CHEMICAL SENSES

Volume 4

CHEMICAL SENSES

Series

Volume 1: Receptor Events and Transduction in Taste and Olfaction, *edited by Joseph G. Brand, John H. Teeter, Robert H. Cagan, and Morley R. Kare*

Volume 2: Irritation, *edited by Barry G. Green, J. Russell Mason, and Morley R. Kare*

Volume 3: Genetics of Perception and Communication, *edited by Charles J. Wysocki and Morley R. Kare*

Volume 4: Appetite and Nutrition, *edited by Mark I. Friedman, Michael G. Tordoff, and Morley R. Kare*

CHEMICAL SENSES

Volume 4
Appetite and Nutrition

EDITED BY
**MARK I. FRIEDMAN
MICHAEL G. TORDOFF
MORLEY R. KARE**

*Monell Chemical Senses Center
Philadelphia, Pennsylvania*

Marcel Dekker, Inc.   New York • Basel • Hong Kong

Library of Congress Cataloging-in-Publication Data

Appetite and nutrition / edited by Mark I. Friedman, Michael G. Tordoff, Morley R. Kare.
      p.   cm.-- (Chemical Senses; v. 4)
  Proceedings of a conference sponsored by the Monell Chemical Senses Center.
  Includes bibliographical references and index.
  ISBN 0-8247-8371-9 (acid-free paper)
  1. Chemical senses--Congresses. 2. Appetite--Congresses.
3. Hunger--Congresses. I. Friedman, Mark I.
II. Tordoff, Michael Guy. III. Kare, Morley Richard.
IV. Monell Chemical Senses Center. V. Series: Chemical senses (New York, N. Y.); v. 4.
QP455.A66   1991
612.3'9--dc20   91-15817
                                                                       CIP

This book is printed on acid-free paper.

Copyright © 1991 by MARCEL DEKKER, INC. All Rights Reserved

Neither this book nor any part may be reproduced or transmitted in any form or by any means, electronic or mechanical, including photocopying, microfilming, and recording, or by any information storage and retrieval system, without permission in writing from the publisher.

MARCEL DEKKER, INC.
270 Madison Avenue, New York, New York 10016

Current printing (last digit):
10 9 8 7 6 5 4 3 2 1

PRINTED IN THE UNITED STATES OF AMERICA

MORLEY RICHARD KARE
1922–1990

Morley was born in the city of Winnipeg in the Canadian province of Manitoba, on March 7, 1922, and was buried there on August 1, 1990. He received a bachelor's degree in science and agriculture from the University of Manitoba, a master's degree in science and agriculture from the University of British Columbia in Vancouver, a doctorate in Zoology from Cornell University in 1952, and two honorary degrees, one from the University of Pennsylvania in 1972 and a D.Sc. degree from the University of Manitoba in 1983.

He served as an officer in the Canadian army during World War II (1943–1946), and as a volunteer in the Israeli army in 1947. In 1951 he married Carol Abramson and they had two children, Susan and Jordin. He was extremely proud of and devoted to his wife and children. Morley served on the faculties of the Universities of British Columbia, Cornell, North Carolina State, and Pennsylvania. During his tenure he was the recipient of a large number of honors and awards, ranging from those honoring his research in sensory and comparative physiology to recognition of his contributions to oenology by the American Institute of Wine and Food in 1988. His skill in oenology was known to many who received a gift sample of his own home-brewed framboise or a sampling of his knowledge exemplified with tastings from his large collection of wines.

A second avocation of his was stamp collecting. He would often ask others to save or trade stamps. On occasion, while giving a friend a ride, he would suddenly stop at the post office. Apologizing for the delay, he would explain that a new commemorative stamp had just been printed and he didn't want to miss the opportunity to add it to his collection.

Morley was a prolific scientific writer. He authored or coauthored approximately 50 book chapters and 200 scientific publications. A scrutiny of the coinvestigators listed on these publications attests to his ability to bring together individuals from different disciplines to do research.

He organized or coorganized many international conferences on the chemical senses, behavior, and neuroscience. He had a zeal and a contagious excitement for teaching. He was especially heartened when investigators and research directors from government, industry, and academia were stimulated by research on the chemical senses that was underway or needed to be done.

His multidisciplinary perspective and convictions about the scientific and practical importance of taste and smell culminated with the founding of the Monell Chemical Senses Center in 1968. Morley's staunch convictions, his vitality, and his persistence served him well in obtaining and maintaining financial resources from a variety of joint ventures, for its development. In spite of a number of setbacks over the years, he succeeded in building the center without losing his wry sense of humor. The center developed from an initial group of five investigators to nearly fifty today.

If Morley was asked about the origins of his scientific interests, he would probably credit two of his academic mentors on the faculty at Cornell University: Professor H. Dukes, of the Veterinary School, and Professor Robbie McCleod, from the Department of Psychology. Dr. Dukes stimulated his interest in comparative physiology, whereas Dr. McCleod encouraged the development of research in the senses of taste and smell.

Morley was strongly dedicated to maintaining and developing the Monell Center, so it was characteristic of him to continue working and being optimistic until a short time before his death on July 30, 1990. He was a productive scientist, a loyal friend to many, especially dedicated to helping young investigators getting started in their careers, a proponent of fairness, and a zealous and compassionate individual.

Owen Maller
Haverford, Pennsylvania

Series Introduction

The Monell Chemical Senses Center celebrated its twentieth anniversay in 1988. Founded as a multidisciplinary organization dedicated to research in all aspects of the chemical senses, the Monell Center is today the only such institution of its kind and has become the focal point for chemosensory research. While the center has always had strong programs of research in many of the traditional areas of chemosensory science, it has also nurtured and helped to expand the scope of less traditional approaches to the study of the chemical senses. With this in mind, the center commemorated its twentieth year by hosting four international conferences on specialized topics within the framework of the chemical senses.

The proceedings of the four conferences are published by Marcel Dekker, Inc. as the series *Chemical Senses*. The first volume, *Receptor Events and Transduction in Taste and Olfaction,* contains the proceedings of the first international symposium that highlighted recent advances in understanding of receptor mechanisms in taste and olfaction. The second volume, *Irritation,* contains the proceedings of the second international symposium, dedicated exclusively to the developing field of common chemical sensitivity. The impact of genetics on the chemical senses was the topic of the third conference, which is covered in the third volume, *Genetics of Perception and Communication.* The proceedings of the final conference, which focused on recent research into the interdependence of the chemical senses and nutrition, appear in the fourth volume, *Appetite and Nutrition.*

The editors of these volumes are confident that the next twenty years will be even more fruitful than the past twenty and that as our research disciplines mature, this anniversary year of the Monell Chemical Senses Center will be recognized as a major turning point in our understanding of the mechanisms, processes, and functions of the chemical senses.

Joseph G. Brand, Mark I. Friedman, Barry G. Green,
Charles J. Wysocki, and Morley R. Kare

Foreword

Of all the research establishments with which I am familiar, the Monell Chemical Senses Center is unique in its focus on the chemical senses—notably taste and smell. This focus reminds us that the five senses we too often take for granted (sight, hearing, touch, taste, and smell) are transducing systems that make it possible for us to interact with our environment and, indeed, survive. It may not be too far-fetched to suggest a rough analogy between the various receptors on the membranes of cells and senses. This is because receptors permit the individual cell to communicate with the body's milieu intérieur while, at a much more complex level of organization, the senses, including their physiological substrates, permit the body to communicate extensively with its milieu extérieur. Both structures, cell and body, are, in effect, surrounded by a medium composed of a complex and variable mixture of ingredients, of which only a relatively small number are pertinent to the needs of the cell or the overall organism. Thus, mechanisms must and do exist to identify and deal with the ingredients of interest. As regards the cell, we might be talking about fuel sources, such as glucose or fatty acids, or chemical messengers, such as insulin or norepinephrine. For the body, taken as a whole, we might be talking about the smell of frying bacon, the sight of an apple, or the aroma of a vintage wine.

The cell, which lives in a stable and predictable environment, has little choice but to accept the fuels, nutrients, and hormonal instructions provided by its milieu; however, the organism, particularly as it becomes more complex and versatile, can decide what it will and will not consume on the basis of an inherited gustatory sense in combination with learned behavior. As pointed out during this conference, monophagous animals such as anteaters depend heavily on genetically programmed behavior in search of their prey. In contrast, polyphagous animals, such as chimpanzees, supplement their genetic information on food with information obtained through experience and social learning. This capability permits the animal to survive in an unstable or unpredictable environment.

If we reflect on the matter, the point is fairly obvious: Cell membrane receptors have meaning because the milieu intérieur provides chemicals useful to the cell that are designed to bind to and act via specific receptors. Our olfactory apparatus has meaning because the ambient environment contains items of importance to the body that can be detected, identified, or evaluated by virtue of the fact that they give rise to a specific odor. In other words, the meaning of the senses is best apprehended in terms of the process by which senses interact with their objects. However, it would be a mistake to forget that, as far as cells are concerned, the number of their receptors and their affinity for specific hormones may be influenced by conditions within the cell. By the same token, the concentration of certain hormones in the circulation that bathes the cell also may affect the cell's receptors. For example, a high concentration of circulating insulin may result in the so-called down-regulation of insulin receptors. Similarly, the ways in which senses function are influenced by conditions within the body. The affective responses to the information the senses provide the brain are subject to modulation by the body's nutritional status. In this volume, Stricker and Verbalis, Gary Beauchamp and his coauthors, and Jay Schulkin all discuss various aspects of the problem of determining the neural, metabolic, and nutritional substrates of salt-seeking behavior, and also the bases of human salt preference.

In light of the foregoing discussion, it may be worth commenting on the dynamics of the experience of a flavor or an odor. In this regard, Alfred North Whitehead, Harvard's philospher of science in the 1930s, may be of some help to us. According to Whitehead (1967), "nature is a structure of evolving processes. The reality is the process." To paraphrase him, it is nonsense to ask if a salty taste or a sweet taste is real. The taste is "ingredient in the process of realization."

There is a poignant echo of this concept of Whitehead's in Chapter 2 by Mark Friedman, entitled "Metabolic Control of Calorie Intake." In the scheme of food intake control suggested by Friedman, it is not the calories in food that are monitored, but rather the realization of these calories through their biochemical oxidation.

Of course, in order to function as scientists, we must work with abstractions like salty taste, sweet taste, and bitter taste, which is to say that we need to think in terms of clear-cut entities that the mind can manipulate. But, to cite Whitehead again:

> The disadvantage of exclusive attention to a group of abstractions, however well-founded, is that, by the nature of the case, you have abstracted from the remainder of things. In so far as the excluded things are important in your

FOREWORD

experiences, your modes of thought are not fitted to deal with them. You cannot think without abstractions; accordingly, it is of the utmost importance to be vigilant in continually revising your *modes* of abstraction.

It seems to me that as we study the chemical senses and the various behaviors that relate to these senses—particularly ingestive or appetitive behavior—it is extremely important to keep in mind that the senses serve as the major interface between the organism and its environment. One definition of the term "interface" describes it as a place where interaction occurs between two systems or processes. In the case of the senses, this interaction has profound philosophical implications because it is the only mechanism by which nature can become conscious of itself. From the scientific standpoint, this definition of interface tells us that to understand a phenomenon like taste or smell in depth we must look at it as a dynamic process involving the interaction of the animal organism (an actively regulated system) with its environment (a system that has considerable stability but gives no evidence of being actively regulated). In other words, we must examine the relation between the physiology of the organism and the environment in which it functions.

In recognition of this need, the Monell Center has made a major commitment to studying the continuum that embodies components, physiological and environmental, that interface at the level of the chemical senses. Thus, in recent years the center has embarked on a substantial and rapidly growing research program in nutrition and appetitive behavior. The International Conference on Appetite, which I was given the privilege of opening, is a striking testimonial to the attention given by biological scientists to the different aspects of the process we call ingestive behavior.

One has only to look at the titles of the chapters compiled in this volume to be impressed by the multiplicity of vantage points from which the phenomenon of appetite has been examined. This multiplicity of perspectives reflects the vitality of the field, its complexity, and the fascination it holds for investigators representing a variety of disciplines.

Theodore B. VanItallie
Professor Emeritus of Medicine
Columbia University College of Physicians and Surgeons
at St. Luke's/Roosevelt Hospital Center
New York, New York

REFERENCE

Whitehead, A. (1967). *Science and the Modern World*. Lowell Lectures, 1925. The Free Press, Macmillan, New York, pp. 59–72.

Preface

This book is the most recent in an informal series, started in 1967, that examines the relationship between the chemical senses and nutritive processes (Kare and Maller, 1967; Kare and Maller, 1977, Kare and Brand, 1986). The emphasis in this volume is on behavior; the factors that guide the selection of nutrients and control their intake. In this context, appetite is seen as the product of interactions between the sensations of taste and smell and the internal bodily signals that reflect changing nutritional status. The appetite for food, salt, water, and alcohol can all be viewed in terms of this interplay between external and internal sensory events. How this interaction is shaped by evolution and individual experience, how it is expressed at different stages in life, and how the nervous system integrates the information from different sensory sources are major themes in this book.

Contributors to the first book on chemical senses and nutrition addressed many of the same issues that are covered here, more than 20 years later. That these basic issues remain unresolved makes it clear that we are dealing with complex processes and difficult scientific questions. The work presented in the present volume makes it clear that much progress has been made. The conference in 1966, on which the first book was based, was motivated by a need to bring together researchers in the chemical senses and those in nutrition who usually had very little scientific contact. Although the need for such cross-disciplinary fertilization remains, the conference on which this book is based was prompted less by a need to forge links between diverse scientific interests than to provide a forum for discussion and integration of ideas by scientists who, for the most part, already have a multidisciplinary approach to what is now an active area of investigation. Progress in this research area is also evident in the detailed knowledge of the physiological, behavioral, and neural mechanisms of appetite described in this book and in the range and breadth of human studies that are reviewed.

The conference that formed the basis for this book was held at the Monell Chemical Senses Center on December 7–9, 1988. This International Conference on Appetite was last in a series of four symposia that were held to commemorate the Monell Center's twentieth anniversay. We hope that the tenor and the tempo of the meeting are captured in the edited transcripts of the question-and-answer periods that followed individual papers and in the general discussion periods that took place at the end of each conference session. Drs. Robert Bolles, L. Arthur Campfield, Paul McHugh, Timothy Morck, and Harvey Weingarten, who served as session chairs, facilitated these lively discussions. Whereas the individual chapters deal with the same material presented at the conference, they are not based on transcripts of the original talks but rather provide broader and more detailed coverage of the topic.

Financial support for the conference was generously provided by the International Life Sciences Institute—Nutrition Foundation, and the U.S. Food and Drug Administration.

The editors are grateful to Ms. Jodie Carr for her assistance with the symposium arrangements and Ms. Linda Robinson for transcribing the proceedings. We also thank Dr. Joseph Brand, Mr. Frederick Sandler, Dr. Danielle Reed, Dr. Rhonda Deems, Ms. Kellie Watson, Ms. Kathy Breslin, and Mr. Henry Lawley for their help during the conference.

Mark I. Friedman
Michael G. Tordoff
Morley R. Kare

REFERENCES

Kare, M. R. and Maller, O. (1967). *The Chemical Senses and Nutrition.* Johns Hopkins Press, Baltimore.

Kare, M. R. and Maller, O. (1977). *The Chemical Senses and Nutrition.* Academic Press, New York.

Kare, M. R. and Brand, J. G. (1986). *Interaction of the Chemical Senses with Nutrition.* Academic Press, New York.

Contents

Series Introduction   *v*
Foreword Theodore B. VanItallie   *vii*
Preface Mark I. Friedman, Michael G. Tordoff, and Morley R. Kare   *xi*
Contributors   *xvii*

PART I   REGULATORY ASPECTS OF APPETITE

1. Biological Bases of Sodium Appetite: Excitatory and Inhibitory Factors   3
   Edward M. Stricker and Joseph G. Verbalis
2. Metabolic Control of Calorie Intake   19
   Mark I. Friedman
3. Nutritional Determinants of Alcohol Intake in C57BL/6J Mice   39
   R. Thomas Gentry

Part I Discussion   61

PART II   HEDONIC ASPECTS OF APPETITE

4. Fat and Sugar: Sensory and Hedonic Aspects of Sweet, High-Fat Foods   69
   Adam Drewnowski

5.	Human Salt Appetite *Gary K. Beauchamp, Mary Bertino, and Karl Engelman*	85
6.	Sensory and Postingestional Components of Palatability in Dietary Obesity: An Overview *Michael Naim and Morley R. Kare*	109
7.	Hunger, Hedonics, and the Control of Satiation and Satiety *John E. Blundell and Peter J. Rogers*	127
8.	Open-Loop Methods for Studying the Ponderostat *Michel Cabanac*	149

Part II Discussion   171

PART III   ACQUIRED PREFERENCES

9.	Social Factors in Diet Selection and Poison Avoidance by Norway Rats: A Brief Review *Bennett G. Galef, Jr.*	177
10.	Primate Gastronomy: Cultural Food Preferences in Nonhuman Primates and Origins of Cuisine *Toshisada Nishida*	195
11.	Innate, Learned, and Evolutionary Factors in the Hunger for Salt *Jay Schulkin*	211
12.	Metabolic Basis of Learned Food Preferences *Michael G. Tordoff*	239
13.	Protein- and Carbohydrate-Specific Cravings: Neuroscience and Sociology *David A. Booth*	261

Part III Discussion   277

PART IV   APPETITE DURING LIFE STAGES

| 14. | Suckling: Opioid and Nonopioid Processes in Mother–Infant Bonding
Elliott M. Blass | 283 |
| 15. | Children's Experience with Food and Eating Modifies Food Acceptance Patterns
Leann L. Birch | 303 |

16.	Changes in Appetitive Variables as a Function of Pregnancy *Judith Rodin and Norean Radke-Sharpe*	325
17.	Changes in Taste and Smell Over the Life Span: Effects on Appetite and Nutrition in the Elderly *Susan S. Schiffman and Zoe S. Warwick*	341
18.	Disturbances of Thirst and Fluid Balance in the Elderly *Barbara J. Rolls and Paddy A. Phillips*	367

Part IV Discussion — 383

PART V NEURAL INTEGRATION

19.	Neuroanatomical Bases of Cephalic Phase Reflexes *Terry L. Powley and Hans-Rudolf Berthoud*	389
20.	Neuroendocrine Activity During Food Intake Modulates Secretion of the Endocrine Pancreas and Contributes to the Regulation of Body Weight *Anton B. Steffens, Jan H. Strubbe, Anton J. W. Scheurink, and Börk Balkan*	405
21.	Central Nervous System Pathways and Mechanisms Integrating Taste and the Autonomic Nervous System *David F. Cechetto*	427
22.	Metabolic Influences on the Gustatory System *Thomas R. Scott*	445

Part V Discussion — 471

Index — 477

Contributors

Börk Balkan   Department of Animal Physiology, University of Groningen, Haren, The Netherlands

Gary K. Beauchamp   Monell Chemical Senses Center, Philadelphia, Pennsylvania

Hans-Rudolf Berthoud   Department of Psychological Sciences, Purdue University, West Lafayette, Indiana

Mary Bertino   Research and Development Sensory Department, Colgate-Palmolive Company, Piscataway, New Jersey, and Monell Chemical Senses Center, Philadelphia, Pennsylvania

Leann L. Birch   Human Development and Family Studies, University of Illinois at Urbana-Champaign, Urbana, Illinois

Elliott M. Blass   Department of Psychology, Cornell University, Ithaca, New York

John E. Blundell   Biopsychology Group, Department of Psychology, University of Leeds, Leeds, England

David A. Booth   Nutritional Psychology Research Group, School of Psychology, University of Birmingham, Birmingham, England

Michel Cabanac Department of Physiology, Laval University, Quebec City, Quebec, Canada

David F. Cechetto Department of Stroke and Aging, The John P. Robarts Research Institute, London, Ontario, Canada

Adam Drewnowski School of Public Health, University of Michigan, Ann Arbor, Michigan

Karl Engelman Department of Medicine, University of Pennsylvania, Philadelphia, Pennsylvania

Mark I. Friedman Monell Chemical Senses Center, Philadelphia, Pennsylvania

Bennett G. Galef, Jr. Department of Psychology, McMaster University, Hamilton, Ontario, Canada

R. Thomas Gentry Department of Medicine, Mount Sinai School of Medicine, New York, New York

Morley R. Kare† Monell Chemical Senses Center, Philadelphia, Pennsylvania

Michael Naim Department of Biochemistry and Human Nutrition, Faculty of Agriculture, The Hebrew University of Jerusalem, Rehovot, Israel

Toshisada Nishida Department of Zoology, Kyoto University, Kyoto, Japan

Paddy A. Phillips Department of Medicine, University of Melbourne and Austin Hospital, Heidelberg, Victoria, Australia

Terry L. Powley Department of Psychological Studies, Purdue University, West Lafayette, Indiana

Norean Radke-Sharpe Department of Mathematics, Bowdoin College, Brunswick, Maine

Judith Rodin Department of Psychology, Yale University, New Haven, Connecticut

---

†deceased

CONTRIBUTORS

Peter J. Rogers Psychobiology Group, Food Acceptability Department, AFRC Institute of Food Research, Shinfield, Reading, England

Barbara J. Rolls Department of Psychiatry and Behavioral Sciences, The Johns Hopkins University, Baltimore, Maryland

Anton J. W. Scheurink Department of Animal Physiology, University of Groningen, Haren, The Netherlands

Susan S. Schiffman Department of Psychology and Psychiatry, Duke University, Durham, North Carolina

Jay Schulkin Department of Anatomy, University of Pennsylvania, Philadelphia, Pennsylvania

Thomas R. Scott Department of Psychology, University of Delaware, Newark, Delaware

Anton B. Steffens Department of Animal Physiology, University of Groningen, Haren, The Netherlands

Edward M. Stricker Department of Behavioral Neuroscience, University of Pittsburgh, Pittsburgh, Pennsylvania

Jan H. Strubbe Department of Animal Physiology, University of Groningen, Haren, The Netherlands

Michael G. Tordoff Monell Chemical Senses Center, Philadelphia, Pennsylvania

Joseph G. Verbalis Department of Medicine, University of Pittsburgh, Pittsburgh, Pennsylvania

Zoe S. Warwick Department of Psychology, Duke University, Durham, North Carolina

CHEMICAL SENSES

Volume 4

Part I
Regulatory Aspects of Appetite

1
Biological Bases of Sodium Appetite: Excitatory and Inhibitory Factors

Edward M. Stricker and Joseph G. Verbalis

University of Pittsburgh
Pittsburgh, Pennsylvania

I. INTRODUCTION

More than 50 years have passed since Richter (1936) first reported that rats increase their intake of NaCl solution to compensate for the uncontrolled loss of sodium in urine caused by bilateral adrenalectomy and the consequent removal of sodium-conserving mineralocorticoid hormones. Later investigations showed that sodium appetite also can be elicited in intact rats by pharmacologic doses of adrenal mineralocorticoids such as desoxycorticosterone (DOC) or aldosterone, which cause sodium retention (Rice and Richter, 1943; Wolf, 1965), and by treatments such as subcutaneous injection of colloidal solution (Stricker and Wolf, 1966), which redistribute body fluid by leaching plasma from the circulation into the interstitial space. Thus, sodium appetite is not only an appropriate response to sodium deficiency; it may arise even when total body sodium stores are normal or elevated.

All models of sodium appetite to date have involved the production of sodium deficiency, elevated mineralocorticoid levels, and/or redistribution of body sodium associated with plasma volume deficits. Although several of these models have been very well characterized, the biological basis of sodium appetite has remained obscure. In this chapter we propose a new perspective for evaluating various models of sodium appetite and elucidating the mechanisms by which enhanced NaCl consumption is elicited. It was derived largely from studies of rats injected subcutaneously with a colloidal solution of polyethylene glycol (PEG), so this model is emphasized. Nevertheless, the implications are likely to be relevant to other models of sodium appetite as well.

II. PHENOMENON

Subcutaneous injection of a hyperoncotic colloidal solution disrupts local Starling forces, causing a gradual sequestration of protein-free intravascular fluid in a sizable edema. The extent of fluid loss and hypovolemia is proportional to the dose and concentration of the colloidal solution (Fitzsimons, 1961; Stricker, 1968). Fluid accumulation increases progressively over 12–16 hours, and its magnitude can be quite considerable: 5 ml of 30% PEG solution injected into an adult rat causes up to 30% reduction of plasma volume, and twice that dose can produce so much hemoconcentration that the kidneys fail.

This procedure is more effective than acute hemorrhage in reducing blood volume because interstitial fluid that enters the vasculature as a result of rising plasma protein concentration is drawn into the edema as well. Thus, the induced hypovolemia may last for 24 hours or more, long enough for most investigations, until the lymphatic system drains the injected colloid and sequestered fluid from the subcutaneous tissue. As well, because red blood cells remain in the circulation, the complications of anemia are avoided. Finally, because the fluid loss is gradual rather than abrupt, studies of behavior can be conducted without the debilitation associated with acute hypotension. Indeed, blood pressure is maintained at near-normal levels despite the substantial plasma volume deficits induced because the losses are so gradual (Stricker, 1977; Stricker et al., 1987).

After subcutaneous PEG treatment, animals lose protein-free plasma fluid from the circulation. What they need to repair plasma volume is an isotonic NaCl solution, in amounts that exceed the induced volume deficits because some portion of the ingested fluids also will be sequestered into the edema until tissue turgor limits further fluid accumulation. Thereafter ingested fluid will remain largely intravascular and thereby restore plasma volume. Appropriate to these needs, rats allowed access to isotonic NaCl solution will consume it in large amounts and repair plasma volume deficits as rapidly as they are formed (Stricker and Jalowiec, 1970). When separate bottles of water and concentrated NaCl solution are offered instead, rats drink both fluids in such proportion that the mixed solution is slightly more dilute than 150 mEq/L. The kidneys retain virtually all the ingested sodium, and excretion of a sodium-free urine then leaves the animals with the isotonic mixture they require (Stricker, 1981).

Teleologically, rats' behavioral responses to colloid treatment clearly are appropriate to the induced plasma volume deficits. But what are the biological bases underlying these responses? Accumulating evidence suggests that the thirst associated with plasma volume deficits is largely stimulated by low-pressure baroreceptors (Stricker, 1978a). There has been considerable speculation that a portion of the drinking response is mediated by circulating levels of angiotensin, a pressor agent with known dipsogenic properties that is formed in blood when the proteolytic enzyme, renin, is secreted from the kidneys (Fitzsimons, 1969). Activity in the renin–angiotensin system does increase after PEG treatment in parallel with the induced water intake (Leenen and Stricker, 1974). However, PEG-treated rats drink water normally even when the renin–angiotensin system is eliminated by bilateral nephrectomy (Fitzsimons, 1961; Stricker, 1973). Moreover, rats given large electrolytic lesions of the septal area are unusually sensitive to the dipsogenic actions of exogenously administered angiotensin, but they drink water normally after PEG treatment even when plasma volume deficits are as large as 20–25%, and excessive intakes are seen only when the deficits (and presumably the induced increases in plasma angiotensin) are still greater (Stricker, 1978b). Thus, it seems unlikely that the stimulus for hypovolemic thirst in rats normally has a large endocrine component, although under conditions of extreme volume depletion, angiotensin could well contribute more significantly to observed thirst. A similar conclusion has been reached in other species, including man (Abraham et al., 1975; Phillips et al., 1985).

Alternatively, there is neural input to the brain from stretch receptors embedded in the smooth muscle wall at various sites in the cardiovascular system, which are sensitive to local filling or blood pressure. Signals from these baroreceptors are well known to be responsible for pituitary secretion of the antidiuretic hormone, vasopressin (AVP), during hypovolemia (Gauer and Henry, 1963), and the same signals appear to stimulate thirst as well. Consistent with this idea are recent observations that the thirst of PEG-treated rats can be virtually eliminated by inflating a small balloon that had been previously implanted in the atrium of their hearts, thereby preventing detection of the lost plasma volume by local baroreceptors. In contrast, drinking was found to proceed normally in rats when the balloon was inflated but thirst was induced by injection of hypertonic NaCl solution, thereby providing a centrally detected osmoregulatory stimulus to thirst (Kaufman, 1983).

With regard to sodium appetite, it seems noteworthy that the induced increase in NaCl intake after PEG treatment always becomes prominent several hours after thirst appears, regardless of the dose of colloid that is administered or the concentration of NaCl solution that is available (Stricker and Jalowiec, 1970; Stricker, 1981). Once such drinking has begun, the animals consume saline and water steadily until they have taken in enough of both fluids to repair the volume deficit. Water consumption alone does relatively little to repair plasma volume deficits, because water distributes intracellularly as well as extracellularly. Indeed, the osmotic dilution that water ingestion causes inhibits both further water intake and release of AVP, thereby preventing even greater degrees of dilution (Stricker, 1969; Stricker and Verbalis, 1986). In contrast, intake of NaCl is critical in repairing the volume deficit, and consequently it seems paradoxical that thirst should appear first while sodium appetite is delayed. The following series of studies considered the biological basis of sodium appetite, including this delay, from three separate perspectives.

III. RESULTS AND DISCUSSION

A. Excitatory Stimuli

It is possible that the delayed expression of sodium appetite after PEG treatment results from the gradual appearance of some factor(s) that provide the signal for sodium appetite. For example, aldosterone levels are known to increase gradually after PEG treatment, in proportion to the growing plasma volume deficit (Stricker et al., 1979), and perhaps sodium appetite is stimulated when their levels exceed a certain threshold. Alternatively, because activity in the renin–angiotensin system also increases in response to the induced hypovolemia (Leenen and Stricker, 1974), critical levels of angiotensin may provide such a stimulus. Or both hormones may be involved in the stimulation of sodium appetite. If any of these hypotheses is correct, then pretreatments causing alterations in the induced changes in hormone levels should produce parallel changes in sodium appetite.

In fact, the results of such experiments are consistent with this prediction. Thus, for example, much greater and more rapid increases in plasma levels of aldosterone can be obtained by switching the maintenance diet of rats from standard laboratory chow to a sodium-deficient diet prior to PEG treatment (Stricker et al., 1979); under such conditions, the delayed appearance of sodium appetite is eliminated and intake of NaCl solution begins immediately, even before thirst is evident (Stricker, 1981). Conversely, when the enhanced aldosterone response to hypovolemia in sodium-deprived rats is prevented by prior hypophysectomy, sodium appetite again appears only after the normal 5-hour delay (Stricker, 1983). Although these experiments suggest the likely significance of aldosterone as a stimulus for sodium appetite, others have indicated that additional factors are involved as well. For example, sodium appetite appears 5 hours after PEG treatment even when rats are adrenalectomized and given DOC replacement therapy, so that mineralocorticoid levels could not increase (Stricker, 1983). These latter findings suggest that increases in plasma aldosterone levels alone do not normally mediate sodium appetite in rats after PEG treatment.

Other experiments indicated that angiotensin could play an important role in stimulating sodium appetite, perhaps acting in concert with aldosterone. For example, when delivery of angiotensin to the brain was increased pharmacologically, the appearance of sodium appetite in PEG-treated rats was accelerated

(Stricker, 1983; Fitts et al., 1985). Similarly, sodium appetite elicited by PEG treatment was augmented in rats with septal lesions (Stricker, 1984), which are unusually sensitive to the central actions of angiotensin. Moreover, rats treated systemically with DOC drank NaCl solution soon after angiotensin was injected directly into their brains (Fluharty and Epstein, 1983), whereas pharmacological blockade of the renin–angiotensin system was found to abolish salt consumption in sodium-deficient rats (Sakai et al., 1986). Collectively, these results support the possibility that angiotensin and aldosterone act synergistically in the rat brain to elicit sodium appetite during sodium deficiency. Thus, the angiotensin and aldosterone secreted during sodium depletion might play an active role in mediating both the specific behavioral response to sodium need (i.e., NaCl ingestion) and the complementary physiological responses that support blood pressure and promote renal sodium conservation.

On the other hand, there are many experimental protocols in which the putative synergetic action of angiotensin and aldosterone in the brain does not predict the appearance of sodium appetite in rats (see Table 1). For example, it is well known that mineralocorticoids are eliminated by adrenalectomy and that renin secretion

TABLE 1 Relation Between Thirst, Sodium Appetite, and Plasma (p) Levels of Angiotensin, Aldosterone, and Oxytocin in Rats Given Various Treatments

Treatment[a]	Results[b]				
	Thirst	Sodium appetite	pANGIO	pALDO	pOT
PEG + water load		×	×	×	c
NaD/PEG	×	××	×	××	d
Adrenalectomy		××	×		
DOC	×	××		××	
Phentolamine/PEG	××		××	×	××
NaD/IVC lig/PEG	××		××	××	××
Nephrectomy/PEG	×			×	××
Bladder-puncture/PEG	×		×	×	××

[a]Six groups of rats were maintained on standard laboratory chow prior to the indicated treatment, whereas the other two groups were maintained for 2 days on sodium-deficient diet (NaD). Among the rats injected with 30% PEG solution, separate groups were otherwise untreated, made hypotensive by injection with phentolamine or ligation of the inferior vena cava (IVC lig), or made uremic by nephrectomy or bladder puncture.
[b]× = increase, ×× = larger increase.
[c]pOT levels increased gradually after PEG treatment, but sodium appetite appeared only after water was consumed and osmotic dilution inhibited the OT secretion.
[d]pOT levels were blunted for 1–2 hours after PEG treatment, when sodium appetite was prominent. When access to drinking fluids was delayed for 4 hours, however, plasma OT levels were elevated and sodium appetite was again inhibited until the hypovolemic rats drank water.

is suppressed by DOC treatment, yet both treatments effectively elicit sodium appetite in rats. Moreover, when rats are made hypotensive by injection of PEG solution after ligation of the inferior vena cava or prior to administration of the α-adrenergic receptor blocking agent phentolamine, both angiotensin and aldosterone levels increase substantially; but thirst, not sodium appetite, is stimulated (Stricker, 1971; Hosutt and Stricker, 1981). Similarly, uremia produced by bilateral nephrectomy or by puncturing the bladder prevents the sodium-appetite-eliciting effects of subcutaneous PEG, but not its effects on thirst (Stricker, 1971; Stricker et al., 1979). On the basis of these and other findings, we abandoned an exclusive search for excitatory signals whose increased presence would stimulate sodium appetite.

B. Inhibitory Stimuli

Forward movement in an automobile continually in gear might be obtained not only by stepping on the gas but also by releasing the brake. Similarly, sodium appetite might appear after a delay not because excitatory stimuli have increased but because inhibitory stimuli have decreased. In considering the possible identity of such inhibitory stimuli, we were struck by the fact that hypovolemic rats invariably drank water before consuming NaCl solution. Indeed, even when water and saline were withheld for 24 hours after PEG treatment, rats always drank water to the point of producing a 3-5% dilution of plasma osmolality before intake of NaCl solution began (Stricker, 1981). As mentioned, this osmotic dilution inhibited thirst as well as the induced secretion of AVP in PEG-treated rats. We therefore determined whether plasma AVP levels also were unusually low when sodium appetite was induced in rats either by DOC treatment or by adrenalectomy. In both cases basal plasma AVP levels were found to be normal. However, we also measured plasma levels of oxytocin (OT), the other neurohypophyseal hormone, which like AVP is secreted in response to hypovolemia and inhibited by concurrent osmotic dilution of body fluids in rats (Stricker and Verbalis, 1986). Plasma levels of OT were notably low in these animals (Stricker and Verbalis, 1987). Such suppressed levels of plasma OT represented the first common factor known to be present in these three dissimilar models of sodium appetite in rats. Thus, it seemed possible that OT in some way inhibited NaCl solution intake, and that sodium appetite was expressed only when circulating levels of the peptide hormone were very low.

To evaluate this hypothesis, we examined plasma OT levels in the various protocols in which PEG-treated rats showed elevated levels of angiotensin and aldosterone yet did not have a sodium appetite. In each case plasma OT levels were found to be substantially elevated (Stricker et al., 1987; see Table 1). Similarly, when sodium appetite in adrenalectomized rats was inhibited temporarily by injection of hypertonic NaCl, we found an acute increase in plasma OT levels whose duration varied in proportion to the duration over which sodium appetite was inhibited (Stricker and Verbalis, 1987). In fact, we have yet to find a condition in which sodium appetite could be elicited in the presence of stimulated OT secretion. Conversely, OT secretion induced by hypovolemia was blunted by prior maintenance on a sodium-deficient diet, in association with enhanced sodium appetite (Stricker et al., 1987).

Despite such correlational evidence, other experiments indicated clearly that circulating OT did not directly affect the intake of NaCl solution in rats. Continuous administration of synthetic OT from an implanted osmotic minipump, raising plasma OT to physiological levels and beyond, did not inhibit sodium appetite in PEG-treated rats, nor did systemic administration of an OT receptor antagonist enhance NaCl intake (Stricker and Verbalis, 1987).

C. Central Oxytocinergic Neurons

If increased pituitary secretion of OT correlates well with the inhibition of sodium appetite in rats but circulating levels of OT do not by themselves mediate that inhibition, then plasma OT may be a marker for some parallel event that is associated with pituitary OT secretion and is responsible for mediating inhibition of sodium appetite. OT secreted from the neurohypophysis is synthesized in magnocellular neurons in the supraoptic nuclei and paraventricular nuclei (SON and PVN) of the hypothalamus. Adjacent to the magnocellular neurons in the PVN, but not the SON, are parvocellular OT-containing neurons, which project widely throughout the brain to sites including the limbic system and the dorsal motor nucleus of the vagus in the brain stem (Swanson and Sawchenko, 1983). Thus, central oxytocinergic projections appear to be advantageously located to influence ingestive behavior and associated autonomic function. Perhaps these centrally projecting hypothalamic nuclei are stimulated simultaneously with the magnocellular neurons projecting to the pituitary so that the induced activity somewhere in the brain serves to inhibit sodium appetite; conversely, sodium appetite may develop when activity is suppressed in this subset of PVN neurons.

We adopted two strategies to evaluate this hypothesis. One was to identify some response also known to be influenced by the parvocellular oxytocinergic neurons, and to determine whether sodium appetite was inhibited when that response was stimulated. In doing so, we made use of a recent report by Rogers and Hermann (1987) in which an oxytocinergic projection from the PVN to the dorsomotor nucleus of the vagus in the brainstem was observed to decrease gastric motility and emptying. In our studies, two chemical agents, lithium chloride and copper sulfate, both inhibited gastric emptying in rats and increased pituitary OT secretion as well (Verbalis et al., 1986; McCann et al., 1989). These agents also decreased sodium appetite in sodium-deprived adrenalectomized rats and in PEG-treated rats, in the latter case without affecting the induced thirst (Stricker and Verbalis, 1987). Thus, stimulation of centrally projecting oxytocinergic neurons in rats apparently leads to a parallel inhibition of NaCl intake.

The second approach used the reverse strategy and determined whether sodium appetite was potentiated when stimulation of the PVN was blocked. Osmoreceptors or osmoreceptor pathways in the ventral portion of nucleus medianus (vNM) are believed to be involved with stimulation of thirst and the secretion of AVP and OT during osmotic dehydration. Accordingly, lesions of vNM abolish thirst and attenuate neurohypophyseal hormone secretion after injection of hypertonic NaCl solution, whereas the responses to hypovolemia are unimpaired (Gardiner et al., 1985). However, rats with such lesions showed a marked spontaneous appetite for concentrated NaCl solution; they drank 15–35 ml of 0.5 M NaCl daily for at least several months after surgery (Gardiner et al., 1986), an effect as large as the sodium appetite obtained with any known treatment, including adrenalectomy. Examination of these animals indicated that they did not become sodium deficient after the brain lesions, nor was the renin–angiotensin system activated or plasma levels of aldosterone increased. Instead, we suggest that the lesions blocked the normal inhibitory effects on sodium appetite that occur when rats consume concentrated NaCl solution, thus permitting the ingestion of saline to persist unchecked.

A putative role of PVN neurons in the inhibitory control of sodium appetite would appear to be inconsistent with the early report that lactating rats increase their daily intake of NaCl solution (Richter and Barelare, 1938), inasmuch as suckling is an established stimulus of pituitary OT secretion. However, recent experiments in our laboratories have indicated that although magnocellular PVN activity is known to be increased during suckling, parvocellular activity is not (Helmreich et al., 1988). Thus, suckling would not be expected to provide a stimulus that directly affected the proposed central control of NaCl consumption. Even if it did, parvocellular PVN neurons might still mediate inhibition of sodium appetite because lactating animals are inactive when nursing, and thus suckling-induced OT secretion and saline ingestion necessarily occur at different times of the day.

IV. CONCLUSIONS

Rats with plasma volume depletions need water and NaCl to restore circulatory volume. Accordingly, they increase their intakes of both fluids. A schema summarizing the interrelated controls of thirst and sodium appetite is presented in Figure 1. Thirst during hypovolemia appears to be controlled by excitatory neural signals from low-pressure cardiovascular baroreceptors and by inhibitory neural signals from cerebral osmoreceptors. Thus, water intake is not maintained despite the continued presence of hypovolemia due to the progressive osmotic dilution resulting from renal retention of the ingested water. These animals need sodium, both to expand plasma volume and to avoid excessive osmotic dilution, and appropriate to this need, hypovolemic animals also manifest an appetite for salty foods and fluids. Sodium appetite appears to be controlled by an excitatory

BIOLOGICAL BASES OF SODIUM APPETITE

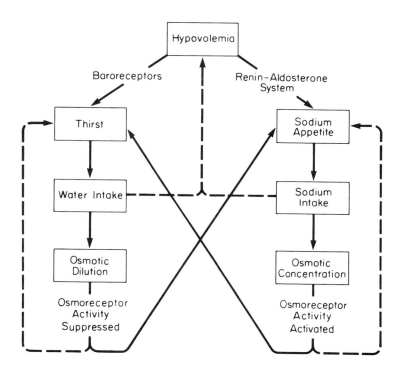

FIGURE 1 Schematic representation of the physiological mechanisms controlling thirst and sodium appetite during hypovolemia. Solid arrows indicate stimulation, dashed arrows indicate inhibition. Hypovolemic rats alternately drink water and concentrated NaCl solution, depending on the current plasma osmolality, and ultimately consume sufficient fluid at an isotonic concentration to repair the volume deficit. However, the water and NaCl intakes are limited when animals have access to only one of the drinking fluids, as a result of activation of the respective inhibitory osmoregulatory pathway. Conversely, neither inhibitory pathway is activated when the rats drink isotonic NaCl solution, and consequently intakes proceed unabated in response to the hypovolemic stimulus. (Adapted from Stricker and Verbalis, 1988.)

stimulus that results at least in part from activity in the renin–angiotensin–aldosterone system, and by an inhibitory stimulus that results from activity in neurons projecting centrally from parvocellular neurons in the PVN. Thus, despite increased secretions of renin and aldosterone, PEG-treated animals will not begin to consume NaCl until osmotic dilution eliminates the parallel increase in activity of central inhibitory neurons, which also is stimulated by hypovolemia. Consumption of concentrated NaCl solution will raise plasma osmolality and thereby inhibit sodium appetite while disinhibiting thirst; subsequent water intake, in turn, will continue until the diluted plasma osmolality once again inhibits thirst while disinhibiting sodium appetite. By alternating their intakes of saline and water in this way, the hypovolemic animals ultimately consume sufficient amounts of both fluids to restore plasma volume.

Similarly, we presume that NaCl consumption is robust after DOC treatment or bilateral adrenalectomy, in part, because activity in central inhibitory pathways is low (as reflected by low plasma OT levels). Thus, the consumption of concentrated NaCl solutions, which usually is self-limiting because increased plasma osmolality stimulates the central inhibitory system, can continue in the adrenalectomized rat because they rapidly lose the ingested NaCl and remain osmotically dilute. Finally, the central inhibitory system seems to be much less easily activated in the volume-expanded, DOC-treated rats than in normal animals, and the increased plasma osmolality resulting from consumption of concentrated NaCl solution disappears quickly because it soon elicits osmoregulatory water intake.

V. FUTURE RESEARCH NEEDS

The brain mechanisms that mediate sodium appetite remain poorly understood. The perspective proposed in this chapter emphasizes the possibility that central oxytocinergic neurons may play some role in this drive, and obviously it will be necessary to evaluate this hypothesis directly. Unfortunately, it probably will be very difficult for concerned investigators to do so by conventional techniques. For example, the concentration of released OT in cerebrospinal fluid is easily measurable but of unknown significance as an index of synaptic OT release at a specific site in brain. Administration of OT into the brain would be expected to influence the intake of NaCl solution only if OT reached appropriate brain structures critical to sodium appetite, but those sites have not yet been determined. Less specific perfusion of the brain with OT would bypass this issue; but the predicted effect, inhibition of NaCl consumption, might be explained by less specific consequences of the treatment. Perhaps the best available approach would involve the intracerebroventricular administration of OT receptor blocking agents, which might be expected to potentiate sodium appetite. Of course, even if the hypothesis is correct, sodium appetite may not be enhanced because the agents never reach the relevant receptors, or because they are more effective at blocking uterine OT receptors (for which they were developed) than cerebral OT receptors. Nevertheless, such experiments should be pursued even though the results obtained must be interpreted cautiously.

It should also be noted that the centrally projecting PVN parvocellular neurons mediating inhibition of sodium appetite need not be oxytocinergic. The PVN is an extraordinarily complex structure containing many different neuropeptidergic cells, any of which might be coactivated in concert with magnocellular OT neurons. Thus, negative results with central manipulations of OT receptors should prompt studies of other possible candidates for PVN neurons mediating inhibition of sodium appetite.

Although a possible role for the PVN in the inhibition of sodium appetite has not been suspected previously, involvement of this structure in mediating satiety for food has been proposed (Leibowitz, 1978; Stricker et al., 1988). Central pathways involved in inhibiting food and NaCl consumption could overlap if increased PVN activity acted to reduce the appetite for osmoles in general. Such activity then would be expressed both as a decreased intake of food, the usual source of osmotic particles, and as a decreased consumption of NaCl solution (see also Wolf et al., 1984). According to this hypothesis, stimuli causing pituitary OT secretion should reduce food intake as well as inhibit sodium appetite. In fact, considerable evidence already is available to support this hypothesis, because each of such nauseogenic agents as lithium chloride and copper sulfate and dipsogenic agents as hypertonic NaCl is well known to reduce food intake in rats. On the other hand, the circumstances in which food and NaCl intakes are stimulated certainly do not always coincide; for example, increased NaCl consumption is observed after bilateral adrenalectomy in rats, but hyperphagia is not. Nevertheless, an integrated hypothesis could be retained if food and NaCl intakes were primarily controlled by separate signals, perhaps related to the availability of metabolic fuels and to the renin–angiotensin–aldosterone system, respectively, but shared some component of a common inhibitory system. This speculation awaits further investigation.

In considering a possible role of central oxytocinergic neurons in the control of food intake, we have used pituitary OT secretion as a peripheral marker of increased neuronal activity in the PVN. Of course, the pituitary secretion of OT presumably serves some systemic physiological function as well. In considering what that function might be, it should be noted that enhanced sodium excretion would best complement inhibition of NaCl intake. In this regard, exogenously administered OT is known to have natriuretic properties in rats (Balment et al., 1980), and recently we have confirmed those findings and extended them to the physiological range (Verbalis et al., 1988). Thus, it seems possible that after an osmotic load in rats, sodium excretion (mediated in part by pituitary OT secretion) and the inhibition of sodium appetite are complementary functions much like renal water conservation (mediated by pituitary AVP secretion) and the stimulation of thirst.

Finally, an important issue that has received little attention to date is represented by the substantial differences in sodium appetite among species. Although a salt craving in humans has been associated with Addison's disease (Wilkins and Richter, 1940), a clinical syndrome of reduced adrenocortical steroid secretion that is equivalent in this respect to the adrenalectomized rat preparation, only about 15–20% of such patients manifest sodium appetite despite marked sodium deficiency (Henkin et al., 1963). Similarly, it seems noteworthy that sodium appetite is uncommon in otherwise healthy humans during clinical states of sodium depletion. When considered together with additional observations that the mechanisms controlling sodium appetite in herbivores like sheep appear to differ from those mediating sodium appetite in rats, as discussed above, it seems likely that there are fundamental differences in the central regulatory systems controlling sodium appetite across species. This may reflect different evolutionary adaptations to differences in the dietary availability of sodium to herbivores and carnivores (Denton, 1982).

In contrast, quite analogous control mechanisms for thirst appear to have evolved in all mammalian species studied in the laboratory to date, including humans, perhaps because the need for water to prevent dehydration in animals is common and relatively independent of their dietary source of calories. As well, the importance of the renin–angiotensin–aldosterone system for blood pressure support and sodium conservation appears to be universal among mammalian species. More comparative studies of sodium appetite, especially including human subjects, are required before these considerations of the interrelated psychobiological mechanisms that control fluid intake and output in animals can be extended to humans.

ACKNOWLEDGMENTS

This work was supported by research grant MH-25140 from the National Institute of Mental Health. The technical assistance of Jen-shew Yen and Marcia Drutarosky is gratefully acknowledged.

DISCUSSION

Thrasher: One situation that seems to contradict the general idea that oxytocin inhibits salt intake is that salt appetite is very prevalent in many species during lactation, which is also a period of time when oxytocinergic pathways are very active. How would you get around that?

Stricker: There are two points I would like to make in answering that important question. First, the oxytocin released into the circulation during lactation is of course from magnocellular neurons in PVN, whereas we are proposing that the parvocellular PVN neurons are involved in the inhibitory control of sodium appetite in rats. We have recently completed experiments suggesting that suckling does not in fact activate the parvocellular neurons in rats. Thus, the pituitary secretion of oxytocin during lactation does not challenge the possibility that parvocellular secretion of oxytocin plays an important role in the control of sodium appetite. The second point to emphasize is that lactation and NaCl consumption are not really coincidental phenomena. That is, the dams do not consume NaCl or food while they nurse the pups. Thus, even if suckling did stimulate the parvocellular neurons, it may not affect ingestive behavior because food and fluid consumption occur at other times of the day, when the dams are not nursing. For these two reasons, we do not find the phenomenon of sodium appetite in lactating rats inconsistent with our hypothesis.

Epstein: Is it possible to put these ideas to a crucial test? It is my understanding that there are now specific receptor blockers for the posterior pituitary peptide, oxytocin. What is the effect of the injection of those materials into the brain of the rat? It should increase the animal's salt intake.

Stricker: We have just finished perfecting a system for actually delivering oxytocin antagonists into the brain. I can tell you that we have done experiments in which the oxytocin antagonist was administered in the periphery, and it has no effect on sodium appetite. But that's not the direct test of the hypothesis; it only eliminates what we never considered in the first place, which is that it was peripheral oxytocin that was involved. The real test is the one that you have identified. Those experiments have not been done, but they'll be done very shortly.

VanItallie: Is there any evidence for a reward or aversion system for salt in brain?

Stricker: I'm not the best person to answer that question. It's certainly true that a salt-deficient animal taking in salt finds that reinforcing, but whether or not that has anything to do with what people describe as reinforcement systems in the brain, I can't answer that.

REFERENCES

Abraham, S. F., Baker, R. M., Blaine, E. H., Denton, D. A., and McKinley, M. J. (1975). Water drinking induced in sheep by angiotensin—A physiological or pharmacological effect? *J. Comp. Physiol. Psychol.* **88**:503–518.

Balment, R. J., Brimble, M. J., and Forsling, M. L. (1980). Release of oxytocin induced by salt loading and its influence on renal excretion in the male rat. *J. Physiol., London,* **308**:439–449.

Denton, D. (1982). *The Hunger for Salt: An Anthropological, Physiological and Medical Analysis.* Springer-Verlag, Berlin.

Fitts, D. A., Thunhorst, R. L., and Simpson, J. B. (1985). Modulation of salt appetite by lateral ventricular infusions of angiotensin II and carbachol during sodium depletion. *Brain Res.* **346**:273–280.

Fitzsimons, J. T. (1961). Drinking by rats depleted of body fluid without increase in osmotic pressure. *J. Physiol., London,* **159**:297–309.

Fitzsimons, J. T. (1969). The role of a renal thirst factor in drinking induced by extracellular stimuli. *J. Physiol., London,* **201**:349–368.

Fluharty, S. J., and Epstein, A. N. (1983). Sodium appetite elicited by intracerebroventricular infusion of angiotensin II in the rat: II. Synergistic interaction with systemic mineralocorticoids. *Behav. Neurosci.* **97**:746–758.

Gardiner, T. W., Verbalis, J. G., and Stricker, E. M. (1985). Impaired secretion of vasopressin and oxytocin in rats after lesions of nucleus medianus. *Am. J. Physiol.* **249**:R681–R688.

Gardiner, T. W., Jolley, J. R., Vagnucci, A. H., and Stricker, E. M. (1986). Enhanced sodium appetite in rats with lesions centered upon nucleus medianus. *Behav. Neurosci.* **100**:531–535.

Gauer, O. H., and Henry, J. P. (1963). Circulatory basis of fluid volume control. *Physiol. Rev.* **43**:423–481.

Helmreich, D. L., Thiels, E., Verbalis, J. G., and Stricker, E. M. (1988). Suckling does not affect gastric motility in rats. *Soc. Neurosci. Abstr.* **14**:629.

Henkin, R. I., Gill, J. R., Jr., and Bartter, F. C. (1963). Studies of taste thresholds in normal man and in patients with adrenal cortical insufficiency: The role of adrenal cortical steroids and of serum sodium concentration. *J. Clin. Invest.* **42**:727–735.

Hosutt, J. A., and Stricker, E. M. (1981). Hypotension and thirst after phentolamine treatment. *Physiol. Behav.* **27**:463–468.

Kaufman, S. (1983). Role of right atrial receptors in the control of drinking in the rat. *J. Physiol., London,* **349**:389–396.

Leenen, F. H., and Stricker, E. M. (1974). Plasma renin activity and thirst following hypovolemia or caval ligation in rats. *Am. J. Physiol.* **226**:1238–1242.

Leibowitz, S. F. (1978). Paraventricular nucleus: A primary site mediating adrenergic stimulation of feeding and drinking. *Pharmacol. Biochem. Behav.* **8**:163–175.

McCann, M. J., Verbalis, J. G., and Stricker, E. M. (1989). LiCl and CCK inhibit gastric emptying and feeding and stimulate OT secretion in rats. *Am. J. Physiol.* **256**:R463–R468.

Phillips, P. A., Rolls, B. J., Ledingham, J. G. G., Morton, J. J., and Forsling, M. L. (1985). Angiotensin II-induced thirst and vasopressin release in man. *Clin. Sci.* **68**:669–674.

Rice, K. K., and Richter, C. P. (1943). Increased sodium chloride and water intake of normal rats treated with desoxycorticosterone acetate. *Endocrinology,* **33**:106–115.

Richter, C. P. (1936). Increased salt appetite in adrenalectomized rats. *Am J. Physiol.* **115**:155–161.

Richter, C. P., and Barelare, B., Jr. (1938). Nutritional requirements of pregnant and lactating rats studied by the self-selection method. *Endocrinology,* **23**:15–24.

Rogers, R. C., and G. E. Hermann. (1987). Oxytocin, oxytocin antagonist, TRH, and hypothalamic paraventricular nucleus stimulation: Effects on gastric motility. *Peptides,* **8**:505–513.

Sakai, R. R., Nicolaidis, S., and Epstein, A. N. (1986). Salt appetite is suppressed by interference with angiotensin II and aldosterone. *Am. J. Physiol.* **251**:R762–R768.

Stricker, E. M. (1968). Some physiological and motivational properties of the hypovolemic stimulus for thirst. *Physiol. Behav.* **3**:379–385.

Stricker, E. M. (1969). Osmoregulation and volume regulation in rats: Inhibition of hypovolemic thirst by water. *Am. J. Physiol.* **217**:98–105.

Stricker, E. M. (1971). Effects of hypovolemia and/or caval ligation on water and NaCl solution drinking by rats. *Physiol. Behav.* **6**:299–305.

Stricker, E. M. (1973). Thirst, sodium appetite, and complementary physiological contributions to the regulation of intravascular fluid volume. In *The Neuropsychology of Thirst,* A. N. Epstein, H. R. Kissileff, and E. Stellar (Eds.). H. V. Winston & Sons, New York, pp. 73–98.

Stricker, E. M. (1977). The renin–angiotensin system and thirst: A reevaluation: II. Drinking elicited by caval ligation or isoproterenol. *J. Comp. Physiol. Psychol.* **91**:1220–1231.

Stricker, E. M. (1978a). The renin–angiotensin system and thirst: Some unanswered questions. *Fed. Proc.* **37**:2704–2710.
Stricker, E. M. (1978b). Excessive drinking by rats with septal lesions during hypovolemia induced by subcutaneous colloid treatment. *Physiol. Behav.* **21**:905–907.
Stricker, E. M. (1981). Thirst and sodium appetite after colloid treatment in rats. *J. Comp. Physiol. Psychol.* **95**:1–25.
Stricker, E. M. (1983). Thirst and sodium appetite after colloid treatment in rats: Role of the renin–angiotensin–aldosterone system. *Behav. Neurosci.* **97**:725–737.
Stricker, E. M. (1984). Thirst and sodium appetite after colloid treatment in rats with septal lesions. *Behav. Neurosci.* **98**:356–360.
Stricker, E. M., and Jalowiec, J. E. (1970). Restoration of intravascular fluid volume following acute hypovolemia in rats. *Am. J. Physiol.* **218**:191–196.
Stricker, E. M., and Verbalis, J. G. (1986). Interaction of osmotic and volume stimuli in regulation of neurohypophyseal secretion in rats. *Am. J. Physiol.* **250**:R267–R275.
Stricker, E. M., and Verbalis, J. G. (1987). Central inhibitory control of sodium appetite in rats: Correlation with pituitary oxytocin secretion. *Behav. Neurosci.* **101**:560–567.
Stricker, E. M., and Verbalis, J. G. (1988). Hormones and behavior: The biology of thirst and sodium appetite. *Am. Sci.* **76**:261–267.
Stricker, E. M., and Wolf, G. (1966). Blood volume and tonicity in relation to sodium appetite. *J. Comp. Physiol. Psychol.* **62**:275–279.
Stricker, E. M., Vagnucci, A. H., McDonald, R. H., Jr., and Leenen, F. H. (1979). Renin and aldosterone secretions during hypovolemia in rats: Relation to NaCl intake. *Am. J. Physiol.* **237**:R45–R51.
Stricker, E. M., Hosutt, J. A., and Verbalis, J. G. (1987). Neurohypophyseal secretion in hypovolemic rats: Inverse relation to sodium appetite. *Am. J. Physiol.* **252**:R889–R896.
Stricker, E. M., McCann, M. J., Flanagan, L. M., and Verbalis, J. G. (1988). Neurohypophyseal secretion and gastric function: Biological correlates of nausea. In *Biowarning Systems in the Brain,* H. Takagi, Y. Oomura, M. Ito, and M. Otsuka (Eds.). University of Tokyo Press, Tokyo, pp. 295–307.
Swanson, L. W., and Sawchenko, P. E. (1983). Hypothalamic integration: Organization of the paraventricular and supraoptic nuclei. *Annu. Rev. Neurosci.* **6**:269–324.
Verbalis, J. G., McHale, C. M., Gardiner, T. W., and Stricker, E. M. (1986). Oxytocin and vasopressin secretion in response to stimuli producing learned taste aversions in rats. *Behav. Neurosci.* **100**:466–475.
Verbalis, J. G., Mangione, M., and Stricker, E. M. (1988). Oxytocin is natriuretic at physiological plasma concentrations. *Soc. Neurosci. Abstr.* **14**:628.
Wilkins, L., and Richter, C. P. (1940). A great craving for salt by a child with corticoadrenal insufficiency. *J. Am. Med. Assoc.* **114**:866–868.
Wolf, G. (1965). Effect of desoxycorticosterone on sodium appetite of intact and adrenalectomized rats. *Am. J. Physiol.* **208**:1281–1285.
Wolf, G., Schulkin, J., and Simson, P. E. (1984). Multiple factors in the satiation of salt appetite. *Behav. Neurosci.* **98**:661–673.

2
Metabolic Control of Calorie Intake

Mark I. Friedman

Monell Chemical Senses Center
Philadelphia, Pennsylvania

I. INTRODUCTION

Forty years ago, E. F. Adolph showed that rats maintain calorie intake in the face of changes in dietary caloric density by altering the amount of food consumed (Adolph, 1947). On the basis of these results, Adolph concluded that "animals eat for calories." The implication is that there is a mechanism that monitors ingested calories and controls food intake accordingly.

Although the constancy of calorie intake is often emphasized, there are a number of conditions in which it changes markedly. Homeostatic demands cause calorie intake to increase—for example, when the expenditure of energy increases as a result of exercise, lactation, or cold ambient temperatures. Calorie intake may also be disturbed, as in the case of hyperphagia associated with obesity. Finally, adjustments in food intake in response to diet dilution or nutrient loading are not always so precise. For example, Adolph (1947) found that calorie intake is maintained less accurately when the diet is diluted with nonnutritive bulk than when it is diluted with water.

Departures from a steady, precisely maintained level of calorie consumption have been explained in various terms. Changes in the level of calorie intake attendant with muscular activity, lactation, and thermoregulation may be responses to other regulatory signals controlling food intake which modify or override the mechanism to monitor calorie intake. Disturbances in calorie intake, such as overeating or anorexia, have been attributed to a defective neural mechanism that controls food intake. Hyperphagia has also been thought to result from the influence of nonregulatory controls of feeding based on environmental factors or the palatability of food. The imprecision of calorie intake control in response to diet dilution or nutrient preloading could be due to inherent system errors, or it may reflect the influence of other controls of food intake.

In this chapter I attempt to show how both the constancy and precision of calorie intake on one hand, and the variability and imprecision on the other, can be understood parsimoniously in terms of a control of food intake that is based on a signal generated in the oxidation of metabolic fuels. This explanation shifts the emphasis away from the food and its caloric density to the processing of metabolic fuels, which can be derived from foodstuffs as well as from endogenous reserves. The source of calorie intake control thus lies not in the energy value of food, but in the biochemical realization of that energy through fuel oxidation.

II. POSTABSORPTIVE CONTROL OF CALORIE INTAKE

Calorie content may be assessed at a number of places along the path that food takes from the mouth to its ultimate metabolic fate. The taste and smell of food do not appear to provide critical information about dietary calories because rats that feed only by self-administered intragastric injections of food maintain calorie intake in the face of dilution of their liquid diet (Epstein, 1967). Rather, feedback about the caloric value of food is obtained further downstream.

A. Gastrointestinal Signals

McHugh and Moran (1985) have observed precise, compensatory changes in food intake after intragastric calorie loads in rhesus monkeys. Their results point to a mechanism of calorie intake control based on a signal from the stomach that reduces food intake in proportion to the degree of gastric fill. Stomach fullness is dependent on the rate of gastric emptying, which is in part determined by the caloric value of the gastric contents; the lower the caloric density of gastric contents, the faster they empty. However, because their experiments are conducted using animals that are fed once daily during a restricted period, it is not clear whether such a gastric control operates under normal free feeding conditions (see Friedman, 1990). Gastrectomy or gastric vagotomy does not disrupt daily food intake or the compensatory response to dietary dilution, although these procedures may alter the size and frequency of meals (Snowdon and Epstein, 1970; Tsang, 1938). These results suggest that the calorie content of food is not assessed in the stomach under ad libitum feeding conditions. Factors beyond the stomach appear to be involved.

Intestinal factors, which may play a role in feeding behavior, apparently are not involved in the control of calorie intake either. The gut peptide hormone cholecystokinin (CCK) can reduce food intake in the short term when administered by injection; repeated administration with each meal does not alter daily food intake, however, because the smaller meal size is compensated for by an increased meal frequency (West et al., 1984). Furthermore, it is not clear how the peptide could signal calories because it is released in response to only certain energy-yielding substrates (i.e., fats). Perhaps the most striking demonstration that intestinal factors are not critical for calorie intake control is the case of parabiotic rats with crossed intestines (Canbeyli and Koopmans, 1984). These animals, which are joined surgically by skin with little exchange of large molecules (e.g., hormones), individually maintain and defend calorie intake over the long term even though ingested nutrients are largely absorbed by their partner. It would seem that we need to follow ingested calories into the postabsorptive compartment to find the mechanism that detects them.

B. Postabsorptive Signals

Rats will reduce calorie intake in a compensatory fashion after intragastric nutrient loads even when time is allowed for gastrointestinal clearance (Booth, 1972). McHugh and Moran have shown that monkeys will compensate precisely during their daily feeding period for intragastric loads given on the previous day, long after the loads would have emptied from the stomach (McHugh and Moran, 1978). Rats and humans reduce daily voluntary energy intake in compensation for calories given intravenously (Nicolaidis and Rowland, 1976; Friedman et al., 1986a). Compensation is especially precise when the mix of macronutrients given intravenously is similar to that consumed in the normal diet. Apparently, signals generated in the postabsorptive metabolism of foodstuffs control calorie intake.

It should be emphasized that whereas the taste and smell of food or the degree of gastric fill may not provide critical feedback in the control of calorie intake, these factors can serve as cues that signal the caloric value of food through learned association with the postabsorptive effects of food (see Tordoff, Chapter 12, this volume). It is possible that under some circumstances these conditioned cues suffice to maintain energy intake, such as when the caloric density of the diet is constant and therefore predictable. Nevertheless, the value of the cues will change when the postabsorptive payoff from the food changes. What appears to be learned, and what is ultimately critical, is the caloric value of the food, a value that is appreciated postabsorptively, and not in the mouth, nose, or stomach.

Fats and carbohydrates are the primary sources of calories in the diet, and one might expect the postabsorptive metabolism of these macronutrients to be involved in the control of calorie intake. Signals associated with lipid and glucose metabolism have long been thought to be involved in the control of food intake (see Friedman and Stricker, 1976; Friedman, 1990). The amount of body fat has been considered to be the source of a long-term signal, although this signal has yet to be identified (Harris and Martin, 1884). In any case, it is unlikely that the size of body fat stores could provide information to control calorie intake because compensatory adjustments in food intake in response to diet dilution can occur well before any appreciable change in the size of body fat depots would be evident. In addition, obese animals maintain calorie intake in the face of diet dilution like normal animals do (e.g., McGinty et al., 1965).

There is substantial evidence that changes in the metabolism of glucose and fatty acids alter food intake (see Friedman and Tordoff, 1986). However, it seems unlikely that signals produced from either glucose or fatty acid utilization alone would control calorie intake, because only part of the available calories would be represented. Recent studies using selective substrate analogues show that combined inhibition of glucose and fatty acid utilization produces a synergistic increase of food intake in rats (Friedman and Tordoff, 1986). This interactive effect suggests the existence of a mechanism that integrates information from the metabolism of glucose and fatty acids. An integrated signal such as this could better reflect the availability of calories in the postabsorptive compartment.

C. Metabolizable Calories

The effects of these metabolic inhibitors of glucose and fatty acid metabolism on food intake depend on the source of dietary calories (Friedman et al., 1986; Scharrer and Langhans, 1986; Tordoff et al., 1988). Thus, inhibition of fatty acid metabolism increases food intake more when fat is the main energy-yielding substrate in the diet, whereas glucose analogues are more effective when carbohydrates are the main energy source. These results serve as another example of the use of alternative fuels in the control of food intake (see Friedman and Stricker, 1976). Such interchangeability implies that some common feature of the different metabolic fuels influences food intake. One thing metabolic fuels have in common, although to different degrees, is their caloric value. The energy content of metabolic fuels could be the common currency that underlies the capacity to maintain calorie intake when one source of energy is exchanged for another.

To realize the caloric value of food fuels, the fuels must be metabolized; simple availability in blood is not enough. The effect of substrate analogues on feeding indicates that the source of the metabolic signal is intracellular because the site of action of these agents is in the metabolic pathways within cells. The importance of metabolizable energy in the control of calorie intake is also demonstrated by the marked differences between normal and diabetic rats to nutrient-specific dilution of their diet (Friedman, 1978). Normal rats increase food intake to the same extent whether the caloric density of their diet is decreased by reducing either the carbohydrate or the fat in equicaloric amounts through substitution with nonnutritive bulk. This is the typical response to diet dilution, and it illustrates the use of alternative fuels in the control of calorie intake. In contrast, diabetic rats do not alter food intake when the carbohydrate content of the diet is decreased, whereas they increase food intake markedly in response to an equicaloric reduction in fat content. This hyperphagia is a distinguishing characteristic of diabetic rats fed typical low-fat commercial diets.

These results cannot be understood in terms of a mechanism that monitors the caloric value of ingested food because normal and diabetic rats ate the same food. Instead, these findings indicate that the capacity to adjust food intake in response to diet dilution is based on a signal related to the extraction of energy from the food, because normal and diabetic rats differ in their ability to derive calories from different energy sources in the diet. Normal rats can utilize both carbohydrates and fats for energy production, and therefore they respond similarly to equicaloric losses in these two fuel sources. In contrast, whereas diabetic rats readily metabolize fats for energy, they do not readily utilize carbohydrates. Indeed, studies showing that diabetic rats increase food intake when their high-fat diet is "diluted" with carbohydrates to the same extent as they do when it is diluted with cellulose indicate that dietary carbohydrate is little more than filler for these animals (Friedman, 1978). Thus, to refine Adolph's conclusion, animals eat for metabolizable calories, not for calories in food.

III. THE METABOLIC STIMULUS AND RECEPTOR

If the extraction of energy from foodstuffs generates a signal that controls food intake, then that signal is likely to originate in the postabsorptive metabolism of food fuels because this is where food calories are ultimately realized. But where in metabolism might the energy derived from metabolic fuels provide a signal that controls food intake? And where in the body might such a signal be generated?

A. The Stimulus of Fuel Oxidation

It would seem that for something as vitally important as caloric homeostasis, a mere index or correlate for metabolizable energy would not do. Whereas the taste and smell of food, gastric fill, substrate blood levels, or utilization of specific fuels could serve as proxies for the caloric value of food, the only foolproof way for the body to ascertain a food's energy content is by an intracellular "analysis." At an intracellular level, it is in the pathways of oxidative phosphorylation that calories in food as well as from endogenous sources are realized. It is at the level of fuel oxidation that calories can be turned into a signal that the nervous system reads to control feeding behavior and calorie intake.

Research from a variety of sources (see, e.g., Booth, 1972; Friedman and Stricker, 1976; Nicolaidis and Even, 1985; Friedman and Tordoff, 1986) provides a picture of a stimulus for food intake control that is associated with changes in fuel oxidation.

1. The stimulus is generated during intramitochondrial oxidative phosphorylation. It is therefore produced during the complete oxidation of fuels to a biochemically useful form of energy, namely adenosine triphosphate (ATP), as opposed to incomplete or uncoupled oxidation, which produces heat.

2. The stimulus is independent of the type of fuel used for oxidation. Although fats, carbohydrates, and amino acids provide substrates for energy produc-

tion, the signal is tied to a metabolic process common to the utilization of these different substrates, not to the presence or metabolism of a specific fuel. Which fuel is being oxidized is therefore less important than how much fuel is oxidized.

3. The stimulus is independent of the source of metabolic fuel. Fuels for oxidation may come directly from food in the gastrointestinal tract or from "food" that has been stored previously in bodily reserves (i.e., adipose tissue lipids and liver glycogen). Because the stimulus is at the end of the metabolic sequence for energy production, it is indifferent to the original source of that energy. Whether the source of oxidizable fuel is exogenous or endogenous is therefore less important than whether the overall supply is adequate, insufficient, or in surplus.

4. The oxidative stimulus for feeding is generated during normal fluctuations in fuel metabolism. This includes metabolic shifts that occur during the diurnal cycle and, in the shorter term, from meal to meal. In addition, changes in fuel metabolism that take place over seasonal and reproductive cycles may also produce an oxidative signal that modifies food intake. Although changes in fuel oxidation do not need to be extreme to provide a stimulus controlling food intake, the signal may also be generated under emergency conditions such as during severe hypoglycemia induced by insulin injection.

5. The oxidative stimulus is inversely related to food intake. Increases in oxidation decrease food intake, whereas reductions in oxidation increase it. The status of the oxidative signal, like that of food intake, is described relative to another circumstance or to a baseline or steady-state condition along a time dimension.

Viewing the metabolic control of food intake in terms of oxidation brings order to a rather fragmented picture of the metabolic control of food intake. This is depicted in Figure 1, which shows how a variety of metabolic pathways that have been implicated in the control of food intake can be organized and understood in terms of their relationship to the pathways of intramitochondrial oxidation. Instead of seeing responses to experimental manipulation of each metabolic pathway as a reflection of an individual control of food intake, one can see how different treatments might all act via their effect on oxidation, a final common pathway in the process.

Crucial questions remain about the nature of oxidative stimulus for feeding, and the most important deals with the central issue of which specific biochemical event or events constitute the signal. Nevertheless, the idea that changes in oxidative metabolism control food intake has considerable theoretical appeal. The model is parsimonious because it integrates earlier theories of food intake that were based on separate signals from glucose and fat metabolism, and it provides a mechanism by which changes in the utilization of glucose and lipid can interact to control feeding. The theory also has broad explanatory power, providing an understanding of many phenomena associated with feeding behavior (see Friedman and Stricker, 1976; Friedman, 1990), including the control of calorie intake, as discussed below.

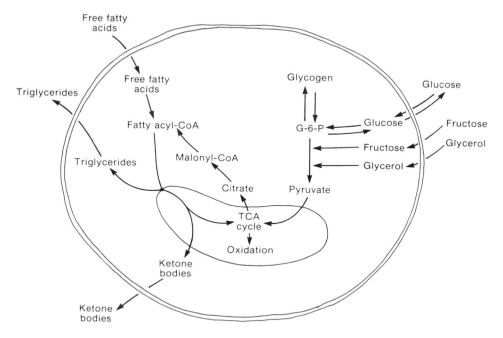

FIGURE 1 Relationship of various metabolic pathways and substrates implicated in the control of food intake to intramitochondrial oxidative phosphorylation: CoA, coenzyme A; G-6-P, glucose-6-phosphate; TCA cycle, tricarboxylic acid cycle.

B. The Hepatic Receptor

For changes in fuel oxidation to affect food intake, they must be detected by the nervous system. Fuel oxidation occurs everywhere in the body. In which tissue does the signal of oxidation originate? Because different tissues have distinctive metabolic capabilities, we might look to the metabolic map shown in Figure 1 for a clue. If, as pictured, the different metabolic processes that affect feeding are located within one cell type, what emerges is a cell that looks a great deal like a hepatocyte. Specifically, it is a cell in which fatty acids can be oxidized, esterified to form triglycerides, and synthesized from glucose and three-carbon precursors. In addition, the cell is capable of metabolizing fructose and glycerol, and storing glycogen. This interpretation based on metabolic data is consistent with empirical evidence that a metabolic stimulus controlling food intake is produced in the liver.

There are four main lines of evidence for a hepatic site of origin of the metabolic signal.

1. Hepatic–portal vein infusions of metabolic substrates, substrate analogues, and drugs are more effective than systemic infusions in altering food intake (see Russek and Racotta, 1986; Tordoff and Friedman 1986). In the case of

glucose, the hepatic–portal infusions reduce food intake under otherwise everyday feeding conditions when substrate concentrations and delivery rates that fall within the physiological range are used (Tordoff and Friedman, 1986). This indicates that the liver participates normally in the control of feeding.

2. Manipulations of metabolism that are specific to the liver affect food intake (see Tordoff and Friedman, 1988). By virtue of its transport or enzymatic capabilities, the liver is able to metabolize certain substrates that other tissues handle in a very minimal way, if at all. Thus, administration of fructose, which does not cross the blood–brain barrier and is metabolized primarily in liver, reduces food intake (Tordoff and Friedman, 1988). Similarly, because of the high K_m of hepatic glycerol dehydrogenase, the liver is most capable of metabolizing the large doses of glycerol, which reduce feeding (Ramirez and Friedman, 1982). Some metabolic inhibitors that increase food intake, such as the fructose analogue 2,5-anhydro-D-mannitol (2,5-AM), or the fatty acid analogue methyl palmoxirate (MP), also appear to have a hepatic site of action (Friedman and Tordoff, 1986; Tordoff et al., 1988).

3. Changes in hepatic metabolism parallel behavioral effects of experimental treatments (see Russek and Racotta, 1986). Studies comparing the effects of hepatic–portal versus jugular substrate infusions have shown that the portal infusions that alter food intake selectively alter hepatic metabolism (Tordoff and Friedman, 1988), and, in the case of glucose, the behavioral effects of different concentrations coincide with effects in liver. Such independent confirmation of the intended effect of treatments on the liver has also been reported for oral fructose administration (see Tordoff, this volume, Chapter 12) and injection of 2,5-AM and MP (Friedman and Tordoff, 1986; Tordoff et al., 1988). However, whereas certain changes in liver metabolism may parallel those in feeding under a particular experimental condition, no such metabolic change has been found that correlates with alterations in food intake across conditions.

4. Cutting hepatic nerves alters ad libitum food intake and the response to experimental treatments. Section of the hepatic branch of the vagus nerve alters the diurnal rhythm of feeding (Friedman and Sawchenko, 1984), abolishes the suppression of food intake by fructose and other metabolites (Friedman and Granneman, 1983; Langhans et al., 1985), and eliminates the feeding response to administration of 2,5-AM (Tordoff and Friedman, unpublished observations) and the fatty acid inhibitor mercaptoacetate (Langhans and Scharrer, 1987). These findings implicate the liver in the day-to-day control of feeding and point to a hepatic site of action of the various experimental manipulations. However, because the hepatic vagus contains both afferent and efferent nerve fibers, it is not possible to determine whether the effects of nerve section are due to interruption of sensory information from the liver or to interference with the generation of the signal by alteration in liver metabolism.

How information about liver metabolism is communicated to the brain to control food intake is a matter for speculation. A neural connection is suggested by the effects of nerve section although, again, these results do not necessarily implicate the afferent nerve fibers. Changes in hepatic metabolism could elicit secretion of a bloodborn signal, but there is no evidence for such a mechanism. Other tissues, particularly the brain, may also monitor fuel metabolism in the control of feeding. Although cerebral sensors of metabolism may be involved in the emergency feeding response to severe fuel depletion, there is no direct evidence that they participate in the normal control of calorie intake (see Epstein et al., 1975). Indeed, brain lesions that disrupt the feeding response to energetic emergencies do not prevent the behavioral compensation for caloric dilution of the diet (Carlisle and Stellar, 1969). Unless there are specialized brain cells with metabolic characteristics of hepatocytes, the weight of the evidence falls toward a metabolic signal for food intake control based in the liver.

IV. FUEL OXIDATION AND CONTROL OF CALORIE INTAKE

Before describing in detail how a stimulus of fuel oxidation can account for the different and sometimes paradoxical aspects of calorie intake—that is, its constancy, change, disruption, and imprecision—several important general features of the model need to be highlighted:

1. As discussed above, the stimulus from oxidation is indifferent to the source of fuels. In other words, at the level of metabolism where the signal is generated, there is no distinguishing between calories (fuels) coming directly from food in the gastrointestinal tract and those derived from internal stores such as adipose tissue and glycogen. Therefore, at any given moment, oxidative metabolism in the receptor cell is dependent on exogenous fuels from food, on endogenous fuels mobilized from storage, or any combination thereof.

2. Oxidation is not the only fate of metabolic fuels (Fig. 2). Fuels can be stored, primarily as fat in adipose tissue and to a lesser extent as glycogen in liver. Metabolic fuels may also be lost. In the context of the model being developed here, "lost" fuels are those that are oxidized in an uncoupled fashion for heat production, are excreted in feces or urine (e.g., as in diabetic glycosuria), or are used for energy production in nonreceptor cells (e.g., muscle, brown adipose tissue). Under steady-state or equilibrium conditions, the flux of fuels and their partitioning between routes of oxidation, storage, and loss remains constant. When the supply of fuels is held constant, changes in the flux of fuels via one route will affect the flux through the other routes. How fuels are partitioned among these three paths for disposal (i.e., oxidation, storage, and loss) thus can determine the level of fuel oxidation and, in turn, food intake.

If food is not the only immediate source of oxidizable fuels, and oxidation in receptor cells is not the only fate of energy-yielding substrate, then it follows that the number of calories ingested does not necessarily have to track changes in dietary calorie content or caloric requirements. Food intake may or may not change when dietary energy density varies or when energy requirements are

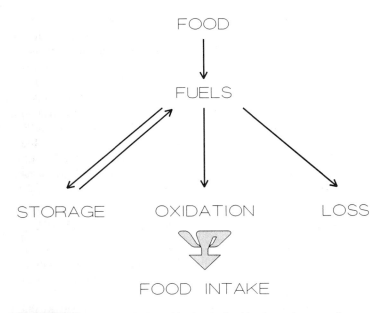

FIGURE 2 Role of fuel partitioning in food intake under normal, unperturbed (steady-state) conditions. Note that fuels may be derived from food or from endogenous stores (adipose tissue lipid or liver glycogen). Loss refers to fuels lost in urine or feces, oxidized in an uncoupled fashion for thermogenesis, or oxidized by nonreceptor tissues (see text).

altered; this is because, in addition to calories in food, there are calories in body reserves that can satisfy energy needs. Similarly, when the energy content of food is constant, energy intake may change because shifts in the partitioning of fuels alter the flux through oxidative pathways in the critical receptor tissue. At times, changes in energy consumption may seem appropriate in terms of energy balance while at others they may seem counter to it. This apparent uncoupling between calorie intake and energy balance occurs because ingested fuels are only one source of substrate for oxidative metabolism and fuels eaten may not be metabolized through oxidative pathways. Thus, even though the signal from oxidation controls calorie intake—indeed, even if it were the only controlling stimulus—it is not necessarily a faithful index of calorie intake.

From this perspective, "regulation of calorie intake" is a misnomer. Calorie consumption is not regulated, but rather is controlled, along with energy expenditure, in the service of energy homeostasis. Changes in calorie intake can be regulatory in the sense that they can be appropriate to homeostatic needs and requirements, even though calorie intake is not regulated in the sense of being held at some predetermined level. If anything is regulated, it is energy production, which is essential for life. Eating can be viewed as a means to this end, to the extent that the number of calories consumed is governed by changes in the oxidation of fuels that produces this energy. In other words, to refine Adolph's conclusion further, animals eat for oxidation.

A. Constancy of Calorie Intake and the Illusion of Imprecision

The constancy of calorie intake under normal, unperturbed conditions can be understood simply by the maintenance of a steady state as shown in Figure 2. When the flux of fuels is regular and the partitioning between routes of disposal is in equilibrium, the stimulus generated by fuel oxidation and, in turn, food intake, remain the same.

The compensatory increase in food intake seen in response to diet dilution can be considered to be a response to a decreased flux of energy-yielding substrate through critical oxidative pathways. Less fuel is available for oxidation because less is consumed. Figure 3, which illustrates the response to dilution, depicts the situation before compensation occurs and, like the figures that follow, should be viewed as a change from the steady state shown in Figure 2. A similar model could

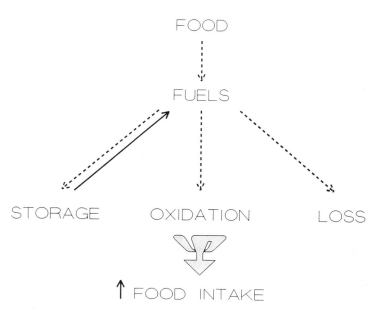

FIGURE 3 Role of fuel partitioning in the compensatory increase in food intake in response to dietary caloric dilution. Dashed line denotes a decrease in fuel flux from steady state as shown in Figure 1. The decrease in fuel flux through oxidative pathways leads to an increase in food intake.

account for the reduction of food intake after caloric preloads, except that the flux of fuels would be increased.

Diet dilution is shown here as reducing the flux of fuels through all three routes of disposal, but that need not be the case. For example, flux through oxidative pathways could be maintained through increased mobilization of stored fuel at the expense of body fat. In this case, there would be no change in food intake, or less than might be predicted from the change in dietary caloric density, and calorie intake would decline. In a similar vein, extra calories given by forced feeding could be disposed of primarily through storage with little effect on fuel oxidation, thereby resulting in a smaller decrease of food intake than might be expected from the amount of the calories administered.

A change in food intake insufficient to compensate for caloric dilution or caloric load is often seen as an example of dysregulation or evidence that the mechanism controlling calorie intake is imprecise. According to the model developed here, these interpretations misidentify the regulated variable and misplace the source of the signal that controls food intake. As discussed above, calorie production, not intake, is regulated, and it is calorie production from fuels in food as well as endogenous sources, not just from food, that provides the stimulus controlling food intake. Thus, what may appear as a failure of regulation or loss of control from the experimenter's point of view, may simply be a reflection of the fact that events inside the animal (i.e., fuel partitioning and oxidation), not the experimenter's treatments, ultimately determine food intake. The approach requires that the control of eating behavior be relocated from the realm of experimental manipulations, such as diet dilution or calorie loading, and placed inside the animal, where it belongs.

B. Change in Response to Homeostatic Demands

Calorie intake usually changes along with the bodily demand for energy. Normally, energy needs and intake increase in the cold (when thermogenesis is required to maintain body temperature), under conditions of sustained exercise (when calories are needed for muscular activity), and during lactation (when energy is diverted to milk to feed young). Reversal of any of these conditions results in a decline in energy requirements and, in turn, a decrease in calorie intake. Within the context of the model for food intake control presented here, metabolic fuels used to meet the demand for energy under these various conditions are lost; that is, lost in the sense that, once used, they are not available for oxidation in tissues, where their oxidation is detected to control food intake. However, as illustrated in Figure 4, increased use of fuels via these pathways can affect food intake by altering the availability of energy-yielding substrates to oxidative pathways that provide a signal for feeding.

An increased demand for energy does not have to be met solely by an upward adjustment in food intake. Fat fuels stored in adipose tissue can be mobilized to help meet the demand. Such a strategy could have survival value: for example,

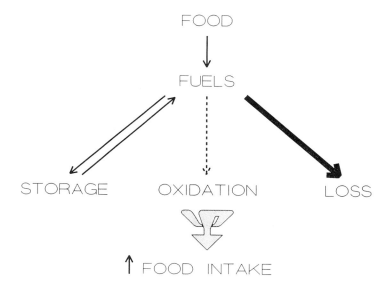

FIGURE 4 Role of fuel partitioning in the compensatory increase in food intake during exercise, lactation, or thermogenesis. Thicker line denotes increased and dashed lines indicate decreased fuel flux relative to normal steady state as shown in Figure 1 (see Fig. 1 and text for description of loss pathway). The decrease in fuel flux through oxidative pathways leads to an increase in food intake.

when foraging for food in the cold or leaving the nest of young presents risks to the individual or its offspring. That food intake does not totally compensate for changing energy demands and that body weight (fat) is lost should not be interpreted as indicating a lack of precise intake control or a failure of homeostasis. After all, what is body fat for if not to help sustain energy requirements in time of need? What is important is that those needs be met efficiently, and there are both behavioral and physiological adaptations that can be flexibly recruited to do so.

C. Disturbances in Calorie Intake

Calorie intake is considered to be disturbed when it changes in the absence of any obvious reason, such as cold ambient temperature or increased muscular activity, and is usually accompanied by marked changes in body weight, primarily due to gain or loss of body fat. Such disturbances can be understood in terms of the model under discussion by considering the impact of body fat deposition and mobilization on metabolic fuel availability.

Hyperphagia associated with obesity is accompanied inevitably by a metabolic state that fosters the storage of body fat. However, increased fuel deposition is not caused by overeating alone. In every animal model of obesity that has been examined, there is a shift in fuel metabolism toward fat storage that occurs primary to and independent of the change in food intake (see Friedman and Stricker, 1976; Friedman, in press). Overeating during the development of obesity can thus be seen as a response to the redirection of fuels away from pathways of oxidation toward those of fat synthesis and storage (Fig. 5). In other words, hyperphagia not only contributes to increased fat deposition, it is also caused by it. In essence, the behavioral response to excessively storing fuels as fat is the same as that to losses of fuels for heat production or muscular exercise (Fig. 4), except in the case of obesity fuels are "lost" to adipose tissue.

There may be any number of metabolic mechanisms that could serve to direct fuels toward storage instead of oxidation (e.g., elevated insulin levels, increased activity of lipoprotein lipase). However, regardless of how this shift in partitioning is achieved, the net effect on oxidation and, in turn, food intake, would be the same. Obesity may occur without any appreciable increase in food intake. According to the model, this would be accomplished if fuels that are otherwise lost (e.g., in thermogenesis) were instead stored as fat with no net change in fuel oxidation through the pathways that provide the signal controlling intake.

Overeating of certain diets is often seen as a response to increased food

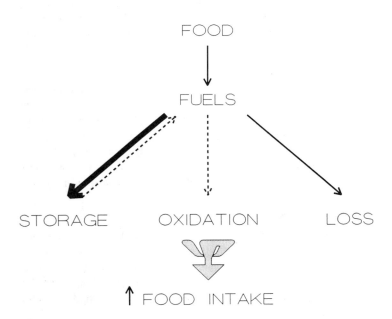

FIGURE 5 Role of fuel partitioning in hyperphagia during the development of obesity. Thicker line denotes increased and dashed lines indicate decreased fuel flux relative to normal steady state as shown in Figure 1. The decrease in fuel flux through oxidative pathways leads to an increase in food intake.

palatability and, as such, as an example of how nonregulatory controls of feeding can override normal homeostatic mechanisms of energy balance. The lack of direct evidence that palatability can drive overeating for more than the short term along with substantial evidence pointing to the critical role of diet composition (see Naim and Kare, this volume, Chapter 6) suggests that this form of overeating may also be due to shifts in the partitioning of fuels toward storage. Diets high in fat and calories, which are especially effective in promoting hyperphagia, are also high in carbohydrates, which would tend to promote the storage of dietary fat (see Ramirez et al., 1989). In a sense, overeating of such diets is like the response to diet dilution except that the reduction in the caloric value of the food occurs internally as energy-yielding substrates are diverted from oxidative pathways where the food energy would be realized. It would appear to be unnecessary to postulate another control of food intake to account for dietary hyperphagia, just as it seems unnecessary to do so to deal with the increase in intake that occurs under conditions of increased energy demand.

Just as increased fuel deposition may lead to hyperphagia by diverting fuels from the fate of oxidation, excessive mobilization of fat fuels from adipose tissue could decrease food intake by supplying fuels to oxidative pathways (Fig. 6).

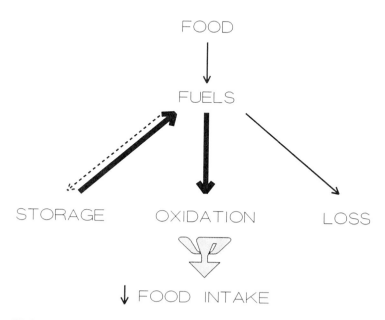

FIGURE 6 Role of fuel partitioning in the hypophagia associated with excessive mobilization of body fat reserves. Thicker line denotes increased and dashed lines indicate decreased fuel flux relative to normal steady state as shown in Figure 1. The increase in fuel flux through oxidative pathways leads to a decrease in food intake.

Reduced calorie intake under conditions of fat mobilization is seen experimentally when treatments that cause animals to overeat and become obese are terminated or when insulin replacement treatment in diabetic rats is terminated (see Ramirez and Friedman, 1983). Clinically, such a situation might obtain in acute diabetic cachexia. Anorexia associated with other diseases might be due to oxidation of mobilized fuels. In this regard, it is intriguing that cachectin, which is found in pathologies marked by anorexia and wasting, is a potent lipolytic agent (see Beutler, 1988).

Hyperphagia and anorexia are often thought to result from a defective mechanism controlling food intake. The model presented here suggests that these disturbances in calorie intake may occur not because the mechanism controlling food intake is broken, but rather because it works too well. It detects internal shifts in fuel metabolism that affect the critical process of energy production and adjusts food intake appropriately. From a clinical point of view, this sensitivity can create real problems for those who are overeating or anorectic. Understanding the nature of the critical signal of oxidation and the factors that can affect it through fuel partitioning should help to guide the development of strategies to correct disturbances in calorie intake.

V. CONCLUSIONS

There are other phenomena that would lend themselves to the type of analysis used above, including the rhythms in calorie intake that occur during seasonal, estrous, and diurnal cycles, or the compensatory hyperphagia that occurs during refeeding after a fast. In these and other cases, like the ones discussed above, the analysis would focus on the status of fuel oxidation. The parsimony inherent in the model used here suggests that, in fact, it is not necessary to view calorie intake in terms of its constancy, change, and disturbances, because all these aspects are manifestations of the same mechanism of control. This mechanism is based on a signal from fuel oxidation that is indifferent to whether fuels are derived from food or from internal stores, and is sensitive to internal shifts in the partitioning of those fuels among pathways of storage, loss, and oxidation. According to this view, calorie intake is not regulated or monitored directly, but is controlled by, and operates in the service of, fuel oxidation. Constancy, changes, disturbances, and even apparent imprecision of calorie intake thus reflect the state of fuel flux through oxidative pathways—the status of a common stimulus—rather than the operation of multiple controls of food intake or a defective mechanism of control.

ACKNOWLEDGMENTS

The author thanks Drs. Israel Ramirez and Michael Tordoff for their helpful comments on the manuscript.

DISCUSSION

Weingarten: I have no difficulty with the idea that the rat has the impressive ability to regulate calories within a 24-hour period; nor, for that matter, that the liver is involved with this. I have difficulty understanding how exactly the liver is involved. If you look at the Adolph studies, the preload studies, and many of your studies, they imply, as Jean Mayer said, that the rat has a metabolic memory. The rat can integrate caloric intake over a 24-hour period to match its output. In a lot of your studies, it seems to me you present the oxidative stimulus as being a stimulus for the initiation of eating. Now I'd like to know whether you think the ability of the animal to integrate energy intake over 24 hours sits in the liver, or whether the liver is primarily involved with kicking in the meal, and the memory resides somewhere else.

Friedman: First of all, I think you can argue given this kind of model that you don't necessarily need to make distinctions between long term and short term or initiation and termination, which is perhaps a more general issue. I think, from a metabolic point of view, that the liver is in an excellent position to integrate metabolic information. It does that in the service of fuel metabolism, and it is possible that it generates a signal that reflects these metabolic integrations and controls food intake.

Steffens: You present some very interesting data, and I believe oxidation is an important factor regarding regulation of food intake. You show some nice results about a free fatty acid analogue that increases food intake. But mostly it is assumed that the brain is involved in the regulation of food intake, and the brain is not able to use free fatty acids. At this point I have a small problem. What is the brain doing with that free fatty acid analogue? Or do you think that it is really the liver that's doing the whole job?

Friedman: When I listed the criteria for hepatic control of food intake, one of them had to do with treatments that are relatively specific to the liver. Methyl palmoxirate, I think, is one of those treatments, and there's evidence to that effect. I think that the brain is getting information from various sources including the liver, the stomach, the mouth, and the nose, and integrating that information. You can make a very strong argument that the evidence for a metabolic signal controlling food intake based in the liver is much better than the evidence that it's based in the brain. The main problem with evidence for a metabolic receptor in the brain is that the treatments that have been used to demonstrate this are not specific to the brain. If you put something in the ventricles or in brain tissue, and it affects food intake, you also affect peripheral metabolism or physiology. The issue for me is, Can you affect the brain without affecting peripheral metabolism physiology and see a change in food intake? In terms of localization, I think the portal–jugular comparisons argue strongly for a hepatic site of reception.

Campfield: I would like to go to the point you made about the oxidative stimulus. It would help us all to know what to measure. In the experiment you outlined of animals in the cold, we all know they eat more. What shall we measure to test this hypothesis?

Friedman: If we knew specifically what the stimulus is, I could tell you what to measure. What we're trying to do by using these metabolic manipulations is to focus on what that stimulus is and how to investigate it. At this point it's theoretical. The power of this comes in a theoretical light: in how to ask that question and how to understand this array of phenomena we're faced with, to say nothing of the growing list of controls of food intake. That's where the value is; but I agree with you, we'd also like to know what the stimulus is and what to measure.

Campfield: Finding the stimulus is another issue. Calling it oxidation implies that oxidation has been measured, and I think that's the issue. What can we measure in any way to correlate with the stimulus? I think that's the challenge to the theory.

Friedman: I would agree with you. I don't mean to say that it's oxidation per se that's being monitored, but it's some aspect of the process of oxidative phosphorylation that is generating the signal.

VanItallie: One of the things that bothers me about this interesting notion is the time factor. In other words, you look at the regulation of food intake in human beings. There's no relationship between food intake and energy expenditure on any given day. If you look at it over a period of a week, there's a pretty good relationship in active young people between food intake and energy expenditure. What I'm asking you is, How could one account for this apparent delay, between what might be going on at any given time in oxidation, or some correlate of oxidation, and the fact that there is a latent period before the adjustments take place? Is there a coupling with some conditioning systems, for example?

Friedman: I would ask what's happening to the fuels at a given time. If the metabolic system can buffer these changes that might occur in response to changes in diet or energy demand by drawing on internal fuels, then it appears that there's not regulation. Again it gets back to the issue of what is being regulated: Is it something associated with oxidative phosphorylation, ingested calories, or energy balance overall? That's my first question. It's certainly true that there are other things that will affect how much food is consumed in the short term, which may have no relationship to this metabolic stimulus.

REFERENCES

Adolph, E. F. (1947). Urges to eat and drink in rats. *Am. J. Physiol.* **151**:110–125.
Beutler, B. (1988). Cachexia: A fundamental mechanism. *Nutr. Rev.* **46**:369–373.
Booth, D. A. (1972). Postabsorptively induced suppression of appetite and the energostatic control of feeding. *Physiol. Behav.* **9**:199–202.
Canbeyli, R. S., and Koopmans, H. S. (1984). The role of jejunal signals in the long-term regulation of food intake in the rat. *Physiol. Behav.* **33**:945–950.
Carlisle, H. J., and Stellar, E. (1969). Caloric regulation and food preference in normal, hyperphagic, and aphagic rats. *J. Comp. Physiol. Psychol.* **69**:107–114.
Epstein, A. N. (1967). Feeding without oropharyngeal sensations. In *The Chemical Senses and Nutrition*, M. R. Kare and O. Maller (Eds.). Johns Hopkins Press, Baltimore, pp. 263–280.
Epstein, A. N., Nicolaidis, S., and Miselis, R. (1975). The glucoprivic control of food intake and the glucostatic theory of feeding behavior. In *Neural Integration of Physiological Mechanisms and Behavior*, G. J. Mogenson and F. R. Calaresu (Eds.). University of Toronto Press, Toronto.
Friedman, M. I. (1978). Hyperphagia in rats with experimental diabetes mellitus: A response to a decreased supply of utilizable fuels. *J. Comp. Physiol. Psychol.* **92**:109–117.
Friedman, M. I. (1990). Making sense out of calories. In *Handbook of Behavioral Neurobiology*, Vol. 10, *Neurobiology of Food and Fluid Intake*, E. M. Stricker (Ed.). Plenum Press, New York, pp. 513–529.
Friedman, M. I. (in press). Body fat and the metabolic control of food intake. *Int. J. Obesity*.
Friedman, M. I., and Granneman, J. (1983). Food intake and peripheral factors after recovery from insulin-induced hypoglycemia. *Am. J. Physiol.* **244**:R374–R382.
Friedman, M. I., and Sawchenko, P. E. (1984). Evidence for hepatic involvement in the control of ad libitum food intake in rats. *Am. J. Physiol.* **247**:R106–R113.
Friedman, M. I., and Stricker, E. M. (1976). The physiological psychology of hunger: A physiological perspective. *Psychol. Rev.* **83**:409–431.
Friedman, M. I., and Tordoff, M. G. (1986). Fatty acid oxidation and glucose utilization interact to control food intake in rats. *Am. J. Physiol.* **251**:R840–R845.
Friedman, M. I., Gil, K. M., Rothkopf, M. M., and Askanazi, J. (1986a). Postabsorptive control of food intake in humans. *Appetite*, **7**:258 (abstract).
Friedman, M. I., Tordoff, M. G., and Ramirez, I. (1986b). Integrated metabolic control of food intake. *Brain Res. Bull.* **17**:855–859.
Harris, R. B. S., and Martin, R. J. (1984). Lipostatic theory of energy balance: Concepts and signals. *Nutr. Behav.* **1**:253–275.
Langhans, W., and Scharrer, E. (1987). Evidence for a vagally mediated satiety signal derived from hepatic fatty acid oxidation. *J. Auton. Nerv. Sys.* **18**:13–18.
Langhans, W., Egli, G., and Scharrer, E. (1985). Selective hepatic vagotomy eliminates the hypophagic effect of different metabolites. *J. Auton. Nerv. Sys.* **13**:255–262.
McGinty, D., Epstein, A. N., and Teitelbaum, P. (1965). The contribution of oropharyngeal sensations to hypothalamic hyperphagia. *Anim. Behav.* **13**:413–418.
McHugh, P. R., and Moran, T. H. (1978). Accuracy of the regulation of caloric ingestion in the rhesus monkey. *Am. J. Physiol.* **235**:R29–R34.
McHugh, P. R., and Moran, T. H. (1985). The stomach: A conception of its dynamic role

in satiety. In *Progress in Psychobiology and Physiological Psychology,* J. M. Sprague and A. N. Epstein (Eds.). Academic Press, New York, pp. 197–232.

Nicolaidis, S., and Even, P. (1985). Physiological determinant of hunger, satiation and satiety. *Am. J. Clin. Nutr.* **42**:1083–1092.

Nicolaidis, S., and Rowland, N. (1976). Metering of intravenous versus oral nutrients and the regulation of energy balance. *Am. J. Physiol.* **231**:661–668.

Ramirez, I., and Friedman, M. I. (1982). Glycerol is not a physiological signal in the control of food intake in rats. *Physiol. Behav.* **29**:921–925.

Ramirez, I., and Friedman, M. I. (1983). Metabolic concomitants of hypophagia during recovery from insulin-induced obesity in rats. *Am. J. Physiol.* **245**:E211–E219.

Ramirez, I., Tordoff, M. G., and Friedman, M. I. (1989). Dietary hyperphagia and obesity: What causes them? *Physiol. Behav.* **45**:163–168.

Russek, M., and Racotta, R. (1986). Possible participation of oro-, gastro-, and enterohepatic reflexes in preabsorptive satiation. In *Interaction of the Chemical Senses with Nutrition,* M. R. Kare and J. G. Brand (Eds.). Academic Press, New York, pp. 373–393.

Scharrer, E., and Langhans, W. (1986). Control of food intake by fatty acid oxidation. *Am. J. Physiol.* **250**:R1003–R1006.

Snowdon, C. T., and Epstein, A. N. (1970). Oral and intragastric feeding in vagotomized rats. *J. Comp. Physiol. Psychol.* **71**:59–67.

Tordoff, M. G., and Friedman, M. I. (1986). Hepatic portal glucose infusions decrease food intake and increase food preference. *Am. J. Physiol.* **251**:R192–R196.

Tordoff, M. G., and Friedman, M. I. (1988). Hepatic control of feeding: Effect of glucose, fructose, and mannitol infusions. *Am. J. Physiol.* **254**:R969–R976.

Tordoff, M. G., Flynn, F. W., Grill, H. J., and Friedman, M. I. (1988a). Contribution of fat metabolism to "glucoprivic" feeding produced by fourth ventricular 5-thio-D-glucose. *Brain Res.* **445**:216–221.

Tordoff, M. G., Rafka, R., DiNovi, M. J., and Friedman, M. I. (1988b). 2,5-Anhydro-D-mannitol: A fructose analogue that increases food intake in rats. *Am. J. Physiol.* **254**:R150–R153.

Tsang, Y. C. (1938). Hunger motivation in gastrectomized rats. *J. Comp. Physiol. Psychol.* **26**:1–17.

West, D. B., Fey, D., and Woods, S. C. (1984). Cholecystokinin persistently suppresses meal size but not food intake in free-feeding rats. *Am. J. Physiol.* **246**:R776–R787.

3
Nutritional Determinants of Alcohol Intake in C57BL/6J Mice

R. Thomas Gentry

Mount Sinai School of Medicine
New York, New York

I. INTRODUCTION

One of the ultimate goals of basic research on alcoholism is to answer the question: Why do alcoholics drink? By analogy to other drugs of addiction, it is assumed that the motivation to consume alcohol is related to its physiological effects. Furthermore, we expect that the requisite effects and acquired behavioral patterns are not unique to humans, but can be modeled in animals.

A. Alcoholism as a Motivational Disease

Frequently alcoholism is defined in terms of its consequences: physical dependence, hepatic cirrhosis, or disrupted commitments to family or work. But the essential characteristic is the overwhelming drive to self-administer alcohol despite its devastating consequences.

Alcoholics no longer have a choice, in the ordinary sense, of whether to drink or not; they need to drink to function. Perhaps suggestive is the National Institute for Alcohol Abuse and Alcoholism (NIAAA) slogan: "If you need a drink to be social, that's not social drinking." Alcoholics spend much of their time thinking about alcohol, seeking alcohol, hoarding alcohol, in much the same way any of us would have our thoughts dominated by food, if we were starving. The analogy to homeostatic appetitive behaviors is obvious.

But addictive drinking is not the same as normal appetitive behavior. The motivation is specific for alcohol. Medical therapies designed to reduce the appetite for alcohol are continuously being evaluated, but no pharmacological agent or dietary regimen has ever been found to dramatically attenuate the drive to drink in the way, for example, methadone satisfies the need for opiates (Nyswander and Dole, 1967; Dole, 1988).

Furthermore alcoholism is acquired. People may inherit a susceptibility to alcoholism (see, e.g., Cloninger et al., 1981; Cloninger, 1987), but no one is born an alcoholic. The addictive state requires years of experience drinking alcohol. The initial reasons for drinking apparently give way to the potent and specific drive of an addiction.

Thus, there are at least two issues: How does someone progress from social drinking to alcoholic drinking? What are the biological determinants of the addictive state? This chapter addresses the possible relationships between alcoholism and appetitive behavior derived from what we have learned about alcohol consumption in C57BL/6J (C57) mice.

B. Animal Models of Alcoholism

C57 mice are one of the few strains of experimental animals that prefer alcohol (i.e., a 10% v/v aqueous solution of ethanol) to water. Other animals that prefer alcohol to water include the P rats developed by Li, Lumeng, and their colleagues in Indianapolis (Lumeng et al., 1977; Li et al., 1986, 1987), the AA rats developed at the ALKO laboratories in Helsinki, Finland (Eriksson, 1968, 1972), and the UChB strain from the University of Chile, Santiago (Mardones, 1960; Mardones and Segovia-Riquelme, 1983). In these cases alcohol preference was genetically bred by mating high drinkers. C57 mice were not bred for alcohol preference, but were a well-established inbred strain only later discovered to prefer alcohol to water (Mirone, 1957; McClearn and Rodgers, 1959).

C. Operational Definitions of Alcohol-Specific Intake

It was Curt Richter who first suggested that alcohol drinking should be studied as a two-bottle choice: alcohol versus plain water. For the alcohol-preferring strains such as C57 mice, the median preference ratio is 0.70 or higher; that is, 70% of their total fluid intake is 10% alcohol when they are offered alcohol and water ad libitum. It was also Richter who pointed out that ethanol is a significant source of energy and is consumed very much as a food by rats (Richter, 1941).

Much later, Lester and Freed (1973) emphasized the distinction between nutritionally and pharmacologically motivated drinking, and suggested that the study of alcoholism required that experimental animals have unlimited access to an adequate source of food. From that time on, most investigators have assumed that if animals have a two-bottle choice and excess food, then alcohol intake is motivated by its pharmacological properties. In defense of this assumption, examples can be presented in which treatments (e.g., injection of various drugs) can decrease alcohol intake without affecting food intake.

Despite these precautions, several points suggested to us that most of the alcohol consumption observed in C57 mice might very well be nutritionally motivated.

1. While ad libitum access to lab chow ensures that animals are not obliged to drink alcohol for its calories, it does not exclude the possibility that they may choose to consume both lab chow and alcohol as foods.

2. Pharmacological agents that decrease alcohol intake in animals also suppress food intake at higher doses. Many of them are well-known anorexics such as amphetamine, fluoxidine, fenfluramine, and naloxone. The issue is whether dose-dependent sensitivity represents evidence for motivational specificity. It may be more parsimonious to conclude that alcohol is a weakly accepted food. The greater sensitivity of alcohol appetite in animals could be attributable either to its relative unpalatability or to the fact that it provides only calories and not the variety of required nutrients provided by chow. Obviously, the discovery of a drug that is relatively safe and specifically suppresses alcohol consumption would be of extraordinary theoretical and clinical importance.

3. The most reliable way to increase alcohol consumption to pharmacologically relevant rates in animals involves either food deprivation or manipulating the palatability of alcohol. These procedures confound the motivation for drinking (Freed and Lester, 1970; Lester and Freed, 1972), and neither is necessary for the maintenance of alcoholic drinking among humans.

4. By far the most effective method to suppress alcohol consumption in animals is to offer them a palatable solution of sucrose as an alternative. In this instance the intake of alcohol is almost zero (see, e.g., Lester and Greenberg, 1952), and it is very low even in alcohol-preferring strains such as C57 mice (Gentry and Dole, 1987). If the animals' motivation for drinking alcohol were pharmacological, then why should they stop drinking when given sugar-water as an alternative?

D. What Is the Motivation for Drinking in C57 Mice?

With these caveats in mind, we set out to determine why C57 mice drink alcohol, asking of them the same question we would eventually like to address for alcoholics. To do this, we characterized the circadian pattern of alcohol intake, the blood levels of alcohol achieved during ad libitum drinking, the temporal relationship between drinking alcohol and eating food, and the response of alcohol intake to various dietary alternatives including sucrose solutions.

II. CHARACTERISTICS OF ALCOHOL CONSUMPTION IN C57 MICE

A. Alcohol Intake

Consistent with previous observations (e.g., McClearn and Rodgers, 1959; Rodgers, 1967), we documented a median alcohol preference ratio of 0.89 after 6 weeks of free-choice drinking (Gentry et al., 1983). Clearly, C57 mice prefer 10% alcohol to water and alcohol is reinforcing for these animals; but the question remains, Why is alcohol reinforcing?

The dose of alcohol consumed by C57 mice is much higher than that of human alcoholics when expressed as a proportion of body weight. Mean intake is typically about 9 g per kilogram of body weight per day, with a large degree of variation ranging from 2 to 16 g/kg in individual mice. But even the lowest rate of drinking is the dose equivalent of 12 oz. of 86 proof whisky for a 70 kg man. The difference, of course, is that mice metabolize alcohol much more rapidly than do humans. We estimate the metabolic capacity for ethanol oxidation in C57 mice drinking alcohol is in the range of 20–30 g/kg per day. Probably the most relevant index of alcohol intake is the percentage of total calories (Lieber, 1982). For mice, the average quantity of alcohol consumed represents only 10% of their total caloric intake and ranges no higher than 20%; this in contrast to human alcoholics, who can consume 50% or more of their total caloric intake as ethanol.

The preference for alcohol is not immediately manifested but gradually increases to maximum after 6 weeks exposure to the alcohol–water choice. The preference remains high during the animals' active growing stage, then slowly decreases over the rest of the animals' lifetime.

B. Circadian Pattern of Drinking

Using lick detectors to monitor when mice drink alcohol revealed a strong circadian pattern of intake, with half of the total daily intake occurring in the first 6 hours of the 12-hour dark period, and another 25–30% consumed during the remaining hours of the dark period (Dole et al., 1983; Gentry et al., 1983). The intake of alcohol follows the same circadian pattern as the ingestion of food and water (Dole et al., 1985), and all three ingestive behaviors are correlated with the nocturnal activity cycle characteristic of rodents.

C. Plasma Concentrations of Ethanol at Midnight

To determine the plasma concentrations of ethanol produced by voluntary (i.e., free-choice) drinking, we regularly took blood samples at midnight as a preliminary step to most experiments. Individual animals exhibited concentrations ranging as high as 116 mg/dl, but the mean plasma concentration was only 25 ± 2 mg/dl ($N = 120$), even though the sampling time immediately followed the 6-hour period of highest intake. Fully a third of all animals had levels below 5 mg/dl despite a minimal rate of intake of 2 g/kg. Curiously, there was only a very modest correlation between each animal's daily intake of alcohol and the plasma concentration at midnight ($r = 0.24$, $p < 0.05$, with 118 degrees of freedom).

D. Plasma Concentrations of Alcohol After Bouts of Drinking

Data collected by computerized monitoring of licks at one-minute intervals demonstrated that the intake of alcohol is highly episodic, suggesting that plasma concentrations of alcohol produced by ad libitum drinking may fluctuate dramatically. To evaluate this possibility, the data acquisition computer was programmed to signal experimenters when a large bout of drinking had just occurred. Blood samples were then taken as quickly as possible, and at 10-minute intervals for 40 minutes. The results indicated that concentrations after bouts of drinking were considerably higher than those taken at midnight. Every animal tested had pharmacologically relevant concentrations of 10 mg/dl or more when samples were taken at time points immediately following drinking, and individual measurements ranged as high as 165 mg/dl.

A computer model of alcohol pharmacokinetics was designed to estimate the minute-to-minute plasma concentrations during an entire 24-hour period, using lick data as input and the four measured concentrations to calculate the rate of ethanol elimination. The results confirmed that the levels of blood alcohol were extremely variable in association with the episodic intake. Every mouse exhibited peaks in the range of 100–175 mg/dl several times each night, and occasionally there were peaks exceeding 200 mg/dl. Between these peaks, plasma concentrations of ethanol fell to zero or near zero before the next bout of drinking. Concentrations were essentially zero for fully half of the total 24 hours (Dole and Gentry, 1984). This analysis provided an explanation for the low correlation between daily intake and midnight levels, since each mouse had an independent sequence of drinking bouts and pauses unrelated to the arbitrary time point of midnight. Furthermore, there was no evidence that mice regulate their intake of alcohol either to maintain any particular level or to avoid transient peaks that were sufficiently elevated to induce intoxication. Alcoholics, in contrast, typically maintain more consistently elevated blood alcohol concentrations. An analysis of human data provided by Enoch Gordis (personal communication, cited in Dole and Gentry, 1984) demonstrated that four alcoholic men permitted to drink alcohol ad libitum maintained concentrations of at least 50 mg/dl, 24 hours per day.

E. Temporal Relationship with Feeding Behavior

Simultaneous monitoring of food, water, and alcohol intake permitted analysis of the temporal relationship between these ingestive behaviors. In general, bouts of food and alcohol intake were temporally related, with cross-correlations ranging from $r = 0.68-0.89$, $p < 0.001$ (Dole et al., 1985). While the imperfect correlations suggest that not all intake of alcohol is prandial, the temporal coincidence between alcohol and food intake was at least as high as the correlation between food intake and water drinking. Independent analysis of the cross-correlations between the intake of 5% sucrose and food yielded similar correlations. These results suggest a common motivational determinant of alcohol intake and food consumption.

F. Effect of Diet on Alcohol Consumption

In early studies we had observed that mice given the recommended diet for breeding (Purina 5005, 12% fat) consumed less alcohol than did mice given standard rodent chow (Purina 5001, 4.5% fat). To characterize the effect of diet on alcohol intake and preference, mice were fed the standard diet, then switched to various alternatives on a weekly basis. When mice were offered fat in the form of Crisco in addition to the standard chow, their intake of alcohol was decreased by approximately two-thirds, and the same effect was observed when the standard chow was mixed 50/50 with sucrose (Dole et al., 1985). When mice had access to these palatable diets, they drank more water than alcohol. Thus, the alcohol preference characteristic of this strain depends on the particular diet they are eating.

G. Response of Alcohol Intake to Sucrose Challenges

As noted by Lester and Greenberg (1952), the dramatic decrease in alcohol intake when animals are given access to solutions of sucrose suggests that the motivation may be nutritional rather than pharmacological. However, it could be argued that the motivation for drinking alcohol is pharmacological, but simply much weaker than the animals' appetite for sugar-water. And the suppression of alcohol intake is not the direct result of a motivational competition, but the consequence of a ceiling effect manifested when the total volumetric intake approaches the maximum animals can tolerate.

To examine the effects of total volumetric intake on the consumption of alcohol, a group of mice were first offered the standard choice of alcohol and water, then challenged with two different concentrations of sucrose in a third bottle during subsequent weeks (Gentry and Dole, 1987). Adding the choice of 5% sucrose had the expected effect. The intake of alcohol was decreased from a mean of 2.3 ± 0.2 ml/day to 0.1 ± 0.1 ml/day. The 5% sucrose was consumed at the rate of 10.5 ± 2.0 ml/day, consistent with the possibility that alcohol intake was suppressed indirectly by a volumetric maximum. However, when the third bottle was changed from 5% to 10% sucrose, the total volumetric intake (virtually all sucrose) increased to 19.6 ± 0.4 ml/day, demonstrating a capacity to drink nearly twice as much fluid. Our interpretation of these results is that C57 mice could have consumed both 10.5 ml of 5% sucrose and 2.3 ml of alcohol if the motivation were independent or if alcohol were consumed for its pharmacological properties.

H. Conclusions Regarding Motivation

The results with C57BL/6J mice do not permit the rejection of what should be the null hypothesis: that the intake of the flavored/caloric substance, alcohol, is motivated by its nutritional properties. While pharmacologically motivated drinking cannot be ruled out, the implication seems to be that it is of minor importance. Furthermore, the observed characteristics of alcohol drinking lack the potency and specificity expected of an addictive behavior. In comparison, the appetite for alcohol among human alcoholics is inelastic. The motivation to drink is much stronger than a "preference." It is more analogous to the necessity of demand for food. Even the most efficacious treatment programs all too often fail, and temporary success is frequently followed by recidivism.

It should be pointed out, however, that other researchers in the field have concluded that pharmacologically motivated drinking and alcoholism occur in animals, and readers are referred to Li et al. (1987) and Meisch (1977).

III. RELEVANCE OF ALCOHOL INTAKE TO ALCOHOLISM

While the results suggest that C57 mice drink alcohol more for its calories or taste than for its pharmacological properties, this conclusion does not mean that alcohol intake in these animals is irrelevant to the study of human alcoholism. First, C57 mice and several strains of rats exhibit chronic alcohol consumption that is voluntary and results in pharmacological effects, characteristics that make it a potential model for the chronic self-exposure to alcohol prerequisite to alcoholism in humans. Second, it could be argued that the brief peaks in blood alcohol after bouts of drinking are precisely the most relevant psychotropic stimuli. Finally, even with a motivation that lacks the specificity and potency of alcoholism, examples can be provided in which animals exhibit inappropriate alcohol satiety, examples in which free-choice drinking produces pathological consequences as severe as those characteristic of human alcoholism.

A. Modeling the Preaddictive State

Despite caution in attributing alcoholism, or pharmacologically motivated drinking, to animals, it is clear that C57BL/6J mice and alcohol-preferring rats voluntarily consume sufficient quantities of ethanol to experience brief periods of intoxication (Dole and Gentry, 1984), as well as other physiologic effects such as metabolic tolerance (Gentry et al., 1985; Lumeng and Li, 1986) and physical dependence (Waller et al., 1982a). Certainly humans drinking equivalent amounts of alcohol would be "at risk" for alcoholism.

Since the addictive state is acquired only after extensive self-exposure to alcohol, a preaddictive state is inherent in the development of alcoholism. Thus, while C57 mice may not exhibit the final stages of human alcoholism, they do exhibit chronic, voluntary intake, which is a necessary part of the etiological process. It is also possible that they represent a model of alcohol abuse as distinguished from alcoholism (Dole, 1986). Finally, the existing animal models of alcohol intake probably offer the best opportunity to investigate the process that mediates the final step to alcohol addiction.

B. Peaks in Blood Alcohol as Reinforcing

As described above, the episodic pattern of ad libitum drinking in C57 mice results in dramatic fluctuations of blood alcohol. We have pointed out that this pattern of alcohol concentrations, which is not characteristic of humans, demonstrates that mice do not regulate their blood alcohol levels within a wide range between zero and 150 mg/dl or more (Dole and Gentry, 1984). There is ample evidence, however, to suggest that the behavioral and subjective effects of alcohol are most sensitive to rising concentrations of blood alcohol, a phenomenon referred to as acute tolerance (Hurst and Bagley, 1972; LeBlanc et al., 1975; Vogel-Sprott, 1979; Haubenreisser and Vogel-Sprott, 1983; Radlow and Hurst, 1985). Thus, it could be inferred that the episodic pattern of drinking combined with a rapid rate of ethanol elimination optimizes the frequency and magnitude of behavioral effects without the accumulation of debilitating alcohol concentrations in the blood. If the reinforcing properties of alcohol exhibit acute tolerance, then C57 mice regularly experience the most relevant stimuli several times each night.

The qualification that must be appended to this interpretation is that there is no evidence to confirm that these peaks in blood alcohol reinforce drinking. They are temporally correlated with other ingestive behaviors and they are not sought by mice given 5% sucrose as an alternative.

C. Examples of Alcohol Intemperance in Animals

A frequent (but not universal) characteristic of alcoholism is drinking to the point of inebriation despite considerable tolerance. Whether excessive drinking is a consequence of an overwhelming drive to drink or a lack of appropriate satiety is yet to be determined. Under certain conditions, "intemperate" drinking can be observed in animals and thus provides potential models for the determination of the mechanisms involved.

Schedule-induced polydipsia greatly augments the intake of alcohol in a variety of animals (Lester, 1961; see Falk and Tang, 1988, for a review). Falk et al. (1972), for example, used an intermittent food schedule to maintain high rates of intake in Holtzman rats, a strain that usually consumes small amounts of alcohol. With optimal conditions, these rats exhibited obvious inebriation and physical dependence, and some died as a consequence. While the excessive intake of alcohol is related to food deprivation and the intermittent access to food (Freed and Lester, 1970), the animals were not obliged to drink. They could have stopped drinking at any point and avoided the debilitating intoxication. The model forces the consideration of the role of adjunctive behavior in human addiction including alcoholism.

Nonoral routes of administration can increase the rate of alcohol intake in a variety of species including primates who avoid the taste of ethanol (see, e.g., Winger and Woods, 1973). Most notably, Waller et al. (1984) demonstrated that P rats would learn to lick at an artificial flavor to obtain intragastric infusions of ethanol without food or water deprivation. They found that the dose of ethanol self-administered was a function of the concentration of ethanol in the infused solution. When the concentration was raised to 40%, these animals exhibited self-intoxication. The mean blood alcohol level was 231 mg/dl after episodes of intake. Since these blood concentrations were considerably higher than those observed in P rats during free-choice drinking (Li and Lumeng, 1977; Murphy et al., 1983), the authors concluded that intragastric infusion circumvents the slightly aversive taste of alcohol even in this strain selected for alcohol preference over water.

Since reinforcement was demonstrable in the absence of orosensory cues for alcohol, Waller et al. (1984) ascribed the motivation for alcohol preference in P rats to "a postabsorptive, pharmacological mechanism (presumably mediated by the central nervous system)." But regardless of the motivation, the model exhibits self-intoxication with apparent correspondence to acute alcohol toxicity and occasional death in humans.

In a recent review Li et al. (1987) addressed the issue of why P rats limit oral intake of alcohol when blood levels exceed 50 mg/dl (Waller et al., 1982b), yet exhibit occasional peaks as high as 218 mg/dl after scheduled access to alcohol (Murphy et al., 1986). They suggested that blood alcohol concentrations continue to rise after intake ceases because of delayed gastrointestinal absorption, a phenomenon referred to as "overshoot." Thus it is possible that intragastric infusion of ethanol, particularly at the 40% concentration, maximizes the "overshoot" in blood alcohol and offers an experimental model of acute intoxication in humans.

Inhibition of the rate of ethanol metabolism does not increase alcohol intake but reduces the capacity of animals to eliminate what they do drink, and results in self-intoxication despite unlimited access to food and water (Gentry et al., 1983; Gentry, 1985a). C57 mice were first acclimated to the alcohol–water choice for 6 weeks, then given daily injections of 4-methylpyrazole (4MP, an inhibitor of hepatic alcohol dehydrogenase: Lester et al., 1968; Li and Theorell, 1969; Blomstrand and Theorell, 1970) at a dose that suppressed the rate of alcohol elimination by 80% for 6–8 hours. We found that 4MP-treated animals continued to choose alcohol despite significantly elevated concentrations of plasma alcohol at midnight. But the 4MP was effective only during the dark period, and circulating blood alcohol returned to zero each day.

In the next experiment, 4MP was administered by subcutaneously implanted pumps (Alzet osmotic minipumps) for 2 weeks (Gentry, 1985b). After 1 week of chronic 4MP infusion and free-choice drinking, plasma concentrations of ethanol were determined at four time points separated by 6-hour intervals. Mean concentrations ranged from 156 ± 33 mg/dl at midday to 254 ± 31 mg/dl at midnight, even though blood samples were taken at arbitrary times (i.e., not correlated with bouts of drinking). All the 10 mice treated with 4MP and offered the alcohol versus water choice exhibited intoxication.

During the second week the self-intoxication became more severe, yet the mice continued to drink alcohol. Five of the 10 experimental subjects lost 20% of their original weight and became hypothermic and unresponsive. Plasma concentrations in these mice exceeded 400 mg/dl. All five were successfully rescued by feeding them a sucrose solution with a hand-held pipette, warming them with a heat lamp, and removing the alcohol bottle. Meanwhile, the control mice treated with 4MP and given only water to drink exhibited normal weight gain and no signs of toxicity. Thus, the toxic consequences were considered to be "self-induced" because one of the necessary components (i.e., alcohol) was freely self-administered in a choice situation.

In subsequent studies, 4MP-treated mice readily avoided alcohol and consumed only 5% sucrose when it was made available in a third bottle during the second week (Gentry, 1989). Our interpretation of this finding is that self-intoxication results from a deficit in the appropriate limitation of alcohol intake, or satiety, rather than from a potentiated motivation to drink alcohol.

Recent experiments indicate that the exposure to alcohol before 4MP treatment is a necessary component of the self-intoxication phenomenon. Alcohol-naive mice were first given Alzet pumps containing 4MP, then offered the alcohol-versus-water choice for the first time. They quickly learned to avoid alcohol and consumed only water (Gentry, 1989). This suggests that tolerance, or adaptation to the taste of alcohol, could contribute to self-intoxication in 4MP-treated mice. Another interpretation is that prior exposure to alcohol trained the animals to consider alcohol "safe" and inhibited their ability to learn the association between its taste and the toxic consequences potentiated by 4MP. Thus, the self-induced intoxication observed in 4MP-treated mice may represent an example of "learned safety" (Kalat and Rozin, 1973) so effective as to be life-threatening. While the relative effects of tolerance, taste adaptation, and learned safety need to be evaluated, the results emphasize the importance of experiential factors and suggest possible mechanisms by which chronic self-exposure to alcohol contributes to the etiology of human alcoholism.

IV. A METABOLIC THEORY OF ALCOHOLISM
A. Background

The motivation among alcoholics for drinking alcohol is not known. We must consider the possibility that the nutritional or metabolic consequences of alcohol play a role in addictive drinking.

Many addictive drugs interact with appetite and to varying degrees usually suppress food intake. Recent findings emphasize that the rate of drug self-administration depends on the nutritional state of animals, with food deprivation increasing the intake of etonitazene, amphetamine, cocaine, pentobarbital, methohexital, ketamine, and phencyclidine (see Carroll and Meisch, 1984, for a review). Intracranial self-stimulation (ICSS) experiments reveal a commonality between food deprivation, ethanol, and other addictive drugs. ICSS is augmented by each, either by increasing the rate of responding or lowering the electrical threshold for responding (see, e.g., Margules and Olds, 1962; Hoebel, 1969; Wise, 1980). The implication seems to be that drug addiction is not created de novo but results from the modification of innate motivational systems such as hunger.

Alcohol differs from other addictive drugs in that it lacks the structural specificity to act directly on neurotransmitter receptors. Its effects on the nervous system and the reinforcement of behavior must be indirect—mediated, for example, by "fluidization" of plasma membranes (Littleton, 1983; Goldstein, 1984) or interference with the chain of events responsible for the transduction of neurochemical signals.

As an alternative to a neurological site of action, it is possible that the metabolic or nutritional consequences of alcohol may provide the reinforcing effects responsible for addictive drinking. Alcohol contains approximately 7 kcal/g and is administered in very large doses compared to other drugs (approximately 3 orders of magnitude more than opiates or cocaine) so that alcoholics consume 1000–2000 kcal/day as ethanol.

Consistent with a nutritional hypothesis, epidemiological studies reveal a coincidence between alcoholism and appetitive disorders such as bulimia, particularly in women (Beary et al., 1986; Killen et al., 1987). And Cuellar et al. (1987) have described the occurrence of alcoholic-like hepatitis as the consequence of bulimia. These observations suggest a connection between alcoholism (or perhaps some types of alcoholism) and an abnormal regulation of appetite. Anecdotal evidence also suggest that recovering alcoholics develop carbohydrate appetite or "sweet tooth." Yung et al. (1983) found that successful sobriety is correlated with an unusually high intake of free sugars, confirming the adage among members of Alcoholics Anonymous that candy may help to suppress the urge to drink. The authors emphasize, however, that it is not clear whether the carbohydrate appetite is a cause of alcoholism or a consequence of chronic alcohol drinking.

While sugar may assist in maintaining abstinence, it is clear that alcoholism is not the same as hunger. The appetite for alcohol is relatively specific. Alcoholics have unlimited access to a variety of nutrients and still seek alcohol. Unlike laboratory rodents, palatable solutions of sucrose do little to abate their need for alcohol.

B. The Role of the Liver

If the energy content of ethanol contributes to the motivation for addictive drinking, there must be an explanation for the specificity of the alcohol appetite. One possibility is that chronic exposure to alcohol alters the functional detection of energy in foods so that ethanol becomes a preferred source of calories. The site of this dysfunction is likely to be the liver. The effects of chronic alcohol exposure on the liver are profound, ranging from acute disruptions in redox states of enzymatic cofactors to gross morphological changes (see Lieber, 1982, for a review). Furthermore, research on the physiology of caloric reinforcement suggests that the energy content of foods is monitored in hepatocytes (Friedman et al., 1986). Both the function and the anatomical location of the liver make it uniquely qualified to monitor energy flow and regulate appetite.

Consistent with a central role of the liver in alcohol intake are several reports indicating that models of hepatotoxicity in animals result in increased consumption of alcohol. Chronic exposure to toluene (Geller et al., 1983) or toluene plus ethanol (Pryor et al., 1985) significantly increases the subsequent self-administration of alcohol. Similarly, portacaval shunts produce hepatotoxicity characterized by alterations in ingestive behavior (Martin et al., 1981; Martin, 1983), including the enhancement of ethanol consumption in rats (Martin et al., 1982, 1985). Finally, Starzl et al. (1988) have reported a remarkably low incidence of recidivism among alcoholics receiving liver transplants. While there are alternative explanations for this observation, it is compatible with the hypothesis that a diseased liver is responsible for alcoholic drinking.

C. Alcohol-Specific Appetite: A Nutritional Hypothesis

As a working hypothesis then, it is suggested that chronic alcohol consumption induces a state of hepatic dysfunction that makes alcohol the preferred source of energy, and then alcoholics learn to consume alcohol by mechanisms analogous to those by which animals learn to consume flavors associated with the delivery of glucose to the liver (Tordoff and Friedman, 1986).

D. The Unique Metabolism of Ethanol

The particular hepatic changes that may make ethanol preferred to other sources of energy are speculative. As a suggestion, the altered availability of cofactors or an enzymatic deficiency may interfere with the metabolism of carbohydrates and fats, converting them into a signal for reinforcement. The oxidation of ethanol could bypass this deficiency, since its metabolism is more direct, yielding energy at its first step of metabolism (to acetaldehyde), then rapidly being converted to acetyl-CoA for further oxidation in the tricarboxylic acid cycle. Perhaps working in concert with an attenuated signal from other energy sources, the induction of the microsomal ethanol oxidizing system (MEOS: Lieber and DeCarli, 1968) caused by chronic alcohol consumption could alter the metabolism of alcohol to maximize its reinforcing properties.

Both MEOS and alcohol dehydrogenase (ADH) oxidize ethanol to acetaldehyde. Usually, ADH accounts for the majority of this rate-limiting step in alcohol oxidation. After chronic alcohol feeding, however, MEOS activity is increased sufficiently to result in a faster rate of total alcohol oxidation (Lieber and DeCarli, 1972), with no induction of ADH (Videla et al., 1973; Kalant et al., 1975). The activity of the ethanol-inducible cytochrome P-450 associated with microsomes increases by severalfold as a consequence of chronic exposure to alcohol (see, e.g., Ardies et al., 1987). One of the differences between MEOS and ADH is that the microsomal system requires NADP rather than NAD as the hydrogen acceptor, and while the reoxidation of NADH is coupled with oxidative phosphorylation, NADPH is not. The result is that MEOS activity generates heat without conservation of chemical energy (Lieber, 1982). This microsomal pathway for the oxidation of ethanol to heat is consistent with the observation that ethanol calories do not fully count in the maintenance of body weight under controlled conditions (Richter, 1941; Pirola and Lieber, 1972). While deficiencies in food absorption and other changes may contribute to this effect, energy wastage in the form of excess heat production is specifically implicated by higher rates of oxygen consumption in alcoholics (Tremolières and Carré, 1961) and rats chronically fed alcohol (Pirola and Lieber, 1975, 1976). Thus, the relevant conclusions are that ethanol oxidation by MEOS generates more heat than by other means and that the proportion of ethanol converted to heat is augmented by chronic exposure to alcohol.

Friedman and Stricker (1976) have concluded that hepatic oxidative metabolism integrates the utilization of all fuels to regulate hunger and satiety. The caloric content of ethanol should be monitored by the same mechanism. If the behaviorally relevant signal is the immediate production of energy, then oxidation of ethanol by MEOS should be especially potent in the induction of satiety and the reinforcement of associated tastes (Tordoff and Friedman, 1986). Furthermore, the capacity of MEOS to contribute to hepatic oxidative metabolism is increased severalfold by exposure to alcohol. Thus, ethanol has unique metabolic characteristics that may make it the preferred source of energy (and reinforcement) under conditions of hepatic dysfunction, conditions that are themselves generated by chronic (preaddictive) drinking.

E. Testing the Hypothesis

The question then is, Why do some rats and mice drink alcohol chronically, yet fail to acquire the hypothesized metabolic appetite for alcohol in preference to other sources of energy? Assuming that liver pathology is a necessary condition, the answer is already known. Rodents simply cannot drink enough alcohol to attain the severity of hepatotoxicity seen in humans and baboons (e.g., Lieber, 1982). To test the hypothesis, alternative means of inducing the appropriate liver dysfunction will be required. Possibilities include toluene or carbon tetrachloride poisoning. Additionally, 2-deoxyglucose and methyl palmoxirate could be utilized to limit the oxidation of glucose and fatty acids in a procedure analogous to that used by Friedman and Tordoff (1986). Induction of MEOS can be achieved by forcing alcohol intake, or by other means, before offering animals an ad libitum access to alcohol and water. Finally, alcohol metabolism by MEOS, at the expense of ADH, can be manipulated by chronic treatment with 4MP. Ultimately, confirmation of an alcohol-specific appetite will depend on the maintenance of alcohol intake and self-intoxication even when animals have simultaneous access to 5 or 10% sucrose.

V. SUMMARY AND CONCLUSIONS

Addressing the question "Why do alcoholics drink?" and achieving progress toward medical treatment of alcoholism are sufficiently important to justify simultaneous pursuit of all legitimate animal models and tenable hypotheses.

Our studies with C57BL/6J mice suggest the reevaluation of animal models of alcoholism in terms of an appetitive behavior. Analysis of the characteristics of alcohol intake in this strain forces us to conclude that they probably consume alcohol for its nutritional characteristics, with no evidence of the specificity or potency of appetite expected of addictive drinking. Despite this caution in attributing alcoholic motivation to animals, there exist a variety of models of alcohol drinking relevant to human alcoholism. Some models exhibit chronic self-exposure to alcohol suggestive of the preaddictive stage of human drinking. Other procedures result in conspicuous self-intoxication and serve as potential paradigms of alcohol intemperance in humans. The appropriate conclusion is to use the available models to construct and test hypotheses regarding the etiology and consequences of human drinking, while avoiding premature satisfaction in our models. It is suggested that animal models that simulate the motivation of addictive drinking are yet to be discovered.

Finally, the possibility is considered that the nutritional consequences of ethanol contribute to addictive drinking. A novel hypothesis is presented based on the unique metabolic characteristics of ethanol that could result in alcohol becoming the preferred source of caloric energy.

The challenge remains to discover the physiological and experiential determinants of alcohol addiction. What motivates alcoholics to drink despite grave consequences and ultimately death? How does social drinking evolve into addictive drinking? Answers to these questions would have profound consequences on the efficacy of medical treatment for alcoholism.

ACKNOWLEDGMENTS

The author acknowledges the technical assistance of Mr. Seth Coplan, Ms. Adrienne Chin, and Dr. Robert Lim, Jr., and is particularly indebted to Drs. Enrique Baraona, Vincent Dole, Mark Friedman, and Charles Lieber for their suggestions and theoretical contributions. This research has been supported by a grant from the National Institute for Alcohol Abuse and Alcoholism (AA-06635).

DISCUSSION

Schneider: What neurotransmitters in the brain are different in these animals?

Gentry: In C57 mice versus other nondrinking mice, not very much that I know about. T. K. Li, working with his P rats versus NP rats, has identified some serotonergic differences.

Blass: A possible additional interpretation of your fascinating sucrose data is that sucrose may cause opioid release. This appears to be the case in infant rats of the Sprague-Dawley strain and certain adult rats. If that is the case, then you have two competing pharmacological effects as opposed to two competing nutritional effects as the explanation.

Gentry: OK, let me tell you a little about that. Of course, people have tested the effects of opiates on alcohol intake; [this is] one of the favorite topics in our field of probably several thousand papers. Opiate agonists and opiate antagonists both decrease alcohol intake. But, also look at it quantitatively. If you give an opiate antagonist or agonist, you can decrease alcohol intake, but it doesn't decrease alcohol intake by very much—10–20%. If it decreases much more than that, you're causing severe illness. If, on the other hand, all you have to do is add a 5 or 10% solution of sucrose in the cage—they're not sick, there's no evidence of illness—the alcohol intake hasn't gone down by 10 or 20%, it's gone down by 99%.

VanItallie: Would these animals, after they've been on alcohol for a period of time, show an increased disposition to work to get the alcohol? In other words, would they work pretty hard to get the alcohol on a demanding reinforcement schedule?

Gentry: Yes, animals will work to get alcohol. They won't work too hard, however, and in fact they'll work a lot harder for a saccharin solution or sucrose solution, and given the choice they prefer one of those. But, there's no doubt the alcohol is reinforcing. It's a mild reinforcer, but it is reinforcing; you can get animals to work for alcohol.

Thrasher: In the last experiment, where you actually got them to mimic the human alcoholic by altering the metabolism of the alcohol, would the sucrose still turn off the alcohol intake?

Gentry: Well, of course I've tried that. When we saw animals actually dying in our cages after slowing down the metabolism, the first thing we said was "Maybe this is it! Maybe this 4-methylpyrazole has created alcoholism like we've never seen before in these animals in this laboratory." Well, all we had to do in the very next experiment was drop a bottle of 5% sucrose in the same cage. The animals didn't drink any alcohol, they didn't show any blood alcohol concentrations, they showed no illness, they just drank the sugar. They weren't alcoholic, and yet they were having regulatory problems. Obviously part of the process of addiction.

Torii: Do you have any idea why animals ingest food and alcohol?

Gentry: Well, the notion is that there's a common motivational fluctuation. When they become hungry for food, there's also the associated intake of water and, in this case, the associated intake of alcohol, which, by the way, combines all characteristics of water, food—that is, calories—and taste. In other words, we can take all our data and it fits the assumption that it's nutritional intake.

Schulkin: You began your lecture by suggesting that hunger could be construed as an addiction. Could you elaborate on that?

Gentry: I think that would be stretching the definition of addiction. What we mean by addiction is something that is acquired and something that is usually bad for you. If drinking alcohol didn't destroy your liver and destroy your marriage, it wouldn't be nearly so much a concern.

McHugh: We do have another pharmacological antagonistic against psychological or behavioral effects of alcohol, the Squibb compound (4513). Have you used that in your animals to see whether that would stop them from drinking?

Gentry: Yes, obviously. When that was published, I think everybody in the field bought some and tried it. It decreases alcohol intake, just like about hundreds of other things I can find on the shelf, and also decreases food intake. The evidence, furthermore, that it is a specific antagonist of alcohol has virtually fallen apart. It appears to be more like a benzodiazapene reverse agonist; that is, it's stimulating the very factors that alcohol is inhibiting.

REFERENCES

Ardies, M. C., Lasker, J. M., and Lieber, C. S. (1987). Characterization of the cytochrome P-450 monooxygenase system of hamster liver microsomes: Effect of prior treatment with ethanol and other xenobiotics. *Biochem. Pharmacol.* **36**:3613–3619.

Beary, M. D., Lacey, J. H., and Merry, J. (1986). Alcoholism and eating disorders in women of fertile age. *Br. J. Addict.* **81**:685–689.

Blomstrand, R., and Theorell, H. (1970). Inhibitory effect on ethanol oxidation in man after administration of 4-methylpyrazole. *Life Sci.* **9**:63–640.

Carroll, M. E., and Meisch, R. A. (1984). Increased drug-reinforced behavior due to food deprivation. In *Advances in Behavioral Pharmacology,* Vol. 4, T. Thompson, P. B. Dews, and J. E. Barrett (Eds.). Academic Press, New York, pp. 47–88.

Cloninger, C. R. (1987). Neurogenetic adaptive mechanisms in alcoholism. *Science* **236**:410–416.

Cloninger, C. R., Bohman, M., and Sigvardson, S. (1981). Inheritance of alcohol abuse: Cross-fostering analysis of adopted men. *Arch. Gen. Psych.* **38**:861–868.

Cuellar, R. E., Tarter, R., Hays, A., and Van Thiel, D. H. (1987). The possible occurrence of "alcoholic hepatitis" in a patient with bulimia in the absence of diagnosable alcoholism. *Hepatology,* **7**:878–883.

Dole, V. P. (1986). On the relevance of animal models to alcoholism in humans. *Alcoholism: Clin. Exp. Res.* **10**:361–363.

Dole, V. P. (1988). Implications of methadone maintenance for theories of narcotic addiction. *JAMA,* **260**:3025–3029.

Dole, V. P., and Gentry, R. T. (1984). Toward an analogue of alcoholism in mice: Scale factors in the model. *Proc. Natl. Acad. Sci., USA,* **81**:3543–3546.

Dole, V. P., Ho, A., and Gentry, R. T. (1983). An improved technique for monitoring the drinking behavior of mice. *Physiol. Behav.* **30**:971–974.

Dole, V. P., Ho, A., and Gentry, R. T. (1985). Toward an analogue of alcoholism in mice: Criteria for the recognition of pharmacologically motivated drinking. *Proc. Natl. Acad. Sci., USA,* **82**:3469–3471.

Eriksson, K. (1968). Genetic selection for voluntary alcohol consumption in the albino rat. *Science,* **159**:739–741.

Eriksson, K. (1972). Alcohol consumption and blood alcohol in rat strains selected for their behavior towards alcohol. *Finnish Found. Alcohol Stud.* **20**:121–125.

Falk, J. L., and Tang, M. (1988). What schedule-induced polydipsia can tell us about alcoholism. *Alcoholism: Clin. Exp. Res.* **12**:577–585.

Falk, J. L., Samson, H. H., and Winger, G. (1972). Behavioral maintenance of high concentrations of blood ethanol and physical dependence in the rat. *Science,* **177**:811–813.

Freed, E. X., and Lester, D. (1970). Schedule-induced consumption of ethanol: Calories of chemotherapy? *Physiol. Behav.* **5**:555–560.

Friedman, M. I., and Stricker, E. M. (1976). The physiological psychology of hunger: A physiological perspective. *Psychol. Rev.* **83**:409–431.

Friedman, M. I., and Tordoff, M. G. (1986). Fatty acid oxidation and glucose utilization interact to control food intake in rats. *Am. J. Physiol.* **251**:R840–R845.

Friedman, M. I., Tordoff, M. G., and Ramirez, I. (1986). Integrated metabolic control of food intake. *Brain Res. Bull.* **17**:855–859.

Geller, I., Hartmann, R. J., and Gause, E. M. (1983). Effect of exposure to high concentrations of toluene on ethanol preference of laboratory rats. *Pharmacol. Biochem. Behav.* **19**:933–937.

Gentry, R. T. (1985a). Voluntary consumption of ethanol and its consequences in C57 mice treated with 4-methylpyrazole. *Alcohol*, **2**:581–587.

Gentry, R. T. (1985b). An experimental model of self-intoxication in C57BL/6J mice. *Alcohol*, **2**:671–675.

Gentry, R. T. (1989). Self-intoxication in C57BL/6J mice: Why do mice treated with 4-methylpyrazole continue to voluntarily drink alcohol? *Alcoholism: Clin. Exp. Res.* **13**:303 (Abstr.).

Gentry, R. T., and Dole, V. P. (1987). Why does a sucrose choice reduce the consumption of alcohol in C57BL/6J mice? *Life Sci.* **40**:2191–2194.

Gentry, R. T., Rappaport, M. S., and Dole, V. P. (1983). Elevated concentrations of ethanol in plasma do not suppress voluntary ethanol consumption in C57BL mice. *Alcoholism: Clin. Exp. Res.* **7**:420–423.

Gentry, R. T., Chin, A., Jacobs, R., and Dole, V. P. (1985). Pharmacokinetics of ethanol in C57BL/6J mice. *Alcoholism: Clin. Exp. Res.* **9**:196.

Goldstein, D. B. (1984). The effects of drugs on membrane fluidity. *Annu. Rev. Pharmacol. Toxicol.* **24**:43–64.

Haubenreisser, T., and Vogel-Sprott, M. D. (1983). Tolerance development in humans with task practice on different limbs of the blood–alcohol curve. *Psychopharmacology*, **81**:350–353.

Hoebel, B. G. (1969). Feeding and self-stimulation. *Ann. N.Y. Acad. Sci.* **157**:758–778.

Hurst, P. M., and Bagley, S. K. (1972). Acute adaptation to the effects of ethanol. *Q. J. Stud. Alcohol.* **33**:358–378.

Kalant, H., Khanna, J. M., and Endrenyi, L. (1975). Effect of pyrazole on ethanol metabolism in ethanol-tolerant rats. *Can. J. Physiol. Pharmacol.* **53**:416–422.

Kalat, J. W., and Rozin, P. (1973). "Learned safety" as a mechanism in long-delay taste aversion learning in rats. *J. Comp. Physiol. Psychol.* **83**:198–207.

Killen, J. D., Taylor, C. B., Telch, M. J., Saylor, K. E., Maron, D. J., and Robinson, T. N. (1987). Evidence for an alcohol–stress link among normal weight adolescents reporting purging behavior. *Int. J. Eating Disord.* **6**:349–356.

LeBlanc, A. E., Kalant, H., and Gibbins, R. J. (1975). Acute tolerance to ethanol in the rat. *Psychopharmacology*, **41**:43–46.

Lester, D. (1961). Self-maintenance of intoxication in the rat. *Q. J. Stud. Alcohol.* **22**:223–231.

Lester, D., and Freed, E. X. (1972). The rat views alcohol—Nutrition or nirvana? In *Biological Aspects of Alcohol Consumption*, Vol. 20, O. Forsander and K. Eriksson (Eds.). The Finnish Foundation for Alcohol Studies, Helsinki, pp. 51–57.

Lester, D., and Freed, E. X. (1973). Criteria for an animal model of alcoholism. *Pharmacol. Biochem. Behav.* **1**:103–107.

Lester, D., and Greenberg, L. A. (1952). Nutrition and the etiology of alcoholism: The effect of sucrose, saccharin and fat on the self-selection of ethyl alcohol by rats. *Q. J. Stud. Alcohol.* **13**:553–560.

Lester, D., Keokosky, W. Z., and Felzenberg, F. (1968). Effect of pyrazoles and other compounds on alcohol metabolism. *Q. J. Stud. Alcohol.* **29**:449–454.

Li, T.-K., and Lumeng, L. (1977). Alcohol metabolism of inbred strains of rats with alcohol preference and nonpreference. In *Alcohol and Aldehyde Metabolizing Systems*, R. G. Thurman, J. R. Williamson, H. Drott, and B. Chance (Eds.). Academic Press, New York, 1977, pp. 625–633.

Li, T.-K., and Theorell, H. (1969). Human liver alcohol dehydrogenase: Inhibition by pyrazole and pyrazole analogs. *Acta. Chem. Scand.* **23**:892–902.

Li, T.-K., Lumeng, L., McBride, W. J., and Murphy, J. M. (1987). Rodent lines selected for factors affecting alcohol consumption. *Alcohol Alcohol., Suppl.* **1**:91–96.

Li, T.-K., Lumeng, L., McBride, W. J., Waller, M. B., and Murphy, J. M. (1986). Studies on an animal model of alcoholism. In *NIDA Monograph 66*, pp. 41–49.

Lieber, C. S. (1982). *Medical Disorders of Alcoholism: Pathogenesis and Treatment*. Vol. XXII in the series *Major Problems in Internal Medicine*. Saunders, Philadelphia.

Lieber, C. S., and DeCarli, L. M. (1968). Ethanol oxidation by hepatic microsomes: Adaptive increase after ethanol feeding. *Science*, **162**:917–918.

Lieber, C. S., and DeCarli, L. M. (1972). The role of hepatic microsomal ethanol oxidizing system (MEOS) for ethanol metabolism in vivo. *J. Pharmacol. Exp. Ther.* **181**:279–287.

Littleton, J. M. (1983). Tolerance and physical dependence on alcohol at the level of synaptic membranes: A review. *J. R. Soc. Med.* **76**:593–601.

Lumeng, L., and Li, T.-K. (1986). The development of metabolic tolerance in the alcohol-preferring P rats: Comparison of forced and free-choice drinking of ethanol. *Physiol. Biochem. Behav.* **25**:1013–1020.

Lumeng, L., Hawkins, T. D., and Li, T.-K. (1977). New strains of rats with alcohol preference and nonpreference. In *Alcohol and Aldehyde Metabolizing Systems*, Vol. 3, R. G. Thurman, J. R. Williamson, H. R. Drott, and B. Chance (Eds.). Academic Press, New York, pp. 537–544.

Mardones, J. (1960). Experimentally induced changes in the free selection of ethanol. *Int. Res. Neurobiol.* **2**:41–76.

Mardones, J., and Segovia-Riquelme, N. (1983). Thirty-two years of selection of rats by ethanol preference: UChA and UChB strains. *Neurobehav. Toxicol. Teratol.* **5**:171–178.

Margules, D. L., and Olds, J. (1962). Identical "feeding" and "rewarding" systems in the lateral hypothalamus of rats. *Science*, **135**:374–375.

Martin, J. R. (1983). Alterations in ingestive behavior following experimental portacaval anastomosis in rats. *Physiol. Behav.* **30**:749–755.

Martin, J. R., Baettig, K., and Bircher, J. (1981). Exaggerated consumption of saccharide solutions following experimental portacaval anastomosis in rats. *Behav. Neural. Biol.* **32**:54–69.

Martin, J. R., Bircher, J., and Porchet, H. (1982). Enhanced ethanol consumption following portal–systemic shunting in rats. *Experientia*, **38**:814.

Martin, J. R., Porchet, H., Buhler, R., and Bircher, J. (1985). Increased ethanol consumption and blood ethanol levels in rats with portacaval shunts. *Am. J. Physiol.* **248**:G287–G292.

McClearn, G. E., and Rodgers, D. A. (1959). Differences in alcohol preference among strains of mice. *Q. J. Stud. Alcohol.* **20**:691–695.

Meisch, R. A. (1977). Ethanol self-administration: Infrahuman studies. In *Advances in Behavioral Pharmacology*, Vol. I, T. Thompson and P. B. Dews (Eds.). Academic Press, New York, pp. 35–84.

Mirone, L. (1957). Dietary deficiency in mice in relation to voluntary alcohol consumption. *Q. J. Stud. Alcohol.* **18**:552–560.

Murphy, J. M., McBride, W. J., Lumeng, L., and Li, T.-K. (1983). Monoamine and metabolite levels in CNS regions of the P line of alcohol-preferring rats after acute and chronic ethanol treatment. *Physiol. Biochem. Behav.* **19**:849–856.

Murphy, J. M., Gatto, G. J., Waller, M. B., McBride, W. J., Lumeng, L., and Li, T.-K. (1986). Effects of scheduled access on ethanol intake by the alcohol-preferring (P) line of rats. *Alcohol*, **3**:331–336.

Nyswander, M. E., and Dole, V. P. (1967). The present status of methadone blockade treatment. *Am. J. Psychiatr.* **123**:1441–1442.

Pirola, R. C., and Lieber, C. S. (1972). The energy cost of the metabolism in drugs, including ethanol. *Pharmacology*, **7**:185–196.

Pirola, R. C., and Lieber, C. S. (1975). Energy wastage in rats given drugs that induce microsomal enzymes. *J. Nutr.* **105**:1544–1548.

Pirola, R. C., and Lieber, C. S. (1976). Energy wastage in alcoholism and drug abuse: Possible role of hepatic microsomal enzymes. *Am. J. Clin. Nutr.* **29**:90–93.

Pryor, G. T., Howd, R. A., Uyeno, E. T., and Thurber, A. B. (1985). Interactions between toluene and alcohol. *Pharmacol. Biochem. Behav.* **23**:401–410.

Radlow, R., and Hurst, P. M. (1985). Temporal relations between alcohol concentration and alcohol effect: An experiment with human subjects. *Psychopharmacology*, **85**:260–266.

Richter, C. P. (1941). Alcohol as a food. *Q. J. Stud. Alcohol.* **1**:650–662.

Rodgers, D. A. (1967). Alcohol preference in mice. In *Comparative Psychopathology*, J. Zubin and H. Hunt (Eds.). Grune & Stratton, New York, pp. 184–201.

Starzl, T. E., Van Thiel, D., Tzakis, A. G., Iwatsuki, S., Todo, S., Marsh, J. W., Koneru, B., Staschak, S., Stieber, A., and Gordon, R. D. (1988). Orthotopic liver transplantation for alcoholic cirrhosis. *JAMA*, **260**:2542–2544.

Tordoff, M. G., and Friedman, M. I. (1986). Hepatic portal glucose infusions decrease food intake and increase food preference. *Am. J. Physiol.* **251**:R192–R196.

Tremolières, J., and Carré, L. (1961). Etudes sur les modalités d'oxydation de l'alcool chez l'homme normal et alcoolique. *Rev. l'Alcool.* **7**:202–227.

Videla, L., Bernstein, J., and Israel, Y. (1973). Metabolic alterations produced in the liver by chronic ethanol administration: Increased oxidative capacity. *Biochem. J.* **134**:507–514.

Vogel-Sprott, M. D. (1979). Acute recovery and tolerance to low doses of alcohol: differences in cognitive and motor skill performance. *Psychopharmacology*, **61**:287–291.

Waller, M. B., McBride, W. J., Lumeng, L., and Li, T.-K. (1982a). Induction of dependence on ethanol by free-choice drinking in alcohol-preferring rats. *Pharmacol. Biochem. Behav.* **16**:501–507.

Waller, M. B., McBride, W. J., Lumeng, L., and Li, T.-K. (1982b). Effects of intravenous ethanol and of 4-methylpyrazole on alcohol drinking in alcohol-preferring rats. *Pharmacol. Biochem. Behav.* **17**:763–768.

Waller, M. B., McBride, W. J., Gatto, G. J., Lumeng, L., and Li, T.-K. (1984). Intragastric self-infusion of ethanol by ethanol-preferring and -nonpreferring lines of rats. *Science,* **225**:78–80.

Winger, G. D., and Woods, J. H. (1973). The reinforcing property of ethanol in the rhesus monkey: I. Initiation, maintenance and termination of intravenous ethanol-reinforced responding. *Ann. N.Y. Acad. Sci.* **215**:162–175.

Wise, R. A. (1980). Action of drugs of abuse on brain reward systems. *Pharmacol. Biochem. Behav.* **13** (Suppl. 1):213–223.

Yung, L., Gordis, E., and Holt, J. (1983). Dietary choices and likelihood of abstinence among alcoholic patients in an outpatient clinic. *Drug Alcohol Depend.* **12**:355–362.

Part I: Discussion

McHugh: I will start with a few general comments about what we've heard today. There were a number of common themes that came up in the talks. Each of our speakers was talking not just about the effects of events, but about regulation and control tied to sodium, to food, to fluid hemodynamics, and the like. In the process of hearing them, a natural set of questions would come up that would be common to all of them, such as whether antagonists would work, what part of the body was producing the signals, and what parts of the body were carrying the signals. Related to all of that, came the sense that what used to be a relatively straightforward kind of work—the kind of work that Adolph and Richter so much enjoyed, showing these events in the active animals, the unanesthetized animals, showing these kinds of controls in their behavior—is, along with the phenomena, still there, but the methods of explanation are far more detailed and far more complicated.

We go from someone like Ed Stricker, who tells us that he sees an important inhibition of an inhibitor, and then he tells us it's not what you're seeing, but up there in the brain. That's brave Ed Stricker talk: he is willing to make a prediction about something that he's only measuring at a distance, and we wish him well in that hypothesis.

My interests in this area have been with Mark Friedman in trying to find where the things that relate to calories might happen. What intrigued me about Mark's work is the remarkable effects that he shows, and at the same time my uncertainty of exactly whether he's looking at controls or effects. And, likewise how bold he is to say we're getting down to a single thing, and I'm still out there waffling around about the brain taking multiple inputs to lead to the final behavior.

The last paper this morning was equally fascinating in what it revealed; but at the level of measurement and phenomenon, it is still talking to us about whether we're dealing about effects or controls.

So the morning was fascinating and I'm prepared now to let anyone else or the speakers who I've insulted respond to my questions. Ed, would you like to begin?

Stricker: I'm happy to have a chance to talk. I don't feel insulted. There are several things I want to say to embellish on what I had said this morning in response to questions that came up after my talk and during the break, so I can save some time by addressing them right now.

Just to deal with your comments, Paul, it's true that we're measuring oxytocin in the periphery and speculating about oxytocin in the brain. There hasn't been a good technique to actually measure oxytocin directly in the brain, although it's certainly clear that there are parvocellular neurons that project. That's not a hypothesis. It's also clear that those neurons have functional effects; they are not just there for people to trace. One of the things that's also clear is that the parvocellular neurons are active under many circumstances in which magnocellular neurons are active, suckling being the exception rather than the rule. Furthermore, when we look at one of the functions of the paraventricular nucleus, which is known to be mediated through oxytocinergic neurons, namely gastric motility, there is a perfect correlation between inhibition of food intake, inhibition of salt intake, and inhibition of gastric motility. So even though we haven't seen increases in oxytocin by direct measurement, it's not so adventuresome to imagine that there are increases in the brain of this transmitter in correlation with what we can see more readily.

The experiment that Alan suggested is a good experiment as long as you assume two things, which we're perfectly willing to assume at the outset because it's the best we can do. First of all, that when you give an oxytocin antagonist into the brain, it actually gets to the place where you want it to get to and blocks the effects that it's suppose to be blocking. Once you put it into the brain, it's hard to know where it goes, and it's also hard to know where it should go because we still don't know where the receptors are that might be mediating the salt appetite. So that's an assumption. The second assumption, of course, has to do with the definition of an oxytocin antagonist that was derived from its ability to block receptors in the uterus. It's an interesting hypothesis to say "Because it blocks in the uterus it's also going to block the oxytocin receptors in the brain." Maybe it will, but maybe it won't. It's the kind of experiment which, if it worked and increased salt intake, we would say "terrific." If it didn't, then it may not be so conclusive because there are two other embedded assumptions which we simply have no way of evaluating. However, the general approach I think is the appropriate one, which is to give an antagonist and try to increase behavior. If you put

PART I: DISCUSSION 63

something in the brain and decrease behavior, then you're open to the possibility that what you're doing is less specifically affecting that behavior. Goodness knows, you can put all sorts of stuff in the brain and decrease ingestive behavior and that may not be so interesting as something put into the brain that increases it.

Powley: To complement that approach, another, albeit dirty, experiment would be to intercept those axons in the brain or reversibly anesthetize the parvocellular neurons. Have you done that?

Stricker: We have been spending the past several months attempting to make paraventricular lesions or knife cuts of paraventricular projections. It is the biggest lesion I ever made without success, and it's hard to know actually what we've done. Sometimes the lesion is where we want it; most of the time it's not. Even when they're there, they intercept all sorts of things. So, if the wiring is more complicated than I'm suggesting, which is a simple inhibitory pathway which you can intercept and nothing else relevant is happening, then we haven't succeeded. Goodness knows what the pathways are. On the basis of what I've seen in the past few months, I'm not optimistic that this approach is going to bear fruit.

Blass: Perhaps you can help shed some light as to the type of mechanism that you seem to be dealing with, if in fact it is inhibitory. Is oxytocin, or whatever oxytocin is implying, inhibiting behavior in general, or is it specific to salt appetite? Either of those answers would be followed up by a "Why?" Why oxytocin? Why a general inhibitor? Why a specific inhibitor of salt? The animal doesn't seem to be penalized for ingesting salt, especially under the circumstances where there is at least a relative need.

Stricker: Thank you for asking that question. It's a good question, and I'll be happy to deal with both parts of it. What you're asking is a general overview and that couldn't have been embedded in a talk on salt appetite, but I'll be happy to tell you about it now.

I don't think that the oxytocin projections are specific for salt appetite at all. The same treatments that inhibit salt appetite always inhibit food intake as well. So I see this as part of a system that affects ingestive behavior, not salt appetite specifically, but not ingestive behavior too broadly. That is to say that it doesn't affect water intake at all or sugar intake at all. I think that what it does affect is the noncaloric contributions of food. We've been focusing on the calories that are embedded in food and what happens when you vary the caloric content or even the specific nutrients that compose the calories. But in addition to food as a source of calories, food is the source of osmoles. There are noncaloric components of food intake, and I think those are components that the PVN, through these oxytocinergic projections, is controlling. I think that's why salt appetite is inhibited, but I also think that's why agents that activate oxytocin release that don't affect calories are so effective in decreasing feeding. Nausea, for example, is a wonderful way of decreasing food intake, that is to say it's tremendously effective—way more effective than satiety is in reducing food intake—and it's terrific for increasing oxytocin in the brain. Dehydration also decreases food intake in a noncaloric way and increases oxytocin, and a tremendous gastric distention is another one that increases oxytocin release and decreases food and salt intake as well. Not only does it decrease intake of salt and the intake of food, and increase oxytocin, but it also decreases gastric motility. So I see the PVN as an integrator of inhibitory effects on osmolar intake with coordinated effects on endocrine secretion and on autonomic function as well.

Now if you ask why the PVN, I would say in part because the hypothalamus, for reasons that people have recognized for a generation, is a central player in homeostasis. If you ask the question "Where does integration take place?" hypothalamus ought to be the first place that you think of. There is visual information as well as taste information that runs through the hypothalamus, so that it gets a lot of the sensory input that it needs. If you look at the parvocellular projections, they go all over the brain in the same way that locus coeruleus projects all over the brain. This is a system that looks like it is wired up to do many things simultaneously. It's not surprising to me that a system with that kind of anatomy would have these kinds of diverse integrative functions.

McHugh: Mark, would you like to respond to my insult?

Friedman: There are two points that you raised that I will respond to. One is the issue of a single versus a multiple control. I certainly don't mean to imply that the metabolic stimulus is the only factor that determines food intake or calorie intake. Your work with Tim Moran, and Deutsch's work, suggest that under certain circumstances gastric fill may play a role, and may even provide another model for studying caloric intake. I think the question is, Under which conditions do these different controls operate, and what is their contribution under different conditions?

The other point you raised, which I think is a very general issue in all of this kind of work, is the distinction between controls and effects. I remember in graduate school having it beat into my head that we needed to make a distinction between controls and regulation, and that's what I was referring to in my talk. The issue that you raised about the distinction between controls and effects gets at the fundamental problem that to understand the controls, we have to create effects. We have to intervene in the system; otherwise we're reduced to looking at correlations which, although interesting, are not informative unless we know ahead of time what we want to correlate. And we know that by doing experiments where we intervene and where we're basically looking at treatment effects. It's an issue of developing a language of talking about all of this that I think is very important to address. I'm glad you raised it.

PART I: DISCUSSION

Powley: You finessed the whole issue of how the liver communicates with the brain. Could you just expand on that a little.

Friedman: There are two basic ways. One is that the communication could be via a neural route, which is one of the reasons we cut the nerves. There's some evidence to indicate, to the extent that the hepatic vagus specifically innervates the liver, that communication could occur via the hepatic vagus. One of the reasons I put hepatic vagotomy or hepatic nerve section last on the list of evidence is I think that it is sufficiently nonspecific that it does not provide a really strong case. Nerve section is more compelling when combined with other lines of evidence. The other possibility, for which there is no evidence for or against, is that there may be a humoral signal. This has not been explored with respect to a hepatic signal and that's an area that ought to be looked at.

Steffens: Nicolaidis has this hypothesis that in the brain there is a representation of the whole periphery. What are your ideas about that?

Friedman: As you know, Mayer did the same thing. When he suggested the glucostat was insulin-sensitive, there was no evidence that insulin actually worked in the brain, let alone that there was any there. So he was actually putting a muscle cell in the brain. One way to look at this—and the reason I kept referring to a cell that looks a hepatocyte—is that we can put a hepatocyte in the brain, too, if you want. I'd rather keep it where it belongs! It's possible that there may be cells in the brain that are homunculi of what is going on in the periphery, but I'm not sure how strong the evidence is for it.

Part II
Hedonic Aspects of Appetite

4
Fat and Sugar: Sensory and Hedonic Aspects of Sweet, High-Fat Foods

Adam Drewnowski

University of Michigan
Ann Arbor, Michigan

I. INTRODUCTION

Increasing prevalence of obesity in Western societies has been linked to changing dietary habits and increasing consumption of palatable foods containing refined sugars and fat. Nutritional surveys indicate that the average adult consumes between 65 and 100 grams of fat and an estimated 95 grams of total sugars per day (Block et al., 1985; Glinsman et al., 1986). Fats and added sugars account for almost 50% of total calories in the typical American diet (Fig. 1). The chief dietary sources of fats are meat, vegetable oils, and milk, cheese, and other dairy products. Milk, fruit, and fruit juices are the main respective sources of natural sugars, lactose and fructose. Added sugars include sucrose in bakery and confectionery products and high-fructose corn syrup in carbonated soft drinks (Block et al., 1985; Glinsman et al., 1986).

Dessert-type foods constitute a special category. Foods such as chocolate, ice cream, cookies, pastries, and frozen desserts are typically composed of two chief ingredients: sugar and fat. Such foods, uniformly described as highly palatable by children and adults, are among the most preferred in the American diet. Candy bars are a popular snack food among children and adolescents. Ice cream and chocolate are a common target of food cravings reported by college-age females. Ice cream, cookies, and other sweet desserts figure in reports of uncontrollable

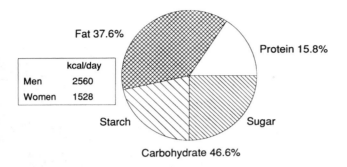

FIGURE 1 Nutrient composition of the U.S. diet: data for men and women ages 19–50 years. (From U.S.D.A. Nationwide Food Consumption Survey, 1985).

eating binges occurring in obesity and bulimia nervosa. The "carbohydrate-craving" phenomenon associated with antidepressant treatment and with seasonal affective disorder (SAD) also seems to involve chocolate, cookies, and other sweet, fat-rich desserts (Paykel et al., 1973; Rosenthal et al., 1987).

Widespread preferences and cravings for palatable desserts attest to the importance of hedonic aspects of appetite. However, relatively few laboratory studies have explored the sensory and hedonic aspects of sugar/fat mixtures (Drewnowski and Greenwood, 1983; Drewnowski, 1987a,b). Past studies on taste responsiveness have tended to equate food palatability with sweetness, focusing on the perception and preferences for sweet solutions (Weiffenbach, 1977). Only recently, attention has shifted to sensory perception of dietary fats and the role fats play in determining food acceptance (Tuorila and Pangborn, 1988; Drewnowski et al., 1989b).

II. SENSORY EVALUATION STUDIES

Laboratory studies on sweet taste responsiveness have generally addressed issues of sensitivity (threshold), perceived intensity, and stimulus acceptability or hedonic preference (Moskowitz et al., 1974). Sensory intensity, measured by magnitude estimation or category scaling procedures, follows the standard logarithmic equation:

$$I = a + b(\log C)$$

where C is the stimulus concentration. In contrast, hedonic preference functions for sweetness can be highly variable. Some subjects prefer, whereas others dislike, sugar solutions of increasing sweetness intensity. The average response typically follows an inverted-U shape function as preferences first increase with increasing sugar concentration, reaching a maximum at an ideal point or "break-

SENSORY AND HEDONIC ASPECTS OF FAT AND SUGAR

point" at around 8-10% sucrose. Preferences then decline as intensely sweet stimuli are judged to be increasingly less pleasant (Type 1 response). However, some individuals including most children continue to prefer solutions of increasing sugar concentration (Type 2 response). Studies on a variety of sweet foods and beverages have also shown that higher sucrose breakpoints are generally observed with solid relative to liquid foods (Moskowitz et al., 1974). Ice cream, for example, contains around 15% sugar, whereas the sugar content of cake frostings may be as high as 75-80%.

Sensory preferences for sweet solutions are generally regarded as the hedonic component of appetite. Sweet taste responsiveness among children, adolescents, and adults is thought to predict food preferences and therefore food consumption. However, there are major problems with this approach. First, sugar solutions in water may not be the ideal sensory stimulus, since most calorie-rich "sweets" are solids rather than liquids (Drewnowski et al., 1989b). Second, such foods often contain fat as the second major ingredient and the principal source of calories. Preferences for sweet taste in such foods may well be modulated by their fat content. Third, taste preferences for sweet solutions have rarely been linked with food intake measures under laboratory conditions, let alone with patterns of food consumption in real life (Mattes, 1985). Although taste preferences are a key factor in determining food acceptability, other psychological and attitudinal variables are also likely to play a major role.

It is unclear whether individuals or population groups can be distinguished by their sweet taste responsiveness. Attempts to link the pleasure response to sweet solutions with measures of body weight status have been largely unsuccessful (Thompson et al., 1976). Studies using sucrose solutions, sweetened Kool-Aid, and chocolate milkshakes found no consistent relationship between sweet taste preferences and overweight (Drewnowski, 1987a). Large-scale consumer studies found no relationship between body weight and hedonic preferences for increasing concentrations of sugar in such foods as apricot nectar, canned peaches, lemonade, or vanilla ice cream. In studies comparing obese and normal-weight subjects, individual variability of hedonic response was far greater than any between-group differences.

Studies on sensory preferences and body weight status have recently focused on the role of fat in relation to food acceptance (Drewnowski et al., 1985; Pangborn et al., 1985). Since the oral perception of fats involves mouthfeel, texture, and olfaction (Pangborn and Dunkley, 1964; Cooper, 1987), it poses a challenge to the investigator. It is not always clear what oral sensations contribute to the overall perception of fat content. Consequently, sensory assessment employs a wide range of attribute scales. The question then arises which of the many possible attributes are most closely linked to stimulus fat content. Past sensory evaluation studies of such foods as peanut butter, margarine, and mayonnaise have established that mouthfeel attributes of thickness, smoothness, and creaminess are all linked to product fat content (Kokini et al., 1977; Cussler et al., 1979). In liquid dairy products, where fat is contained in emulsified globules, the perception of fatness is largely guided by stimulus smoothness, thickness, or viscosity (Pangborn and Dunkley, 1964; Mela, 1988). Elsewhere, the mouthfeel of beverages has been described in such terms as thick, creamy, heavy, syrupy, and viscous. The terminology of food texture (Jowitt, 1974) contains a wide variety of such terms.

Sensory assessment of fats in solid foods is even more problematic. Fats can be an integral part of the food itself (e.g., marbling in meat) and are often used during food preparation. Fats are responsible for the texture, flavor, and aroma of many foods, and they endow foods with a wide range of textural characteristics (Schneeman, 1986). Again, it is often unclear what oral sensations contribute to the perception of fat in foods, since no single sensory attribute can be unambiguously linked to fat content. Among the attributes that have been associated with fat in foods are such terms as hard, soft, juicy, chewy, greasy, viscous, slippery, creamy, crisp, crunchy, and brittle (Szczesniak, 1971). It should be noted that many attributes in the terminology of food texture are not hedonically neutral. While creaminess may be a desirable property in many foods, greasiness generally is not.

Systematic evaluation of food texture has long been based on a classic body of work, the General Foods Texture Profile (Brandt et al., 1963). This method approached texture evaluation as a dynamic analysis of the mechanical, geometrical, fat, and moisture aspects of foods, occurring along the dimension of time—from first bite through complete mastication. The primary mechanical characteristics of food were defined as hardness, cohesiveness, adhesiveness, and viscosity, while secondary ones were brittleness, chewiness, and gumminess. Geometrical characteristics were defined as those related to the shape, size, and orientation of food particles (e.g., gritty, grainy, coarse), while mouthfeel characteristics were related to the perception of moisture or fat (wet, oily, greasy).

The Texture Profile introduced a number of anchored rating scales, each referenced to an array of physical standards (Szczesniak et al., 1963). For example, the viscosity scale (shown in Table 1) represented eight degrees of viscosity, each one defined by a commercially available standard. It can be seen that more than half the standards are foods containing different amounts of fat. Similarly, the adhesiveness scale included such standards as Crisco oil, cream cheese, and peanut butter. Szczesniak et al. (1963) provided rating scales for six different textural properties of food: hardness, fracturability, chewiness, gumminess, adhesiveness, and viscosity. Each scale represented a wide range of textures in order to encompass many different foods.

The relationship between perceived fat content and food acceptance is fairly complex. In dairy products, elevated fat content generally serves as an index of product quality (Cooper, 1987). Consumers value richness or smoothness of ice cream, while the quality of heavy cream is judged by its thickness or pour. In other cases, consumers show negative attitudes toward foods with a high perceived fat content (Shepherd and Stockley, 1985). Further studies suggest that sensory

TABLE 1 Viscosity Scale from the General Foods Texture Profile

Rating	Product	Brand or manufacturer	Sample size	Sample temperature (°F)
1	Water	Crystal Spring	½ tsp	Room
2	Light cream	Sealtest Foods	½ tsp	45–55
3	Heavy cream	Sealtest Foods	½ tsp	45–55
4	Evaporated milk	Carnation Co.	½ tsp	45–55
5	Maple syrup	F. H. Leggett & Co.	½ tsp	45–55
6	Chocolate syrup	Hershey Co.	½ tsp	45–55
7	Mixture: ½ cup mayonnaise and 2 tablespoons heavy cream	Best Foods–Hellman's and Sealtest Foods	½ tsp	45–55
8	Condensed milk	Borden Foods Magnolia brand	½ tsp	45–55

Source: Brandt et al. (1963).

preferences for high-fat foods need not be analytical. Under some circumstances, respondents may prefer high-fat foods without being aware of their elevated fat content.

A recent study (Drewnowski et al., 1989b) illustrates some of the problems in the sensory assessment of liquid and solid foods. In that study, 25 young men and women tasted and rated the sweetness, creaminess, and fat content of liquid and solid dairy products of varying sugar and fat content. The range of fat levels varied between 0.1 and 52%, weight by weight, and the samples were sweetened with 0, 5, 10, or 20% sucrose. Liquid dairy products included sweetened skim milk, whole milk, half and half, and heavy cream. Solid food samples were made up of cottage cheese, cream cheese, or the two blended together to produce mixtures of comparable fat content (0.1–52%). The sweetened blends were then spread on slices of white bread, rolled "jelly roll" fashion and cut into small slices to produce the solid food units (Drewnowski et al., 1989b).

The subjects reliably estimated sweetness intensity of both liquid and solid food. However, their ability to assess the creaminess and fat content of solids as opposed to liquids was greatly impaired. Despite specific instructions, some subjects were unable to track the increasing fat content of solid foods. Mean estimates of perceived fat content as a function of stimulus fat in liquids and solids are summarized in Figure 2. Unlike stimulus intensity judgments, hedonic preference ratings for liquids and solids were not appreciably different. As shown in Figure 3, preferences for high-fat samples did not suffer because of difficulties in estimating stimulus fat content. On the contrary, best-liked fat levels were consistently higher in solids than in liquids. In contrast, preferences for sweet taste were

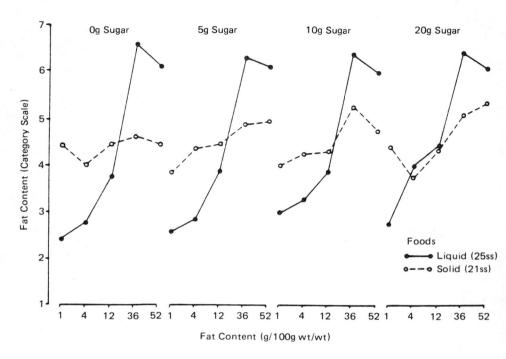

FIGURE 2 Mean estimates of stimulus fat as a function of fat level at each level of sucrose. Fat content is expressed as grams per 100 g. (Drewnowski et al., 1989b. Reprinted with permission.)

approximately constant across food media. Subjects who liked intensely sweet milkshakes also liked intensely sweet solid food samples. Both Type 1 and Type 2 responses were obtained for fat as well as for sugar.

Preferences for fat in foods may be to a large degree system-specific. As the present data indicate, sensory response to milk beverages does not predict the response to fat in solid foods. The present data further suggest that preferences for fat in foods may be formed in the absence of an accurate, conscious judgment of stimulus fat content. The reliance on texture cues in sensory assessment of fats means that an illusion of fat content can be created by making the stimulus more viscous, either by gelling or by the use of hydrocolloid thickeners. A newly developed fat substitute, Simplesse, consists of protein microglobules designed to emulate the cooling mouthfeel of emulsified dairy fat.

Stimulus texture and the perception of fat content may thus depend on food ingredients other than fat. In a recent study (Drewnowski and Schwartz, 1989), 50 young women rated the sweetness and fat content of 15 stimuli resembling cake frostings. The samples were composed of sucrose (20–77% wt/wt), polydextrose (a bland, partly metabolizable starch), unsalted butter (15–35% wt/wt), and

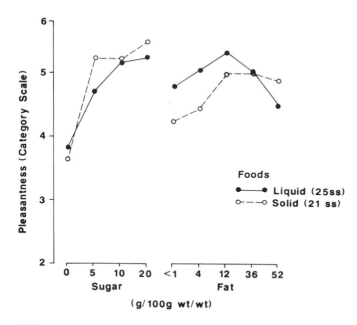

FIGURE 3 Mean hedonic ratings as a function of sucrose and fat content in a range of sweetened dairy products. (Data from Drewnowski et al., 1989b.)

distilled water. As expected, the perception of sweetness intensity was a function of sucrose levels only. In contrast, oral assessment of fat content was largely mediated by product texture. Fatness ratings were a combined function of three ingredients: fat, polydextrose, and water. The addition of sugar caused a sharp decrease in fatness ratings. The sweetest stimuli were also perceived as lowest in fat. This observation may help explain why fat in sweet, fat-rich desserts often behaves as invisible calories, with most of the attention focusing on the other ingredient, sugar. The relationship between food texture and the perception of fat content also has important implication for the incorporation of fat substitutes into reduced-calorie food products.

III. HEDONIC RESPONSE PROFILES

Hedonic preference data provide further information regarding ingredient composition of the best-tasting mixture. Our early studies (Drewnowski and Greenwood, 1983; Drewnowski et al., 1985) addressed the possibility that sensory preferences for sugar versus fat are a biological marker that may help discriminate between clinical patient populations at extremes of body weight.

We conducted a series of sensory evaluation studies using the previously described 20 different mixtures of milk, cream, and sugar. The studies were conducted with anorectic and bulimic patients and with massively obese women enrolled in a program for weight reduction. Additional responses were obtained from groups of age-matched, normal-weight female controls. Subject characteristics are summarized in Table 2.

The subjects rated each stimulus in turn for its sweetness, creaminess, and perceived fat content. They also rated the acceptability of each sample on a 9-point category scale. The use of a mathematical modeling technique known as the response surface method allowed us to predict the shape of hedonic response for a wide range of ingredient levels (Drewnowski and Greenwood, 1983). Figure 4 shows graphic representations of the average hedonic response surface for the group of normal-weight adults. The data are presented as a three-dimensional projection (Fig. 4a) and as corresponding isopreference or isohedonic contours (Fig. 4b).

The response profile to mixtures of sugar and milkfat was strongly interactive. Preference ratings for sweetened skim milk or unsweetened milk or cream were relatively low. Highest ratings were obtained for a mixture estimated as containing 8% sugar and 20% fat. As shown in Table 2, the optimal sugar and fat levels varied among subject groups. Obese women tended to prefer stimuli that were rich in fat but relatively low in sugar. In contrast, anorectic women liked sweetness but showed reduced preferences for the oral sensation of dietary fat.

Expressing preferences for sugar versus fat in terms of an optimal sugar/fat (S/F) ratio allowed us to correlate individual taste responses with values of the body mass index (kg/m^2), a common measure of overweight. As shown in Figure 5, there was a negative relationship between S/F ratios and the degree of overweight. The present data are consistent with clinical reports that obese patients crave fat-rich foods, while anorectic patients show dislike if not an aversion to fats, particularly milk and meat. However, it should be noted that the observed correlation was relatively weak ($r = -0.43$), accounting for only a small proportion of the variance. Although other, more direct measures of metabolic status

TABLE 2 Subject Characteristics and Optimal Stimulus Composition

Subject group	N	Age (years)	Weight (kg)	Sucrose (%)	Fat (%)
Obese	12	38.0	95.8	4.4	34.5
Reduced obese	8	32.7	67.9	10.1	35.1
Normal weight	15	30.1	58.8	7.7	20.7
Normal weight	16	19.1	57.7	9.1	28.7
Bulimic	7	19.4	56.8	15.3	27.9
Anorectic	25	17.2	40.5	12.7	16.5

Source: Drewnowski et al., (1985; 1987).

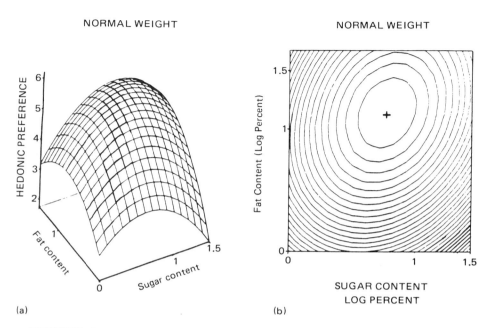

FIGURE 4 Hedonic response surface to sugar–fat mixtures expressed as a three-dimensional projection (a) or as isopreference contours (b). Region of optimal preference is denoted by + sign.

may be better correlated with taste preferences, behavioral factors are likely to be involved as well. Preferences for sugar and fat may be modulated by previous experience, or they may be contingent on attitudinal or social variables, including attitudes toward body weight and dieting.

IV. METABOLIC FACTORS

Some cravings for sweet desserts are thought to be triggered by physiological or metabolic events (Wurtman et al., 1981). For example, cravings for carbohydrates are said to be caused by deficiencies in the metabolism of a central neurotransmitter, serotonin. A majority of obese individuals are thought to suffer from serotonin deficiency and have been classified as carbohydrate cravers (Lieberman et al., 1986). Carbohydrate snacks are said to relieve depression and fatigue (Lieberman et al., 1986), serving effectively as a form of self-medication. The sensory aspects of such snacks have been dismissed as unimportant, and the cravings are said to be specific for a macronutrient, carbohydrate, rather than for any given food item.

Unfortunately, the "carbohydrate-rich" snacks used in most laboratory stud-

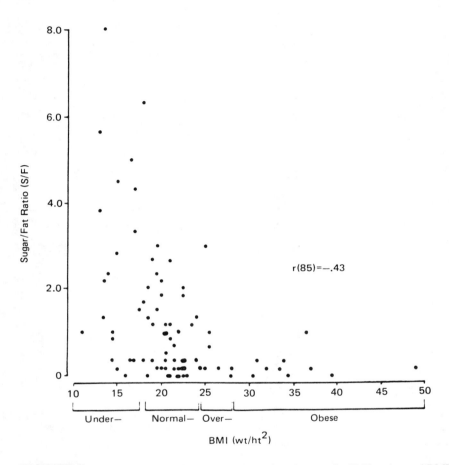

FIGURE 5 Relationship between optimally preferred sugar/fat (S/F) ratios and BMI values of anorectic, normal-weight, and obese females: body mass index (BMI) = kg/m^2. Low S/F ratios denote preferences for fat over sugar in stimulus mixtures; high S/F ratios denote preferences for sugar over fat.

ies were almost without exception mixtures of sugar and fat, including such foods as chocolate candy (Snickers and M&M's), chocolate cupcakes, chocolate chip cookies, and cookies of other types. In other studies, cakes, frozen pastries, and other desserts (Paykel et al., 1973), and even ice cream sundaes with whipped cream, were described as carbohydrate-rich foods. Cravings for these foods are clearly more common than cravings for bland carbohydrates such as potatoes or bread. It would appear that "carbohydrate cravings" as described in the psychological literature are most commonly directed at the palatable mixtures of sugar and fat.

The argument that cravings for chocolate or ice cream are mediated through serotonin metabolism ignores the hedonic aspects of appetite and is therefore unconvincing. A more promising approach that is more sensitive to the hedonic aspects of appetite involves endogenous opioid peptides, pleasure-enhancing molecules manufactured by the human brain. In several studies with rats and mice (Blass, 1987), intakes of sugar and fat have been linked to the endogenous opioid peptide system. Dietary studies have further shown that morphine-injected animals selectively increase fat intake, while the opioid antagonist naltrexone blocks overeating induced by a palatable cafeteria diet.

One preliminary report (Drewnowski et al., 1989a) examined the effects of manipulating the opioid peptide system on sensory preferences for sugar and fat. Nine normal-weight females again rated the sweetness and acceptability of 20 different mixtures of milk, cream, and sugar following infusions of the synthetic opioid agonist butorphanol, or the opioid antagonist naloxone. Preference ratings for the sugar/fat mixtures were reduced following naloxone, while the perception of sweetness intensity was not affected by either drug. The only effects on food intake were observed for sweet, high-fat foods (chocolate and cookies), whose intake was significantly lowered by naloxone. The drop in fat calories was significant.

Of course, metabolic factors influencing food preferences need not be limited to central events. The pleasure response to food may also be influenced by postingestive consequences of previous exposure to the food through a conditioning procedure. In one view, the reinforcing qualities of food are linked to the rate of metabolic flux, the speed with which calories are made available to the organism. This approach is also sensitive to the differential hedonic appeal of different sugar/fat mixtures, since stimulus composition determines the relative rates of oxidation and fat storage.

V. ATTITUDINAL STUDIES

Enhanced sensory preferences for sugar/fat mixtures do not necessarily lead to elevated intakes of all sweet, fat-rich foods. Although taste factors are undoubtedly important, attitudes toward health, body weight, and dieting often override physiological and metabolic signals. Most dramatic examples of interaction between taste factors, attitudes, and behavior are to be found among women with eating disorders (e.g., anorexia nervosa and bulimia).

Although sensory responsiveness studies have shown that anorectic patients like intensely sweet stimuli (Drewnowski et al., 1987), they also view sweet desserts as forbidden foods that have no place in their regular diet. High-calorie sweets are consumed only in the course of uncontrollable eating binges that are followed by self-induced vomiting. There is some evidence that the binges are not caused by a satiety deficit, but rather are triggered by physiological events. Patients sometimes report knowing that a binge is imminent, and there are reports of buying, hoarding, and preparing foods ahead of time.

While eating binges may include sweet, high-fat foods such as cookies, chocolates, or ice cream, eating disorder patients generally show an aversion to dietary fats (Drewnowski et al., 1987). Clinical observations confirm that anorectic women avoid starches, sugars, and especially fats. According to some clinical reports, anorectic women were willing to eat vegetables, lettuce, fresh fruit, cheese, and sometimes eggs, but were disgusted by milk and meat. Although the pattern of food selection in eating disorders has been described as a "carbohydrate phobia," it may be that the reported avoidance of high-calorie foods is linked to an avoidance of dietary fats (Drewnowski et al., 1988).

It is unclear whether fat aversion has a physiological or a psychological basis. One possibility is that the avoidance of fats is the result of prior conditioning. The common targets of food aversion are foods that are novel, of low preference, and characterized by salient sensory characteristics. Since fats absorb the flavors and odors of other foods, they may prove especially aversive to susceptible individuals. Fat/protein mixtures in particular are avoided by pregnant women and by patients undergoing chemotherapy (Midkiff and Bernstein, 1985). Food aversions in eating disorder patients may similarly be linked to nausea and vomiting following eating binges.

Another possibility is that eating disorder patients avoid foods high in calories, that is, foods rich in fat. Eating disorder patients show a rigid attitude structure, allowing themselves to eat only those foods that they also perceive as nutritious and low in calories. Attitudes toward dieting clearly influence hedonic judgment, since vegetables are typically rated as more preferred than ice cream. Indeed many anorectic patients profess to like only salads, vegetables, and fruit, and are disgusted by milk and meat.

Reports of food preference can be subject to cognitive bias. For example, dietary intake assessments of children ages 2–12 typically report that the children derive the bulk of dietary sugars from milk and fresh fruit (Morgan and Zabik, 1981). However, it must be realized that such data are based on information typically provided by the child's mother. Actual observation of foods consumed over 3 days by schoolchildren in Washington D.C. both at home and in school revealed a totally different pattern of intake. The most commonly consumed food was candy, followed by potato chips, popcorn, sweetened Kool-Aid, presweetened cereals, other carbonated beverages, cupcakes, doughnuts, and pies (Davidson et al., 1986).

Hedonic aspects of appetite may also be influenced by the accepted social norms. A survey of food preferences among males in the U.S. Armed Forces (Meiselman et al., 1974) showed milk and grilled steak among the most highly preferred foods. In contrast, foods that had connotations of dieting (e.g., skim milk, diet soda, fruit yogurt) tended to be rated as unacceptable. Apparently, "diet" foods are undesirable in this particular milieu. Social desirability and peer pressure may have a further influence in determining food choice.

ACKNOWLEDGMENT

This work was supported by grants DK37011 and DK38073 from the National Institutes of Health.

DISCUSSION

McHugh: I'm sorry that you're only going to have one question because some of this is perhaps lighthearted. That was wonderful data you showed, Adam, but the first question that came from the last slide is "Didn't those children have a mother?"

Drewnowski: This was observed at school. The idea was to show that children really have patterns of food intake that parents know nothing about.

McHugh: Now the serious question related to your type 1 and type 2 people that you described and their behavior. Are they differentiated in other ways in their taste capacities like their PROP tasting or anything of that sort?

Drewnowski: We looked at a number of things, but the type 1 and type 2 response to sugar, which is the one that has been characterized by Rose Marie Pangborn and Howard Moskowitz. We have looked at just about everything. We looked at body mass indices, we looked at dietary restraint, we looked at all kinds of stuff. There was not really very much there. However if we look at type 1 and type 2 response to fats, we have preliminary evidence—it's very preliminary right now—that women who seemed to like unsweetened fats are the ones who have had a problem with body weight. They are not necessarily overweight right now, but they tend to be more restrained on a restraint scale and they have shown greater weight fluctuations in the past. So it's not their current status but perhaps a lifetime history, previous yo-yo cycles, whatever, that may be the issue here.

REFERENCES

Blass, E. M. (1987). Opioids, sugar and the inherent taste of sweet: Broad motivational implications. In *Sweetness,* J. Dobbing (Ed.). ILSI–Nutrition Foundation Symposium, Springer-Verlag, New York, pp. 115–124.

Block, G., Dresser, C. M., Hartman, A. M., and Carroll, M. D. (1985). Nutrient sources in the American diet: Quantitative data from the NHANES II survey. *Am. J. Epidemiol.* **122**:27–40.

Brandt, M. A., Skinner, E. Z., and Coleman, J. A. (1963). Texture Profile method. *J. Food Sci.* **28**:404–409.

Cooper, H. R. (1987). Texture in dairy products and its sensory evaluation. In *Food Texture,* H. R. Moskowitz (Ed.). Dekker, New York, pp. 251–272.

Cussler, E. L., Kokini, J. L., Weinheimer, R. L., and Moskowitz, H. R. (1979). Food texture in the mouth. *Food Technol.* **33**(10):89–92.

Davidson, F. R., Hayek, L. A., and Altschul, A. M. (1986). Perception and food use of adolescent boys and girls. *Ecol. Food Nutr.* **18**:309–317.

Drewnowski, A. (1987a). Sweetness and obesity. In *Sweetness,* J. Dobbing (Ed.). ILSI–Nutrition Foundation Symposium, Springer-Verlag, New York.

Drewnowski, A. (1987b). Fats and food texture: Sensory and hedonic evaluations. In *Food Texture,* H. R. Moskowitz, (Ed.). Dekker, New York, pp. 217–250.

Drewnowski, A., and Greenwood, M. R. C. (1983). Cream and sugar: Human preferences for high-fat foods. *Physiol. Behav.* **30**:629–633.

Drewnowski, A., Brunzell, J. D., Sande, K., Iverius, P. H., and Greenwood, M. R. C. (1985). Sweet tooth reconsidered: Taste preferences in human obesity. *Physiol. Behav.* **35**:617–622.

Drewnowski, A., Halmi, K. A., Pierce, B., Gibbs, J., and Smith, G. P. (1987). Taste and eating disorders, *Am. J. Clin. Nutr.* **46**:442–450.

Drewnowski, A., Gosnell, B., Krahn, D. D., and Canum, K. (1989a). Opioids affect taste preferences for sugar and fat. Paper presented at the Annual Meeting of the American Psychiatric Association, San Francisco.

Drewnowski, A., Shrager, E. E., Lipsky, C., Stellar, E., and Greenwood, M. R. C. (1989b). Sugar and fat: Sensory and hedonic evaluations of liquid and solid foods. *Physiol. Behav.* **45**:177–183.

Drewnowski, A., Pierce, B., and Halmi, K. A. (1988). Fat aversion in eating disorders. *Appetite,* **10**:119–131.

Drewnowski, A., and Schwartz, M. (1990). Invisible fats: Sensory assessment of sugar/fat mixtures. *Appetite,* **14**:203–217.

Glinsman, W. H., Irausquin, H., and Park, Y. K. (1986). Evaluation of health aspects of sugars contained in carbohydrate sweeteners. *J. Nutr.* **116**(11S):S1–216.

Jowitt, R. (1974). The terminology of food texture. *J. Texture Stud.* **5**:351–358.

Kokini, J. L., Kadane, J. B., and Cussler, E. L. (1977). Liquid texture perceived in the mouth. *J. Texture Stud.* **8**:195–218.

Lieberman, H. R., Wurtman, J. J., and Chew, B. (1986). Changes in mood after carbohydrate consumption among obese individuals. *Am. J. Clin. Nutr.* **44**:772–778.

Mattes, R. D. (1985). Gustation as a determinant of ingestion: Methodological issues. *Am. J. Clin. Nutr.* **41**:672–683.

Meiselman, H. L., Waterman, D., and Symington, L. E. (1974). Armed Forces Food Preferences. Technical Report 75-63-FSL, U.S. Army Natick Development Center, Natick, MA.

Mela, D. J. (1988). Sensory assessment of fat content in fluid dairy products. *Appetite,* **10**:37–44.

Midkiff, E. E., and Bernstein, I. L. (1985). Targets of learned food aversions in humans, *Physiol. Behav.* **34**:839–841.

Morgan, K. J., and Zabik, M. E. (1981). Amount and food sources of total sugar intake by children aged 5 to 12 years. *Am. J. Clin. Nutr.* **34**:404–413.

Moskowitz, H. R., Kluter, R. A., Westerling, J., and Jacobs, H. L. (1974). Sugar sweetness and pleasantness: Evidence for different psychophysical laws. *Science,* **184**:583–585.

Pangborn, R. M., and Dunkley, W. L. (1964). Sensory discrimination of fat and solids-not-fat in milk. *J. Dairy Sci.* **47**:719–725.

Pangborn, R. M., Bos, K. E., and Stern, J. (1985). Dietary fat intake and taste responses to fat in milk by under-, normal-, and overweight women. *Appetite,* **6**:25–40.

Paykel, E. S., Mueller, P. S., and de la Vergne, P. M. (1973). Amitryptyline, weight gain and carbohydrate craving: A side effect. *Br. J. Psychiatr.* **125**:501–507.

Rosenthal, N. E., Genhart, M., Jacobsen, F. M., Skwerer, R. G., and Wehr, T. A. (1987). Disturbance of appetite and weight regulation in seasonal affective disorder. *Ann. N.Y. Acad. Sci.* **499**:216–230.

Schneeman, B. O. (1986). Fats in the diet: Why and where? *Food Technol.* **10**:115–120.
Shepherd, R., and Stockley, L. (1985). Fat consumption and attitudes towards food with a high fat content, *Hum. Nutr.: Appl. Nutr.* **39A**:431–442.
Szczesniak, A. S. (1971). Consumer awareness of texture and other food attributes: II. *J. Texture Stud.* **2**:196–206.
Szczesniak, A. S., Brandt, M. A., and Friedman, H. H. (1963). Development of standard rating scales for mechanical parameters of texture and correlations between the objective and the sensory methods of texture evaluation. *J. Food Sci.* **28**:397–403.
Thompson, D. A., Moskowitz, H. R., and Campbell, R. (1976). Effects of body weight and food intake on pleasantness ratings for a sweet stimulus. *J. Appl. Psychol.* **41**:77–83.
Tuorila, H., and Pangborn, R. M. (1988). Prediction of reported consumption of selected fat-containing foods. *Appetite,* **11**:81–95.
U.S. Department of Agriculture (1985). Nationwide Food Consumption Survey, Continuing Survey of Food Intakes by Individuals. NFCS, CSFII Reports No. 85/1-3. Hyattsville, Md. U.S. Department of Agriculture.
Weiffenbach, J. M., Ed. (1977). *Taste and Development: The Genesis of Sweet Preference,* U. S. Government Printing Office, Bethesda, MD.
Wurtman, J. J., Wurtman, R. J., Growdon, J. H., Henry, P., Lipscomb, A., and Zeisel, S. H. (1981). Carbohydrate craving in obese people: Suppression by treatments affecting serotoninergic transmission. *Int. J. Eating Disord.* **1**:2–15.

5
Human Salt Appetite

Gary K. Beauchamp

Monell Chemical Senses Center
Philadelphia, Pennsylvania

Mary Bertino

Colgate–Palmolive Company, Piscataway, New Jersey, and
Monell Chemical Senses Center
Philadelphia, Pennsylvania

Karl Engelman

University of Pennsylvania
Philadelphia, Pennsylvania

I. INTRODUCTION

An ability to taste sodium presumably evolved as a consequence of the critical nature of this nutrient and its patchy distribution (Dethier, 1977). Many animals that consume plants respond to sodium deficit with a specific salt appetite that is characterized by increased activity and a heightened tendency to orally explore novel aspects of the environment. When sodium is encountered, it is innately recognized as a distinctive taste, and ingestion proceeds to correct the deficit (see Denton, 1982).

That taste is critical in sodium detection was shown first by Richter (1956), who found that adrenalectomized rats that became sodium deficient as a result of the absence of mineralocorticoid hormones were able to survive if they were permitted access to salt. When Richter denervated almost all taste receptors in such animals, they no longer consumed adequate amounts of salt (NaCl), presumably because they could not detect its presence, and they died as a consequence.

How does the taste system operate to ensure increased intake to correct the deficit? This question has been the focus of a vast amount of research during the past 50 years. While substantial progress has been made in understanding the hormonal and central nervous system control for salt appetite in rats and sheep (see, e.g., Denton, 1982; Sakai et al., 1986; Stricker and Verbalis, 1988), an unresolved issue is whether changes in the sense of taste could account for the salt appetite during deficiency.

Richter (1956) reported a decreased salt preference threshold (the lowest concentration of salt that is ingested preferentially to water) in salt-depleted animals. However, adrenalectomized rats could not detect salt (detection threshold) at any lower level than could intact animals (Carr, 1952). Apparently, the decreased preference threshold was not due to an increase in absolute taste sensitivity. In fact, Contreras and colleagues (Contreras and Frank, 1979; Contreras et al., 1984) demonstrated a depressed sensory response to salt in sodium-depleted rats. It has been argued that this reduced sensitivity could account in part for increased intake of higher salt concentrations, since the solutions would taste less salty to the rat. Recently, Jacobs et al. (1988) studied brain stem neural responses in sodium-depleted rats. It was suggested that the taste of salt may change to become more sweetlike, or at least more pleasant, after sodium depletion.

A. Human Salt Appetite

Since sensory experience of rats and sheep must be inferred from results of behavioral or neurophysiological experiments, an alternative is to study taste in humans who are salt-depleted. Here, one can employ psychophysical testing techniques that are capable of objectively evaluating various parameters of taste function and sensory experience.

Very few experimental studies have been published on the sensory effects of sodium depletion in humans. Much of what is available is anecdotal, uncontrolled, and subject to experimental bias, particularly since experimenter and subject often were the same individual, and since studies were conducted at a time when procedures were not well developed and validated. Although it is widely assumed that adult humans develop a salt appetite when sodium deficient, the data are weak (see Beauchamp, 1987; Stricker and Verbalis, 1988). Since many behavioral as well as neurophysiological responses to salt differ among rats, sheep, mice, and humans, it is questionable whether these other animals provide a good model for humans (see, e.g., Denton, 1982; Beauchamp and Bertino, 1984; Barnwell et al., 1986; Sakai et al., 1986; Denton et al., 1988; Rowland and Fregley, 1988).

B. Cultural Variation

Historical and anthropological sources (Multhauf, 1978; Denton, 1982) document the remarkable amount of effort humans have devoted to securing salt. These observations provide circumstantial evidence that salt insufficiency stimulates a kind of salt appetite. Salt has served as a basis for taxation throughout the world, with Indian salt taxes being among the most prominent. Wars have been fought over rights to salt-producing regions, and communities often established salt deposits as communal property to prevent salt monopolies (see, e.g., Wilson, 1937; Block, 1976).

In spite of this, people in some cultures do not consume salt beyond what is found naturally in their food and are reported to have no desire for added salt (see Denton, 1982, for reference). It has been suggested that this occurs generally in people who consume meat, thereby ensuring sufficient sodium consumption, while salt assumes more importance among pastoral, largely plant-eating, cultures (Denton, 1982). Unfortunately, this generalization is not completely accurate; the anthropological literature on salt usage and its variation in different cultures does not admit of a completely satisfactory explanation based on physiology (see Lapicque, 1896; Dastre, 1901; Neumann, 1977).

C. Clinical Reports

The most dramatic and often-cited illustration of a depletion-specific sodium appetite in humans came from the observation of a child with adrenal insufficiency and sodium wasting who demonstrated what appeared to be a specific craving for salt (Wilkins and Richter, 1940; Wilkins et al., 1940). The child died when not allowed to satisfy the craving in the hospital. This and a number of other clinical case studies are presented in Table 1 with annotations. The list, while not exhaustive, describes the most prominent cases in the English language medical literature.

In almost every case among those described in Table 1, the onset of the extreme salt appetite was in childhood, although the underlying medical problem varied. Indeed, there have been very few documented cases of salt appetite with an adult onset, with cases F and L (Table 1) being apparent exceptions. It is generally assumed that salt appetite is a common symptom of adult-onset Addison's disease. As can be seen in Table 1 (case D), 16% (10 out of 64) of patients with adrenal insufficiency are reported to exhibit a salt appetite, with the time of onset of this symptom not reported. And in a widely cited monograph on Addison's disease (Thorn, 1949), a figure of 19% is given for the incidence of salt craving (Table 1, case E). Yet salt appetite or craving is *not* listed as one of the nine signs and symptoms of this disease, and no description of symptoms similar to those of the childhood onset of salt appetite appears. In fact, patients with Addison's disease seem to be equally likely to eat salty foods or licorice to ameliorate their symptoms (see Table 1, case F). Whether the salty foods and/or licorice candy actually tasted especially good to these patients or whether the patients had merely learned that consumption of these items relieved symptoms, as might be the case for a medicine that is taken even though it never really tastes good, cannot be determined from the clinical reports.

TABLE 1 Clinical Reports of Salt Appetite

Case	Subject(s) description (Reference)[a]	Age at onset of salt appetite	Amount of NaCl consumed	Medical diagnosis	Comments
A	A 15-year-old diabetic boy (three other diabetic children referred to—one seemed also to have a salt appetite) (1–3)	Unknown (<15 years)	60–90 g/day	Diabetes	Consumed salt to satisfy "an abnormal craving for salt." Excess salt consumption resulted in elevated blood pressure.
B	A 20-year-old woman (4)	"Present since early childhood"	130–195 g/day (self-report); 138 g/day (clinically determined)	Primary pulmonary arteriosclerosis	First noticed appetite for salt when large enough to climb into chair to reach it. Carried rock salt with her at all times. Refused to go on low-salt diet for more than 4 days. Threshold for salt taste low relative to control values.
C	A 3½-year-old boy (5)	First noticed at 1 year of age	Unknown, probably substantially greater than 20 g/day	Corticoadrenal insufficiency	First began eating pure salt at about 18 months of age. "Salt" among first words learned. Did not like sweets.
D	10 of 64 patients Addison's disease (6)	Unknown	Unknown	Addison's disease	"Increased desire for salt and salty foods." According to Richter (5), one patient covered food with salt and ate very salty foods, salting such items as oranges and lemons. He disliked sweets.
E	19% of patients with Addison's disease (7)	Unknown	Unknown	Addison's disease	Salt appetite *not* listed among nine diagnostic signs and symptoms of disease. Not one of the dozen adult cases described had elevated salt appetite listed as a symptom.

F	A 36-year-old male (8)	Apparently recent	0.25 kg of salty olives each day	Addison's disease	Also exhibited an appetite for licorice sweets, known to contain a palliative for Addison's disease. In another case (9), an adult craved licorice but denied craving salt.
G	The Southwood–Gannon family (n = 20 total; 10)	Unknown: disease onset in childhood	Unknown	Genetically determined periodic paralysis with normalkalemia	No direct evidence of salt appetite. Many family members consumed a large quantity of salt daily, perhaps as a therapeutic agent as attacks are moderated.
H	A 5-year-old male (11)	"The patient had always been a poor eater and craved salt."	Unknown	Bartter's syndrome	Another child was not noted to have salt appetite. Most often clinical descriptions of Bartter's syndrome do not mention salt appetite; most involve observations on adults.
I	An 8-year-old male (12)	Unknown (< 8 years)	Unknown	Pleoconial myopathy	Desired salt; covered food with it and "frequently" ate a teaspoon of salt directly. A 12-year-old sibling also exhibited salt hunger. No relationship among sodium loading, sodium deprivation, aldosterone antagonism, and attacks.
J	A 13-year-old male (13)	"Present all his life."	Unknown	Mitochondrial myopathy	Ate sandwiches of bread and salt and salted bananas. Mother had to forbid him from pouring salt from store containers directly onto food.

TABLE 1 *Continued*

Case	Subject(s) description (Reference)[a]	Age at onset of salt appetite	Amount of NaCl consumed	Medical diagnosis	Comments
K	16 of 43 children with sickle cell disease (14)	Unknown ("as early as 2½ years")	Unknown	Sickle cell hemoglobinopathies	Six had "abnormal" and 10 had "exceedingly increased" salt appetite relative to rest of family. Latter category included children who salted apples, oranges, and peaches, as well as one 2½ year old who salted rock candy, licked off salt, and discarded candy.
L	A 33-year-old female (15)	Adult; recent	"By the shakerfull."	Iron deficiency	Iron replacement therapy resulted in a cessation of salt eating.

[a] Numbers in parentheses designate the following items, found in the References section.

1. McQuarrie (1935).
2. Thompson and McQuarrie (1934).
3. McQuarrie et al. (1936).
4. Darley and Doan (1936).
5. Wilkins and Richter (1940).
6. Thorn et al. (1942).
7. Thorn (1949).
8. Knowles and Asher (1958).
9. Cotterill and Cunliffe (1973).
10. Pokanzer and Keir (1961).
11. Bartter et al. (1962).
12. Shy et al. (1966).
13. Spiro et al. (1970).
14. Grossman et al. (1977).
15. Shapiro and Linas (1985).

Clinical studies of salt taste sensitivity (e.g., threshold), as distinguished from preference or appetite, are also infrequent. The studies of Henkin et al. (1963) found a nonspecific lowered threshold (greater sensitivity) for taste compounds representing salty, sweet, sour, and bitter among patients with adrenal insufficiency. A report of increased taste sensitivity among older children with cystic fibrosis (Henkin and Powell, 1962) has not been supported in subsequent studies (Wotman et al., 1964; Desor and Maller, 1975; Hertz et al., 1975). One patient studied at 20 years of age, who exhibited a salt appetite from early childhood associated with primary pulmonary arteriosclerosis, was reported to have a strikingly low salt threshold (Table 1, case B; Darley and Doan, 1936).

D. Human Experimental Approaches

Few attempts have been made to induce sodium depletion experimentally in human volunteers. McCance (1936) induced sodium loss in 4 subjects over periods of approximately 10 days with a combination of very low salt diets and heavy sweating. This study was a tour de force in terms of rigor and dedication required of experimenter and of subjects; McCance was included in both groups. Induced sweat was collected in rubber sheets and all urine and stools were collected for analysis. To collect insensible perspiration, special underclothes were worn, which were then washed in distilled water to extract the salt. As estimated by chloride loss measured in sweat, urine, stools, and clothes, subjects lost up to 22 g of total sodium (out of about 100 g of total body sodium for a 70 kg man). No taste tests were conducted. However, McCance reported that foods (and even cigarettes) seemed tasteless, and that anorexia and nausea were prominent symptoms. McCance found that rinsing his mouth with salt water was refreshing and restorative of taste, but he was unable to find changes in salivary electrolyte composition. It is notable that no obvious specific cravings for salt were noted.

Yensen (1959) used a procedure similar to that of McCance to induce sodium loss in two individuals. The subjects were reported to have adapted to the diet rather well with the exception of the low sodium bread, which they found unpleasant. "Apart from the feeling at meal times that some of the food would probably 'taste nicer with some salt,' no cravings for salt were felt by either subject" (p. 233). Significant increases in salt taste sensitivity (lower thresholds) were found during the periods of sodium depletion. However, rather uncontrolled methods were employed to evaluate threshold (see Henkin et al., 1963). Mildly salty solutions were reported to taste more salty when subjects were depleted, but no data were presented to support this.

De Wardener and Herxheimer (1957) induced sodium depletion by having 2 subjects (themselves) drink huge amounts of water for a period of 12 days. Although subjects were fed a diet low in sodium, they could use salt ad libitum from a shaker. One subject was in substantial negative sodium balance throughout this period, while the second was only in slight negative balance. For days 2–4, food was reported to be tasteless unless large amounts of salt were added. Salt intake from shakers increased for both subjects. Following a transient increase in salt taste sensitivity, threshold returned to baseline prior to normalization of sodium balance. These observations suggested to the authors that there was an increased desire for salt of central origin.

Stinebaugh et al. (1975) evaluated salt taste sensitivity in fasting patients. The patients exhibited substantial negative sodium balance over the course of the fast, but there were no changes in salt taste threshold.

E. Summary

Little is known about the sensory effects of sodium depletion in humans. While anecdotal and clinical reports suggest that an appetite or craving may develop, this may be largely restricted to sodium depletion occurring during infancy or early childhood. Early salt restriction may be particularly likely to induce salt appetite, perhaps by altering some receptor population or CNS structures that are labile at this time (Sakai et al., 1986). Additionally, salt appetite has not been reported in the few available experimental studies. There are suggestions from these studies that there is an increased desire for salty foods, not salt itself. However, no specific tests were performed to study salt appetite. The studies that seem to demonstrate increased sensitivity for salt during depletion used taste assessment methods now known to be imprecise.

II. TASTE AND HUMAN SODIUM DEPLETION: A NEW STUDY

We have recently evaluated salt taste perception (threshold, intensity scaling, preference, food desires) among normal subjects during sodium depletion (Beauchamp et al., 1990). Due to ethical considerations, our depletion regimen was not as stringent as some previously used (e.g., McCance, 1936). However, using a combination of low-sodium diets and diuretics, we successfully induced substantial negative sodium balance and investigated the effects of this manipulation on taste using modern psychophysical techniques.

A. Methods

Ten University of Pennsylvania students (6 men, 4 women, mean age = 21.9 years, range: 18–33) participated in the study. All participants were in good physical health as determined by a medical examination. Subjects were told that the purpose of the experiment was to investigate relationships among diet, diuretics, and taste, though the details of the hypotheses tested were not presented to them.

The protocol is outlined in Table 2. Subjects were required to stay in the Clinical Research Center (CRC) of the hospital of the University of Pennsylvania for 10 days. During this period, sodium depletion was accomplished through

TABLE 2 Depletion Study Design

Variable	Pre-depletion		Depletion days (low-Na diet)											Post-depletion	
			1[a]	2	3	4	5	6	7	8	9	10	11[b]		
Diuretic			×	×	×		×		×		×		×		
			×	×	×	×		×		×		×			
Urine collection	×	×	×	×	×	×	×	×	×	×	×	×		×	×
Blood pressure	×	×	×	×	×	×	×	×	×	×	×	×		×	×
Weight	×	×	×	×	×	×	×	×	×	×	×	×			
Hormones			×							×					
Taste tests	×	×						×					×	×	×
Food item questionnaire	×	×						×					×	×	×

[a]Range: 12–14 hours.
[b]Range: 8–12 hours.

administration of diuretics and a diet very low in sodium (≈ 10 mmol/day), prepared by the CRC research diet kitchen. Subjects received one of two diuretics orally. In one group, subjects were administered furosemide (40 mg), whereas the other group received ethacrynic acid (50 mg). Both drugs are high-ceiling loop diuretics, which can produce potassium as well as sodium depletion. Subjects received a total of seven drug administrations, the maximum considered advisable. Drugs were administered daily for the first 3 or 4 days of hospitalization, then on alternate days for the remaining 7 or 8 days.

During each day of the 10-day hospitalization period, upright and supine blood pressures were measured four times and body weight was recorded. Daily complete urines were collected during the 10-day hospitalization period and were analyzed for sodium, potassium, and creatinine. To estimate subjects' customary salt intake on ad libitum diets, two 24-hour urine collections were made 1–3 weeks prior to depletion and again on one ($n = 4$) or two days ($n = 6$), 1–4 weeks following the depletion period. At these times, blood pressure and body weight were also recorded. Plasma levels of aldosterone were obtained for 8 of 10 subjects prior to the depletion period and on day 8 of hospitalization. Renin measurements during the same periods were obtained for only 4 subjects.

Taste tests were conducted on six days throughout the study: twice 1–3 weeks prior to hospitalization, days 6 (or 7 for 3 subjects) and 11 of hospitalization, and 1–4 weeks ($n = 7$) following discharge (Table 2). In the pre- and postdepletion periods, taste testing occurred on the same days that physiological measures were obtained. NaCl and sucrose detection thresholds (the lowest level at which the tastant could be discriminated from the diluent, water) were determined during each test session. In addition, subjects rated the intensity and the pleasantness, of salt in soup and sugar in Kool-Aid and indicated their preferred concentrations using methods previously described (Bertino et al., 1982).

After each taste test, subjects were given a list containing 29 common food items. The list included a variety of foods that varied in degree of sweetness and saltiness. Subjects were instructed to rate the imagined pleasantness of the taste of each food using a category rating scale (0 = least pleasant, 10 = most pleasant). At the bottom of the questionnaire space was provided for them to list any foods that they craved or would particularly like to eat.

B. Results

Detailed results are presented in Beauchamp et al. (1990). Here a brief summary is provided.

The average urinary excretion values for sodium and potassium are shown in Figure 1. Following an initial high level of sodium excretion during the first 1–2 days, there was a substantial decline in sodium excretion during days 3–4 of hospitalization. Excretion of sodium leveled off at approximately the amount in the hospital diet beginning about day 5. During the last five days on the low sodium diet, urinary sodium excretion averaged 9 mmol/day on the days when no diuretics were given and equaled 21.5 mmol/day when the drugs were administered. Serum aldosterone values, obtained on two days (once prior to depletion

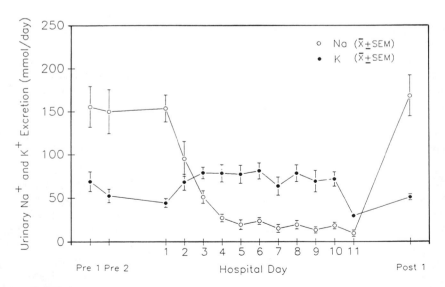

FIGURE 1 Mean (± SEM) 24-hour urinary sodium and potassium excretion values for the predepletion, depletion, and postdepletion periods.

HUMAN SALT APPETITE

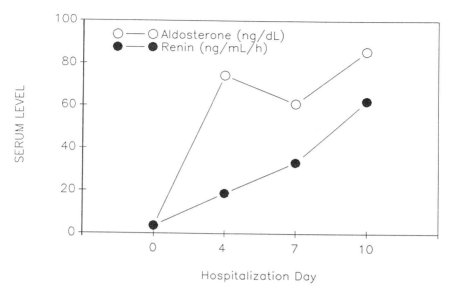

FIGURE 2 Mean values for aldosterone ($n = 8$) and renin ($n = 4$) for the predepletion and depletion periods.

and once on day 8 of depletion) for 8 of the 10 subjects increased markedly, demonstrating significant sodium depletion and volume contraction (Fig. 2).

Initial analyses revealed no difference in taste data between the two tests prior to depletion, between the two depletion tests (days 6 and 11), or between the two tests after repletion. These data were therefore collapsed and three periods were considered: predepletion, depletion, and postdepletion. Additionally, since no significant effects for drug were obtained, data for the two drugs were combined in all subsequent analyses.

No significant differences (ANOVA) were found between the three time periods for NaCl and sucrose thresholds. Examination of individual thresholds, however, show that NaCl thresholds decreased during depletion in 6 of the 10 subjects. Depletion also did not alter judgments of taste intensity.

Breakpoint value, the concentration receiving the greatest pleasantness rating, was used as the primary indicator of hedonic responses for pleasantness scaling. In contrast to threshold and suprathreshold intensity judgments, hedonic judgments were influenced by the sodium depletion regimen. As shown in Figure 3, breakpoints were highest during the depletion period for salt in crackers (7 out of 10 subjects) and soup (8 out of 10 subjects; $p = 0.055$). In contrast, breakpoints for sucrose in Kool-Aid changed in an opposite direction (8 out of 10 subjects; $p = 0.055$).

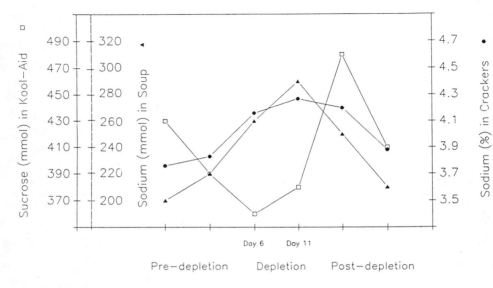

FIGURE 3 Mean breakpoint values for NaCl in crackers and soup and sucrose in Kool-Aid for each taste testing period.

Similarly, but more dramatically, salty foods increased in desirability as determined by the food pleasantness questionnaire. The goal was to evaluate whether subjects' desires for foods changed during depletion and to determine whether such changes were consistently associated with the attribute of food saltiness. To this end, the degree to which each of the 29 foods exhibits attributes of saltiness, sweetness, nutritiousness, fattiness, and fillingness was first independently determined by a naive, nondepleted group of subjects drawn from the same pool as the depletion subjects. These 14 subjects were provided with a list of the 29 food items (Fig. 4) and asked to judge the saltiness, sweetness, nutritiousness, fattiness, and fillingness. These attributes were not independent of each other. Intercorrelations among the values of the food items within the five attributes were then computed. The judged salty and sweet contents were inversely related, while nutritiousness was negatively correlated with saltiness and fattiness. Fattiness was positively correlated with saltiness. All other intercorrelations were nonsignificant.

Food items were ordered from highest to lowest within each of the five categories (e.g., for saltiness, potato chips were judged most salty and apples least salty). To determine whether depletion significantly affected ratings of foods and, more importantly, whether effects were specific to foods high in salt, the rating for each category during depletion was compared with the average of the ratings before and after depletion. Figure 4 presents the judged saltiness of each food item

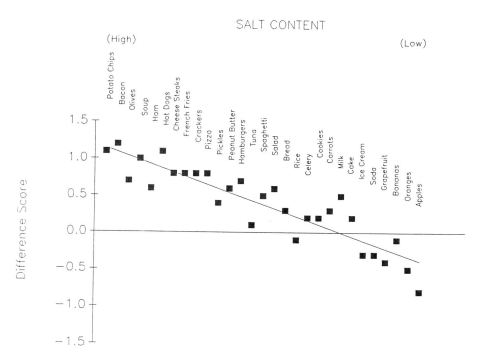

FIGURE 4 Changes during the depletion period in the desirability of foods as a function of their judged saltiness. The foods are ranked from left to right by judged saltiness: potato chips were ranked by a separate group of nondepleted subjects as most salty, salad as moderately salty, and apples as least salty. The difference score represents the difference between the subjects' imagined preference for this food while depleted minus the average imagined preference for that same food prior to and following depletion. Potato chips, bacon, soup, and hot dogs—foods independently judged to be high in saltiness—exhibited an increase in desirability during sodium depletion. In contrast, apples, oranges, and grapefruit—foods judged to be low in saltiness—were less desirable during depletion.

(determined as described above from independent analysis) plotted against the difference score, which reflects differences between pleasantness judgments during depletion compared with those before and after depletion. The correlation between ranked saltiness of the 29 foods and differences in pleasantness as a function of depletion was high ($r = +.86$; $p < 0.0001$), indicating that foods judged high in salt increased most in pleasantness when subjects were sodium depleted.

The other attributes rated (sweetness, fillingness, fattiness, nutritiousness) were treated in a similar manner, and the data for these are presented in Figures 5–

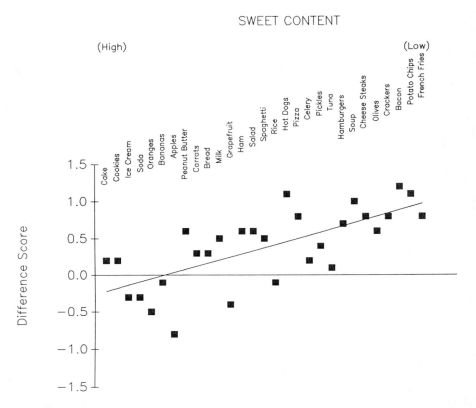

FIGURE 5 Same as Figure 4 except that the foods are arranged from left to right in order of decreasing sweetness.

8. For sweetness, the correlation between ranked sweetness and change in pleasantness was the opposite of that observed for saltiness, as would be expected based on the negative correlation between these two attributes. The correlation for the fattiness was positive but less so than that for saltiness; correlations for fillingness and nutrition were low.

Of the 7 subjects who listed additional foods that they craved, 6 clearly expressed a greater desire for salty foods during the depletion period compared to pre- and postdepletion, with the seventh subject's responses being ambiguous. For example, one subject desired cookies and jelly beans during the pre- and postdepletion periods but wished for chili dogs with cheese during the depletion days. Other foods listed as desired during depletion days included anchovies, soup, pizza, bacon, and French fries, as well as a number of foods not usually

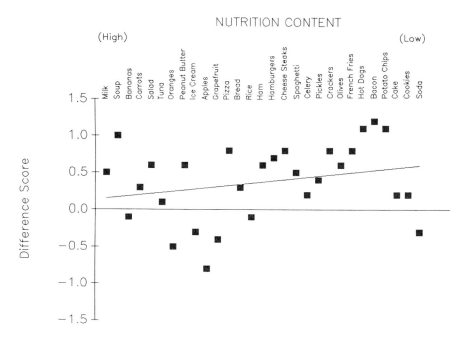

FIGURE 6 Same as Figure 4 except that the foods are arranged from left to right in order of decreasing nutritiousness.

classed as salty. However, only one subject explicitly used the word "salty" to describe the desired foods.

C. Discussion

The level of sodium depletion attained in this study did not significantly alter taste sensitivity as determined by threshold and suprathreshold tests. While there was a trend for thresholds to decrease during the depletion period, in agreement with some previous work, this was not evident in all subjects. Psychophysical scaling of salt intensity was unaffected by sodium depletion. As expected, there was no change in sweet perception.

In direct contrast to these negative results for salt sensitivity, judgments of the pleasantness of salt taste in food and the desirability of salty foods were influenced by the treatment. Breakpoints for salt in soup and crackers increased for the majority of subjects. In contrast, breakpoints for sucrose in Kool-Aid declined during sodium depletion. Data obtained in the food desirability questionnaire were entirely consistent with these sensory–hedonic results. Foods presumed (by

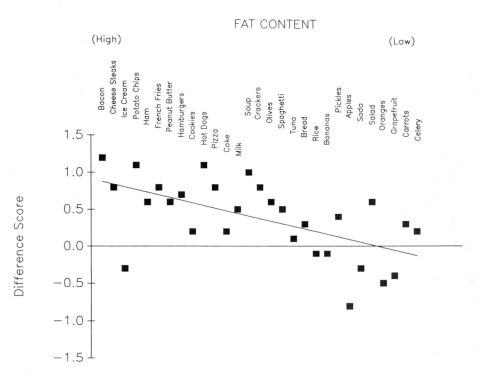

FIGURE 7 Same as Figure 4 except that the foods are arranged from left to right in order of decreasing fattiness.

others) to be high in salt were substantially more desirable during depletion compared to before and after depletion, whereas those high in sweetness exhibited the opposite pattern. Finally, of the subjects listing other foods they craved, most desired saltier foods during the depletion period. It appears that a moderate craving for salty foods was induced by this protocol.

III. SALT APPETITE OR ALTERED SALT PREFERENCE?

Salt appetite refers to a set of behavioral responses of an organism when it has been rendered sodium deficient. Generally, salt-deficient rats increase locomotor activity, become more willing to ingest novel substances, recognize the taste of salt as the missing nutrient, if it is made available, and immediately ingest sufficient quantities of salt to reverse the deficit (Richter, 1956; Denton, 1982). There is an extensive literature on the physiological, endocrine, and central nervous system substrates for this example of regulatory behavior (see Denton, 1982; Sakai et al., 1986; Schulkin, this volume, Chapter 11).

HUMAN SALT APPETITE

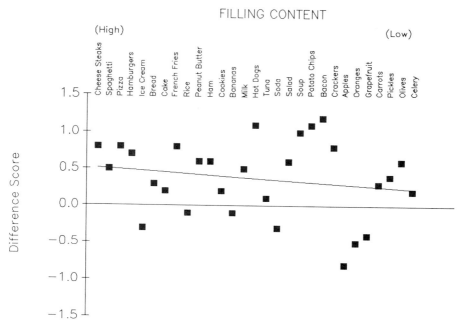

FIGURE 8 Same as Figure 4 except that the foods are arranged from left to right in order of decreasing fillingness.

For several species, including humans, consumption of salt occurs even when there is no known sodium deficit. The causes for this "salt preference," as well as the factors that alter its expression, are less well understood. Furthermore, whether different or identical physiological mechanisms underlie salt appetite and salt preference is not known.

In the study described here, we attempted to induce a salt appetite in human volunteers. The subjects were sodium depleted, and serum levels of hormones of sodium conservation (aldosterone and renin) increased as expected. While no extreme cravings for sodium were evident, increased desire for salty foods was observed.

We cannot yet conclude, however, that sodium *depletion* was responsible for the change in the desirability of salty foods. These changes in salt liking could be due to (a) physiological need for sodium induced by the depletion (i.e., salt appetite) and/or (b) the deprivation of experience tasting salty foods. A number of studies indicate that the second alternative—alterations in saltiness exposure causing changes in salt desirability—does operate. When individuals are placed on lowered sodium diets ($\approx$ 50–75 mmol/day), a shift in preferences is observed such that the optimal level of salt in food declines (Bertino et al., 1982; Teow et al., 1984; Blais et al., 1986). That this hedonic shift may take weeks or months to occur suggests that it is due to the experience with less salty food rather than the absolute decrease in ingested salt. Studies on taste preference following increases in salt intake (Bertino, 1986) provide direct support for this hypothesis that it is the level of sodium consumed that is critical for the change in taste hedonics (review: Beauchamp, 1987). When sodium intake was increased by requiring volunteers to add 10–12 g of extra NaCl to their food every day, thereby increasing their exposure to salty taste, breakpoints shifted upward: subjects' optimal levels of salt in food increased. However, if subjects were required to increase their salt intake without tasting it, by taking salt tablets, no taste preference changes were observed (see Fig. 9). Thus, taste preference changes following moderate increases (doubling) and decreases (halving) of sodium intake appear to be due to changes in the experience of tasting saltiness, not to the absolute amount of salt consumed.

To be determined is whether the increase in desirability of salty foods following the sodium depletion we induced, including reducing sodium intake to less than 10% of the preexperimental level, was due to sodium depletion itself or to the marked decrease in experience with the taste of salty foods that occurred as a consequence of the very low sodium diet (cf., Bertino et al., 1981). To answer this question, studies are needed in which subjects are exposed to the very low sodium regimen but have their sodium losses replaced without experiencing the taste of salt (e.g., by salt tablets). Such studies are under way.

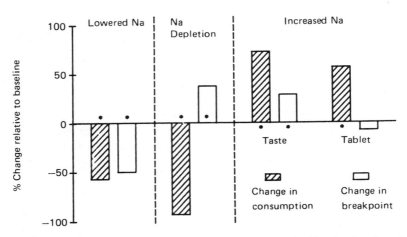

FIGURE 9 Summary of changes in sodium excretion and breakpoints for salt taste in soup from several studies. The data on the left represent a summary of Bertino et al. (1982), the middle bars summarize results from Beauchamp et al. (1990; the study described here), whereas the set of bars on the right summarize data from Bertino et al. (1986). Asterisks indicate significant ($p \leq 0.05$) changes from baseline.

IV. CONCLUSIONS

Human salt appetite, an intense desire for salty-tasting substances, has been observed in several clinical cases, almost all of which involve childhood onset. Reports of previous experimental studies of salt depletion in human adults did not describe strong desires for salt. In a new study outlined here, moderate increases in the desirability of salty foods were observed in subjects fed diets very low in sodium and given diuretics. It remains to be determined whether this change in preference was due to salt depletion and/or deprivation of experience with salty tastes.

ACKNOWLEDGMENTS

This research was supported by National Institutes of Health grant NIH HL-31736 and Clinical Research Center (CRC) grant RR-00040.

DISCUSSION

Grill: With respect to whether or not the sodium-depleted humans were less deprived than rats, I would assume you can do an experiment with rats achieving a similar degree of loss and examine the degree to which it would affect their behavior.

Beauchamp: There's no doubt that our humans were actually less deprived than most rat models. That's true.

Grill: I think it would be interesting to take a look at rats that were similarly deprived. A second point is you speak of species differences, and there are very clear species differences with respect to taste and neurophysiology of taste. Boudreau's data clearly suggest, for example, that there's a sodium/lithium specificity in the response of geniculate ganglion cells, but only in rats and goats and not in carnivores (dogs and cats). Humans clearly seem to perceive potassium compounds as being embodied with the same saltiness as sodium compounds. I was curious to know whether you looked at potassium-containing compounds in your humans that were sodium-depleted.

Beauchamp: In that particular study, no, we didn't. We have looked at it in other studies, and I would say that it's a little more complicated. There is a saltiness in potassium chloride. There's also a bitterness in which there may be great individual differences. We've had trouble getting good data on preference with potassium chloride as a salty stimulus.

VanItallie: What happened to the blood pressure in these people whose salt was withheld? And, to comment on patients with Addison's disease, I had a patient with Addison's disease who did not give an overt history of salt craving, but then it turned out she used to go to the refrigerator every night to eat pickles. That was something she had not done until she got this illness. So I think that the history of salt craving may be masked sometimes because the patients don't necessarily associate a particular food with saltiness. The second issue is that people with Addison's disease may have varying degrees of sodium depletion because it's not an absolute uniform destruction of the adrenal gland and the adrenal cortex. Therefore, it's conceivable that people have varying degrees of loss of salt-retaining hormone.

Beauchamp: First, I'll say one of our subjects actually exhibited a significant decline in blood pressure and we had to reduce the diuretic dosage. The rest of them showed a slight decline in blood pressure during the first 2–3 days of the sodium depletion period, but it normalized by day 7 or 8. The second point, in terms of foods chosen, I would agree with you. In fact, the one Addison's patient who was described there did not consume salt; he consumed half a pound of olives a day. There's a subtle issue, but it may well be that these people are actually treating themselves with foods that are high in salt and that their preference is not for salt but for the flavors associated with those foods. Similarly, licorice seems to be at least as common a desired food item amongst Addison's patients as is salt. They developed what looks like a licorice appetite, presumably as a result of its pharmacological effects.

Stricker: Isn't it possible to take advantage of natural experiments, by which I mean people, who because of their work or play do a lot of sweating, lose a lot of salt? You don't have to induce anything in these individuals, all you have to do is track them and find out what they do after they have lost all this salt and sweat. I assume that athletes in this city during the summer lose a lot of salt, and what do they do to replace it?

Beauchamp: I have several comments to that question. First of all, we certainly have thought of that and are casting about for an appropriate group. The salt-retaining capacities are so good that, at least I've been told, it will be difficult to find such a group that's truly sodium-depleted.

Stricker: If they lose it by sweating rather than in urine.

Beauchamp: If they lose it by sweating and they're not used to sweating a lot, that's the kind of group we'd like to find. We've not been able to find such a group, but that's certainly something we'd like to try.

REFERENCES

Barnwell, G. M., Dollahite, J., and Mitchell, D. S. (1986). Salt taste preference in baboons. *Physiol. Behav.* **37**:279–284.

Bartter, F. C., Pronove, P., Gill, J. R., and McCardle, R. C. (1962). Hyperplasia of the juxtaglomerula and hypokalemic alkalosis. *Am. J. Med.* **33**:811–828.

Beauchamp, G. K. (1987). The human preference for excess salt. *Am. Sci.* **75**:27–33.

Beauchamp, G. K., and Bertino, M. (1985). Rats do not prefer salted solid food. *J. Comp. Psychol.* **99**(2):240–247.

Beauchamp, G. K., Bertino, M., Burke, D., and Engelman K. (1990). Experimental sodium depletion and salt taste in normal human volunteers. *Am. J. Clin. Nutr.* **51**:881–889.

Bertino, M., Beauchamp, G. K., and Engelman, K. (1986). Increasing dietary salt alters salt taste preference. *Physiol. Behav.* **31**:825–828.

Bertino, M., Beauchamp, G. K., Riskey, D. R., and Engelman, K. (1981). Taste perception in three individuals on a low sodium diet. *Appetite,* **2**:67–73.

Bertino, M., Engelman, K., and Beauchamp, G. K. (1982). Long-term reduction in dietary sodium alters the taste of salt. *Am. J. Clin. Nutr.* **36**:1134–1144.

Blais, C., Pangborn, R. M., Borhani, N. O., Ferrell, M. F., Prineas, R. J., and Laing, B. (1986). Effect of dietary sodium restriction on taste responses to sodium chloride: A longitudinal study. *Am. J. Clin. Nutr.* **44**:323–343.

Bloch, M. R. (1976). Salt in human history. *Interdisciplinary Sci. Rev.* **1**:4.

Carr, W. J. (1952). The effect of adrenalectomy upon the NaCl taste threshold in the rat. *J. Comp. Physiol. Psychol.* **45**:377–380.

Contreras, R. J., and Frank, M. E. (1979). Sodium deprivation alters neural responses to gustatory stimuli. *J. Gen. Physiol.* **73**:569–594.

Contreras, R. J., Kosten, T., and Frank, M. E. (1984). Activity in salt taste fibers: Peripheral mechanism for mediating changes in salt intake. *Chem. Senses,* **8**(3):275–288.

Cotterill, J. A., and Cunliffe, W. J. (1973). Self-medication with liquorice in a patient with Addison's disease. *Lancet,* **1**:294–295.

Darley, W., and Doan, C. A. (1936). Primary pulmonary arteriosclerosis with polycythemia: Associated with the chronic ingestion of abnormally large quantities of sodium chloride (halophagia). *Am. J. Med. Sci.* **191**:633–646.

Dastre, M. A. (1901). Salt and its physiological uses. *Annual Report of the Smithsonian Institution,* Part 1. Government Print Office, Washington, DC, pp. 561–574.

de Wardener, H., and Herxheimer, A. (1957). The effect of a high water intake on salt consumption, taste thresholds and salivary secretion in man. *J. Physiol.* **139**:53–63.

Denton, D. (1982). *The Hunger for Salt.* Berlin: Springer-Verlag.

Denton, D., McBurnie, M., Ong, F., Osborne, P., and Tarjan E. (1988). Na deficiency and other physiological influences on voluntary Na intake of BALB/c mice. *Am. J. Physiol.* **255**:R1025–R1034.

Desor, J. A., and Maller, O. (1975). Taste correlates of disease states: Cystic fibrosis. *J. Pediatr.* **87**(1):93–96.

Dethier, V. G. (1977). The taste of salt. *Am. Sci.* **65**:744–751.

Grossman, H., Kennedy, E., McCamman, S., Rice, K., and Hellerstein, S. (1977). Salt appetite in children with sickle cell disease. *J. Pediatr.* **90**:671–672.

Henkin, R. I., and Powell, G. F. (1962). Increased sensitivity of taste and smell in cystic fibrosis. *Science,* **138**:1107–1108.

Henkin, R. I., Gill, J. R., and Bartter, F. C. (1963). Studies on taste thresholds in normal man and in patients with adrenal cortical insufficiency: The role of adrenal cortical steroids and of serum sodium concentration. *J. Clin. Invest.* **42**(5):727–735.

Hertz, J., Cain, W., Bartoshuk, I. M., and Dolan, T. F. (1975). Olfactory and taste sensitivity in children with cystic fibrosis. *Physiol. Behav.* **14**:89–94.

Jacobs, K. M., Mark, G. P., and Scott, T. R. (1988). Taste responses in the nucleus tractus solitarius of sodium-deprived rats. *J. Physiol.* **406**:393–410.

Knowles, J. P., and Asher, R. (1958). Addison's disease with glycyrrhizophilia. *Proc. R. Soc. Med.* **51**(3):178.

Lapique, L. (1896). Sur l'alimentation minérale. *Anthropologie,* **1**:35–45.

McCance, R. A. (1936). Medical problems in mineral metabolism. *Lancet,* **1**:823–830.

McQuarrie, I. (1935). Effects of excessive salt ingestion on carbohydrate metabolism and arterial pressure in diabetic children. *Proc. Staff Meet.* Mayo Clinic, **10**:239–240.

McQuarrie, I., Thompson, W. H., and Anderson, J. A. (1936). Effects of excessive ingestion of sodium and potassium salts on carbohydrate metabolism and blood pressure in diabetic children. *J. Nutr.* **11**:77–101.

Multhauf, R. P. (1978). *Neptune's Gift: A History of Common Salt.* Johns Hopkins University Press, Baltimore.

Neumann, T. W. (1977). A biocultural approach to salt taboos: The case of the southeastern United States. *Curr. Anthropol.* **18**:289–308.

Poskanzer, D. C., and Kerr, D. N. (1940). A third type of periodic paralysis with normokalemia and favorable response to sodium chloride. *Amer. J. Med.* **31**:328–342.

Richter, C. P. (1956). Salt appetite of mammals: Its dependence on instinct and metabolism. In M. Autuoure (Ed.). *L'instinct dans le comportement des animaux et de l'homme.* Masson, Paris.

Rowland, N. E., and Fregley, M. (1988). Sodium appetite: Species and strain differences and role of renin–angiotensin–aldosterone system. *Appetite,* **11**:143–178.

Sakai, R. R., Nicolaidis, S., and Epstein, A. N. (1986). Salt appetite is suppressed by interference with angiotensin II and aldosterone. *Am. J. Physiol.* **251**:R762–R768.

Shapiro, M. D., and Linas, S. L. (1985). Sodium chloride pica secondary to iron-deficiency anemia. *Am. J. Kidney Dis.* **V**(1):67–68.

Shy, G. M., Gonatas, N. K., and Perez, M. (1966). Two childhood myopathies with abnormal mitochondria: I. Megaconial myopathy. II: Pleoconial myopathy. *Brain,* **89**(1):133–158.

Spiro, A. J., Pineas, J. W., and Moore, C. L. (1970). A new mitochondrial myopathy in a patient with salt craving. *Arch. Neurol.* **22**:259–269.

Stinebaugh, B. J., Vasquez, M. I., and Schloeder, F. X. (1975). Taste thresholds for salt in fasting patients. *Am. J. Clin. Nutr.* **28**:814–817.

Stricker, E. M., and Verbalis, J. G. (1988). Hormone and behavior: The biology of thirst and sodium appetite. *Am. Sci.* **76**:261–267.

Teow, B. H., Di Nicolantonio, R., and Morgan, T. O. (1984). Sodium detection threshold and preference for sodium salts in humans on high and low salt diets. *Chem. Senses,* **8**:267.

Thompson, W. H., and McQuarrie, I. (1934). Effects of various salts on carbohydrate metabolism and blood pressure in diabetic children. *Proc. Soc. Exp. Biol. Med.* **31**:907–909.

Thorn, G. W. (1949). *The Diagnosis and Treatment of Adrenal Insufficiency.* Charles C. Thomas, Springfield, IL.

Thorn, G. W., Dorrance, S. S., and Emerson, D. (1942). Addison's disease: Evaluation of synthetic desoxycorticosterone acetate therapy in 158 patients. *Ann. Intern. Med.* **16**(6):1053–1096.

Wilkins, L., and Richter, C. P. (1940). A great craving for salt by a child with corticoadrenal insufficiency. *JAMA,* **114**:866–868.

Wilkins, L., Fleischmann, W., and Howard, J. E. (1940). Macrogenitosomia praecox associated with hyperplasia of the androgenic tissue of the adrenal and death from corticoadrenal insufficiency. *Endocrinology,* **26**:385–395.

Wilson, D. V. (1937). Federal grants-in-aid to the state of Ohio. Unpublished master's thesis, University of Chicago.

Wotman, S., Mandel, I. D., Khotim, S., et al. (1964). Salt taste threshold and cystic fibrosis. *Am. J. Dis. Child.* **108**:372–374.

Yensen, R. (1959). Some factors affecting taste sensitivity in man: II. Depletion of body salt. *Q. J. Exp. Psychol.* **11**:230–238.

6
Sensory and Postingestional Components of Palatability in Dietary Obesity: An Overview

Michael Naim

The Hebrew University of Jerusalem
Rehovot, Israel

Morley R. Kare[†]

Monell Chemical Senses Center
Philadelphia, Pennsylvania

I. INTRODUCTION

Obesity that is primarily induced by feeding certain foods is classified as dietary obesity. In many cases, increased caloric intake caused by specific food items is responsible for the positive energy balance produced. The resulting obesity can often be reversed once these foods have been removed from the diet. The issue of whether hedonic aspects of the food contribute to the overeating phenomenon presents a major nutritional question that has attracted many behavioral scientists. Until recently, most available data on behavioral responses, such as diet preference and intake, were collected during short-term experiments. As a result, short-term characteristics of palatability, which are usually influenced by diet sensory properties, were extrapolated to explain long-term effects of palatability on obesity. Nutritionally controlled studies in recent years, however, indicate the signifi-

[†] deceased

cance of postingestional factors of the food as stimulators of prolonged overconsumption of calories. These factors include high fat content, presence of sugars in solution, and liquid diets. After ingestion by experienced animals, such factors appear to modulate palatability patterns that differ from those observed in naive animals. The objective of this overview is to assess the nutritional significance of dietary sensory and other factors that promote overeating. Attention will be given to the view that common Western human foods present a variety of sources of postingestional signals of appetite.

II. GENERAL CONSIDERATIONS OF DIETARY OBESITY

A positive energy balance can result either from an elevation in energy intake or from a reduction in energy expenditure and heat loss. When such processes continue for extended periods of time, fat deposition and obesity are observed. Studies with experimental animals allow us to classify the obesity syndrome into dietary and genetic causes (Mickelsen et al., 1955; Zucker, 1967; Bray and York, 1979; Bray 1982). The induction of hypothalamic obesity by means of lesions in the ventromedial (VMH) or paraventricular (PVN) hypothalamus provides another useful experimental system (Hetherington and Ranson, 1942; Brobeck et al., 1943; Leibowitz et al., 1981). A major difference between genetic and dietary obesity is that in the genetic form, fat deposition is produced because of a variety of metabolic defects and even under food restriction (Cleary et al., 1980), although such manipulations may modify the extent to which the obesity develops. Changes in diet composition are responsible for the increased fat deposition in dietary obesity.

Hyperphagia and changes in energy expenditure can occur in both types of obesity (Cleary et al., 1980; James et al., 1981). Increased energy intake appears to be a common feature of dietary obesity (Kanarek and Hirsch, 1977; Rothwell and Stock, 1979; Naim et al., 1985); although hyperphagia is not a prerequisite in some strains of rats (Schemmel et al., 1970; Applegate et al., 1984). Undoubtedly, genetic factors are also important in dietary obesity, since similar dietary manipulations can produce obesity in some strains but not in others (Schemmel et al., 1970). Dietary obesity in animals can often be reversed by diet (Sclafani and Springer, 1976; Wade and Bartness, 1985), suggesting that nutritional manipulations could be effective as a major strategy for treating humans susceptible to this type of obesity, whereas in genetic obesity such manipulations are secondary to other treatments.

Evidently, obesity is common in Western societies but not in others (Keys, 1970; Kato et al., 1973; Kaufmann et al., 1982). It is not a major health problem in Japan, but Japanese immigrants to the United States have developed metabolic disorders that are probably due to changes in eating habits (Kato et al., 1973).

III. DIETARY SENSORY FACTORS CAN CONTRIBUTE TO A BALANCED INTAKE OF CALORIES AND ESSENTIAL NUTRIENTS IN NUTRIENT-DEFICIENT SITUATIONS

A. Diet Selection

The innate response to and preference of newborns for the sweet taste of sugars (Desor et al., 1973; Steiner, 1973), and the immediate recognition of sodium by sodium-deficient rats (Nachman and Cole, 1971), demonstrate a significant role for the chemical senses in the selection of essential nutrients. Along with slow acceptance of new foods and aversion to bitter stimuli, which are in many cases toxic constituents, innate responses have important survival value for mammals. Furthermore, animals and humans prefer to select foods with a variety of sensory qualities; the hedonic level for a specific food is decreased during ingestion of a meal, while preference for uneaten food remains the same (Le Magnen, 1956; Rolls et al., 1981a,b; Treit et al., 1983), thereby ensuring a balanced intake of nutrients. In addition, animals use the sensory properties of foods to evaluate physiological consequences produced after food has been ingested. The ability of rats deficient in B vitamins to select diets containing these vitamins, and the development of specific hungers for salt and other essential nutrients such as amino acids, have been documented (Harris et al., 1933; Richter, 1936; Leung et al., 1972).

A key experimental challenge is to evaluate the proportional contribution of food sensory properties, per se, to the regulation of diet selection. Other food components that produce postingestional consequences may lead to a complex learning process. An elegant and perhaps appropriate way to deal with this multivariate question in the short term is to perform sham-feeding experiments (Janowitz and Grossman, 1949; Naim et al., 1978; Weingarten and Watson, 1982). Such an approach, however, is for a variety of reasons not suitable for long-term nutritional studies. Many studies have indicated that if enough time is allowed for the postingestional effects to develop, such effects can dominate food preference behavior. Preferences of humans and rats for low over high concentrations of sucrose solutions can be overturned if blood glucose levels are decreased following insulin injection (Mayer-Gross and Walker, 1946; Jacobs, 1958). In rats, the initial preference for saccharin and the aversiveness toward the bitter tastant sucrose octaacetate (SOA) can be reversed in long-term choice feeding experiments if saccharin is coupled to a diet of poor protein quality and SOA to a diet containing protein of good quality (Naim et al., 1977). Undoubtedly, under situations of nutrient deficiency, the mechanisms above have helped mammals to select nutrients according to physiological needs.

Nevertheless, food availability and previous experience with specific sensory qualities can influence subsequent preferences and eating habits (Rozin, 1976). Indian laborers, whose diet contained sour foods, indicated a strong preference for sour taste compared with Indian medical students who were accustomed to Western cuisine (Moskowitz et al., 1975, 1976). Similarly, the concentration of salt that produced maximum pleasantness was decreased in human subjects who were fed a low-sodium diet for 5 months (Bertino et al., 1982). Therefore, in the long run, diet governs hedonic responses.

B. Caloric Intake

The importance of postingestional effects in the regulation of caloric intake under situations of nutrient deficiency has been demonstrated by the pioneering studies of Epstein, Teitelbaum, and their colleagues, who prevented oral feeding of rats but permitted them to initiate intragastric feeding (Epstein and Teitelbaum, 1962; Epstein, 1967). As shown in Figure 1, rats bearing gastric cannulas could perform a voluntary intragastric feeding by pressing a bar that delivered liquid diet directly into the stomach. As the volume of the delivered food was decreased, animals pressed the bar more frequently to achieve a stable food intake level without

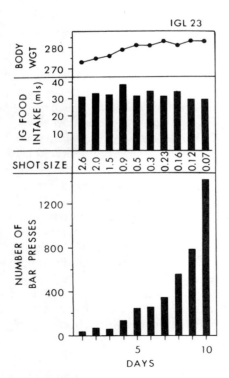

FIGURE 1 Increased number of bar presses in response to decreased gastric shot size by a rat feeding itself intragastrically. (From Epstein, 1967.)

oropharyngeal sensation, and they eventually gained weight normally. Humans can also control caloric intake when offered intragastric meals (Jordan, 1969). However, when the oral feeding was added to the intragastric ingestion, a slight increase in intake was noted. Thus, the need for calories and essential nutrients is a primary controller of intake in emergency situations. Could an aversive-tasting diet overrule such mechanisms? Adulteration of rat diets with the bitter additive quinine produced a reduction in caloric intake (Gentile, 1970), but a bitter stimulus free of pharmacological side effects, such as SOA, was not effective (Naim and Kare, 1977; Kratz and Levitsky, 1978). We were able to induce young rats to consume 10–12% less calories during a 2-week experiment by using four different diets characterized by extremely aversive taste, presented daily in rotation (Naim et al., 1980). It is possible that with such dramatic manipulations, mechanisms of sensory adaptation and habituation were, at least, temporarily overruled.

IV. OVERSUPPLEMENTATION OF PALATABLE FOODS AND OVERCONSUMPTION OF CALORIES; CONTROVERSIAL STUDIES AND THEIR RELEVANCE TO OBESITY

Both animals and humans often overeat when exposed to oversupplementation of specific foods (Sclafani and Springer, 1976; Porikos et al., 1982). Certain Western human foods commonly offered as fast foods and snacks, which are palatable to human and animals, induce overeating and dietary obesity. Conditioned sensory satiety in animals (Le Magnen, 1956), and the sensory-specific satiety in humans (Rolls et al., 1981a), have been extrapolated to explain the high degree of palatability of these foods. Hence, a "cafeteria" model was developed for studying dietary obesity (Sclafani and Springer, 1976; Rothwell and Stock, 1979; Tulp et al., 1982). It has been further hypothesized that the cephalic phase of insulin release, following oral feeding, may initiate ingestive events that contribute to the overeating of "cafeteria" foods (Louis-Sylvestre and Le Magnen, 1980; Berthoud et al., 1981). The "cafeteria" experimental system has stimulated studies that have provided useful physiological and biochemical information related to dietary obesity (Rothwell and Stock, 1979; Himms-Hagan, 1983). Until recently, however, controlled nutritional studies have not pursued these significant findings.

One important role of nutrition is to apply knowledge of metabolic responses to the formulation of dietary recommendations. It would be desirable to elucidate the dietary factors that are responsible for overeating and obesity, which can then be related to metabolic consequences. A variety of high-energy foods varying significantly in composition along with sugar-sweetened milk are usually offered in "cafeteria" experiments. Limitation of this model have been summarized (Moor, 1987). Even if the proportional intake of macro- and micronutrients can be calculated in the "cafeteria" studies, as recently proposed (Rothwell and Stock,

1988), an experimental strategy for the separation of dietary effects responsible for obesity is not feasible.

There are two major drawbacks to the suggestion that food palatability and the presentation of a variety of sensory properties are responsible for the hyperphagia induced by so-called cafeteria foods. One concern is the misinterpretation of the term "food palatability" by many investigators and public health professionals. The second drawback is the extrapolation of short-term effects of diet taste to diet-induced obesity over the long term.

Are Western snack foods palatable? Undoubtedly, a positive answer is appropriate for humans living in Western societies and experimental animals (Sclafani and Springer, 1976; Porikos et al., 1982). Food palatability, however, is a preference variable dependent on both sensory and postingestional factors (Janowitz and Grossman, 1949; Mook, 1963). As discussed in Section III, when enough time is allowed during feeding, these postingestive factors can dominate food preference. In fact, physiological feedback due to such factors can convert initial preference for a specific food to an aversion and vice versa. One may hypothesize, for example, that "fatty taste," which is strongly preferred in Western society (Drewnowski and Greenwood, 1983; Pangborn et al., 1985), is related to possible postingestional effects of fat rather than to its taste per se. Therefore, obtaining a preference profile for various dietary fats with experienced human individuals may not tell us much about the basic mechanisms and factors responsible for such preferences.

V. THE SIGNIFICANCE OF SENSORY FACTORS IN FOODS ON PALATABILITY AND DIETARY OBESITY

Numerous data are available to indicate that the presence of a variety of sensory stimuli in a meal can enhance intake in humans and animals (Le Magnen, 1956; Rolls et al., 1981a,b; Treit et al., 1983). However, there is no evidence that such a variety effect can lead to a long-term increase of caloric intake and obesity. In recent years, our laboratory has initiated studies aimed at investigating this topic in long-term experiments and under situations in which nutrient composition is controlled (Naim et al., 1985, 1986, 1987). Two major objectives have been pursued. The first was to determine whether hyperphagia could be observed in rats fed nutritionally balanced diets supplemented with a variety of appealing flavors and textures. These diets lack the high concentration of fat and sugars that can facilitate physiological pathways of fat deposition (Herman et al., 1970; Reiser and Hallfrisch, 1977; Flatt, 1978). The second objective was to administer the same appealing flavors into high-fat and high-sucrose diets, thus mimicking, to some extent, the consumption of "cafeteria" foods. Dose–preference curves of rats for nutritionally balanced diets containing potent food flavors and texture variety were determined (Naim et al., 1986). These experiments established a catalog of 12 preferred flavors (Table 1) and textures for rats. Subsequently,

TABLE 1 Preference of Rats for Diets Containing Flavors of Common Human Foods[a]

Flavor	Concentration in diet (%, w/w)	Preference index[b]
Peanuts	0.80	0.82
Bread	0.04	0.80
Beef	0.80	0.95
Chocolate	0.40	0.75
Nacho cheese	0.80	0.75
Cheese paste	0.80	0.75
Chicken	0.20	0.95
Cheddar cheese	0.80	0.75
Bacon	0.20	0.78
Salami	0.30	0.75
Vanilla	0.40	0.80
Liver	0.80	0.95

[a]Results are derived from two-choice preference tests between a flavored diet and an unadulterated diet.
[b]Preference index = intake of flavored diet/total intake.
Source: Adapted from Naim et al. (1986).

nutritionally balanced and high-energy diets containing a variety of flavors and textures were tested for their effects on long-term energy intake and fat deposition (Naim et al., 1985). The results indicated that flavor and texture varieties incorporated into nutritionally balanced diets did not induce overconsumption of calories at any time up to 23 days (Fig. 2). During the first 5 days, but not thereafter, the flavored high-fat diets stimulated energy intake. This short and transient elevation in energy intake produced by flavored, high-fat diets (Fig. 2) could not be confirmed in an additional trial (Naim et al., 1985). Body weight gain reflected the intake, and neither low- nor high-fat diets containing flavors induced more fat deposition than did the unadulterated diets (Fig. 3). Thus, neither the appealing flavors and texture nor their variety had any significant effect on caloric intake and fat deposition.

The sweet taste of sugars has been alleged to promote excess intake of calories and deposition of fat. A slight increase, no change, or even reduction in energy intake is produced in rats fed diets containing sucrose (Cohen and Teitelbaum, 1964; Marshall et al., 1969; Reiser and Hallfrisch, 1977). Yet, sucrose and other sugars when offered in solutions along with water and chow stimulate caloric consumption and lead to obesity (Kanarek and Hirsch, 1977; Castonguay et al., 1981; Kanarek and Orthen-Gambill, 1982). One interpretation of these observations is that sugar in solution produces a clearer and more appealing sweet taste

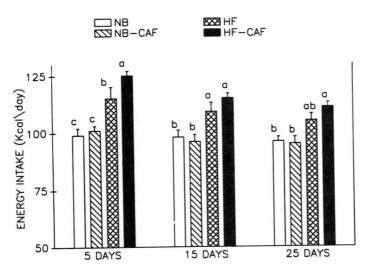

FIGURE 2 Cumulative energy intake of rats fed nutritionally balanced (NB) or high-fat (HF) diets adulterated with food flavors and offered in a multichoice cafeteria (CAF) arrangement. Results are the mean and SEM of 13–15 rats for each treatment. Values not sharing the same superscript letter are different at $p < 0.05$. (Adapted from Naim et al., 1985.)

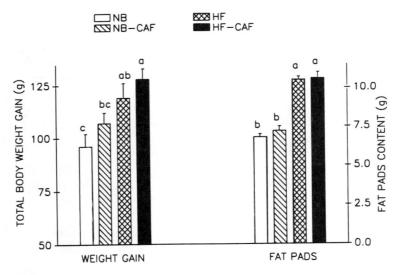

FIGURE 3 Total body weight gain and fat content of fat pads of rats fed nutritionally balanced (NB) or high-fat (HF) diets adulterated with food flavors and offered in a multichoice cafeteria (CAF) arrangement. Results are the mean and SEM of 13–15 rats for each treatment. Values not sharing the same superscript letter are different at $p < 0.05$. (Adapted from Naim et al., 1985.)

than the taste of sugar in the diet, which is probably masked by other constituents. The synergistic effect on intake of solutions containing glucose and saccharin (Valenstein et al., 1967), although shown only in short-term testing, supports the suggestion that sweet taste per se is an important determinant of intake. However, solutions containing the polysaccharide Polycose were also effective in stimulating long-term energy intake, body weight gain, and fat deposition (Sclafani, 1987b). Most important, rats offered solutions adulterated with the less acceptable Polycose–SOA mixture (Sclafani and Vigorito, 1987), along with food and water, consumed comparable amounts of this solution and displayed similar increases in caloric intake, weight gain, and body fat, as did rats offered solutions adulterated with either Polycose of Polycose–saccharin (Fig. 4). In fact, there was no relationship between the degree of preference of rats for sugar-containing solutions and the resulting obesity (Castonguay et al., 1981). Recent studies by

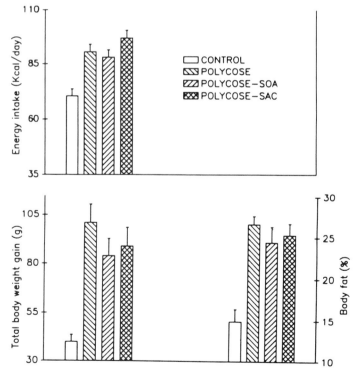

FIGURE 4 Energy intake, body weight gain, and body fat in rats fed chow only (Control), or chow along with either Polycose solution adulterated with sucrose octaacetate (SOA) or Polycose solution adulterated with sodium saccharin (SAC). Results are the mean and SEM of 8 rats for each treatment. (Adapted from Sclafani, 1987b.)

Ramirez (1987c) have demonstrated that rats fed complete liquid diets containing either sucrose or starch as the carbohydrate source consumed more energy, gained more weight, and became fatter than rats fed the same diet in dry form. Furthermore, it has been suggested that altering the viscosity of liquid diets by using xanthan gum as a suspending agent had no reliable effect on gain (Ramirez, 1987b). In summary, studies conducted under nutrient-controlled conditions suggest that flavor variety, texture, and sweet taste play a minor role in the diet-induced, long-term overeating and obesity of experimental animals.

VI. SOURCES OF POSTINGESTIONAL FACTORS THAT CAN BE RESPONSIBLE FOR THE PALATABILITY OF DIET-INDUCED OBESITY

Studies with rats suggest that three major dietary sources may induce overconsumption of calories. These include high-fat diets, sugars or polysaccharides in solution as supplements to food, and liquid diets (Schemmel et al., 1970; Kanarek and Hirsch, 1977; Ramirez, 1987a; Sclafani, 1987b). Apparently, all these stimulatory factors are present in the "cafeteria"-induced obesity diet. Dietary fat contributes 30–44% of total energy consumed in Western societies (Kaufmann et al., 1982). Many food items containing high levels of fat are palatable to humans (Porikos et al., 1982; Drewnowski and Greenwood, 1983; Pangborn et al., 1985). Exposing rats to high-fat diets often leads to overeating and obesity (Schemmel et al., 1970; Naim et al., 1985), and previous studies have suggested that rats may prefer high-fat food (Corbit and Stellar, 1964; Hamilton, 1964; Larue, 1978).

To explore the contribution of sensory and possible postingestional factors of fat to food palatability, naive and relatively young rats (weighing about 110 g) were subjected to short- and long-term two-choice preference tests. Three types of dietary fat were employed (Fig. 5). Although the preference of rats for fat may increase with age (Kanarek, 1985), these experiments and others (Golobov et al., 1986) employing nutritionally controlled diets have demonstrated that naive animals under choice conditions may not show a strong preference for dietary fat. Yet, feeding high-fat diets to rats in no-choice situations may stimulate increased energy intake and fat deposition (Schemmel et al., 1970; Naim et al., 1985). Furthermore, exposing rats to a high-fat diet may subsequently increase the preference for this diet (Golobov et al., 1986). These results suggest, therefore, that experience is needed to establish the preference for high-fat diets. One may hypothesize that humans and most mammals have been more often exposed to nutrient-deficient situations during their phylogenetic development than to energy-condensed foods typically abundant in modern Western societies. In fact, for optimal growth and physiological function, the recommended diets for the rat contains no more than 5% oil (American Institute of Nutrition, 1977), and there is good reason to believe that the optimal fat level for humans can also be below 10%. The feeding of foods high ($> 25\%$) in fat should, therefore, be expected to

PALATABILITY IN DIETARY OBESITY

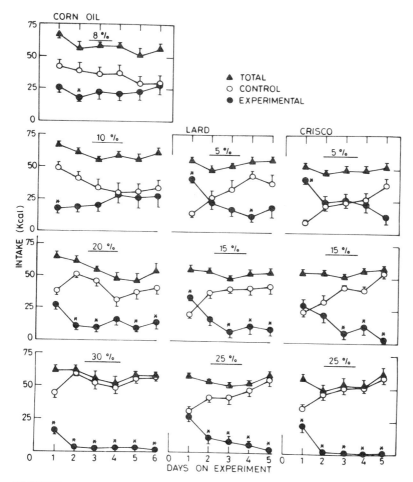

FIGURE 5 Results of two-choice preference tests between a nutritionally controlled diet containing one concentration of either corn oil, lard, or Crisco (experimental diet) versus a nutritionally balanced diet (control diet). Results are the caloric intake from the experimental and control diets, and the total intake from both diets. Values are the mean and SEM of 10–15 rats per group. (From Naim et al., 1987.)

stimulate major metabolic adaptations that may affect diet selection and intake. Thus, one may propose that yet to be defined postingestional signals of fat are responsible for the palatability of fat.

The contributions of carbohydrates and sugars to palatability, intake, and fat deposition have been recently reviewed (Ramirez, 1987a; Sclafani, 1987a). Feeding rats high-carbohydrate or high-sugar diets has little or no effect on caloric intake, although changes in fat deposition and metabolism may occur (Marshall et al., 1969; Reiser and Hallfrisch, 1977). However, both polysaccharide and sugar solutions, when offered along with chow diet, often stimulate intake and fat deposition (Kanarek and Hirsch, 1977; Sclafani, 1987b). As discussed in Section III, many data are available to indicate that sugar-induced intake is not correlated with sweetness. It should be noted that sugar and polysaccharides in solutions and liquid diets (Kanarek and Hirsch, 1977; see also Figs. 4 and 6) had larger effects on weight gain and fat deposition than on caloric intake. This suggests that these diets stimulate weight gain and fat deposition primarily through increased feed efficiency rather than increased caloric intake. Evidently, there is a variety of factors other than taste that can influence the appetite for sugars. Sugars vary in their ability to induce major physiological responses, such as insulin release and brown fat thermogenesis, and the insulin response in humans is larger after

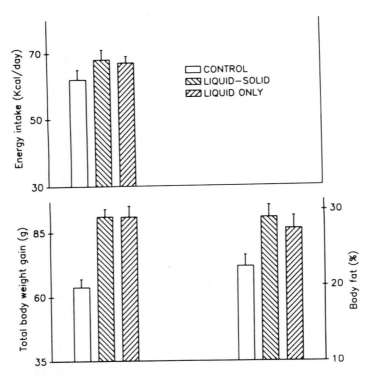

FIGURE 6 Energy intake, body weight gain, and body fat in rats fed a casein starch based diet (Control), or the control diet along with a 32% water suspension of the same diet (Liquid–solid), or the 32% water suspension diet (Liquid only). Results are the mean and SEM of 10 or 11 rats per treatment. (Adapted from Ramirez, 1987a.)

consumption of sugar solutions than after carbohydrate meals (Crapo et al., 1976; Glick et al., 1984). As for dietary fat, the effect of sugars on intake and fat deposition is also dependent on age and perhaps on sex (Sclafani, 1987a; Sclafani et al., 1987). Since complete liquid diets containing starch as the carbohydrate source are more effective in inducing fat deposition in rats than the same diets in dry form (Fig. 6), the critical factor in the obesity induced by sugar solutions may be the "wet source of calories" (Ramirez, 1987a) rather than the sugar per se. The rate of food digestion and absorption (Sclafani, 1987a), and changes in osmotic pressure in the digestive tract (Rogers and Harper, 1965), may be important in these forms of obesity induced by sugar solutions and liquid diets.

VII. CONCLUSIONS AND RESEARCH NEEDS

1. Recent nutritional studies have emphasized both sensory properties of food and other factors responsible for long-term stimulation of energy intake and fat deposition in dietary obesity.

2. However, such studies have not demonstrated a significant role for a variety of appealing flavors, textures, and sweet taste as contributors to the long-term increase in energy intake and fat deposition of rats.

3. The palatability of diets that induce obesity is significantly dependent on, and can be modified by, postingestional factors.

4. The increased energy intake caused by high-fat diets, sugars and polysaccharides in solution, and liquid diets can explain only partly the development of dietary obesity in rats. These diets appear to increase feed efficiency, which further contributes to a positive energy balance.

5. High-fat foods and soft drinks provide a major proportion of daily caloric intake of people living in Western societies. The results above call for further studies to elucidate the mechanism by which high-fat and liquid diets promote energy intake and obesity. Since both animal and human studies reveal variability in individual susceptibility to dietary obesity, future studies should address the involvement of genetic factors. Attention should be given to multidisciplinary studies where nutritional, behavioral, and clinical parameters can be monitored conclusively.

ACKNOWLEDGMENTS

Our studies were supported by the Nutrition Program of the Monell Chemical Senses Center and by a grant from the Schoenbrunn Foundation awarded through the Hebrew University. We acknowledge the collaboration by Drs. J. G. Brand and R. G. Carpenter, Mrs. Yael Shpatz-Golobov, Mrs. Martha Levinson, and Mrs. S. Van Buren. We thank Drs. I. J. Lichton and I. Ramirez for reviewing the manuscript. We also thank Mrs. J. Blescia and Mrs. M. Rockitter for processing the manuscript.

DISCUSSION

Ramirez: The impression you give is that taste is merely a mediator between postingestive effects and food intake. If so, then what are the postingestive effects? Do you have any suggestions?

Naim: What I was trying to do during this talk was to identify the dietary components that are responsible for the phenomenon of increased energy intake in dietary obesity.

REFERENCES

American Institute of Nutrition (1977). Report of the Ad Hoc Committee on Standards for Nutritional Studies. *J. Nutr.* **107**:1340–1348.
Applegate, E. A., Upton, D. E., and Stern, J. S. (1984). Exercise and detraining: Effect on food intake, adiposity and lipogenesis in Osborn–Mendel rats made obese by a high-fat diet. *J. Nutr.* **114**:447–459.
Berthoud, H. R., Bereiter, D. A., Trimble, E. R., Siegel, E. G., and Jeanrenaud, B. (1981). Cephalic phase, reflex insulin secretion. Neuroanatomical and physiological characterization. *Diabetologia*, **20**:393–401.
Bertino, M., Beauchamp, G. K., and Engelman, K. (1982). Long-term reduction in dietary sodium alters the taste of salt. *Am. J. Clin. Nutr.* **36**:1134–1144.
Bray, G. A. (1982). Regulation of energy balance: Studies on genetic, hypothalamic and dietary obesity. *Proc. Nutr. Soc.* **41**:95–108.
Bray, G. A., and York, D. A. (1979). Hypothalamic and genetic obesity in experimental animals: An autonomic and endocrine hypothesis. *Physiol. Rev.* **59**:719–809.
Brobeck, J. R., Tepperman, J., and Long, C. N. H. (1943). Experimental hypothalamic hyperphagia in albino rat. *Yale J. Biol. Med.* **15**:831–835.
Castonguay, T. W., Hirsch, E., and Collier, G. (1981). Palatability of sugar solutions and dietary selection? *Physiol. Behav.* **27**:7–12.
Cleary, M. P., Vasseli, J. R., and Greenwood, M. R. C. (1980). Development of obesity in Zucker obese (fafa) rat in absence of hyperphagia. *Am. J. Physiol.* **238**:E284–E292.
Cohen, A. M., and Teitelbaum, A. (1964). Effect of dietary sucrose and starch on oral glucose tolerance and insulin-like activity. *Am. J. Physiol.* **206**:105–108.
Corbit, J. D., and Stellar, E. (1964). Palatability, food intake and obesity in normal and hyperphagic rats. *J. Comp. Physiol. Psychol.* **58**:63–67.
Crapo, P. A., Reaven, G., and Olefsky, J. (1976). Plasma glucose and insulin responses to orally administered simple and complex carbohydrates. *Diabetes*, **25**:741–747.
Desor, J. A., Maller, O., and Turner, R. E. (1973). Taste in acceptance of sugars by human infants. *J. Comp. Physiol. Psychol.* **84**:496–501.
Drewnowski, A., and Greenwood, M. R. C. (1983). Cream and sugar: Human preference for high-fat foods. *Physiol. Behav.* **30**:629–633.
Epstein, A. (1967). Feeding without oropharyngeal sensations. In *The Chemical Senses and Nutrition*, M. R. Kare and O. Maller (Eds.). The Johns Hopkins Press, Baltimore, pp. 263–280.
Epstein, A., and Teitelbaum, P. (1962). Regulation of food intake in the absence of taste, smell and other oropharyngeal sensations. *J. Comp. Physiol. Psychol.* **55**:753–759.

Flatt, J. P. (1978). The biochemistry of energy expenditure. In *Recent Advances in Obesity Research*, Vol. II, G. A. Bray (Ed.). Newman Publishing, London, pp. 211–228.

Gentile, R. L. (1970). The role of taste preference in the eating behavior of the albino rat. *Physiol. Behav.* 5:311–316.

Glick, Z., Bray, G. A., and Teague, R. J. (1984). Effect of prandial glucose on brown fat thermogenesis in rats: Possible implications for dietary obesity. *J. Nutr.* **114**:286–291.

Golobov, Y., Levinson, M., and Naim, M. (1986). Rats' preference for dietary fat. Paper presented at the Fifth International Congress on Obesity, September, Jerusalem, Israel.

Hamilton, C. L. (1964). Rat's preference for high-fat diets. *J. Comp. Physiol. Psychol.* **58**:459–460.

Harris, L. J., Clay, J., Hargreaves, F., and Ward, A. (1933). Appetite and choice of diet. The ability of the vitamin B deficient rat to discriminate between diets containing the lacking vitamin. *Proc. R. Soc. London (Biol.)* **113**:161–190.

Herman, R. H., Zakim, D., and Stifel, F. B. (1970). Effect of diet on lipid metabolism in experimental animals and man. *Fed. Proc.* **29**:1302–1307.

Hetherington, A. N., and Ranson, S. W. (1942). The spontaneous activity and food intake of rats with hypothalamic lesions. *Am. J. Physiol.* **136**:609–617.

Himms-Hagan, J. (1983). Brown adipose tissue thermogenesis in obese animals. *Nutr. Rev.* **41**:261–267.

Jacobs, H. L. (1958). Studies on sugar preference: 1. The preference for glucose and its modifications by injections of insulin. *J. Comp. Physiol. Psychol.* **51**:304–310.

James, W. P. T., Trayhurn, P., and Garlic, P. (1981). The metabolic basis of subnormal thermogenesis in obesity. In *Recent Advances in Obesity Research*, Vol. III, P. Bjorntorp, M. Cairella, and A. N. Howard (Eds.). Libbey, London, pp. 220–227.

Janowitz, H. D., and Grossman, M. (1949). Some factors affecting food intake of normal dogs and dogs with esophagostomy and gastric fistulae. *Am. J. Physiol.* **159**:143–148.

Jordan, H. A. (1969). Voluntary intragastric feeding: Oral and gastric contributions to food intake and hunger in man. *J. Comp. Physiol. Psychol.* **68**:498–506.

Kanarek, R. B. (1985). Determinants of dietary self-selection in experimental animals. *Am. J. Clin. Nutr.* **42**:940–950.

Kanarek, R. B., and Hirsch, E. (1977). Dietary-induced overeating in experimental animals. *Fed. Proc.* **36**:154–158.

Kanarek, R. B., and Orthen-Gambill, N. (1982). Differential effects of sucrose, fructose and glucose on carbohydrate-induced obesity in rats. *J. Nutr.* **112**:1546–1554.

Kato, H., Tillotson, J., Nichaman, M. Z., Rhoads, G. G., and Hamilton, H. B. (1973). Epidemiologic studies of coronary heart disease and stroke in Japanese men living in Japan, Hawaii and California—Serum lipid and diet. *Am. J. Epidemiol.* **97**:372–385.

Kaufmann, N. A., Friedlander, Y., Halfon, S.-T., Slater, P. E., Dennis, B. H., McClish, D., Eisenberg, S., and Stein, Y. (1982). Nutrient intake in Jerusalem—Consumption in adults. *Isr. J. Med. Sci.* **18**:1183–1197.

Keys, A. (1970). Coronary heart disease in seven countries. The diet. *Circulation*, **41**(Suppl. 1):162–183.

Kratz, C. M., and Levitsky, D. A. (1978). Post-ingestive effects of quinine on intake of nutritive and non-nutritive substances. *Physiol. Behav.* **21**:851–854.

Larue, C. (1978). Oral cues involved in rat's selective intake of fats. *Chem. Sense Flavor*, **3**:1–6.

Leibowitz, S. F., Hammer, N. J., and Chang, K. (1981). Hypothalamic paraventricular nucleus lesions produce overeating and obesity in the rat. *Physiol. Behav.* **27**:1031–1040.

Le Magnen, J. (1956). Hyperphagie provoquée chez le rat blanc par alteration du méchanisme de satiété périphérique. *C.R. Soc. Biol. (Paris)* **150**:32–34.

Leung, P. M. B., Larson, D. M., and Rogers, Q. R. (1972). Food intake and preference of olfactory bulbectomized rats fed amino acid imbalanced or deficient diets. *Physiol. Behav.* **9**:553–557.

Louis-Sylvestre, J., and Le Magnen, J. (1980). Palatability and preabsorptive insulin release. *Neurosci. Biobehav. Rev.* **4**(Suppl. 1):43–46.

Marshall, M. W., Womack, M., Hildebrand, H. E., and Munson, A. W. (1969). Effects of types and levels of carbohydrates and proteins on carcass composition of adult rats. *Proc. Soc. Exp. Biol. Med.* **132**:227–232.

Mayer-Gross, W., and Walker, J. W. (1946). Taste and selection of food in hypoglycemia. *Br. J. Exp. Pathol.* **27**:297–298.

Mickelsen, O., Takahashi, S., and Craig, C. (1955). Experimental obesity: I. Production of obesity in rats by feeding high-fat diets. *J. Nutr.* **57**:541–554.

Mook, D. G. (1963). Oral and postingestional determinants of the intake of various solutions in rats with esophageal fistulas. *J. Comp. Physiol. Psychol.* **56**:645–659.

Moor, B. J. (1987). The cafeteria diet: An inappropriate tool for studies of thermogenesis. *J. Nutr.* **117**:227–231.

Moskowitz, H. R., Kumraiah, V., Sharma, K. N., Jacobs, H. L., and Sharma, S. D. (1975). Cross-cultural differences in simple taste preferences. *Science*, **190**:1217–1218.

Moskowitz, H. R., Kumraiah, V., Sharma, K. N., Jacobs, H. L., and Sharma, S. D. (1976). Effects of hunger, satiety and glucose load upon taste intensity and taste hedonics. *Physiol. Behav.* **16**:471–475.

Nachman, M., and Cole, L. P. (1971). Role of taste in specific hungers. In *Handbook of Sensory Physiology*, Vol. 4, L. M. Beidler (Ed.). Springer-Verlag, Berlin, pp. 337–362.

Naim, M., and Kare, M. R. (1977). Taste stimuli and pancreatic function. In *The Chemical Senses and Nutrition*, M. R. Kare and O. Maller (Eds.). Academic Press, New York, pp. 145–162.

Naim, M., Kare, M. R., and Ingle, D. E. (1977). Sensory factors which affect the acceptance of raw and heated defatted soybeans by rats. *J. Nutr.* **107**:1653–1658.

Naim, M., Kare, M. R., and Merritt, A. M. (1978). Effects of oral stimulation on the cephalic phase of pancreatic exocrine secretion in dogs. *Physiol. Behav.* **20**:563–570.

Naim, M., Brand, J. G., Kare, M. R., Kaufmann, N. A., and Kratz, C. M. (1980). Effect of unpalatable diets and food restriction on feed efficiency in growing rats. *Physiol. Behav.* **25**:609–614.

Naim, M., Brand, J. G., Kare, M. R., and Carpenter, R. G. (1985). Energy intake, weight gain and fat deposition in rats fed flavored, nutritionally controlled diets in a multichoice ("cafeteria") design. *J. Nutr.* **115**:1447–1458.

Naim, M., Brand, J. G., Christensen, C. M., Kare, M. R., and Van Buren, S. (1986). Preference of rats for food flavors and texture in nutritionally controlled, semi-purified diets. *Physiol. Behav.* **37**:15–21.

Naim, M., Brand, J. G., and Kare, M. R. (1987). The preference–aversion behavior of rats for nutritionally controlled diets containing oil or fat. *Physiol. Behav.* **39**:285–290.

Pangborn, R. M., Bos, K. E., and Stern, J. S. (1985). Dietary fat intake and taste responses to fat in milk by under-, normal- and overweight women. *Appetite,* **6**:25–40.

Porikos, K. P., Hesser, M. F., and Van Itallie, T. B. (1982). Caloric regulation in normal weight men maintained on a palatable diet of conventional foods. *Physiol. Behav.* **29**:293–300.

Ramirez, I. (1987a). When does sucrose increase appetite and adiposity? *Appetite,* **9**:1–19.

Ramirez, I. (1987b). Diet texture, moisture and starch type in dietary obesity. *Physiol. Behav.* **41**:149–154.

Ramirez, I. (1987c). Feeding a liquid diet increases energy intake, weight gain and body fat in rats. *J. Nutr.* **117**:2127–2134.

Reiser, S., and Hallfrisch, J. (1977). Insulin sensitivity and adipose tissue of rats fed starch or sucrose diets ad libitum or in meals. *J. Nutr.* **107**:147–155.

Richter, C. P. (1936). Increased salt appetite in adrenalectomized rats. *Am. J. Physiol.* **115**:155–161.

Rogers, Q. R., and Harper, A. E. (1965). Amino acid diets and maximal growth in the rat. *J. Nutr.* **87**:267–273.

Rolls, B. J., Rolls, E. T., Rowe, E. A., and Sweeney, K. (1981a). Sensory-specific satiety in man. *Physiol. Behav.* **27**:137–142.

Rolls, B. J., Rowe, E. A., Rolls, E. T., Kingston, B., Megson, A., and Gunary, R. (1981b). Variety in a meal enhances food intake in man. *Physiol. Behav.* **26**:215–221.

Rothwell, N. J., and Stock, M. J. (1979). A role for brown adipose tissue in diet induced thermogenesis. *Nature,* **281**:31–35.

Rothwell, N. J., and Stock, M. J. (1988). The cafeteria diet as a tool for studies of thermogenesis. *J. Nutr.* **118**:925–928.

Rozin, P. (1976). The selection of foods by rats, humans and other animals. In *Advances in the Study of Behavior,* Vol. 6, J. S. Rosenblatt, R. A. Hinde, E. Shaw, and C. Beer (Eds.). Academic Press, New York, pp. 21–76.

Schemmel, R., Mickelsen, O., and Gill, J. L. (1970). Dietary obesity in rats: Body weight and body fat accretion in seven strains of rats. *J. Nutr.* **100**:1041–1048.

Sclafani, A. (1987a). Carbohydrate taste, appetite, and obesity: An overview. *Neurosci. Biobehav. Rev.* **11**:131–153.

Sclafani, A. (1987b). Carbohydrate-induced hyperphagia and obesity in the rat: Effects of saccharide type, form and taste. *Neurosci. Biobehav. Rev.* **11**:155–162.

Sclafani, A., and Springer, D. (1976). Dietary obesity in adult rats: Similarities to hypothalamic and human obesity syndrome. *Physiol. Behav.* **17**:461–471.

Sclafani, A., and Vigorito, M. (1987). Effects of SOA and saccharin adulteration on polycose preference in rats. *Neurosci. Biobehav. Rev.* **11**:163–168.

Sclafani, A., Hertwig, H., Vigorito, M., and Feigin, M. B. (1987). Sex differences in polysaccharide and sugar preferences in rats. *Neurosci. Biobehav. Rev.* **11**:241–251.

Steiner, J. (1973). The human gustofacial response: Observation of normal and anencephalic newborn infants. In *Fourth Symposium—Oral Sensation and Perception,* J. F. Bosma (Ed.). U.S. Department of Health, Education and Welfare, Bethesda, MD, pp. 254–278.

Treit, D., Spetch, M. L., and Deutsch, J. A. (1983). Variety in the flavor of food enhances eating in the rat: A controlled demonstration. *Physiol. Behav.* **30**:207–211.

Tulp, O. L., Frink, R., and Danforth, E., Jr. (1982). Effect of cafeteria feeding on brown and white adipose tissue cellularity, thermogenesis, and body composition in rats. *J. Nutr.* **112**:2250–2260.

Valenstein, E. S., Cox, V. C., and Kakolewski, J. W. (1967). Polydipsia elicited by the synergistic action of a saccharin and glucose solution. *Science,* **157**:552–554.

Wade, G. N., and Bartness, T. J. (1985). Photoperiod and diet effects on obesity in hamsters. In *Recent Advances in Obesity Research,* Vol. IV, J. Hirsch and T. B. Van Itallie (Eds.). Libbey, London, pp. 37–44.

Weingarten, H. P., and Watson, S. D. (1982). Sham feeding as a procedure for assessing the influence of diet palatability on food intake. *Physiol. Behav.* **28**:401–407.

Zucker, L. M. (1967). Some effects of caloric restriction and deprivation on the obese hyperlipemic rat. *J. Nutr.* **91**:247–254.

7
Hunger, Hedonics, and the Control of Satiation and Satiety

John E. Blundell

University of Leeds
Leeds, England

Peter J. Rogers

AFRC Institute of Food Research
Shinfield, Reading, England

I. CLARIFICATION OF TERMS: HUNGER AND PLEASURE, SATIATION AND SATIETY

One of the most intriguing issues in the study of appetite control concerns the roles of hunger and pleasure in determining the pattern of food consumption. Both hunger and pleasure are terms with apparently high explanatory power, but to be employed usefully and consistently, both need to be logically and empirically anchored. Potentially, each can be used as a research tool to investigate the control of appetite, and as a concept to give greater understanding to descriptions of appetite fluctuations in the home, workplace, and clinic.

One of the first steps is to reach some agreement about the logical status of the terms "hunger" and "pleasure." Considering hunger, the logical status of the term should be clarified, for it is clear that this term is being used in more than one sense. On one hand, hunger is a motivational construct with the logical status of a mediating concept or intervening variable (Macquorquodale and Meehl, 1958).

That is, the term refers to an explanatory principle, which is inferred from other directly observable and measurable events. In this sense, the term helps the understanding of motivational processes. On the other hand, "hunger" may be used to refer to certain conscious sensations or feelings linked to a desire to obtain and eat food. This is the sense in which the layman recognizes the notion of hunger.

It is necessary to emphasize the distinction between hunger the psychological process with relevance for motivational theory, and hunger the conscious sensation. Clearly, hunger as a subjectively perceived sensation may occur during the operation of any of the processes or states of satiation and satiety. It is these hunger sensations that probably exert motivational pressure on behavior, and it is these that researchers attempt to capture by means of rating scales and other devices. When hunger is used to connote a conscious feeling, then it is clearly not the converse of satiety, although one often hears the two terms—hunger and satiety—referred to as if they constituted opposing controls over food intake. When representing a subjective state or cognition, hunger becomes one index or measure of the strength of satiety. Of course, when used as a motivational construct, hunger can be opposed to satiety, at least as long as satiety is regarded as a construct representing a state of inhibition over eating. This state cannot be directly measured but can be inferred from measures such as the length of the interval until the next period of eating or the amount of food consumed in a given period of time.

The term "satiation" is probably best considered as the process leading to the termination of an episode of eating (Blundell, 1979). Consequently, satiation and satiety refer to phenomena occurring either within or between meals, respectively. In this sense, satiation and satiety—as here defined—are the equivalent of intrameal and intermeal satiety as described by Van Itallie and Vanderweele (1981). Accordingly, the term "satiating power of food" (Kissileff, 1984; Blundell et al., 1988a) has two meanings. It can describe the amount of food consumed before a period of eating is terminated (the effect of food on satiation) or—the more common meaning—the inhibition of further intake after consumption has ended (the effect of food on satiety). It is worth noting that these two effects can be dissociated, and certain manipulations (e.g., dietary fiber), according to the circumstances, can influence both satiation and satiety (Blundell and Burley, 1987). Consequently, this distinction between satiation and satiety is not just a semantic nicety but is a logical requirement.

The notion of hedonics—set out in the title to this chapter—is rather loosely defined in psychology, usually with reference to affective aspects of an event or experience. In this particular context it can best be conceived of as the state of pleasure arising from the taste and/or consumption of food. In turn, this pleasure is usually measured by rating the subjective sensation of perceived pleasantness of (or liking for) the taste of food.

Figure 1 sets out a scheme for conceptualizing the ways in which the subjective

HUNGER, HEDONICS, AND SATIATION CONTROL

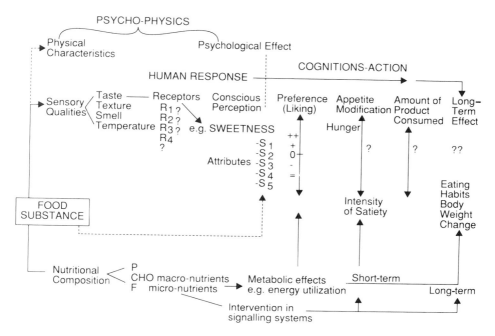

FIGURE 1 Conceptual scheme illustrating the possible place of hedonics (preferences) and hunger in the system mediating the action of physical characteristics and metabolic properties of food upon eating and body weight.

feelings of pleasantness or hunger may mediate in the satiation and/or satiety associated with food consumption. For heuristic reasons the scheme somewhat artificially separates the sensory and metabolic components of food ingestion. For ease of comprehension, the sensory components refer to sensations arising in or around the nose and mouth. The scheme indicates the recognized distinction between the metering of the perceived intensity of a physical stimulus (classical psychophysics) and the recognition of an affective component (do I like it or not?). The assignment of a value to the degree of pleasantness of a stimulus is clearly a cognitive task. The perceived pleasantness is influenced by many factors in addition to the physical qualities of the stimulus; these may include prevailing sociocultural norms for attributing pleasantness to such a stimulus (i.e., the legitimacy of regarding such a stimulus as pleasant) as well as individual experience with similar sensations. Since the scheme also indicates that the evaluated "liking for" a stimulus is influenced by metabolic effects arising from food consumption, the hedonic aspect of food is not a simple parameter. The subjective appreciation of pleasantness can be regarded as a cognitive state built out of the physical aspects of the stimulus, attributions normally applied to such a stimulus, and the prevailing metabolic circumstances.

II. SOME THOUGHTS ON PALATABILITY

At this stage it is relevant to inquire about the relationship between the appreciation of pleasantness and the notion of palatability. Indeed both the definition and the function of palatability are hotly debated in the psychology of appetite. The term "palatability" is widely used as if it possessed high explanatory power, and it is common to see and hear references to increased food consumption arising from highly palatable food or dietary-induced obesity depending on the palatability of the diet. To the casual observer at least, it appears that palatability is being used to connote some property of food that automatically leads to increased consumption.

We propose, however, that palatability is not merely an aspect of the stimulus properties of food such as its taste, smell, flavor, texture, and temperature—on their own, these are simply categories of orosensory stimuli. At another extreme, palatability is not the final common pathway defining food intake, ingestive responses, or food preference. Although these dependent variables may indicate palatability, they are also subject to separate influences (see below).

Rather, food palatability is determined by the result of the integration of orosensory and postingestive stimuli, and consequently it depends on the interaction of food and the organism. Palatability, like hunger (Blundell, 1979), is a hypothetical construct: that is, an explanatory concept that cannot itself be directly observed but is inferred from operationally defined and measurable events (cf. Grill and Berridge, 1985). Accordingly, it is a concept relevant to the understanding of the ingestive behavior of both human and nonhuman animals. Palatability may be experienced as the hedonic (affective) quality of orosensory stimuli; therefore, in humans, pleasantness ratings can usefully indicate hedonic quality, and in turn palatability. Accordingly, the term "palatability," like "hunger," can be conceived of in two ways: either as an intervening variable not directly measurable or, when construed as the perceived pleasantness of food, as a subjective experience that can be objectively monitored by means of a rating. Consequently, in reference to human appetite, palatability appears to be considered to be equivalent to perceived pleasantness of food. Nonetheless, the relationship of pleasantness (or palatability) to actual consumption remains to be determined and should not be prejudged.

The need for the construct defined above, regardless of whether it is called palatability, arises from the demonstration that, whereas hedonic responses to, for example, sweet and bitter stimuli are preprogrammed (Steiner, 1987), powerful conditioned preferences and aversions can be acquired through association of sensory cues and postingestive stimuli (Garcia et al., 1974; Sclafani and Nissenbaum, 1988). As a result of aversive conditioning, a previously accepted taste (even a sweet taste) may come to elicit behavioral responses and an accompanying affective reaction indistinguishable from unconditioned rejection (Booth, 1978; Grill and Berridge, 1985). In the opinion of Booth (1978), "Conditioned aversion is a nasty taste, not merely or at all the refusal to take something that is perceived as dangerous" (p. 566). At the present time the nature of conditioned food preferences (liking) remains largely unexplored (but see Elizalde and Sclafani, 1988); nonetheless, it is probable that the conditioning of a "liking for" foods plays an important role in the overall influence of palatability on food intake and food selection.

III. THE SATIETY CASCADE, PALATABILITY, AND HUNGER

The objective of the present analysis is to examine the relative roles of perceived hunger and perceived pleasantness of food (hedonics) in the control of food consumption. What contribution do these two phenomena make to the initiation of eating, its maintenance and termination, and to the preservation of inhibition over subsequent eating? One way to begin the analysis is to consider the major factors that influence satiety. The conceptualization set out in Figure 2 can be regarded as the satiety cascade. The major mediating processes contributing to the strength of satiety are classified as sensory, cognitive, postingestive (but preabsorptive) and postabsorptive. Obviously, complex physiological mechanisms underlie each of these process categories. To what extent are these mediating processes (and the associated mechanisms) related to fluctuations in the perceived pleasantness of food (here used synonymously with palatability) and perceived hunger? This is an empirical issue, and some answers should be provided by experimental investigation.

A number of questions may be posed. For example, is high palatability a sufficient or necessary condition for the initiation of eating? In other words, is the presentation of a preferred food (rated highly as pleasant) by itself sufficient to trigger eating? The answer must be "no," since there are times when even the most delicious of foods will not be consumed. Is the presentation of a highly preferred food an essential element involved in the initiation of eating? The answer again must be "no," since there are times when even unpleasant tasting food will be consumed. Therefore, palatability is neither a sufficient nor necessary condition for the initiation of eating.

However, it may be more appropriate to ask about the relationship between pleasantness (palatability) and the termination of eating (satiation) since there are at least two theoretical formulations (sensory-specific satiety and negative gustative alliesthesia) in which the decay of the perceived pleasantness of food is regarded as instrumental in bringing eating to a halt. Is it the case that eating (a meal) ceases when the rated pleasantness of food has fallen to zero (or below some threshold value required for the maintenance of consumption)? This issue will be discussed later, but initially it appears extravagant to propose that the perceived

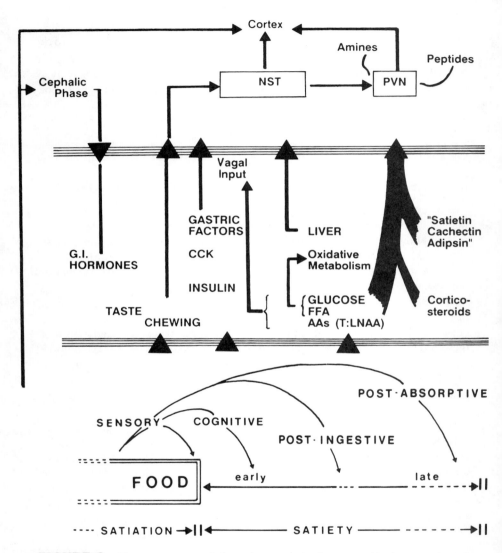

FIGURE 2 The components of the satiety cascade (bottom) with the operation of mediating peripheral physiological mechanisms (middle) and a simple representation of the involvement of elements in the brain (NST, nucleus of the solitary tract; PVN, paraventricular nucleus of the hypothalamus).

quality of a food must change from being highly preferred to revolting before eating ceases.

Similar questions can be posed about the role of hunger and the satiety cascade. Is hunger a necessary or sufficient condition for the initiation of eating? Probably not, since in some instances eating fails to occur in the presence of manifest hunger (during dieting, for example), and it is widely felt that eating can occur when hunger is negligible or at a very low level (see, e.g., Cornell et al., 1989). It appears then that neither hunger nor palatability constitutes a biological imperative that might be responsible for starting and stopping eating. That is, neither hunger nor perceived pleasantness can be guaranteed to predict either the timing of eating episodes or the amount consumed. However, logic insists that some relationship must exist between these parameters; how can it be characterized? As a next step, some theoretical positions will be examined.

IV. A CONSIDERATION OF SENSORY-SPECIFIC SATIETY

Over the past few years a loosely constructed theoretical formulation has been proposed in which a central role is ascribed to the palatability or pleasantness of food in the development of satiation. The concept is also extended to account for satiety and has been given the title of "sensory-specific satiety" (e.g., see Rolls et al., 1981, 1984; Hetherington et al., 1989). The perceived pleasantness of food is invoked in two ways to account for changes in the disposition to eat a particular food. First, it is assumed that, given a choice between two or more foods, subjects will eat more of the food given the higher pleasantness rating. Second, it is argued that consuming a particular food will reduce its perceived pleasantness and therefore render it less likely to be eaten. Although Le Magnen (1985) has stated that "Eating a food in a state of satiety induced by the immediately preceding intake of a sensorial different food is a trivial experience within human meals" (p. 41), the notion of sensory-specific satiety has been promoted as a profound mechanism governing the choice of foods and the pattern of eating.

At the outset it can be noted that the evidence for sensory-specific satiety has been harvested from a highly stylized experimental maneuver in which (usually) the natural rhythm of eating is interrupted by the periodic requirement to taste (and eat) a variety of small food samples and to rate the pleasantness of these. This procedure may take place up to six times within an experimental session (Rolls et al., 1984, 1988), but the effect of this repeated consumption (as opposed to tasting) of food samples does not appear to have been reported. In addition, although adjustments in palatability are central to the credibility of sensory-specific satiety, it should be noted that it is relative (and not absolute) pleasantness that is the axiomatic parameter in controlling the maintenance of consumption. Although the theoretical formulation has emphasized the concept of satiety, the experiments have usually demonstrated a prolongation or reinitiation of eating arising from the presentation of new foods (whose pleasantness has not been diluted by prior consumption).

In other words, the experiments have demonstrated sensory stimulation of appetite—a phenomenon revealed by Le Magnen some 30 years ago (Le Magnen, 1956, 1960). On logical grounds the demonstration of a sensory stimulation of eating (by a different food) does not justify the invocation of a mechanism of "sensory-specific satiety." A further logical issue arises from the alleged embodiment of satiating power exclusively in the sensory qualities of food; for it could be argued that people would continue to feel sated by consuming food that possessed simply sensory qualities but no nutritional value. An examination of the logical and methodological features of sensory-specific satiety could clarify the role of sensory factors (particularly those to which a hedonic dimension can be added) in eating.

The proposal that there are sensory-specific decreases in the pleasantness of food during its consumption has been used to account for the finding that food variety can stimulate intake. Humans and rats, for example, eat more when they are presented with different foods (or the same foodstuff differently flavored) either successively or simultaneously in a test period than if only one of the foods is available for the same length of time (see, e.g., Morrison, 1974; Rolls, 1981; Rolls et al., 1981, 1984; Treit et al., 1983; Rogers and Blundell, 1984). However, although the stimulatory effect of food variety appears to be well established, its explanation may be different from that proposed by Rolls and colleagues.

As indicated above, the direct evidence for (sensory-specific) changes in the palatability of food during eating comes mainly from studies in which human subjects are asked to report on the pleasantness of the taste (and smell, texture, and appearance) of food during and after eating. Superficially, the instruction to rate the taste of a sample of food is clear, and subjects are apparently able to respond without difficulty. The basis on which subjects make their judgments is, however, much less clear. Are they rating how good the food tastes or, for example, how good it is to eat? [Mook (1987) raises the same issue in commenting on Cabanac's alliesthesia experiments.] This is a very important distinction if, as will be argued later, palatability can be distinguished from hunger. The palatability of food (i.e., how good it tastes) may be unaffected by deprivation or a full stomach, whereas the willingness to consume that food will under most circumstances be influenced by the level of hunger or fullness experienced. One way in which this problem might be resolved is to reexamine the effects of eating on the subjective appreciation of food by asking subjects to make the following ratings:

    How pleasant is the taste of this food?
    How pleasant would it be to eat this food?
    How much of this food would you like to eat?

There is no reason to believe that this would force an unnatural distinction between these aspects of eating (it is quite possible for the responses to these questions to change in parallel), but it would require subjects to focus on what may turn out to be different components of subjective experience.

It is said that people spontaneously report that food tastes pleasant when hungry and less so when sated, but this may also confuse the pleasantness of the taste of food with the pleasantness of eating it. In any case, counterexamples such as "It's delicious, but I really can't eat any more" can be provided. In general, anecdotes are not good sources of evidence. More importantly, it is possible that the beliefs implied by such statements will bias responses on self-report rating scales.

The interpretation of subjective ratings presents a number of difficulties, and therefore such data should not be accepted uncritically. The measurement of subjective experiences associated with eating should not, however, be abandoned. In some cases it is possible to test subjective ratings against objective data (e.g., changes in subjective hunger predict changes in food intake in response to the covert manipulation of the energy content of a food preload—see below). It is possible to be optimistic that with appropriate care, subjective ratings can reveal important information about the processes controlling eating and food intake.

Alternatively, facial expressions of human infants in response to taste stimuli (Steiner, 1987), and similarly the taste-elicited, fixed-action patterns (FAP) displayed by rats (Grill and Berridge, 1985) may provide objective data to test the proposal that palatability changes as a function of internal state. Infusion of a sucrose solution into a rat's mouth through a fixed intraoral catheter elicits characteristic FAPs. These decline in number as the amount ingested increases, and the decline is more rapid in animals that have recently fed. A second observation is that aversive reactions (gapes, head shakes, paw shakes) then begin to appear. These results have been interpreted as indicating a shift in the palatability of the sucrose (Grill and Berridge, 1985). The extent to which these various reactions relate to the natural satiation process is, however uncertain. For the intraorally fed rat, the delivery of food is under the experimenter's control: the rat has the option of either ingesting the food, allowing it to drip passively out of its mouth, or actively expelling (rejecting) it. This latter response may be a reaction to the infusion of food beyond the point of full satiety. The freely feeding rat when satiated would no longer take food into its mouth.

It was argued above that the notion that the pleasantness of the taste of food (its palatability) declines with eating should be reexamined. This in turn suggests that an alternative explanation for the stimulatory effect of food variety may be required. One such explanation, which although simple is often overlooked, is that intake may be increased when a variety of foods is offered because of the differential presence of a preferred food(s). For example, in a typical experiment subjects might be offered four different foods in four successive courses (variety condition). In the control condition, the same food might be re-presented for each of the four courses, with separate tests being conducted for the four foods. This means that the most favored food would be available at some point in every test of the variety condition, whereas it is would be present in only one-quarter of the control tests. Note that individual subjects in the experiment may differ as to their most preferred food, and consequently, grouped data may conceal this effect. When a variety of foods is available, individual preferences are more likely to be satisfied. Despite suggestions to the contrary (Rolls et al., 1984), many studies fail to rule this out as a possible explanation of the stimulatory effect of food variety (but see Treit et al., 1983). There is, nonetheless, clear evidence that this is not the only mechanism involved. For example, observations on free-feeding rats showed that *eating* a variety of foods increases intake (Rogers and Blundell, 1984). Analysis of the selection patterns of individual animals showed that the consumption of different foods within the same meal led to an increase in meal size compared with meals in which only one food was eaten. Control data indicated that this was not simply because the likelihood of swapping between foods increased with the length of the meal.

The sensory-specific satiety hypothesis maintains that sensory contact with food is satiating. In general terms, the most obvious alternative to this is that food stimuli have a stimulating effect on eating. Indeed, as noted above, Le Magnen (1956, 1960), who published the first systematic studies on the effects of manipulating food variety, discusses the results as an example of the sensory stimulation of intake (Le Magnen, 1985). Other workers have argued that there is a positive feedback effect on eating arising from tasting food at the start of the meal (see, e.g., Wiepkema, 1971), and our studies on intense sweeteners indicate that sweet tastes increase rather than inhibit appetite (see, e.g., Rogers and Blundell, 1989b). The importance of food stimuli in promoting eating is also demonstrated in a recent study in which the unexpected presentation of food was found to increase desire to eat and to elicit further eating in apparently satiated subjects (Cornell et al., 1989). The mechanisms responsible for the stimulatory effects arising from exposure to food stimuli are largely unknown, although the involvement of cephalic phase autonomic and endocrine responses has been suggested. These findings indicate that further studies examining how palatable food stimuli act to promote ingestion are needed if the effects of food variety are to be fully understood.

Two main deductions arise from this analysis. First, the axiomatic role of rated pleasantness in sensory-specific satiety is more precarious than has hitherto been claimed. The crucial evidence to demonstrate that changes in pleasantness—and *only* changes in pleasantness—provide the mechanism underlying the effect of variety on intake is not available. Second, the prominence of the notion of sensory-specific satiety as an explanatory principle accounting for our understanding of human eating has been overstated. The strength of the term results from its repeated use as an explanation for the outcome of a number of laboratory experiments on rats and humans. Considering the apparently widespread acceptance of this explanatory principle, formal and systematic statements of the theory are notably scarce and weak. In our view, to occupy a place of importance in the theoretical machinery concerning appetite control, the theory of the sensory-specific nature of satiety should be elaborated and formally set out as a set of propositions and formulations that can account for the expression of appetite. Only then will researchers properly be able to evaluate its contribution to our understanding of human appetite.

V. ALLIESTHESIA AND SATIATION

Almost 20 years ago, a seminal article on the physiological role of pleasure appeared in the journal of the American Association for the Advancement of Science (Cabanac, 1971). The article described the phenomenon of alliesthesia and defined the role of subjective appreciation of pleasantness in this phenomenon. At the time, some of the most persuasive evidence for the biological significance of alliesthesia was founded in the experimental study of sweet-tasting solutions and obesity (Cabanac and Duclaux, 1970a,b). One particularly striking finding was that obese subjects, asked to rate the pleasantness of varying intensities of sucrose solutions, failed to display a suppression of rated pleasantness after the ingestion of a glucose load (50 g in 200 ml of water). Unlike lean subjects, therefore, obese individuals did not display negative gustative alliesthesia. Since it was argued that pleasantness (of a stimulus) was an indication of the biological need for that stimulus, this finding was interpreted as indicating that the need for food (indexed by the pleasantness rating) in obese subjects was not diminished by the prior supply of calories. In other words, pleasantness could be used as an assay for the development of satiation, and therefore satiation (or at least one major component of satiation) was weak or absent in obese people. In turn this could explain why obese people ingested more calories than lean individuals.

However, as far as we are aware, the actual food consumption of obese subjects was never measured following either the glucose load or a control volume of water. Therefore the validity of the maintenance of the pleasantness ratings was never put to the test. It would of course be predicted that the glucose load (or food of other types) would fail to suppress subsequent intake just as it failed to suppress pleasantness ratings.

In recent studies we have had the opportunity to repeat the assessment of negative gustative alliesthesia in lean and obese individuals within the context of a larger study in which a large number of additional subjective and objective measures were made (Blundell and Hill, 1988). It was indeed confirmed that, just as Cabanac had claimed, in obese subjects a glucose load did not suppress subsequent pleasantness ratings of sweet-tasting sucrose solutions. However, the glucose load did suppress perceived hunger in both lean and obese subjects, and also reduced food consumption in a later test meal. This dissociation between pleasantness and hunger together with actual measures of energy intake will be referred to later. However, the study mentioned above showed that although the failure of glucose to suppress pleasantness does appear to be a feature of the response repertoire of obese subjects, this effect is independent of either changes in hunger or food consumption. Pleasantness ratings did not predict (or correlate with) food intake (see also later study).

Considering these findings along with previous comments on the mechanisms underlying sensory-specific satiety, we do not feel there is convincing evidence that moment-to-moment adjustments in the palatability of food (pleasantness ratings) are critically involved in the development of satiation, that is, in the process of bringing eating to a halt. However, both logic and common sense demand that palatability exerts some influence over the eating response; what is the nature of this influence? We will return to this matter after a consideration of the role of hunger.

VI. THE IMPORTANCE OF HUNGER?

To what extent is the subjective experience of hunger important in determining the pattern of eating? Does a buildup of hunger give rise to the initiation of eating, and is the rapid decay of hunger by eating responsible for satiation and satiety? For the moment it does not matter that hunger is not a single unified experience; it may derive from a variety of physical sensations (see, e.g., Monello and Mayer, 1965) and the profile of these may vary from individual to individual (see, e.g., Harris and Wardle, 1967). Nor is it necessary here to enter a philosophical discussion on whether hunger constitutes a cause of eating. For the moment it is only necessary to accept that for every person there is a particular constellation of feelings and sensations that is recognized and labeled as hunger. It is not essential yet to agree on the importance of these feelings. To be used as an experimental tool to investigate satiation and satiety, the feeling of hunger must be brought under methodological control. The visual analogue rating scale first apparently used by Silverstone and Stunkard (1968) is now almost universally accepted as a way of measuring the intensity of hunger. One first step in validating the significance of these ratings is to establish the extent to which they are associated with food consumption. This is normally done by calculating coefficients of correlation. Under some conditions—for example, following anorectic drug administration—correlation coefficients are extremely high and range up to 0.7–0.9 (Rogers and Blundell, 1979; Blundell and Rogers, 1980; Goodall et al., 1987). Under other circumstances, such as nutritional loading, coefficients may be in the order of 0.6–0.75 (Hill et al., 1987). A different method of investigation has used diary recording to monitor the profile of eating along with associated ratings of hunger over a period of several days (de Castro and Elmore, 1988). Interestingly, subjective hunger was the best predictor of meal size and was negatively correlated with estimated energy value of stomach contents. The authors suggested that hunger represents an intermediary step in the causal sequence linking prior ingestion with subsequent meal size.

Despite some impressive correlations between hunger ratings and food consumption, certain authors tend to diminish the importance of hunger as an influential factor in the control of ingestion and claim to find little experimental evidence in favor of hunger, although often without providing any statistical verification for the claim. To resolve this issue, it is essential that supporters and opponents of the hunger theory bring the same degree of statistical power to bear on the problem.

However, there may be methodological reasons for the variation of correlation coefficients from study to study, and certainly there are different ways in which the association between hunger and food intake can be adduced. Table 1 illustrates correlations between hunger and subsequent food intake from a study in which sweetened or unsweetened yogurt preloads were eaten by subjects one hour before the presentation of a test meal (Rogers and Blundell, 1989b). The table shows three different ways of treating the hunger rating scores and test meal intakes. We refer to these as absolute values, change (from baseline, i.e., pre-preload) ratings, and calibrated ratings. The absolute method simply works with the raw rating scores and the meal intakes. In the second method the rating score is Δ hunger (i.e., the change in value from the initial baseline rating). The third procedure depends on the presence of a genuine control condition (in this case, the vehicle or

TABLE 1 Correlations Between Hunger Ratings and Food Intake Calculated Using Three Different Methods[a]

Rating	Preload	Pre-preload (baseline)	Post-preload 2 min	Post-preload 60 min	Post-preload 60 min
Absolute[b]	Glucose	.39	.39	.54	.47
	Saccharin	.22	.35	.33	.59*
	Unsweetened (vehicle)	−.04	−.10	.49	.70**
Change from baseline[c]	Glucose		.02	.08	−.05
	Saccharin		.02	.35	.39
	Unsweetened (vehicle)		−.07	.40	.59*
"Calibrated"[d]	Glucose		.39	.75***	.59*
	Saccharin		.50	.66**	.77***

[a] Hunger ratings were made on a 100 mm visual analogue scale word-anchored at either end (As hungry as I've ever felt—Not at all hungry). Food intake was measured in an ad libitum, lunchtime test meal begun 1 hour after eating a 215 ml yogurt preload. Data are for 12 subjects (Rogers and Blundell, 1989b).
[b] Raw scores for hunger and food intake.
[c] Hunger scores are postyogurt minus preyogurt (Δ_{hunger}), and raw scores for food intake.
[d] Hunger scores are treatment minus vehicle calculated from Δ_{hunger}; food intake scores are treatment minus vehicle (from raw scores for food intake).
*$p < 0.05$.
**$p < 0.02$.
***$p < 0.01$ (two-tail).

unsweetened yogurt—see Blundell et al., 1988b). Here the values used to compute the correlation coefficient at each time point are derived from the treatment minus control values for both ratings and test meal intakes. This table of coefficients shows a high degree of consistency, with the highest coefficients provided by the strongest treatment of the data—namely the "calibrated" procedure.

In a series of experiments stretching back over many years and conducted under carefully controlled laboratory conditions, we have invariably found significant correlations between premeal hunger ratings and test meal intakes with coefficients of 0.49–0.86 (Blundell and Rogers, 1980), 0.73 (Hill and Blundell, 1986b), 0.6–0.75 (Hill et al., 1987), 0.65 (Hill and Blundell, 1990b), and 0.74 (Blundell and Hill, 1987). We conclude from our own data and those of others that hunger ratings generally constitute a valid predictor of energy intake, although it is clear that there will be circumstances (physiological, social, or methodological) where this relationship is weakened or lost. However, these circumstances, such as prolonged fasting (Wadden et al., 1985) or disordered eating (Owen et al., 1985) can still provide useful information about the role of hunger.

VII. HUNGER AND THE CALORIC CONTROL OF SATIATION

The observation that ratings of hunger are reliably linked to subsequent food intake does not in itself throw light on the mechanisms that are responsible for this relationship. However, information on this issue can be derived from the examination of the fluctuations in hunger ratings that have been measured during experiments on the satiating efficiency of foods (Kissileff, 1984; Blundell et al., 1988a) or on caloric compensation. Many studies in this domain conform to a preload–test meal formulation in which a portion of some foodstuff, precisely controlled for calories and nutrients (the preload), is consumed at a specified interval before subjects are allowed to eat freely from an abundant supply of food (the test meal). The test meal may be so composed that it can sensitively detect changes in the consumption of total calories and/or the selection of different dietary commodities or nutrients. This experimental procedure has been widely used to evaluate the satiating power of particular food materials including the proportion of protein and carbohydrate (Hill and Blundell, 1986a,b), the energy content and flavor of soups (Kissileff et al., 1984), and the effect of the presence of dietary fiber (Burley et al., 1987). Not surprisingly, these and other studies have shown that various foods differ in their power to induce satiation and maintain satiety. These studies have also been used theoretically to investigate the phenomenon of caloric compensation, that is, the tendency for test meal intakes to be reduced by a calorie value equivalent to that delivered in the preload. One fundamental question addressed by all these studies is this: Are test meal intakes suppressed in proportion to the calories delivered in the preload?

We have investigated this issue in a long series of studies (presently amounting to 14 separate experiments) in which the dimensions of sweetness and energy value of preloads have been separately and independently manipulated. The methodological requirements of such studies have been set out elsewhere (Blundell et al., 1988a,b). The materials used in these studies have included sweet energy-yielding carbohydrates, such as sucrose, glucose, and fructose; nonsweet carbohydrates, such as starch and maltodextrin; high-intensity sweeteners, such as saccharin, aspartame, and acesulfame-K; and potent sweetness inhibitors (see Rogers and Blundell, 1989a for review). These experimental manipulations have been made using a variety of preload vehicles, including water, soups, and yogurts (Rogers et al., 1988; Rogers and Blundell, 1989b). Except for one or two very special experimental circumstances, the outcome of these studies has repeatedly indicated that the presence of calories in the preload is the most important factor leading to a subsequent suppression of intake in the test meal. Our studies have consistently demonstrated that high calorie preloads suppress later food intake to a greater extent than low calorie loads. In addition, because these studies have used a temporal tracking of hunger ratings, the results have disclosed a close parallel between the effect of preload calories on both hunger and energy intake.

The potency and reliability of the association between preload energy value, hunger ratings, and test meal intakes has obliged the adoption of a formulation that may be termed the caloric control of satiation. Even when the presence of calories in the preload was disguised to prevent any conscious recognition of the energy value, the caloric impact was detected by the biological system and metered to result in the adjustment (not always perfect) of the calories consumed in subsequent eating episodes. Since in our studies a minimum of one hour normally elapsed between preload and test meal, the caloric adjustment of intake could have been achieved by a number of mechanisms involving pre- and postabsorptive processes.

In principle this position is similar to other theoretical formulations that have given prominence to the recognition of delivered energy in the control of food intake. These include the energostatic control of feeding (Booth, 1972) and the metabolic hypothesis based on the notion of substrate oxidation (Friedman et al., 1986). Both these propositions emphasize the postabsorptive mechanisms of energy utilization. The position presented here is that the subjective experience of hunger is linked to the availability and metabolism of supplied calories, which in turn influence the pattern of subsequent intake. Accordingly, hunger does not have to be viewed as a "cause" of eating, since changes in experienced hunger and food intake alike may arise from the processing of energy or metabolic fuels. Considering biological influences on motivation, behavioral actions have both a particular direction and a topography; actions are embellished by the formation of symptoms. The subjective experience of hunger can be seen as a prominent symptom generated, in part, by metabolic processing and closely related to actions of food seeking and ingestion. The impact of metabolic processing on conscious experience (subjective hunger) gives a clear biological purpose to this feeling and ensures that behavioral actions will be appropriately supported by cognitive drives.

VIII. THE RELATIONSHIP BETWEEN PALATABILITY AND HUNGER

The discussion above leads to the inevitable conclusion that while palatability and hunger may both exert some control over eating, they do so in quite different ways. Indeed, it appears that perceived pleasantness and perceived hunger are largely independent of each other. Does this mean that they are separately regulated? There is evidence from pharmacological studies that the two phenomena can be dissociated. One approach arises from the postulated linkage between endogenous opioids and hedonic processes. In animal research it has been shown that the opioid antagonist naltrexone suppressed the hyperphagia induced by a "palatable" diet (Apfelbaum and Mandenoff, 1981). In humans, the same antagonist has been reported to depress the perceived pleasantness of a sweet solution, while having no effect on perceived hunger (Fantino et al., 1986). There is additional evidence that naloxone—another opioid antagonist—reduces food intake but has no action on hunger (Trenchard and Silverstone, 1983). Moreover, naltrexone potentiates negative gustative alliesthesia induced by glucose (Melchior et al., 1989). Interestingly, an opposing profile of changes have been induced by the serotoninergic agent of D-fenfluramine (Blundell and Hill, 1988). The administration of this drug had no effect on perceived pleasantness and did not augment glucose-induced alliesthesia. However, this serotoninergic manipulation reduced perceived hunger and intensified the suppressive action of glucose on hunger. Considered together, these studies suggest that opioid mechanisms are involved in mediating hedonic aspects of food stimuli, whereas serotoninergic systems play a mediating role in the expression of hunger.

Perhaps the key question concerns the relationship of palatability and hunger to the occurrence of satiation and the maintenance of satiety. As noted earlier, there exist strong and divided views on this issue. The question was firmly addressed in studies on the satiating efficiency of protein and carbohydrate meals in lean and obese subjects (Hill and Blundell, 1990a,b). During consumption of the fixed-preload meals, subjects were required to rate hunger and the pleasantness of the taste of the food at four stages during the meal. Interestingly, perceived pleasantness of the food did not change noticeably during the course of this meal, whereas hunger ratings declined sharply (Table 2). The suppression of hunger during eating and the maintenance of this suppression during postmeal satiety is consistent with results from many previous studies (e.g., Blundell and Hill, 1987, 1988). However, the stability of the pleasantness ratings of food was surprising in view of the claims made for this parameter in the control of satiety. (However it is worth noting that there exist reports of the failure to detect an alliesthesia-type

TABLE 2 Pleasantness Ratings (100 mm Visual Analogue Scale) Taken at Four Times During the Consumption of a Fixed-Energy Lunchtime Meal (473 kcal)[a]

Lunchtime rating no.	Carbohydrate/ placebo	Carbohydrate/ drug	Protein/ placebo	Protein/ drug
	Lean Subjects (body mass index = 21)			
1	69.9	59.2	77.0	67.1
	(7.5)	(8.4)	(5.0)	(7.4)
2	63.1	56.8	71.4	66.4
	(9.6)	(10.2)	(4.4)	(6.7)
3	58.6	50.1	73.7	68.0
	(9.9)	(11.3)	(4.4)	(7.5)
End	62.4	55.6	78.1	66.7
	(10.4)	(11.5)	(4.9)	(8.5)
	Obese Subjects (body mass index = 38.0)			
1	89.0	82.1	89.0	89.1
	(3.3)	(5.7)	(3.3)	(3.7)
2	85.5	79.0	89.5	87.5
	(9.6)	(6.2)	(3.4)	(4.2)
3	86.0	81.5	86.9	89.0
	(4.6)	(6.5)	(4.5)	(5.1)
End	86.1	83.9	87.5	89.9
	(5.2)	(7.0)	(4.5)	(5.3)

[a]The lunch was high (63%) in carbohydrates on two occasions and high (64%) in protein on two occasions. Subjects were tested under placebo or drugged conditions.

reaction with real foods; see, e.g., Wooley, 1976.) We therefore conclude that while the feeling of hunger is closely related to the metabolic processing of food and the development of satiety, pleasantness of taste is relatively independent of this action.

How should the role of pleasantness in controlling food intake be understood? Logic and common sense suggest that it must have some influence. One way to reconcile a belief in the importance of pleasantness with the negative findings reported above is to enforce the distinction, mentioned earlier, between the pleasantness of the *taste* of food and the pleasantness of *consuming* that same food. If this distinction were accepted, then the potency of one aspect of pleasantness (of consumption) as a factor in satiation could be sustained while the other form of pleasantness (of taste) would not be threatened by attempts to link it with metabolic processing or behavior (e.g., eating).

However, the pleasantness of taste (what is termed "palatability") could modulate food intake via hunger. Experimental evidence supports this view. First, it has been demonstrated that highly preferred food both augments hunger during eating and leads to a more rapid restoration of hunger during late-phase satiety (Hill et al., 1984). Second, it has also been shown that the more rapid return of hunger and desire to eat following consumption of preferred food is accompanied by faster occurrence of the (preprandial) hypoglycemia that precedes the next meal (Chabert et al., 1988). This proposal embodies a modulatory influence of pleasantness on consumption while assigning a more fundamental function to hunger in guiding the pattern of eating.

IX. SYNCHRONICITY OF PHYSIOLOGY, BEHAVIOR, AND COGNITIONS

A scientific approach to the study of appetite suggests that changes in physiology, behavior (eating), and cognitive states (e.g., hunger) be synchronized to provide a cohesive basis for biological functioning. Many experiments are therefore based on the expectation of finding a tight coupling between these three domains. However, in research on anxiety it has been demonstrated that there may be dissociation between autonomic indices (physiology), avoidance (behavior), and subjective anxiety (see, e.g., Epstein and Fenz, 1965). Moreover, in our recent cultural climate it is clear that uncoupling within the appetite system can be brought about by the strategy called dieting.

It is tempting to approach the study of hunger and hedonics as biologists attempting to understand a homeostatic system refined over millions of years of evolution. However, it is generally agreed that the currently prevailing nutritional habitat (at least in technologically advanced societies) has undergone radical alteration within the past 100 years. These changes include the provision of an abundant food supply, altered proportions of macronutrients, and the introduction of a vast array of tastes, flavors, and textures combined unsystematically with the nutritional components of foods. In turn, these factors are embraced within an aggressive culture advocating the attainment of arbitrarily defined standards of appetite and body shapes. The extent to which functional biological principles continue unimpeded to gain expression is debatable. Consequently, it may be unrealistic to seek truly synchronous relationships among the domains of physiology, behavior, and cognitions. Some slippage may be expected, and this would reflect the operations of a system in a state of some turbulence. In studies on appetite it may be worth reflecting on the extent to which researchers are measuring (a) the operating principles of an antique biological system, (b) a biological system biased by prevailing social and personal states, or (c) a biological system undermined by cultural and nutritional forces.

DISCUSSION

Kissileff: Are you aware of the criticism that David Booth has leveled against this four-design protocol?

Blundell: No.

Kissileff: He claims that you can't really call it four cells because of differences in cognitive perceptions between the cells. Even though you might manipulate calories and sweetness, the subject may see these manipulations as four entirely different kinds of food; their palatability might be different, there might be differences in other factors like osmolarity. I wondered if you had any thoughts, but since you hadn't seen the criticism you might not have thought about it.

Blundell: No, but I can think about it very quickly. We certainly know that cognitively, subjects perceive no difference between these preloads. We take very severe measures and great care to ensure that the deception is not broken open by the subjects. So I'm confident about that. The osmolarity is interesting, and I was thinking about it when Gary Beauchamp gave his talk. That's something we don't control for, but then neither do other people using weaker designs. You can certainly criticize the strong design in that it is not as strong as it possibly could be, but it is certainly stronger than most other designs. If these criticisms are going to stick, then they have to be applied even more firmly to weaker, two-cell designs.

Hirsch: Do you imply [sic] from this that the palatability of food at the end of a meal would have to be greater than that at the start of a meal?

Blundell: Not necessarily, but you might ask that question of the next speaker. I did pose the question whether it's necessary for palatability during the meal to reach a low value in order to suppress eating. Is that either a sufficient or necessary condition? Some of the data that I did show, and some I didn't show, indicate that eating still stops when palatability has not changed at all; it's still as high at the end as it was in the beginning. So palatability is certainly not a necessary condition for stopping eating.

REFERENCES

Apfelbaum, M., and Mandenoff, A. (1981). Naltrexone suppresses hyperphagia induced in the rat by a highly palatable diet. *Pharmacol. Biochem. Behav.* **15**:89–91.

Blundell, J. (1979). Hunger, appetite and satiety—Constructs in search of identities. In *Nutrition and Lifestyles*, M. Turner (Ed.). Applied Science, London, pp. 21–42.

Blundell, J. E., and Burley, V. J. (1987). Satiation, satiety and the action of fibre on food intake. *Int. J. Obesity*, **11**(Suppl. 1):9–25.

Blundell, J. E., and Hill, A. J. (1987). Serotoninergic modulation of the pattern of eating and the profile of hunger–satiety in humans. *Int. J. Obesity*, **11**(Suppl. 3):141–155.

Blundell, J. E., and Hill, A. J. (1988). On the mechanism of action of dexfenfluramine: Effect on alliesthesia and appetite motivation on lean and obese subjects. *Clin. Neuropharmacol.* **11**(Suppl. 1):S121–S134.

Blundell, J. E., and Rogers, P. J. (1980). Effects of anorexic drugs on food intake, food selection and preferences, hunger motivation and subjective experiences: Pharmacological manipulation as a tool to investigate human feeding processes. *Appetite,* **1**:151–165.

Blundell, J. E., Rogers, P. J., and Hill, A. J. (1988a). Hunger and the satiety cascade—Their importance for food acceptance in the late 20th century. In *Food Acceptability,* D. M. H. Thompson, (Ed.). Elsevier, Amsterdam, pp. 233–250.

Blundell, J. E., Rogers, P. J., and Hill, A. J. (1988b). Uncoupling sweetness and calories: Methodological aspects of laboratory studies on appetite control. *Appetite,* **11**:54–61.

Booth, D. A. (1972). Postabsorptively induced suppression of appetite and the energostatic control of feeding. *Physiol. Behav.* **9**:199–202.

Booth, D. A. (1978). Acquired behaviour controlling energy intake and output. *Psychiatr. Clin. North Am.* **1**:545–579.

Burley, V. J., Leeds, A. R., and Blundell, J. E. (1987). The effect of high and low fibre breakfasts on hunger, satiety, and food intake in a subsequent meal. *Int. J. Obesity,* **11**(Suppl. 1):87–93.

Cabanac, M. (1971). The physiological role of pleasure. *Science,* **173**:1103–1107.

Cabanac, M., and Duclaux, R. (1970a). Obesity: Absence of satiety aversion to sucrose. *Science,* **168**:496–497.

Cabanac, M., and Duclaux, R. (1970b). Specificity of internal signals in producing satiety for taste stimuli. *Nature,* **227**:966.

Chabert, M., Verger, P., and Louis-Sylvestre, J. (1988). Pre-prandial glycemic phenomenon is precipitated by more palatable food ingestion. *Proceedings of the First European Congress on Obesity,* Stockholm, June 5–6, p. 16.

Cornell, C. E., Roding, J. E., and Weingarten, H. (1989). Stimulus-induced eating when satiated. *Physiol. Behav.* **45**:695–704.

De Castro, J. M., and Elmore, D. K. (1988). Subjective hunger relationships with meal patterns in the spontaneous feeding behaviour of humans: Evidence for a causal connection. *Physiol. Behav.* **43**:159–165.

Elizalde, G., and Sclafani, A. (1988). Starch-based conditioned flavour preference in rats: Influence of taste, calories and CS-US delay. *Appetite,* **11**:179–200.

Epstein, S., and Fenz, W. D. (1965). Steepness of approach and avoidance gradients in humans as a function of experience: Theory and experiment. *J. Exp. Psychol.* **70**:1–12.

Fantino, M., Hosotte, J., and Apfelbaum, M. (1986). An opioid antagonist, naltrexone, reduces preference for sucrose in humans. *Am. J. Physiol.* **251**:R91–R96.

Friedman, M. I., Tordoff, M. G., and Ramirez, I. (1986). Integrated metabolic control of food intake. *Brain Res. Bull.* **17**:855–859.

Garcia, J., Hankins, W. G., and Rusiniak, K. W. (1974). Behavioural regulation of the *milieu interne* in man and rat. *Science,* **185**:824–831.

Goodall, E., Trenchard, E., and Silverstone, T. (1987). Receptor blocking drugs and amphetamine anorexia in human subjects. *Psychopharmacology,* **92**:484–490.

Grill, H. J., and Berridge, K. C. (1985). Taste reactivity as a measure of the neural control of palatability. *Prog. Psychobiol. Physiol. Psychol.* **2**:1–61.

Harris, A., and Wardle, J. (1987). The feeling of hunger. *Br. J. Clin. Psychol.* **26**:153–154.

Hetherington, M., Rolls, B. J., and Burley, V. J. (1989). The time-course of sensory-specific society. *Appetite,* **12**:57–68.

Hill, A. J., and Blundell, J. E. (1986a). Macro-nutrients and satiety: The effects of a high protein or a high carbohydrate meal on subjective motivation to eat and food preferences. *Nutr. Behav.* **3**:133–144.

Hill, A. J., and Blundell, J. E. (1986b). Model system for investigating the actions of anorectic drugs: Effect of D-fenfluramine on food intake, nutrient selection, food preferences, meal patterns, hunger and satiety in healthy human subjects. In *Advances in Biosciences.* Pergamon Press, Oxford, pp. 377–389.

Hill, A. J., and Blundell, J. E. (1990a). Comparison of the action of macronutrients on the expression of appetite in lean and obese subjects. In *The Psychobiology of Human Eating Disorders,* New York Academy of Sciences, pp. 529–531.

Hill, A. J., and Blundell, J. E. (1990b). Sensitivity of the appetite control system in obese subjects to nutritional and serotoninergic challenge. *Int. J. Obesity,* **14**:219–233.

Hill, A. J., Magson, L. D., and Blundell, J. E. (1984). Hunger and palatability: Tracking ratings of subjective experience before, during and after the consumption of preferred and less preferred food. *Appetite,* **5**:361–371.

Hill, A. J., Leathwood, P. J., and Blundell, J. E. (1987). Some evidence for short-term calorie compensation in normal weight human subjects: The effects of high and low energy meals on hunger, food preference and food intake. *Hum. Nutr: Appl. Nutr.* **41A**:244–257.

Kissileff, H. R. (1984). Satiating efficiency and a strategy for conducting food loading experiments. *Neurosci. Biobehav. Rev.* **8**:129–135.

Kissileff, H. R., Gruss, L. P., Thornton, J., and Jordan, H. A. (1984). The satiating efficiency of foods. *Physiol. Behav.* **32**:319–332.

Le Magnen, J. (1956). Hyperphagia provoquée chez le rat blanc par l'alteration du mécanisme de satiété périphérique, *C.R. Soc. Biol.* **150**:32–34.

Le Magnen, J. (1960). Effets d'une pluralité de stimuli alimentaires sur le determinisme quantitatif de l'ingestion chez le rat blanc. *Arch. Sci. Physiol.* **14**:411–419.

Le Magnen, J. (1985). *Hunger.* Cambridge, Cambridge University Press.

Macquorquodale, K., and Meehl, P. E. (1948). On the distinction between hypothetical constructs and intervening variables. *Psychol. Rev.* **55**:95–109.

Melchior, J.-C., Fantino, M., Rozen, R., Igoin, L., Rigaud, D., and Apfelbaum, M. (1989). Effects of a low dose of naltrexone on glucose-induced alliesthesia and hunger in humans. *Pharmacol. Biochem. Behav.* **32**:117–121.

Monello, L. F., and Mayer, J. (1967). Hunger and satiety sensations in men, women, boys, and girls. *Am. J. Clin. Nutr.* **20**:253–261.

Mook, D. G. (1987). *Motivation: The Organization of Action.* New York, Norton.

Morrison, G. R. (1974). Alternations in palatability of nutrients for the rat as a result of prior testing. *J. Comp. Physiol. Psychol.* **86**:56–61.

Owen, W. P., Halmi, K. A., Gibbs, J., and Smith, G. P. (1985). Satiety responses in eating disorders. *J. Psychiatr. Res.* **19**:279–284.

Rogers, P. J., and Blundell, J. E. (1979). Effect of anorexic drugs on food intake, and the microstructure of eating in human subjects. *Psychopharmacology,* **66**:159–165.

Rogers, P. J., and Blundell, J. E. (1984). Food patterns and food selection during the development of obesity in rats fed a cafeteria diet. *Neurosci. Biobehav. Rev.* **8**:441–453.

Rogers, P. J., and Blundell, J. E. (1989a). Evaluation of the influence of intense sweeteners on the short-term control of appetite and caloric intake: A psychobiological approach. In *Progress in Sweeteners,* T. H. Grenby (Ed.). Elsevier Applied Science, New York, pp. 267–289.

Rogers, P. J., and Blundell, J. E. (1989b). Separating the actions of sweetness and calories: Effects of saccharin and carbohydrates on hunger and food intake in human subjects. *Physiol. Behav.* **45**:1093–1099.

Rogers, P. J., Carlyle, J., Hill, A. J., and Blundell, J. E. (1988). Uncoupling sweet taste and calories: Comparison of the effects of glucose and three high intensity sweeteners on hunger and food intake. *Physiol. Behav.* **43**:547–552.

Rolls, B. J. (1981). Palatability and food preference. In *The Body Weight Regulatory System: Normal and Disturbed Mechanisms,* L. A. Cioffi, W. P. T. James, and T. B. Van Itallie (Eds.). New York, Raven Press, pp. 271–278.

Rolls, B. J., Rolls, E. T., Rowe, E. A., and Sweeney, K. (1981). Sensory-specific satiety in man. *Physiol. Behav.* **27**:137–142.

Rolls, B. J., Van Duijenvoorde, P. M., and Rolls, E. T. (1984). Pleasantness changes and food intake in a varied four-course meal. *Appetite,* **5**:337–348.

Rolls, B. J., Hetherington, M., and Burley, V. J. (1988). Sensory stimulation and energy density in the development of satiety. *Physiol. Behav.* **44**:727–733.

Sclafani, A., and Nissenbaum, J. W. (1988). Robust conditioned flavor preference produced by intragastric starch infusions in rats. *Am. J. Physiol.* **225**:R672–R675.

Steiner, J. E. (1987). What the neonate can tell us about umami. In *Umami, A Basic Taste,* Y. Kawamura and M. R. Kare (Eds.). New York, Dekker, pp. 97–123.

Treit, D., Spetch, M. L., and Deutsch, J. A. (1983). Variety in the flavor of food enhances eating in the rat: A controlled demonstration. *Physiol. Behav.* **30**:207, 211.

Trenchard, E., and Silverstone, J. T. (1983). Naloxone reduces the food intake of human volunteers. *Appetite,* **4**:43–50.

Van Itallie, T. B., and Vanderweele, D. A. (1981). The phenomenon of satiety. In *Recent Advances in Obesity Research,* Vol. III, P. Bjorntorp, M. Cairella, and A. N. Howard (Eds.). Libbey, London, pp. 278–289.

Wadden, T. A., Stunkard, A. J., Brownell, K. D., and Day, S. C. (1985). A comparison of two very-low calorie diets: Protein-sparing-modified fast versus protein-formula-liquid diet. *Am. J. Clin. Nutr.* **41**:533–539.

Weipkema, P. R. (1971). Positive feedbacks at work during feeding. *Behaviour,* **39**:266–273.

Wooley, S. C. (1976). Psychological aspects of feeding. In *Appetite and Food Intake,* T. Silverstone (Ed.). Abakon, Berlin, pp. 331–354.

8
Open-Loop Methods for Studying the Ponderostat

Michel Cabanac

*Laval University
Quebec City, Quebec, Canada*

I. INTRODUCTION
A. Steady State Versus Regulation

A system in *steady state* such as the one shown in Figure 1a receives from the environment a constant flow of energy and/or matter and returns to the environment an equal flow of energy or matter (Burton, 1939). In this process, energy balance in the long term is equal to zero, but entropy increases in the environment because energy through the system evolves from a state of higher to a state of lower free energy. This is used by the system to maintain the stability of such tensive variables as temperature, pressure, and ionic concentrations, or the water level of Figure 1 (Cabanac and Russek, 1982).

A steady state is able to regulate its level against transient perturbations. If, for example, a transient increase in inflow raises the level of the water in the tank shown in Figure 1, the elevated pressure will transiently increase the outflow, and the level will return to its original value exponentially.

A steady state can also resist a continuous perturbation. A leak in the wall of the tank of Figure 1 will lower the water level. The pressure within the tank will also drop, and in turn so will the outflow through the output faucet. The level will stabilize to a new steady state when inflow has become equal to outflow plus leak. Therefore a steady state not only regulates against transient perturbations but also resists continuous perturbations.

A steady state is not a regulated system, however. Stability of a settling point is by no means a criterion sufficient to define a regulated system. A *regulated system*

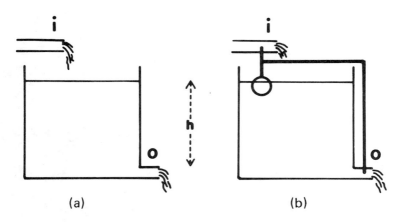

FIGURE 1 Steady state versus regulation. (a) The container receives and loses equal flows (i and o) of water and remains in steady state with water level (h) constant. (b) The container is also a steady state but, in addition, a sensor (float) measures h, a negative feedback loop controls i, and a positive feed-forward loop controls o; the water level is regulated. (From Cabanac and Russek, 1982.)

consists of a steady state equipped with a sensor of the regulated variable, a feedback loop controlling the inflow negatively, and/or a feed-forward loop controlling the outflow positively (Fig. 2). Therefore, a regulated system possesses all the properties of a steady state and adds to them new properties due to the existence of the regulatory loops. These new properties are mainly an increase in response of inflow and/or outflow to perturbations, resulting in a quicker return to the regulated value after a transient perturbation of the system and lesser deviation from the regulated value during a continuous perturbation. The value of the regulated variable is therefore more stable and constant than is the settling point of a steady state. The *set point* is the value at which the system tends to maintain the regulated variable. In Figure 1, the set point is defined by the length of the shaft between the float and the input and output faucets. Finally the words "regulation" and "control" are not interchangeable (see Brobeck, 1965). In the tank of Figure 1, the water level is regulated by controlling the input and output faucets.

It may not be easy to differentiate a regulated system from a steady state. One way is to measure how the system responds to a global perturbation. The existence of a response opposed to the perturbation is the property of a regulated system. Thus, the time course of the regulated variable, and that of the controlled regulatory response, evolve in opposite directions, and their movements, plotted against time, provide mirror images. In certain cases, however, perturbation takes place in the input or output faucet, as, for example, when there is an increased flow in

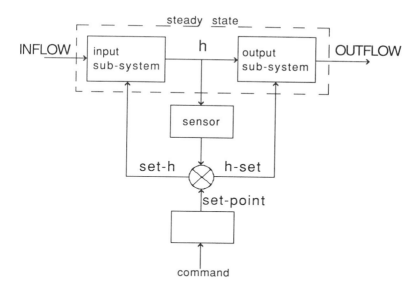

FIGURE 2 Block diagram of Figure 1b and of regulated systems. The set point exists in Figure 1a as the built-in length of shafts between float, and inflow and outflow faucets. (From Cabanac and Russek, 1982.)

the input faucet. In such cases, it is impossible to know the prime mover of synchronous evolutions of the water level and of inflow/outflow of water. The open-loop model permits us to answer that question.

B. The Open-Loop Model

In the open-loop model, the perturbation actuates the negative feedback loop, but the corrective response is prevented from acting on the regulated variable. This situation was studied from the theoretical point of view by Von Euler (1964), who measured the rectal temperature reached by rabbits when their hypothalamic temperature was clamped at a constant value with local thermodes. Thus, the panting response produced by the warming of the hypothalamus could not lower the hypothalamic temperature but produced an open-loop hypothermia with a gain of 8 : 1 (1.6°C hypothermia for 0.2°C hypothalamic heating).

The open-loop model was used by Claude Bernard (1855) to study drinking behavior in the horse and dog. He opened a fistula in the esophagus; thus the ingested water could not reach the stomach and the animals drank endlessly "until they were stopped by fatigue." This elegant technique is being used in rats to study the determinism of hunger and the laws of food intake (Mook, 1989; Sclafani, 1989), but has not been applied yet to study the regulation of body weight.

C. The Ponderostat

The foregoing analysis applied to food intake and body weight reveals that the former is not regulated but rather is controlled to regulate the latter, if body weight is the object of regulation rather than the result of a mere steady state. It has been hypothesized that body mass, or a variable closely correlated to body mass, such as body fat content, is constant over the adult life span because it is regulated (Hervey, 1969). This hypothesis has been the subject of a debate.

A very strong argument in favor of regulated body weight is the fact that body weight changes, resulting from hypothalamic lesions (Keesey and Powley, 1975) or from a circannual cycle (Mrosovsky and Fisher, 1970; Mrosovsky and Powley, 1977), are defended by the animal's food intake. The argument in favor of regulation is not drawn from the measurement of body weight alone but from the fact that food intake in the circannual cycle or after lateral hypothalamus lesions were adjusted to defend body weight. Periods of limited access to food were followed by compensatory overfeeding when body weight was below set point but not when body weight was above set point. Such a behavior is pathognomonic of regulation. An opposite interpretation however was proposed (Wirtshafter and Davis 1977; Davis and Wirtshafter, 1978) shortly after these data were published. As an alternative, Wirtshafter and Davis (1977) hypothesized that body weight remains constant at, or near, a settling point, the adjustment taking place without a set point.

Is it useful, 10 years later, to reopen the controversy? Keesey and Powley (1986) admitted that the application of the concepts of physiological regulation to body weight and body energy has not won wide acceptance. Examples from the recent literature confirm this statement. Le Magnen (1984) considered the stability of body weight to result from the equilibrium between filling and emptying the adipose reservoir. That is the description of a steady state as defined above. The same model is implied when the constancy of body weight is described as resulting only from energy balance in obesity (Himms-Hagen, 1984) and in anorexia nervosa (Bernstein and Borson, 1986). Energy imbalance, of course, must occur if any change in body weight is to be observed, but the concept of regulation implies that energy balance is not the final cause but rather only a means to achieve the goal of body weight equal to its set value (see Cabanac, 1975).

This chapter will examine the studies in which a regulatory response has been measured in such a way as to open the circularity of denouncing energy balance as a cause of body weight stability. Since regulation is a steady state with stability improved by a negative feedback loop and a positive feed-forward loop, the main difference between regulation and steady state is the existence of a regulatory response to a perturbation. A review of the literature shows that there exist three types of regulatory response on the inflow of energy that do not influence food intake and thus can be measured in subjects at various body weights; they are gustatory alliesthesia and saliva secretion in humans, and hoarding behavior in rats. One method that results in increased food intake and body weight is the addition of artificial sweeteners to food. These studies also are reviewed below because they exemplify a typical open-loop situation.

Finally one regulatory response on the outflow of energy has also been measured in rats at various body weights.

II. ALLIESTHESIA

The pleasure aroused by alimentary stimuli in fasted subjects turns into displeasure during satiation (Cabanac, 1971; Esses and Herman, 1984; Fantino, 1984). Thus, postingestive signals are taken into account by the brain to adjust the pleasure to the usefulness of the alimentary stimulus.

The same relationship of pleasure with an internal signal was found in the case of thermal sensation. This point is emphasized here for its implications. In the case of temperatures, the set point for body temperature regulation is proved to move up and down. During fever, the set point for temperature regulation is up and pleasure defends this elevated set point. During the nycthemeral and the ovarian cycles, alliesthesia follows the oscillating set point. This identifies the internal signal responsible for alliesthesia, not as deep body temperature only, but rather as the difference between actual and set point temperatures.

The same rationale pleads in favor of a set point in the case of body weight regulation. Negative alliesthesia with alimentary stimuli contributes to limit quantitatively the amount of ingested food. When subjects are deprived of food for several weeks and lose several kilograms of body weight, negative alliesthesia after a gastric load tends to be diminished or to disappear (Cabanac et al., 1971; Herman et al., 1987). Thus, the size of any meal and total food intake will tend to increase and body weight will tend to return to predieting value. Such an evolution is identical to that found with temperature sensation during an episode of hypothermia. In both cases pleasure defends not any actual deep body temperature or body weight but set points for these variables.

The concept of a ponderostat finds an application in the case of obesity. Alliesthesia disappears during the period of onset of obesity or "dynamic obesity," when presumably the patient's body weight is lower than set point. Alliesthesia in patients is identical to that of healthy controls during "static" obesity, when presumably actual body weight is equal to the elevated set point (Guy-Grand and Sitt, 1974, 1975). This typical pattern of response can be found also with more or less intensity in obese patients under restricted diet, probably in relation to the intensity of the restriction (Underwood et al., 1973; Rodin et al., 1976; Gilbert and Hagen, 1980), as well as in rats (Mook and Cseh, 1981). The discrepancies from this pattern reported in the literature may be attributed to a cultural bias against sugar and sweet sensations (Meiselman, 1977; Enns et al., 1979; Frijters and Rasmussen-Conrad, 1982; Drewnowski, 1985; Tuorila-Ollikainen and Mahlamaki-Kultanen, 1985; Drewnowski et al., 1988).

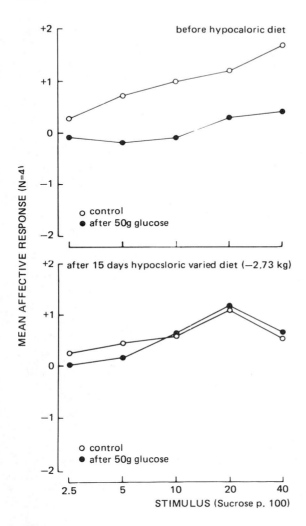

FIGURE 3 Weight loss, palatability (weighted as follows: +2, very pleasant; +1, pleasant; 0, indifferent; −1, unpleasant; −2, very unpleasant), and alliesthesia. *Top left:* Mean affective responses to increasingly sweet solutions given before and 60 minutes after receiving 50 g glucose intragastrically. *Bottom left:* Same experiment but the subjects had lost a mean 2.7 kg by voluntarily reducing their food intake for 15 days, with a diet otherwise normal. *Right:* Same experiment on four different subjects. The body weight loss (lower plot) was obtained spontaneously by feeding the subjects exclusively with bland food ad libitum. (From Cabanac and Rabe, 1976.)

THE PONDEROSTAT: OPEN-LOOP STUDY METHODS 155

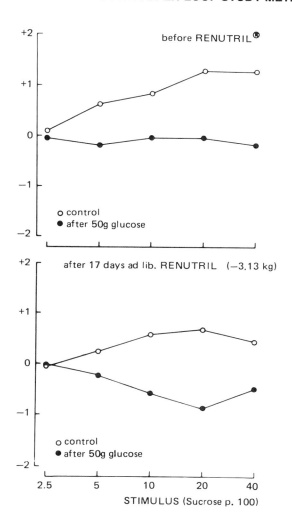

Finally, the analogy with temperature regulation can be found in the influence of peripheral sensory input. Temperature regulation can be described as a proportional regulation with an adjustable set point; one of the main factors in the adjustment of the set point is the peripheral sensory input (Hammel, 1968). In the case of body weight, the ponderostat seems to respond in the same way as the thermostat. When subjects were chronically fed with unpalatable food, their body weight tended to decrease (Cabanac and Rabe, 1976). Their capacity to respond to a gastric load of glucose with negative alliesthesia remained intact (Fig. 3). Such a result can be expressed theoretically in the same way as any other regulatory process by plotting alliesthesia on the ordinate and body weight as the independent variable (Fig. 4). According to this hypothesis, the set point for body weight regulation can be moved upward by chronic feeding with sensory-rich and palatable foods, and downward by chronic feeding with sensory-poor and bland foods.

III. SALIVA SECRETION

Salivation was first measured as an involuntary index of appetite (Wooley and Wooley, 1973) and then was related for the first time to body weight by Wooley et al. (1975). One hour after a high-calorie preload, the amount of saliva secreted on the sight of palatable food was measured in obese and in control subjects. Salivary response was less in the nonobese than in the obese, which suggests that the obese were below set point.

In the following years, the amount of saliva secreted on the sight or thought of food was used merely as an index of appetite (Booth and Fuller, 1981; Wooley

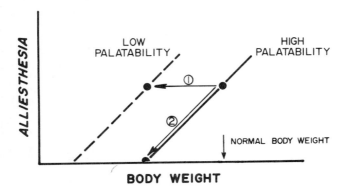

FIGURE 4 Schematic representation of the ponderostat hypothesis with alliesthesia proportional responses: the amount of satiety explored through alliesthesia after 50 g of intragastric glucose is a function of body weight. This function is adjusted by the sensory input of daily food. The subjects of Figure 3 followed arrows 1 (Fig. 3, right), and 2 (Fig. 3, left). (From Cabanac and Rabe, 1976.)

and Wooley, 1981; Booth et al., 1982). Correlation of salivation with the body needs can be found, however, in the fact that the flow of saliva secreted tended to be higher in food-deprived subjects than in nondeprived subjects (Franchina and Slank, 1988). It was shown that salivation can differ from taste judgments (Hyde and Pangborn, 1987) and from ratings of hunger symptoms, preferred foods, and satiety (Wardle, 1987).

It was only recently that the open-loop nature of the salivary response to food was used to study the ponderostat. The explicit relation with weight regulation can be found in the measurements made on normal college students (Legoff and Spigelman, 1987) and on patients with eating disorders (Legoff et al., 1988). The experiment featured the collection of saliva secreted in response to food odors in nonsatiated subjects. College students were classified into four categories: normal-weight nondieters, overweight nondieters, normal-weight dieters, and overweight dieters. The results on these subjects (Fig. 5) show that salivation was stimulated in dieters (both normal-weight or overweight) and depressed in nondieters (both normal-weight and overweight). This result supports the existence of the ponderostat. As seen from that point of view, the enhanced salivary response shows that dieters were below their body weight set point and the inhibited salivary response shows that nondieters were presumably at set point.

Support of the set point theory was also brought by the study of patients with eating disorders (Legoff et al., 1988). In response to food odors, bulimic patients salivated more and anorectics less than controls (Fig. 6). Such a result suggests that bulimic patients are below their body weight set point, and anorectics at or

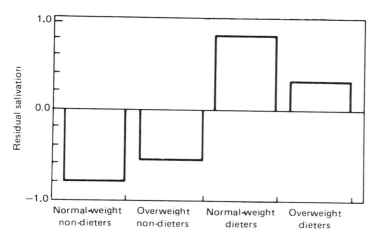

FIGURE 5 Saliva secretion, dieting, and body weight set point. Mean residual salivation values for overweight and normal-weight dieters (restrained) and nondieters (unrestrained). (From Legoff and Spigelman, 1987.)

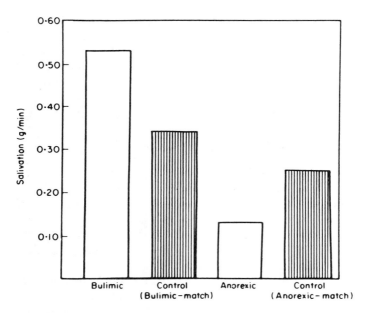

FIGURE 6 Saliva secretion and body weight set point. Mean salivary response to food odors above baseline for bulimics, anorectics, and matched controls at first testing. (From Legoff et al., 1988.)

above set point. Such an interpretation would explain the binge sessions of bulimics by their strong motivation to eat, being below set point, and the robustness of the low body weight of anorectic patients by their being at set point.

IV. ARTIFICIAL SWEETENERS

Saccharin and other artificial sweeteners uncouple sweetness and calories. This open-loop property has been studied in both humans and rats. In humans, it has been shown that preloads sweetened with saccharin, contrary to a sweet glucose load, reduced neither hunger (Rogers et al., 1988) nor the following food intake (Blundell et al., 1988). As a result of this absence of postingestive satiety, chronic users of artificial sweeteners seem to be more prone to increase their food intake: the rate of their weight gain over a one-year period was significantly greater than that of nonusers of artificial sweeteners, irrespective of initial relative weight (Stellman and Garfinkel, 1986).

In rats, the influence of saccharin on food and fluid intake was studied experimentally and theoretically (Mook, 1974). More recently, however, Mook and Cseh (1981) studied the influence of body weight manipulation on the appeal of sweet solutions on rats. They showed that a decrease of body weight, obtained by diet restriction, was followed by an increase of both glucose and saccharin intake. Reciprocally, body weight elevation, obtained with repeated gavage, was followed by a decrease of glucose and saccharin intake.

The experiment above clearly represents an open look in the case of saccharin, since the absence of calories in saccharin prevents any feedback on body weight. As they stand, these results confirm the ponderostat hypothesis inasmuch as they can be almost superimposed on those of alliesthesia experiments in humans.

V. HOARDING OF FOOD BY RATS

The evidence drawn from experiments on rats has the advantage of eliminating any hint of bias in the subjects. Laboratory rats do not hoard pellets when food is continuously available (Wolf, 1939; Bindra, 1948). However, when they are placed on schedules in which food is restricted, they start to hoard (Morgan et al., 1943; Blundell and Herberg, 1973). Herberg and associates (1975) have demonstrated that this behavior is related not to deprivation but rather to body weight decrease: rats hoard food when their body weight is lower than set point. This behavior can be considered to be a regulatory response with a magnitude proportional to the decrease of body weight (Fantino and Cabanac, 1980). The method is open-loop provided the food hoarded is withdrawn from the rat at the end of the session. The intersect with the x-axis of the regression line of mass hoarded versus body weight provides both the rat's body weight when fed ad libitum and the set point (Fig. 7).

The method has been applied to explore the influence of ambient temperature (Fantino and Cabanac, 1984) and cost to reach food (Cabanac and Swiergiel, 1989). None of these ambient perturbations influenced the set point. On the other hand, the method was applied to female rats and was able to detect a change of about 30 g in the set point from estrus to diestrus in rats weighing about 350 g (Fig. 8). Dexfenfluramine, another internal factor, also was shown to be able to reduce the set point for body weight regulation (Fantino et al., 1987).

VI. THE POSITIVE FEED FORWARD: OUTFLOW OF ENERGY

Global energy metabolism is adjusted to the inflow of food. This relation had been demonstrated in the case of starvation, where the energy expenditure decreases with the duration of starvation (Kleiber, 1961). The modulation of basal metabolic rate occurs when food intake is chronically reduced or increased (Apfelbaum, 1974). On first examination, this could be simply a typical steady state as defined above: inflow and outflow of energy would be adjusted to be equal, and thus would result in high body weight when food intake is high and low body weight when food intake is low. The introduction of the notion of a ponderostat seems, here, to be a useless complexity. However, it is possible to manipulate body weight independently from energy metabolism (Keesey, 1988). In a series

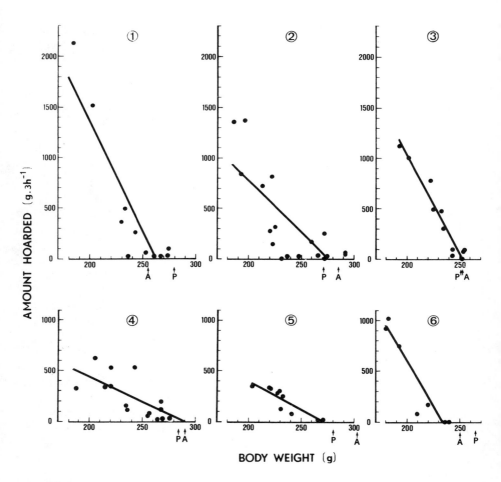

FIGURE 7 Experimental representation of the ponderostat hypothesis with food hoarding proportional response: amount of food pellets hoarded by six rats in repeated 3-hour sessions plotted against the rats' body weights. The experimental set points (intersect of regression lines with x-axis) can be compared to rats' body weight before (A = ante) and after (P = post) the period of body weight reduction and subsequent recovery. (From Fantino and Cabanac, 1980.)

of elegant experiments, Keesey et al. have demonstrated that energy expenditure does not follow energy intake passively, as would be the case in a steady state. On the contrary, energy expenditure participated in the defense of the set point for body weight regulation when body weight was manipulated.

It was proposed by Powley and Keesey (1970) that the decrease of body weight occurring after lesions of the lateral hypothalamus is not simply due to aphagia,

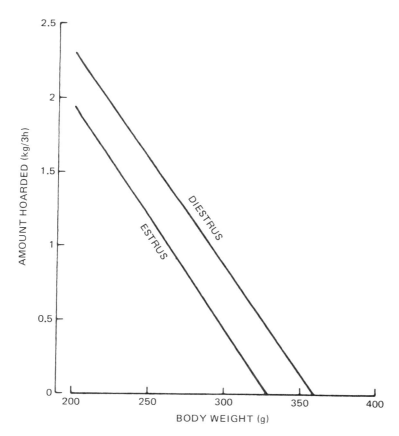

FIGURE 8 Adjustable set point of body weight. Mean regression of mass of food hoarded in 3-hour sessions by eight female rats, against body weight. "Estrus" and "diestrus" refer to the endocrine state of the animals, which continued to ovulate during the food deprivation period. (From Fantino and Brinnel, 1986.)

but rather to a resetting of the ponderostat at a lower value of set point. The measurement of the resting metabolic rate of operated rats supported the hypothesis (Keesey et al., 1984; Corbett et al., 1985). They spent energy at a rate appropriate to the reduced metabolic mass they chronically maintained, and they adjusted expenditure in ways that resisted further reduction of body weight. When their weight was raised (Fig. 9) to the level of nonlesioned rats, they became hypermetabolic.

Another way to manipulate the body weight of rats consists of feeding the animals with food of enriched palatability (Sclafani and Springer, 1976). The obesity produced tends to persist even after the animals have been returned to chow (Rolls et al., 1980). Here again the measurement of energy expenditure of

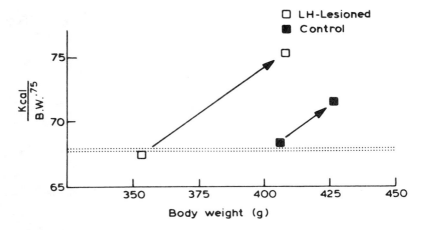

FIGURE 9 Thermogenesis and the ponderostat: relation of resting metabolic rate, in kilocalories per three-quarters body weight per day (kcal/BW$^{0.75}$ day), to body weight (g) in lateral hypothalamus (LH)-lesioned and nonlesioned rats. Nutrament was used to stimulate the intake of half the lesioned and half the nonlesioned rats, and to produce the weight gains indicated by the arrows. The associated increase in resting metabolism was significantly greater than expected (dotted line) for this increase in body weight. (From Keesey, 1988.)

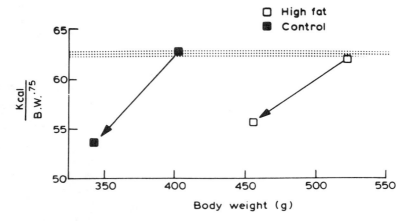

FIGURE 10 Palatability and adjustable set point of body weight: relation of resting metabolic rate (kcal/BW$^{0.75}$ day) and body weight (g) in rats fed either a standard or high fat diet for 6 months. The food intake of half the rats fed the standard diet and half the rats fed the high-fat diet was restricted to produce the weight losses indicated by the arrows. The associated decline in resting metabolism was significantly greater than expected (dotted line) for this decrease in body weight. (From Keesey, 1988.)

rats rendered and maintained obese with a diet of enriched palatability showed that they maintained a lower metabolism than would have been expected from their mass (Corbett et al., 1986). In addition, when these rats had limited access to food, their weight dropped, but their metabolic rate also diminished in a way suggesting the defense of a higher set point than controls (Fig. 10).

VII. CONCLUSIONS

The evidence provided by the present review is conclusive regarding the ponderostat and indicative regarding the pathology of body weight.

Open-loop methods allow the manipulation of body weight without prejudice to the measured response. This was the case with gustatory alliesthesia, salivary secretion, and hoarding behavior, as well as the measurement of resting metabolic rate. The results of all experiments converge to support the ponderostatic hypothesis. There is no indication yet on the nature of the actual variable linked to body weight, or on the mechanism of regulation, but there remains little doubt as to the existence of this regulation. This, of course, implies the existence of a set point.

An important consequence of this general conclusion concerns the pathology of body weight. In a regulated system, when a perturbation modifies inflow, the system regulates by modifying outflow accordingly. Reciprocally, when pathology perturbs outflow, the system regulates by the control of inflow. As a result, it is difficult to modify a regulated variable. It takes overwhelming steady perturbations to change a regulated variable.

On the other hand, a perturbation in the set point reflects on the system with little modifications on inflow and outflow, once the actual variable is equal to its set value.

These considerations are enlightening in the case of the ponderostat. The available evidence shows that during health, both inflow and outflow of energy are controlled. Any long-term change in body weight is therefore likely to be due not simply to excessive food intake or reduced metabolic rate but rather to a modified set point. Once body weight has reached its new set point, the subject maintains its energy balance without much trouble.

The evidence reviewed here indicates that obesity is a case of high set point for body weight regulation. Anorexia is likely to be a resetting at a low value, whereas bulimia might be a permanent fight of the subject to reduce body weight while set point remains normal.

DISCUSSION

Stricker: Because this isn't your work, I hope you'll feel free to comment on it.

Cabanac: Of course!

Stricker: Yes. With regard to the bulimics and the anorexics, because I think those are provocative conclusions that you reached, you would agree of course that the autonomic nervous system can be affected by the psychology of the subject. Salivary flow, being influenced by the autonomic nervous system, might well reflect their thoughts about their weight rather than a ponderostat. So why are you so energetic in proposing this?

Cabanac: Very good question. I can answer. You have the same situation in temperature regulation. Emotional fevers are fevers. Before examinations, and in lectures as well, it has been repeatedly demonstrated that before the speech, body temperature goes up. It is a resetting that you can prevent by Aspirin. And after the speech, you are flushed; you return to your original set point. If psychological influence can act on temperature regulation, why should it be impossible on body weight regulation? It's not an argument against the regulation, it just shows that the human is a complex system. And the animal as well is complex.

Stricker: There is an alternative explanation for those findings.

Cabanac: OK, go ahead, shoot.

Stricker: In using your initial semantic distinctions between a steady state and the regulation, I think it's easy to think of body weight as a steady state rather than a regulated state, the rest of your comments not withstanding. That is, if you look at the biology of adipose tissue regulation or adipose tissue maintenance—I have to be careful about what I say—there are several variables that influence how many calories will be stored. Adipose tissue doesn't have an infinite capacity to store triglycerides, of course; the more it stores, the less it can continue to store.

Cabanac: Some people weigh 200 or 300 kilograms. That's an infinite storage.

Stricker: Yes, there are individuals like that, but there are also individuals who become diabetic when they become really fat, presumably reflecting the fact that there are, at least in them, limits. It's certainly true that the more triglycerides that are stored in adipose tissue, the harder it is to get further triglycerides and that slower storage takes place. I have always thought of that as an interplay among various variables—how much is stored, how many insulin receptors you have, what is the state of sympathetic tone, how much substrate is outside the tissue. This is an interplay that can be changed following a lesion, following changes in pancreatic secretion of insulin. But I would say that's no different from what you have said in the beginning with regard to the flow of water into your vessel and the flow of water out. You're adjusting the magnitude—the intensity—of these variables and therefore changing the height. I wouldn't have called that regulation; I would have called that a steady-state condition in which you're changing

THE PONDEROSTAT: OPEN-LOOP STUDY METHODS 165

the variables. Now I don't want to say that the points I'm making prove something. I'm just saying that there is a point of view different from the point of view that you've raised.

Cabanac: May I add a comment that has nothing to do with what you say? I wish just to pinpoint that John Brobeck had made a wonderful definition of regulation versus control. Regulation is the process that keeps constant the level within the container. Control is the action on the inflow and the outflow. Therefore, inflow and outflow responses are controlled to regulate the water level.

Blass: I was thinking of a Brobeck article as you were presenting your work. I was wondering, especially in your example of hoarding behavior, how one would conceptualize that as being a part of a negative feedback system in the sense that this activity of hoarding, under the two conditions that you demonstrated hoarding, doesn't affect the animal in any meaningful way; it's simply an activity that's engaged in by an animal in very restricted circumstances. Given that, and the facility with which this is incorporated into negative feedback systems, I don't quite understand why one would utilize that as an example of negative feedback. To the extent that one could not utilize that as an example of negative feedback, I would seek additional interpretations. Would that hold with some other examples?

Cabanac: The definition of negative feedback is algebraic. It's a response that goes in the opposite direction to the change of the regulated variable—opposite sign. Therefore, if you have an elevated body weight you should have no hoarding. If you have a decreased body weight you should have increased hoarding, which is what you get.

Blass: But the effector system, as I understood it, should in turn influence the variable that is giving rise to the other signal. That's not what happened in the example you presented.

Cabanac: I don't understand the question then.

Blass: I just wanted to share my understanding of the behavior which you are considering to be controlled. Being incorporated into this regulated set point system, it was my understanding that the behavior itself is going to affect the error signal that is giving rise to that behavior. We saw examples of that this morning in drinking, salt appetite, feeding, and so on. One does not see that easily in the case of hoarding behavior.

Cabanac: In other words, what you're saying is that to hoard food does not modify the total energy balance of the animal. This is precisely why this is an open-loop situation, because the hoarding does not immediately modify body weight. But if you leave the food there, the next day it's gone. Therefore, it's not instantaneous, it's a delayed effect.

Blass: So you incorporate that into the closed-loop system.

Cabanac: Of course, yes.

VanItallie: I spend a lot of time trying to reduce the weights of people who are overweight for various reasons. One of the ways that can be done is by increasing their level of physical activity. And if they do increase the level of physical activity, they will lose weight. If they maintain that level, they continue to maintain the weight lost. Another way might be to feed them a low fat, high complex-carbohydrate diet. Sometimes that results in weight loss. If these people are at a set point because they're actively regulated that way, then I would expect that there's some force in the organism that's trying to combat these changes. I find some difficulty in finding those in the organism, and, second, it seems to me more parsimonious just to say that you change the equilibrium conditions and the level has gone down.

Another problem that I have with the set point concept, although it has appealing characteristics, is, for example, that women in the United States below the poverty line have almost double the prevalence of obesity than women above the poverty line. Now does being below the poverty line change your set point? Certainly, there are probably genetic reasons why some people are fatter than others, and that suggests a set point because there's a genetic role. But it seems to me there's an overlay—that environmental conditions, which affect physical activity level or the nature of the diet, manipulate these weights. I would hate to think that every time I'm doing something to change the environment, that I'm changing the set point. It seems to me that is not a parsimonious explanation for what I'm doing.

Cabanac: There are three questions. First, the Manhattan study, second the exercise, and then the philosophy of regulation and set point. The Manhattan study: well, there is no evidence that the people on the upper layers of the society are above or below set point. This has not been measured, therefore it's impossible to answer. They may have the same set points as the others but resist because there is social pressure, and therefore they are underweight compared to their set point. Nobody knows.

Second, exercise: it's not known whether exercise—chronic exercise, training—changes the set point or not. It's known that there is a decrease in body weight. This has been shown in athletes, people who train, in ballerinas, but it's not known whether it's a change in set point or, again, a passive system. My guess is that there is a change in set point because regulatory processes are integrated responses. You cannot artificially change the body weight of a person if the set point is not changed. Regulation is there. If you manipulate the inflow, the outflow will correct the variable. This is why Mark Friedman said that obesity may be due to increasing palatability and a passive increase in body weight. I don't believe that, because regulatory processes are extremely powerful. In fact, it has been discovered that along with the effect of the cafeteria diet on intake there is also an effect on the outflow of energy. It's totally integrated.

REFERENCES

Apfelbaum, M. (1974). Influence of level of energy intake on energy expenditure in man: Effects of spontaneous intake, experimental starvation, and experimental overeating: In *Obesity in Perspective*, G. A. Bray (Ed.). Government Printing Office, Washington, DC, pp. 145–155.

Bernard, C. (1855). Leçons de physiologie expérimentale, Vols. I and II. Editions de l'Ecole, Paris.

Bernstein, I. L., and Borson, S. (1986). Learned food aversion: A component of anorexia syndromes. *Psychol. Rev.* **93**:462–472.

Bindra, D. (1948). The nature of motivation for hoarding food. *J. Comp. Physiol. Psychol.* **41**:211–218.

Blundell, J. E., and Herberg, L. J. (1973). Effectiveness of lateral hypothalamic stimulation, arousal, and food deprivation in the initiation of hoarding behaviour in naive rats. *Physiol. Behav.* **10**:763–767.

Booth, D. A., and Fuller, J. (1981). Salivation as a measure of appetite: A sensitivity issue. *Appetite*, **2**:370–372.

Booth, D. A., Mather, R. P., and Fuller, J. (1982). Starch content of ordinary foods associatively conditions human appetite and satiation, indexed by intake and eating pleasantness of starch-paired flavours. *Appetite*, **3**:163–184.

Brobeck, J. R. (1965). Exchange, control and regulation. In *Physiological Controls and Regulations*, W. S. Yamamoto and J. R. Brobeck (Eds.). Saunders, Philadelphia, pp. 1–13.

Burton, A. C. (1939). The properties of the steady state compared to those of equilibrium as shown in characteristic biological behavior. *J. Cell. Comp. Physiol.* **14**:327–349.

Cabanac, M. (1971). Physiological role of pleasure. *Science*, **173**:1103–1107.

Cabanac, M. (1975). Temperature regulation. *Annu. Rev. Physiol.* **37**:415–439.

Cabanac, M., and Rabe, E. F. (1976). Influence of monotonous food on body weight regulation in humans. *Physiol. Behav.* **17**:675–678.

Cabanac, M., and Russek, M. (1982). *Régulation et contrôle en Biologie*. Presses de l'Université Laval, Québec.

Cabanac, M., and Swiergiel, A. (1989). Rats eating and hoarding as a function of body weight and cost of foraging. *Am. J. Physiol.* **257**:R952–R957.

Cabanac, M., Duclaux, R., and Spector, N. H. (1971). Sensory feed-backs in regulation of body weight; is there a ponderostat? *Nature*, **229**:125–127.

Corbett, S. W., Wilterdink, E. J., and Keesey, R. E. (1985). Resting oxygen consumption in over- and underfed rats with lateral hypothalamic lesions. *Physiol. Behav.* **35**:971–977.

Corbett, S. W., Stern, J. S., and Keesey, R. E. (1986). Energy expenditure in rats with diet-induced obesity. *Am. J. Clin. Nutr.* **44**:173–180.

Davis, J. D., and Wirtshafter, D. (1978). Set-point or settling points for body weight? A reply to Mrosovsky and Powley. *Behav. Biol.* **24**:405–411.

Drewnowski, A. (1985). Food perception and preferences of obese adults: A multidimensional approach. *Int. J. Obesity,* **9**:201–212.

Drewnowski, A., Pierce, B., and Halmi, K. A. (1988). Fat aversion in eating disorders. *Appetite,* **10**:119–131.

Enns, M. P., Van Itallie, T. B., and Grinker, J. A. (1979). Contribution of age, sex and degree of fatness on preferences and magnitude estimations for sucrose in humans. *Physiol. Behav.* **22**:999–1003.

Esses, V. M., and Herman, C. P. (1984). Palatability of sucrose before and after glucose ingestion in dieters and nondieters. *Physiol. Behav.* **32**:711–715.

Fantino, M. (1984). Role of sensory input in the control of food intake. *J. Auton. Nerv. Syst.* **10**:347–359.

Fantino, M., and Brinnel, H. (1986). Body weight set-point changes during the ovarian cycle: Experimental study of rats using hoarding behavior. *Physiol. Behav.* **36**:991–996.

Fantino, M., and Cabanac, M. (1980). Body weight regulation with a proportional hoarding response in the rat. *Physiol. Behav.* **24**:939–942.

Fantino, M., and Cabanac, M. (1984). Effect of a cold ambient temperature on the rat's food hoarding behavior. *Physiol. Behav.* **32**:183–190.

Fantino, M., Faion, F., and Rolland, Y. (1987). Effect of dexfenfluramine on body weight set-point: Study in the rat with hoarding behaviour. *Appetite,* **7**(Suppl.):115–126.

Franchina, J. J., and Slank, K. L. (1988). Effects of deprivation on salivary flow in the apparent absence of food stimuli. *Appetite,* **10**:143–147.

Frijters, J. E. R., and Rasmussen-Conrad, E. L. (1982). Sensory discrimination, intensity perception and affective judgement of sucrose sweetness in the overweight. *J. Gen. Psychol.* **107**:233–247.

Gilbert, D. G., and Hagen, R. L. (1980). Taste in underweight, overweight and normal-weight subjects before, during and after sucrose ingestion. *Addictive Behav.* **5**:137–142.

Guy-Grand, B., and Sitt, Y. (1974). Alliesthésie gustative dans l'obésité humaine. *Nouv. Presse Med.* **3**:92–93.

Guy-Grand, B., and Sitt, Y. (1975). Gustative alliesthesia: Evidence supporting the ponderostatic hypothesis for obesity. In *Recent Advances in Obesity Research,* Vol. I, A. Howard (Ed.). Newman, London, pp. 238–241.

Hammel, H. T. (1968). Regulation of internal body temperature. *Annu. Rev. Physiol.* **30**:641–710.

Herberg, L. J., Franklin, K. B. J., and Stephens, D. N. (1975). The hypothalamic set-point in experimental obesity. In *Recent Advances in Obesity Research,* Vol. I, A. Howard (Ed.). Newman, London, pp. 235–237.

Herman, C. P., Polivy, J., and Esses, V. M. (1987). The illusion of counter-regulation. *Appetite,* **9**:161–169.

Hervey, G. R. (1969). Regulation of energy balance. *Nature,* **222**:629–632.

Himms-Hagen, J. (1984). Thermogenesis in brown adipose tissue as an energy buffer. *New Engl. J. Med.* **311**:1549–1558.

Hyde, R. J., and Pangborn, R. M. (1987). Taste modifiers: Parotid reflexes can differ from taste judgments. *Chem. Senses,* **12**:577–589.

Keesey, R. E. (1988). The relation between energy expenditure and the body weight setpoint: Its significance to obesity. In *Handbook of Eating Disorders,* Part 2, Burrows, Beaumont, and Casper (Eds.). Elsevier, New York, pp. 87–102.

Keesey, R. E., and Powley, T. L. (1975). Hypothalamic regulation of body weight. *Am. Sci.* **63**:558–565.

Keesey, R. E., and Powley, T. L. (1986). The regulation of body weight. *Ann. Rev. Psychol.* **37**:109–133.

Keesey, R. E., Corbett, S. W., Hirvonen, M. D., and Kaufman, L. N. (1984). Heat production and body weight changes following lateral hypothalamic lesions. *Physiol. Behav.* **32**:309–317.

Kleiber, M. (1961). *The Fire of Life.* Wiley, New York, p. 454.

Legoff, D. B., and Spigelman, M. N. (1987). Salivary response to olfactory food stimuli as a function of dietary restraint and body weight. *Appetite,* **8**:29–35.

Legoff, D. B., Leichner, P., and Spigelman, M. N. (1988). Salivary responses to olfactory food stimuli in anorexics and bulimics. *Appetite,* **11**:15–25.

Le Magnen, J. (1984). Is regulation of body weight elucidated? *Neurosci. Biobehav. Rev.* **8**:515–522.

Meiselman, H. L. (1977). The role of sweetness in the food preference of young adults. In *Taste and development,* J. W. Weiffenbach (Ed.). U.S. Department of Health, Education, and Welfare (now Health and Human Services), Bethesda, MD, pp. 269–279.

Mook, D. G. (1974). Saccharin preference in the rat: Some unpalatable findings. *Psychol. Rev.* **81**:475–490.

Mook, D. G. (1989). Oral factors in appetite and satiety. *Ann. N.Y. Acad. Sci.* **575**:265–280.

Mook, D. G., and Cseh, C. L. (1981). Release of feeding by the sweet taste in rats: Influence of body weight. *Appetite,* **2**:15–34.

Morgan, C. T., Stellar, E., and Johnson, O. (1943). Food deprivation and hoarding in rats. *J. Comp. Psychol.* **35**:275–295.

Mrosovsky, N., and Fisher, K. C. (1970). Sliding set points for body weight in ground squirrels during the hibernation period. *Can. J. Physiol.* **48**:241–247.

Mrosovsky, N., and Powley, T. L. (1977). Set points for body weight and fat. *Behav. Biol.* **20**:205–223.

Powley, T. L., and Keesey, R. E. (1970). Relationship of body weight to the lateral hypothalamic feeding syndrome. *J. Comp. Physiol. Psychol.* **70**:25–36.

Rodin, J., Moskowitz, H. R., and Bray, G. A. (1976). Relationship between obesity, weight loss, and taste responsiveness. *Physiol. Behav.* **17**:591–597.

Rolls, B. J., Rowe, E. A., and Turner, R. C. (1980). Persistent obesity in rats following a period of consumption of a mixed high energy diet. *J. Physiol. (London)* **298**:415–427.

Sclafani, A. (1989). Diet-induced overeating. *Ann. N.Y. Acad. Sci.* **575**:281–291.

Sclafani, A., and Springer, D. (1976). Dietary obesity in adult rats: Similarities to hypothalamic and human obesity syndromes. *Physiol. Behav.* **17**:461–471.

Stellman, S. D., and Garfinkel, L. (1986). Artificial sweetener use and one year weight change among women. *Prev. Med.* **15**:195–202.

Tuorila-Ollikainen, H., and Mahlamaki-Kultanen, S. (1985). The relationship of attitudes and experiences of Finnish youths to their hedonic responses to sweetness in soft drinks. *Appetite,* **6**:115–124.

Underwood, P. J., Belton, E., and Hulme, P. (1973). Aversion to sucrose in obesity. *Proc. Nutr. Soc.* **32**:93A–94A.

Von Euler, C. (1964). The gain of the hypothalamic temperature regulating mechanisms. *Prog. Brain Res.* **5**:127–131.

Wardle, J. (1987). Hunger and satiety: A multi-dimensional assessment of responses to caloric loads. *Physiol. Behav.* **40**:577–582.

Wirtshafter, D., and Davis, J. D. (1977). Set points, settling points, and the control of body weight. *Physiol. Behav.* **19**:75–78.

Wolfe, J. B. (1939). An exploratory study of food storing in rats. *J. Comp. Psychol.* **28**:97–108.

Wooley, O. W., and Wooley, S. C. (1981). Relationship of salivation in humans to deprivation inhibition, and the encephalization of hunger. *Appetite,* **2**:331–350.

Wooley, S. C., and Wooley, O. W. (1973). Salivation to the sight and thought of food: A new measure of appetite. *Psychosom. Med.* **35**:136–142.

Wooley, O. W., and Wooley, S. C., and Woods, W. A. (1975). Effect of calories on appetite for palatable food in obese and nonobese humans. *J. Comp. Physiol. Psychol.* **89**:619–625

Part II Discussion

Hirsch: This question is to Dr. Beauchamp. It appeared in one of the slides that there was an inverse relationship between sweet and salt preference. Is this the case? And if they're related, in what way are they related?

Beauchamp: Let me describe the feedback loops that I see. In this particular study there was a negative correlation between the perceived sweetness of particular foods and the perceived saltiness of particular foods. The issue whether there is a relationship between, say, a person's salt preferences and sweet preferences is one that this particular study, or none of our studies, really deal with. In fact, our evidence seems to suggest those two things are not necessarily related.

Scott: The first point is that it may not be such a matter of concern that the rat's neurophysiological response to saltiness appears to be increased through salt deprivation because that may be only an epiphenomenon of a qualitative change. The question is: In your assaying of subjects' preferences when they are salt deprived or relatively salt deprived as opposed to replete, did you ever ask about the qualitative experience of salt under those conditions?

Beauchamp: What you're referring to is a very interesting issue, which I'm sure you will clarify in your talk, and it is mentioned in our paper. One possibility is that people who are salt-depleted judge the salty foods as being sweet, and I think you have some data that might be consistent with that. I didn't find out about your work until we actually designed and had done our study. To the extent that we do more work, we will ask questions about the perceived sweetness of various salty foods.

Thrasher: I want to make a comment to Dr. Beauchamp. You have in your final slide there, I think, that one of the possibilities for the failure in humans to show salt appetite was a species difference, and you didn't like it.

Beauchamp: No, I liked it.

Thrasher: You liked it. OK, I was just going to say that my experience with dogs, and it's limited pretty much to dogs, is that you cannot make them drink salt—any kind of salt solution, be it isotonic or whatever—with any of the manipulations that cause salt appetite in rats. You can put angiotensin into their head chronically, enough to cause them to drink their body weight in water for 7 days in a row, but they will not touch or they will step over the salt bowl to get to the water. You can load them with aldosterone; nothing happens. You can deplete them. No matter what they do, we cannot get a dog to drink salt when he needs it. On the other hand, if he's dehydrated and needs water, he will drink water or an equal amount of isotonic saline, or twice normal saline in equal volume, but he's making up his water deficit. If you give him the choice, he'll always pick the water. So I really do believe species differences are important.

Beauchamp: If I misled you into saying I didn't believe there were species differences to get you to say that publicly, I'm pleased because in fact I've been making similar arguments also with cats. I'm glad to hear a definitive statement on the dog situation.

Plata-Salaman: You talk about Addison's syndrome. Addison's syndrome may be primary or secondary depending on the ACTH deficits or not, and the primary might be produced by several reasons. Is there any evidence for a selective craving for salt in the etiology of the Addison's syndrome, because sometimes it may not be complete and sometime it is only partial?

Beauchamp: I think that was the same point that Dr. Van Itallie was making. I have no expertise in this area at all. There's nothing that I found in the literature to deal with that.

Weingarten: I cannot resist and will take moderator's prerogative to make one final comment. It's more of a personal comment on the tone of both sessions today. As an absolutely devoted "regulationist" who is interested in the problem of how hedonics drive intake and affect body weight, I think it's interesting that we force the dichotomy between hedonics and regulation. It's also clear that there are certain accoutrements to the regulatory people and certain accoutrements to

PART II DISCUSSION

the hedonic people. The regulatory people are very concerned about physiological mechanism; it is not as well represented among people who do analysis of hedonics. The people who do hedonics get involved in a whole host of semantic arguments, which frankly at times makes me want to go run into the lab and just do an experiment.

The other thing that is interesting to me is that, if you look at the field and how the field has looked at energy and water balance, and how they viewed regulation, the regulatory people for the most part have absolutely fixated on the concept of the negative feedback loop. There is very little discussion of how we can effect the regulation of energy or water balance from the regulatory tradition from anything but the perspective of negative feedback loops. It's very heartening to see there is clearly more than one way to make a regulation. We know from analysis of other physiological systems—there are feed-forward systems, there are adaptive control systems—that invoking those concepts is incredibly productive in our analysis of the physiological control of other behaviors. That is a conceptualization that is, interestingly, part of the hedonic tradition, but is not well regarded or certainly has not been well exploited in the regulatory tradition.

Part III
Acquired Preferences

9
Social Factors in Diet Selection and Poison Avoidance by Norway Rats: A Brief Review

Bennett G. Galef, Jr.

*McMaster University
Hamilton, Ontario, Canada*

I. INTRODUCTION

Study of the behavior of individual omnivores as they choose among foods of differing nutritive value has been an active area of research in psychobiology for more than 50 years. During that half-century, considerable progress has been made in identifying behavioral processes that contribute to an individual's ability to compose a nutritionally balanced, safe diet by choosing appropriately among a number of different substances. Congenital flavor preferences (Young, 1959, 1968), specific hungers (Richter, 1956), learning about the positive and negative consequences of eating various foods (Garcia and Koelling, 1966; Zahorik and Maier, 1969; Rozin, 1976; Booth, 1985), patterned sampling of foods (Rozin, 1969), and hesitancy to eat unfamiliar foods (Barnett, 1958) have all been implicated in adaptive diet selection by animals.

In general, redundancy in the processes that can lead to the accomplishment of a goal indicates that success in achieving that goal makes an important contribution to fitness, and animals are usually very proficient at achieving those goals that increase their probability of survival and reproduction. Hence, the observed redundancy in the behavioral processes that influence individual omnivores to choose adaptively among potential foods suggests, in itself, that such choices tend to be made wisely. It is, consequently, not surprising to find that many psychobiologists, not specialists in the study of diet selection, have formed the general impression that individual omnivores are very good at deciding which foods to eat and which to avoid eating.

This generally held view, that animals acting independently are proficient at selecting appropriate substances to eat, poses a problem for discussions of the role of social influence in diet choice by animals. If, as the literature suggests, the development of adaptive patterns of diet choice by animals can be understood fully in terms of the responses of individuals acting in isolation, then discussion of social influences on diet selection is an unnecessary elaboration. That is, there is little point in talking about social contributions to the development of adequate feeding repertoires in omnivores if each omnivore is perfectly capable of deciding for itself what to eat.

The main argument of the present chapter is that animals acting individually are, in fact, not nearly so good at selecting nutritionally balanced, safe diets as most discussions of food choice published during the past half-century would lead the unwary to conclude. Because individuals are far from perfect in their choice of substances to eat, socially acquired information can often play a critical role both in permitting naive rats to identify nutritive substances and in allowing the naive to determine which, if any, of the substances that they have eaten are toxic and should not be eaten again.

In Sections II and III I focus on data that suggest that individual animals have considerably greater difficulty than generally is appreciated both in selecting balanced diets and in avoiding ingestion of lethal quantities of toxins. In Section IV, I review briefly some of the evidence that indicates that interaction of naive individuals with more experienced others can help the naive both to select nutritive foods and to identify potential toxins.

The discussion of the literature on diet choice that comprises Sections II and III is not intended to suggest that earlier explanations of food choice by free-living animals are in any sense wrong. Congenital flavor preferences (Young, 1959, 1968), responses to novelty (Barnett, 1958), specific hungers (Richter, 1956), and learning about the positive or negative consequences of eating various substances (Garcia and Koelling, 1966; Zahorik and Maier, 1969; Rozin, 1976; Booth, 1985) are each important contributors to the making of adaptive food choices. Rather, it is my view that previous discussions of how free-living omnivores come to select the foods they eat are incomplete. These discussions are incomplete because they fail to take sufficiently into account the effects of social influence on diet choice, though, of course, passing mention of possible social influences on adaptive patterns of food selection is as old as the field itself (see, e.g., Dove, 1935; Richter, 1942–1943).

II. HOW GOOD ARE ANIMALS AT SELECTING A NUTRITIONALLY ADEQUATE DIET?

In a series of classic studies conducted in the 1930s and 1940s, Curt Richter (1942–1943) introduced naive, nondeprived, adult rats into a "cafeteria" setting in which each subject was presented with an array of relatively pure nutrients from which to compose a diet. Richter found that subjects in his cafeteria feeding situation selected substances to eat and drink with great efficiency. His subjects grew both faster and with lower caloric intake than did control subjects eating the McCullum diet (a nutritionally adequate diet compounded by nutritionists). The efficiency of Richter's subjects in self-selecting foods in a cafeteria led Richter (1942–1943) to speak of a "total self-regulatory" capacity, allowing rats to precisely control their intake of various micro- and macronutrients so as to optimize their efficiency of resource utilization.

The observed ability of animals to select a balanced diet by self-selecting from an array of relatively pure constituents in a laboratory cafeteria came as no surprise. Richter argued that the result was predictable from the simple observation that omnivores survive in natural habitat: "The survival of animals and humans in the wild state in which the diet had to be selected from a great variety of beneficial, useless, and even harmful, substances is proof of this ability [to make dietary selections which are conducive to normal growth and reproduction]" (Richter et al., 1938, p. 734).

Unfortunately, Richter's assertion is too broad. The observation that omnivores survive outside the laboratory provides no evidence that they can construct nutritionally adequate diets from purified dietary components. In fact, survival outside the laboratory tells one little about the ability of omnivores to select the foods they need. Perhaps free-living omnivores can survive only in a relatively restricted range of environments where palatable, nutrient-rich foods are abundant. One cannot deduce a general ability to select foods with great efficiency from the fact of survival in nature.

Whether logically compelling or not, the success of Richter's subjects in his cafeteria feeding situation, taken together with the suggestion that this success was only a limited demonstration of the ability of omnivores to select appropriate substances to ingest in natural environments (after all, Richter's cafeteria contained no useless or harmful substances), had profound influence. For decades, the notion that both animals and humans had the ability to totally self-regulate their nutrient intake guided both the design of experiments and the interpretation of data.

The power of Richter's conception of total self-regulatory behavior to hold the scientific imagination is illustrated by publication, as recently as 1987, in the prestigious *New England Journal of Medicine,* of a paper (Story and Brown, 1987) that simply states that 50 years ago, Clara Davis, an early nutritionist, neither showed nor claimed to show that children could self-select a well-balanced diet from a cafeteria of purified nutrients. Davis (1928, 1939) presented children with an array of highly nutritious foods among which to choose. Her subjects could hardly have failed to select adequate diets so long as they showed some variability in their food choices. As Davis stated explicitly in 1939 (and Richter seems to have failed to appreciate), the success or failure of subjects in a cafeteria feeding situation depends on the particular array of foods offered to them to choose among (Galef and Beck, 1990).

Even in the 1940s, when Richter was completing his studies of diet selection by rats in cafeteria feeding situations, there was reason to question the generality of his finding of efficient diet choice by rats faced with a cafeteria of foods. The majority of early studies of diet selection failed to confirm Richter's observation of efficient self-selection by rats (Lát, 1967). Richter attributed the negative findings in the literature of his day to "the complex nature of the natural foods or food mixtures offered for choice" (Richter et al., 1938, p. 176); it might have been more accurate to attribute the relatively infrequent, great success of rats in some cafeteria situations to the provision of particularly felicitous combinations of foods for rats to choose among. Epstein (1967), for example, has suggested that availability of multiple sources of protein and the presence of a carbohydrate of low palatability were responsible for the success of rats in the particular cafeteria Richter used in his studies of total self-selection.

The range of environments in which individual omnivores can succeed in self-selecting adequate diets was and is far more restricted than Richter realized. If the number of foods among which rats must choose is greater than two or three (Harris et al., 1933), if the consequences of eating particular substances are not felt for many hours (Harris et al., 1933), if important nutrients (particularly proteins) are available only in a relatively unpalatable form (Kon, 1931; Scott, 1946), then the ability of individual rats to self-select a balanced diet from among "a great variety of beneficial, useless, and even harmful substances" is not impressive. The data are not presently and have never been consistent with the view that individual adult animals are exceptionally capable of composing nutritionally adequate diets by choosing among an array of foods.

One might wish to argue that, although not perfect at composing balanced diets, adult rats are still pretty good at choosing foods to eat. It's really a matter of opinion as to whether the performance of a subject surviving in a cafeteria situation is "good" or "bad," since no a priori probability of success has ever been calculated. There can be no question, however, that weanling rats perform abysmally in cafeteria feeding situations. Weanlings fail to choose adequate diets even in situations where adults do reasonably well. For example, Tribe (1954, 1955) presented fifteen 100 g, female rats with seven foods. Thirteen solved the problem and grew almost normally; only 2 of 10 weanling rat pups survived in the same situation. Scott et al. (1948) offered twenty 12- to 15-week-old rats a choice among but four foods. Thirteen of these 20 sexually mature individuals gained weight, and all lived; only 9 of 31 weanlings survived in the same situation. Kon (1931) offered four 28-day-old rats three foods to choose among, supplemented by hand-fed vitamins; two of the four weanlings died and one gained no weight for 7 weeks.

Such failures of weanlings to self-select adequate diets in situations far simpler than those one might expect them to face outside the laboratory seem to me to be of particular importance in understanding the behavioral processes responsible for the development of adequate feeding repertoires in natural circumstances. It is at weaning that young rats, or other young mammalian omnivores, must undertake the potentially arduous task of developing de novo a diet of independently acquired foods adequate to support growth and development. An adult may be challenged from time to time by failure of one or another of the resources on which it has come to depend. Every weanling must respond to the withdrawal of its major source of sustenance and the need rapidly to develop a nutritionally adequate, safe diet composed of substances selected from among a plethora of available ingestables. Yet, rat pups consistently fail to solve diet-selection problems in the laboratory that must be considerably simpler than those they face in natural circumstances. This failure of weanling rats to self-select a balanced diet suggests that the cafeteria feeding experiment, the presumed laboratory analogue of diet selection in nature, fails to capture some important aspect of the process of learning to select a balanced diet as such learning occurs outside the laboratory.

III. CAN INDIVIDUAL RATS AVOID TAKING LETHAL QUANTITIES OF TOXIC BAITS?

Rats selectively associate tastes with gastrointestinal malaise (Garcia and Koelling, 1967). They tolerate very considerable delays between tasting a novel substance and experiencing illness and still form an aversion to the taste of the novel substance (Garcia et al., 1966). In the literature, both these proclivities have been discussed frequently as adaptive specializations of general Pavlovian conditioning processes modified in the service of poison avoidance (Seligman, 1970; Rozin and Kalat, 1971; Shettleworth, 1983; but see Logue, 1979; Revusky, 1977; Domjan, 1980; Domjan and Galef, 1983).

If rats are adaptively specialized to learn to associate tastes with toxicosis, then by implication rats should be good at avoiding ingestion of lethal quantities of any toxic substances they encounter. Indeed, review of the psychological literature suggests that because each rat has available a variety of behavioral tactics that can decrease its probability of ingesting lethal quantities of poisons, killing free-living rats by poisoning them is a formidable task. Consequently, it was somewhat surprising to me to find that although total extermination of a rat population by poisoning is difficult, it is not unusual for professional exterminators to kill 80–90% of a target population by introducing a poison bait into a rat-infested area (Meehan, 1984). Often, even quite unsophisticated application of a poison bait will suffice to kill the majority of a target population of rodents. For example, Chitty (1954), working during the war years at Oxford, found that poison baits that were introduced without prebaiting into the home ranges of censused colonies of wild rats typically caused the death of 75% of colony members. Chitty had no evidence that surviving rats in his studies had learned to avoid the poison bait; perhaps some individuals found the bait unpalatable, were exceptionally reluctant to eat unfamiliar food, or simply failed to encounter the poison. Poison-avoidance learning may be even less successful in protecting rats against introduced poisons than Chitty's (1954) data suggest.

By increasing the probability that rodents will consume a lethal quantity of poison before toxicosis is experienced (e.g., by using palatable poison baits, by using rodenticides that are both lethal in small amounts and have delayed onset of symptoms, by "prebaiting," etc.), it is possible to create an environment in which individual rodents have low probabilities of survival. Of course, if one's goal is complete extermination of a population, escape of even one pregnant female means failure of the extermination effort.

Ingestion of lethal doses of toxins by animals is also observed in the case of poisons other than those specifically composed by humans to kill pests. Naturally occurring toxins, like man-made poisons, kill large numbers of mammals. For example, each year the American cattle industry loses an estimated 3–5% of its total herd to poisonous plants, particularly halogeton, larkspur, lupine, and locoweed (James, 1978). Naturally occurring poison baits, like their man-made counterparts, are not easily avoided by animals that encounter them.

In sum, the abilities of animals either to self-select a balanced diet or to avoid eating toxins are not so highly evolved as to preclude a meaningful contribution of socially acquired information to decisions about what to eat and what to avoid eating. There exists, at the least, the logical possibility of important social contributions to the development of adaptive patterns of selection and rejection of foods by animals foraging outside the laboratory.

IV. SOCIAL SOLUTIONS TO THE PROBLEM OF SELECTING NUTRITIONALLY ADEQUATE, SAFE DIETS

One invariant feature of the environment in which each mammal develops, if it survives to weaning age, is the presence of a conspecific adult, a dam, who by her very existence and reproductive success has demonstrated both the nutritional adequacy and safety of her diet. No matter how slow or inefficient individual adults might be at learning either to avoid toxins or to identify sources of requisite nutrients, a juvenile could be reasonably sure that its dam (or any other surviving adult it encountered) had not eaten seriously injurious quantities of debilitating substances and had located sources of all those nutrients necessary for growth and survival. Most important, a juvenile could be reasonably sure that sympatric adults had composed a nutritionally adequate, safe diet in the same geographical area in which the juvenile had to achieve nutritional independence. "Eat what adults are eating, do not eat what adults are not eating" could serve as a useful rule of thumb to juvenile members of any species living in a reasonably stable environment (Boyd and Richerson, 1985).

Of course, the fact that young omnivores could benefit from using adult conspecifics as guides to what to eat and what not to eat does not mean that they do so. It is an empirical question whether young mammals can, in fact, exploit their elders as sources of information about foods.

For 20 years, my students, co-workers, and I have been studying the ways in which naive, developing individuals can incorporate the behavior of more knowledgeable others into their own behavioral repertoires. We have found, in a broad range of circumstances, that the food choices of young rats are profoundly influenced by the food choices of the adults with whom the young interact. In the present section, I review briefly some of the major findings of this research program to illustrate the many ways in which social interactions can facilitate the development of adaptive dietary repertoires by weanlings (for more complete reviews see Galef, 1977, 1982, 1985a, 1986a, 1989a, 1989b; Galef and Beck, 1990).

The original impetus for examining the possibility that weanling Norway rats might use information garnered from adult conspecifics in developing their own feeding repertoires arose from field observations made by Fritz Steiniger, an expert in the control of rodent pests. Steiniger (1950) reported that if he continued to use one poison bait in an area for several months, acceptance of that bait declined dramatically. He observed, in particular, that young rats, born to adults that had learned to avoid a bait, rejected the bait without even sampling it themselves. The young fed exclusively on alternative, safe foods available to their respective colonies.

This avoidance by young wild rats of a food that the adults of their colony had learned to avoid turned out to be a very robust phenomenon, easily captured in the laboratory. In our first experiments (Galef and Clark, 1971a), we established groups of male and female, adult, wild rats (*Rattus norvegicus*) in 1×2 m enclosures. Water was continuously available in the enclosure, and food was present for 3 hours/day in two food bowls placed about 1 m apart. Each food bowl contained one of two, nutritionally adequate diets, discriminable from the other in taste, smell, texture, and appearance.

Adults of each colony were trained to eat one of the foods presented each day and to avoid the other by the simple expedient of introducing a nausea-inducing agent (LiCl) into samples of one of the two foods. The adults rapidly learned not to eat the poisoned food and, most important, for several weeks thereafter, avoided eating the previously poisoned food when offered uncontaminated samples of it.

Our experiments began when litters of pups, born to our trained colonies, left their nest sites to feed on solid food for the first time. We observed both adults and pups throughout daily 3-hour feeding periods on closed-circuit television and recorded the number of times pups in each colony ate from each food bowl. After members of a litter of pups had fed on solid food for 2 weeks, we transferred them to new enclosures where, now without the adults of their colony, the pups were again offered a choice between uncontaminated samples of the two diets. In this second situation, we could directly measure the amount of each diet eaten by pups during daily feeding periods simply by weighing each of the two food bowls offered to the pups before and after each feeding period.

The results of our manipulations were exceptionally clear. Only 1 of 247 rat pups we watched feed for 2 weeks with adults trained to avoid one diet and eat another ever ate even a single bite of the food that the adults of its colony had been trained to avoid. The other 246 pups fed exclusively on whichever diet the adults of their colony were eating. Furthermore, after transfer to enclosures separate from the adults of their colony, pups continued, for several days, to eat only the food that the adults of their colony had eaten and to avoid the alternative that the adults of their colony had learned to avoid. Taken together, these results demonstrate, as Steiniger's (1950) field observations suggested, that juvenile rats can and will use adults with whom they interact as sources of information about which foods to eat and which to avoid.

My co-workers and I have spent much of the past 20 years identifying and describing four independent ways in which the behavior of adult rats can influence diet selection by their young. Below, I review each of these four modes of social influence briefly before turning to consideration of evidence that such social influence can lead rats in complex environments away from toxins and toward nutritionally balanced, safe diets.

A. Modes of Social Influence of Adult Rats on Their Young

1. Presence of Adults at a Feeding Site

One of the simplest and, perhaps, less interesting ways in which adult rats can induce naive young to eat one food rather than another is for the adults to eat at one location rather than at another. The presence of adults at a feeding site attracts young to that site and causes them to feed on whatever food is present there. For example, four times as much food was eaten by both 19- and 25-day-old, domesticated rat pups from a food bowl with an anesthetized female rat draped over its rim than from an identical bowl 1 m away (Galef, 1981). Similarly, Clark and I observed each of nine, individually marked, wild rat pups eat their very first meal of solid food in a large enclosure; the young animals ate that meal both while an adult was eating and at the same food bowl from which the adult was eating, not from a second food bowl 1.5 m away (Galef and Clark, 1971b).

2. Deposition of Residual Olfactory Cues

The results of a number of studies indicate that adult rats mark both the area around a food source and a food that they are eating with residual olfactory cues that make the marked site or marked food more attractive to juveniles seeking food than identical unmarked foods or sites. Galef and Heiber (1976) restricted either mothers and their young or groups of virgin female rats to one end of a 2 × 1 m cage for several days. Then, while these stimulus animals were absent from the cage, food-deprived juveniles were tested individually for 1-hour periods with identical bowls of food at each end of the cage. The young took 70–90% of the food they ate from the end of the enclosure that had been soiled by other rats. Young rats also ate more food from a bowl that an experimenter had surrounded with rat excreta than from a clean food bowl in an unsoiled area (see also Galef, 1981). Galef and Beck (1985), similarly, showed that rats spontaneously mark a feeding site they are visiting, making it more attractive to other rats than an identical, unmarked site.

Though effective in biasing the diet preferences of young in adaptive directions, indirect behavioral mechanisms for communication of food selection such as those described above and in Section IV.A.1 are not very sophisticated. Instances in which adult rats communicate directly to their young information as to what foods should be eaten and avoided are, perhaps, of greater interest.

3. Flavor Cues in Mother's Milk

The results of two sets of studies (Galef and Henderson, 1971; Galef and Sherry, 1973) in my laboratory, as well as a variety of findings from other laboratories (see, e.g., Martin and Alberts, 1979) are consistent with the hypotheses (a) that the milk of a lactating rat contains cues reflecting the flavor of her diet, and (b) that flavor cues in mother's milk influence weanlings' selection of solid foods to eat.

Some of the most convincing evidence of the existence of flavor cues in mother's milk that reflect the flavor of mother's diet came from studies in which rat pups nursing from a female rat fed one diet were made ill by injection of LiCl after having been hand-fed a small quantity of milk that had been expressed manually from a second lactating female eating a different diet. Tests at weaning showed that pups treated in this way had developed a substantial aversion to the diet fed to the lactating female from which the manually expressed milk had been taken (Galef and Sherry, 1973). In the other studies, Galef and Henderson (1972) found that rat pups raised by mothers eating diet B and fostered for 6 hours/day for 18 days to a lactating female eating diet A showed an enhanced preference for diet A at weaning, relative to pups fostered daily to maternal, nonlactating females. Further evidence of transmission of diet cues through mother's milk has been provided by artificially introducing a flavor into the milk of a mother rat (e.g., by intraperitoneal injection) and, thus, causing her pups to exhibit an enhanced preference at weaning for foods of the introduced flavor (Le Magnen and Tallon, 1968; Martin and Alberts, 1979).

4. Olfactory Cues on the Breath of Adult Rats

Galef and Wigmore (1983) and Posadas-Andrews and Roper (1983) discovered independently that after a naive rat (an observer) interacted for a few minutes with a conspecific that had previously eaten some food (a demonstrator), the observer would show a substantially enhanced preference for the food that its demonstrator had previously eaten. We have made considerable progress in understanding the messages passing from demonstrator to observer that allow demonstrators to influence their respective observers' food choices.

To summarize the results of a complex series of studies, the data suggest that both olfactory cues (Galef and Wigmore, 1983) escaping from the digestive tract of demonstrator rats and the smell of bits of food clinging to their fur are sufficient to allow observers to identify the foods that their respective demonstrators ate (Galef and Stein, 1985). However, simple exposure of observers to the smell or taste of a food is not, in itself, sufficient to enhance an observer's preference for that food (Galef et al., 1985). Observers' preferences for foods are altered by experience of the smell of a food in combination with olfactory cues emerging from the anterior end of a living demonstrator rat (Galef and Stein, 1985). These demonstrator-produced semiochemicals, when experienced by a naive rat at the same time that it experiences the smell of a food, alter the subsequent diet preference of the observer. The semiochemicals produced by demonstrators that increase the preference of observers for a food are probably volatile sulfur compounds (e.g., carbon disulfide, a chemical present on rat breath) that, when added to a food, increase the preference of both rats and mice for that food (Bean et al., 1988; Galef et al., 1988; Mason et al., 1988).

B. Uses of Socially Acquired Information

The evidence reviewed briefly above indicates that rats can be influenced in their choices of either feeding sites or foods by the feeding behavior of other rats. Below, I am concerned with evidence that information garnered by naive rats from more knowledgeable conspecifics can be used by the naive to find nutritionally adequate foods, to avoid toxins, and to forage more efficiently than would be possible in the absence of socially-acquired information.

1. What to Eat

In a recent experiment, Beck and Galef (1988) presented individual weanling rats with a choice among four distinctively flavored foods. Three of these foods contained inadequate levels of protein (4.4%) and one, the least palatable of the four, had ample protein (17.5%) to support normal growth. We found, as had others before us (Kon, 1931; Scott and Quint; 1946; Scott et al., 1948; Tribe, 1954, 1955), that young subjects performed poorly in such a situation. None of our juvenile subjects was able to develop a preference for the protein-adequate food in 6 days; each pup lost weight, and if we had not terminated the experiment when we did, all probably would have died. Weanling rats faced with the same diet selection problem while in the presence of adults previously trained to eat the protein-rich alternative, grew rapidly in the experimental situation.

Analyses of the ways in which trained rats influenced the food preferences of naive, adolescent rats (150–175 g) juveniles showed that the naive rats were influenced in their food choices by the foods eaten by their respective demonstrators, not by the place where their respective demonstrators ate. In one experiment, we separated naive rats from their demonstrators with a screen partition and, choosing among four food cups (one of which contained the protein-rich diet that the demonstrator was eating and the other three of which each contained a different protein poor diet), allowed each naive rat to interact through the partition with a demonstrator that was given protein-rich diet to eat from a food cup placed adjacent to its observer's food cup, containing a protein-poor diet. The observers ate as much protein-rich diet as did observers whose demonstrators ate protein-rich diet from a food cup placed adjacent to the observer's food cup containing protein-rich diet.

In a situation in which individual, naive rats found it impossible to select a nutritionally adequate diet from among more palatable, but inadequate, alternative diets, the naive rats used information acquired from more knowledgeable conspecifics to identify the adequate diet.

2. Identification and Avoidance of Toxins

New recruits to a population (recent immigrants or naive juveniles) not only must find and ingest nutritionally adequate diets, they also must avoid ingesting any toxic substances they encounter. In the search for needed nutrients, a naive individual might have to sample broadly among unfamiliar, ingestible substances. By sampling one unfamiliar substance at a time and waiting long enough between meals of unfamiliar foods to evaluate independently the postingestional consequences of each (Rozin and Kalat, 1971), a naive individual could evaluate the toxicity of each unfamiliar substance it ingested. There is, however, little evidence that rats actually sample among several unfamiliar foods to permit their independent evaluation. A growing body of evidence suggests, to the contrary, that even wild rats offered several unfamiliar foods to eat often sample most of them during a single, initial bout of feeding on the unfamiliar foods (Barnett, 1956; Beck et al., 1988). Use of information about the foods that others are eating could provide an alternative route for identification of toxic substances by naive individuals, even if the naive individual had eaten more than one unfamiliar substance before falling ill.

Galef (1986b, c, 1987), found that naive rats were less likely to form aversions to foods that other rats had eaten than they were to form aversions to totally unfamiliar foods. Naive "observer" rats, which had the opportunity to interact with a recently fed "demonstrator" rat before eating two unfamiliar foods and becoming ill, learned an aversion to whichever food their respective demonstrators had not eaten (Galef, 1986b, 1987). Furthermore, a substantial portion of rats that had formed an aversion to a diet, as a result of association of that diet with toxicosis, abandoned their aversions following exposure to other rats that had eaten the averted diet (Galef, 1985; 1986c). If, as seems likely, it is usually the case that an unfamiliar substance that others are eating is less likely to be toxic than an unfamiliar substance that others are not eating, then social influences on taste-aversion learning could be an important source of information as to whether illness was food related and which of several recently eaten foods was most probably illness inducing (Domjan and Galef, 1983, Galef, 1986b, 1987).

3. Where to Eat

Galef et al. (1987) found that rats that were familiar with a maze would spontaneously follow trained rats through the maze to food. They also found that rats trained to follow leader rats through a maze were more likely to follow leaders that had just eaten a familiar, safe food than to follow leaders that had just eaten a food that the potential followers had been trained to avoid. Hungry rats exhibited both a readiness to follow others to feeding sites and an ability to select others to follow on the basis of the desirability of the foods those others had been eating.

Furthermore Galef and Wigmore (1983) have shown that rats that were familiar with the locations at which particular foods were sometimes to be found, but did not know which of several foods was currently available, could use information garnered from a recently fed conspecific to decide where to look for food. After interacting with an "informer" rat that had just eaten cinnamon-flavored diet, a subject rat went to the location where it had previously learned that cinnamon-flavored diet was to be found. After interacting with an "informer" rat that had just eaten cheese-flavored diet, the same subject went to the location where it had previously learned that cheese-flavored diet was to be found. Rats familiar with the location of food patches within their home ranges can find out from others which foods were available and can use that information to orient their foraging trips in profitable directions (Galef, 1983).

V. CONCLUSION

The preceding review leads to two conclusions. First, that rats and, by extrapolation, other omnivores, are not as efficient either at selecting balanced diets or at avoiding repeated ingestion of toxins as is generally believed. Second, that information acquired from others can improve the performance of native rats faced with the need to select a nutritionally adequate diet while avoiding ingestion of poisons. Successful others can serve the naive animal as useful sources of information about where to eat, what to eat, and what to avoid eating.

In discussing the results of her previously mentioned, classic studies of diet selection by human infants, Clara Davis (1939, p. 261) concluded that "the results of the experiment . . . leave the selection of the foods to be made available to young children in the hands of their elders, where everyone has always known it belongs." A similar conclusion can be drawn from the present review of diet selection by rats. Naive, young rats are not generally capable of making appropriate food choices. The selection of foods on which to subsist is better made by their more experienced elders. Elders that either had the misfortune to eat substantial quantities of toxic substances or failed to compose a nutritionally adequate diet die and are not available to serve as behavioral models for juveniles. Other adults, successful in diet selection, remain in a population and are available to shape the ingestive behavior of succeeding generations.

In effect, living adults increase the availability of the foods they are eating to the young with whom they interact. Living adults attract pups to feeding sites and expose the juveniles to olfactory cues that bias the diet preferences of the young. Adult rats, like adult humans, induce their naive young to eat safe, nutritious foods and reduce the probability of exposure of their young to potentially deleterious substances.

Although, in benign environments, naive individuals probably can compose adequate diets, in less congenial circumstances the naive are often not able to identify either toxic or nutritionally adequate diets fast enough to survive. By acting as though adult others are more likely to be eating nutritious, safe substances than useless substances or toxic ones, the naive can facilitate the development of their own adaptive feeding repertoires. Information acquired from successful others may make the difference between survival and failure for naive weanlings searching for adequate diets among the "great variety of beneficial, useless, and even harmful substances" to be found in natural circumstances.

ACKNOWLEDGMENTS

Preparation of this chapter was greatly facilitated by funds received from the Natural Sciences and Engineering Research Council of Canada and the McMaster University Research Board. I thank Mertice Clark, Harvey Weingarten, and Laurel McQuoid for their careful reading of earlier drafts of the manuscript.

DISCUSSION

Rozin: I'm amazed at the power of this effect. It seems that a rat is more affected by one contact with a demonstrator than by one personal experience with a diet being, say, safe or unsafe; so familiarity would have less of an effect than one short exposure to a demonstrator. Now, I'm puzzled as to the adaptive evolutionary significance. It's obvious why it's useful to learn from conspecifics, but it's not obvious why your own experience should be of less value than a roughly equivalent exposure to another animal with all the errors that might occur in the filtration of that to you. My first question is, Why is this so powerful with respect to individual experience? Second, though I agree that you're certainly showing a greater role of social transmission in animals than any of us had thought, it is also true that your own work shows how narrowly defined social transmission is in rats; that is, the context in which social learning occurs is so tightly limited to an exposure in the presence of a particular chemical. In the case of humans, "instruction" is an elaborate social structure to convey things rather than accidental exposure over time. I'm sure you'll agree with that.

Galef: Let me answer the second point first because it's easier to deal with. That just seems to me to be anthropocentric. Rats have this very broad ability to use all kinds of chemical information, whereas we're limited to a single verbal channel in communicating to one another about what to eat. The first question is more difficult. My own view is that wild rats are, as you know, very reluctant to eat unfamiliar foods. What I'm looking at are the situations in which the animal has no information, and is much better off assuming that those who are alive have done the right thing in the past than it is in trying to figure out for itself what's going on. You see, I'm looking always at a situation of the animal being somehow induced to eat unfamiliar things. That's a very dangerous situation for rats; they're very reluctant to do it, as you know. I assume in those situations, indeed evolutionarily, they've been better off attending to live others, because the live others have obviously never made a really serious mistake.

Scott: But you're saying "find an almost dead other," that is, a deeply anesthetized rat.

Galef: It's hard to get into that right now. I believe that the importance of poison avoidance in the life of rats has been exaggerated and that the critical issue for animals is in finding needed nutrients, not in avoiding ingesting toxins.

Kissileff: Is the conditioning stimulus in these experiments always an odor? Will a taste ever work?

Galef: I believe a taste will work. We're going to try to do that next year. We have preliminary data that indicates if a rat is anosmic and interacts with the demonstrator, the observer can still determine what food the demonstrator ate. Possibly, by licking the mouth of the demonstrator.

Kissileff: Wouldn't it be simpler to use a taste as the conditioning stimulus instead of an odor? Then you would be sure that you had completely removed all the olfactory cues.

Galef: Yes, that procedure might work.

Stricker: If the demonstrator had had access to a food and then had been poisoned before giving the demonstrator to the observer and giving the observer that food, would the observer then avoid or eat the food that the demonstrator had consumed?

Galef: You bring up the one great counterintuitive outcome in these experiments. No matter how hard we work at it, and we've tried for three years, we find that a sick demonstrator induces just as much of a preference for a food in its observer as does a well demonstrator. Returning to Dr. Scott's question, that's one of the reasons why I think this system is designed to tell animals what to eat, and not to tell them what not to eat. That suggests to me that this is a more important problem in the lives of the rats than avoiding toxins.

REFERENCES

Barnett, S. A. (1956). Behaviour components in the feeding of wild and laboratory rats. *Behaviour,* 9:24–42.

Barnett, S. A. (1958). Experiments on "neophobia" in wild and laboratory rats. *Br. J. Psychol.* 49:195–201.

Bean, N. J., Galef, B. G., Jr., and Mason, J. R. (1988). At biologically significant concentrations, carbon disulfide both attracts mice and increases their consumption of bait. *J. Wildl. Manage.* 52:502–507.

Beck, M., and Galef, B. G., Jr. (1988). Social control of diet selection: How do animals find nutritious food? Paper presented at the April meeting of the Eastern Psychological Association, Buffalo, NY.

Beck, M., Hitchock, C., and Galef, B. G., Jr. (1988). Diet sampling by wild Norway rats (*Rattus norvegicus*) offered several unfamiliar foods. *Anim. Learn. Behav.* 16:224–230.

Booth, D. A. (1985). Food-conditioned eating preferences and aversions with interoceptive elements: Conditioned appetites and satieties. *Ann. N.Y. Acad. Sci.* 443:22–41.

Boyd, R., and Richerson, P. J. (1985). *Culture and the Evolutionary Process.* University of Chicago Press, Chicago.

Chitty, D. (1954). The study of the brown rat and its control by poison. In *Control of Rats and Mice,* Vol. 1, D. Chitty (Ed.). Oxford University Press (Clarendon), London, pp. 160–299.

Davis, C. M. (1928). Self-selection of diet by newly weaned infants: An experimental study. *Am. J. Dis. Child.* 36:651–679.

Davis, C. M. (1939). Results of the self-selection of diets by young children. *Can. Med. Assoc. J.* 41:257–261.

Domjan, M. (1980). Ingestional aversion learning: Unique and general process. In *Advances in the Study of Behavior,* Vol. 11, J. S. Rosenblatt, R. A. Hinde, C. Beer, and M.-C. Busnel (Eds.). Academic Press, New York, pp.000–000.

Domjan, M., and Galef, B. G., Jr. (1983). Biological constraints on instrumental and classical conditioning: Retrospect and prospect. *Anim. Learn. Behav.* 11:151–161.

Dove, W. F. (1935). A study of individuality in the nutritive instincts and of the causes and effects of variations in the selection of food. *Am. Nat.* 69:469–544.

Epstein, A. N. (1967). Oropharyngeal factors in feeding and drinking. In *Handbook of Physiology,* Vol. 1, *Alimentary Canal,* C. F. Code (Ed.). American Physiological Society, Washington, DC, pp. 197–218.

Galef, B. G., Jr. (1977). Mechanisms for the social transmission of food preferences from adult to weanling rats. In *Learning Mechanisms in Food Selection,* L. M. Barker, M. Best, and M. Domjan (Eds.). Baylor University Press, Waco, TX, pp. 123–150.

Galef, B. G., Jr. (1981). The development of olfactory control of feeding site selection in rat pups. *J. Comp. Physiol. Psychol.* 95:615–622.

Galef, B. G., Jr. (1983). Utilization by Norway rats (*R. norvegicus*) of multiple messages concerning distant foods. *J. Comp. Physiol. Psychol.* 97:364–371.

Galef, B. G., Jr. (1985a). Studies of social learning in Norway rats: A brief review. *Devel. Psychobiol.* **15**:279–295.

Galef, B. G., Jr. (1985b). Socially induced diet preference can partially reverse a LiCl-induced diet aversion. *Anim. Learn. Behav.* **13**:415–418.

Galef, B. G., Jr. (1986a). Olfactory communication among rats of information concerning distant diets. In *Chemical Signals in Vertebrates,* Vol. IV, *Ecology, Evolution, and Comparative Biology,* D. Duvall, D. Muller-Schwarze, and R. M. Silverstein (Eds.). Plenum Press, New York, pp. 487–505.

Galef, B. G., Jr. (1986b). Social identification of toxic diets by Norway rats ($R.$ $norvegicus$). *J. Comp. Psychol.* **100**:331–334.

Galef, B. G., Jr. (1986c). Social interaction substantially modifies learned aversions, sodium appetite, and both palatability and handling-time induced dietary preference in rats ($R.$ $norvegicus$). *J. Comp. Physiol. Psychol.* **100**:432–439.

Galef, B. G., Jr. (1987). Social influences on the identification of toxic foods by Norway rats. *Anim. Learn. Behav.* **15**:327–332.

Galef, B. G., Jr. (1989a). An adaptationist perspective on social learning, social feeding, and social foraging in Norway rats. In *Contemporary Issues in Comparative Psychology,* D. A. Dewsbury (Ed.). Sinauer, Sunderland, MA, pp. 55–79.

Galef, B. G., Jr. (1989b). Innovation in the study of social learning in animals: Developmental and biological perspectives. In *Developmental Psychobiology: Current Methodological and Conceptual Issues,* H. N. Shair, G. A. Barr, and M. A. Hofer (Eds.). Oxford University Press, New York.

Galef, B. G., Jr., and Beck, M. (1985). Aversive and attractive marking of toxic and safe foods by Norway rats. *Behav. Neurol. Biol.* **43**:298–310.

Galef, B. G., Jr., and Beck, M. (1990). Diet selection and poison avoidance by mammals individually and in social groups. In *Handbook of Neurobiology,* Vol. 11, E. M. Stricker (Ed.). Plenum Press, New York, pp. 329–349.

Galef, B. G., Jr., and Clark, M. M. (1971a). Parent–offspring interactions determine time and place of first ingestion of solid food by wild rat pups. *Psychonomic Sci.* **25**:15–16.

Galef, B. G., Jr., and Clark, M. M. (1971b). Social factors in the poison avoidance and feeding behavior of wild and domesticated rat pups. *J. Comp. Physiol. Psychol.* **75**:341–357.

Galef, B. G., Jr., and Heiber, L. (1976). The role of residual olfactory cues in the determination of feeding site selection and exploration patterns of domestic rats. *J. Comp. Physiol. Psychol.* **90**:727–739.

Galef, B. G., Jr., and Henderson, P. W. (1972). Mother's milk: A determinant of the feeding preferences of weaning rat pups. *J. Comp. Physiol. Psychol.* **78**:213–219.

Galef, B. G., Jr., Kennett, D. J., and Stein, M. (1985). Demonstrator influence on observer diet preference: Effects of simple exposure and the presence of a demonstrator. *Anim. Learn. Behav.* **13**:25–30.

Galef, B. G., Jr., Mason, J. R., Preti, G., and Bean, N. J. (1988). Carbon disulfide: A semiochemical mediating socially-induced diet choice in rats. *Physiol. Behav.* **42**:119–124.

Galef, B. G., Jr., Mischinger, A., and Malenfant, S. A. (1987). Further evidence of an information centre in Norway rats. *Anim. Behav.* **35**:1234–1239.

Galef, B. G., Jr., and Sherry, D. F. (1973). Mother's milk: A medium for the transmission of cues reflecting the flavor of mother's diet. *J. Comp. Physiol. Psychol.* **83**:374–378.

Galef, B. G., Jr., and Stein, M. (1985). Demonstrator influence on observer diet preference: Analysis of critical social interactions and olfactory signals. *Anim. Learn. Behav.* **13**:31–38.

Galef, B. G., Jr., and Wigmore, S. W. (1983). Transfer of information concerning distant food in rats: A laboratory investigation of the "information centre" hypothesis. *Anim. Behav.* **31**:748–758.

Garcia, J., Ervin, F. R., and Koelling, R. A. (1966). Learning with prolonged delay of reinforcement. *Psychonomic Sci.* **5**:121–122.

Garcia, J., and Koelling, R. A. (1966). Relation of cue to consequence in avoidance learning. *Psychonomic Sci.* **4**:123–124.

Harris, L., Clay, J., Hargreaves, F., and Ward, A. (1933). The ability of vitamin B deficient rats to discriminate between diets containing and lacking the vitamin. *Proc. R. Soc. B,* **113**:161–190.

James, L. F. (1978). Overview of poisonous plant problems in the United States. In *Effects of Poisonous Plants on Livestock,* R. F. Keeler, K. R. Van Kampen, and L. F. James (Eds.). Academic Press, New York, pp. 3–5.

Kon, S. K. (1931). LVIII. The self-selection of food constituents by the rat. *Biochem. J.* **25**:473–481.

Lát, J. (1967). Self-selection of dietary components. In *Handbook of Physiology,* Vol. 1, *Alimentary Canal,* C. F. Code (Ed.). American Physiological Society, Washington DC, pp. 367–386.

Le Magnen, J., and Tallon, S. (1968). Préférence alimentaire du jeune rat induite par l'allaitment maternel. *C.R.S. Soc. Biol.* **162**:387–390.

Logue, A. W. (1979). Taste aversion and the generality of the laws of learning. *Psychol. Bull.* **86**:276–296.

Martin, L. T., and Alberts, J. R. (1979). Taste aversions to mother's milk: The age-related role of nursing in acquisition and expression of a learned association. *J. Comp. Physiol. Psychol.* **93**:430–445.

Mason, J. R., Bean, N. J., and Galef, B. G. Jr. (1988). Attractiveness of carbon disulfide to wild Norway Rats. *Proceedings of the 13th Vertebrate Pest Conference,* **13**:47–95.

Meehan, A. P. (1984). *Rats and Mice: Their Biology and Control.* Brown, Knight and Truscott, Tonbridge, Kent.

Posadas-Andrews, A., and Roper, T. J. (1983). Social transmission of food preferences in adult rats. *Anim. Behav.* **31**:265–271.

Revusky, S. (1977). The concurrent interference approach to delay learning. In *Learning Mechanisms in Food Selection,* L. M. Barker, M. Best, and M. Domjan (Eds.). Baylor University Press, Waco, TX, pp. 319–370.

Richter, C. P. (1942–1943). Total self regulatory functions in animals and human beings. *Harvey Lect. Ser.* **38**:63–103.

Richter, C. P. (1956). Salt appetite of mammals: Its dependence on instinct and metabolism. In *L'instinct dans le comportement des animaux et de l'homme.* Masson, Paris, pp. 577–629.

Richter, C. P., Holt, L., and Barelare, B. (1938). Nutritional requirements for normal growth and reproduction in rats studied by the self-selection method. *Am. J. Physiol.* **122**:734–744.

Rozin, P. (1969). Adaptive food sampling patterns in vitamin deficient rats. *J. Comp. Physiol. Psychol.* **69**:126–132.

Rozin, P. (1976). The selection of foods by rats, humans and other animals. In *Advances in the Study of Behavior,* Vol. 6, J. S. Rosenblatt, R. A. Hinde, E. Shaw, and C. Beer (Eds.). Academic Press, New York, pp. 21–76.

Rozin, P., and Kalat, J. W. (1971). Specific hungers and poison avoidance as adaptive specializations of learning. *Psychol. Rev.* **78**:459–486.

Scott, E. M. (1946). Self-selection of diet: I. Selection of purified components. *J. Nutr.* **31**:397–406.

Scott, E. M., and Quint, E. (1946). Self-selection of diet: IV. Appetite for protein. *J. Nutr.* **32**:293–301.

Scott, E. M., Smith, S., and Verney, E. (1948). Self-selection of diet: VII. The effect of age and pregnancy on selection. *J. Nutr.* **35**:281–286.

Seligman, M. E. P. (1970). On the generality of the laws of learning. *Psychol. Rev.* **77**:406–418.

Shettleworth, S. J. (1983). Function and mechanism in learning. In *Advances in Analysis of Behavior,* Vol. 3, M. Leila and L. Harzem (Eds.). Wiley, Chichester, pp. 1–37.

Steiniger, von F. (1950). Beitrage zur Soziologie und sonstigen Biologie der Wanderratte. *Z. Tierpsychol.* **7**:356–379.

Story, M., and Brown, J. E. (1987). Do young children instinctively know what to eat? The studies of Clara Davis revisited. *N. Engl. J. Med.* **316**:103–106.

Tribe, D. (1954). The self-selection of purified food constituents by the rat during growth, pregnancy and lactation. *J. Physiol.* **124**:64.

Tribe, D. (1955). Choice of diets by rats: The choice of purified food constituents during growth, pregnancy, and lactation. *Br. J. Nutr.* **9**:103–109.

Young, P. T. (1959). The role of affective processes in learning and motivation. *Psychol. Rev.* **66**:104–125.

Young, P. T. (1968). Evaluation and preference in behavioral development. *Psychol. Rev.* **75**:222–241.

Zahorik, D., and Maier, S. (1969). Appetitive conditioning with recovery from thiamine deficiency as the unconditioned stimulus. *Psychonomic Sci.* **17**:309–310.

10
Primate Gastronomy: Cultural Food Preferences in Nonhuman Primates and Origins of Cuisine

Toshisada Nishida

Kyoto University
Kyoto, Japan

I. INTRODUCTION

This paper is concerned with the similarity between human food practices and those of higher primates. First I will show that human primates share their gustatory sense with their close relatives, chimpanzees. Second, I will show the evidence that human food culture has its counterpart in nonhuman primates by illustrating, in particular, the transmission of food preferences in nonhuman primates. I will limit most of my discussion to chimpanzees, macaques, and baboons, because, like humans, they are typical omnivores and they have also been studied in many different locations.

II. GUSTATORY SENSE

Three pieces of evidence demonstrate that the gustatory sense of chimpanzees is similar to that of human beings.

First, chimpanzees cause damage to crops. Chimpanzees of Mahalé, for example, feed on sugarcane, corn stalks, and fruits of the banana, mango, lemon, and guava trees (Nishida and Uehara, 1983).

Second, the local people compete with chimpanzees for the same food items from the natural environment. For example, chimpanzees of Mahalé feed on 41 of 67 plant food types that are regarded as edible by the local people (Tongwe tribe) (Nishida, 1974, and unpublished data).

Third, I experimentally tasted some items from the food list of chimpanzees, some of which even local people do not regard as edible. Although my experimental feeding was not systematic, I have obtained some idea of the gustatory preferences of wild chimpanzees. Table 1A shows 16 major and 25 important food items of Mahalé chimpanzees, selected on a basis of long-term observations of feeding behavior (Nishida and Uehara, 1983, and unpublished data). Twenty-three of 41 food items were tasted. The taste was classified into 14 categories (Table 1B), and the palatability was quantified and scored on a 1–5 scale from "excellent" to "intolerable" (Table 1C). The result was as follows: The most important characteristic of the taste of chimpanzee foods was "sweet and sour," or simply "sweet." The average rating of palatability was 3.5 on a 5-point scale. This means that all the tested foods were more than "acceptable" to me. Other, less important food items I tasted informally were also good to eat.

III. PRIMATE "CUISINE"

Humans display diversified food culture: E. Rozin (1973, cited by P. Rozin, 1977) called "cuisine" culturally transmitted food practices that determine (a) basic foods (staple and secondary foods), which provide the principal sources of calories and other nutrients, (b) the methods of preparation of these foods, (c) the flavorings added to the basic foods, and (d) a variety of limitations on mixing of certain types of foods, taboos, and rules concerning eating.

Nonhuman primate species also display diversified food habits within their own geographic distribution. In addition to Rozin's categories, two more aspects of cuisine are relevant for exploration: (a) dietary selection (what constitutes the diet in a local population) and (b) the discrimination of food from medicine.

The most easily observable result of cultural transmission processes is the difference in modes of behavior between different local populations of a species, which is uncorrelated with gene or resource distribution (Galef, 1976). Using this technique I will describe local differences of cuisine between local populations.

A. Dietary Selection

1. Chimpanzees (*Pan troglodytes*)

Between Gombé and Mahalé, Tanzania, at least 14 plant food types that are eaten at one locale might not be eaten at the other, although identical plants are available at both areas (Nishida et al., 1983). Here are some examples:

a. Oil palm. The chimpanzees at Gombé eat the fruits, pith of leaf fronds, flowers, and wood of the oil palm. On the other hand, those at Mahalé do not eat

PRIMATE GASTRONOMY

TABLE 1A Taste and Palatability to a Human of Food Items of Wild Chimpanzees

Species	Part eaten[a]	Taste	Palatability
Major food			
Aframomum alboviolaceum	P		
Baphia capparidifolia	LS	SB	3
Cordia millenii	F	SS	3
Ficus capensis	F	SS	4
Ficus exasperata	F		
Ficus exasperata	LS	NT	2
Ficus urceolaris	LS		
Ficus vallis-choudae	F		
Garcinia huillensis	F	SO	5
Harungana madagascarienisis	F	LS	4
Pennisetum purpureum	P		
Pseudospondias microcarpa	F	SA	3
Pterocarpus tinctorius	LS		
Pycnanthus angolensis	F	SU	2
Saba florida	F	OS	4
Sterculia tragacantha	LS		
Important food			
Aframomum mala	P		
Antidesma venosum	F	LS	3
Azanza garckeana	F	SS	3
Baphia capparidifolia	BL	SB	3
Blepharis buchneri	LS	NT[b]	3
Brachystegia bussei	B	LS	3
Canthium crassum	F	SS	5
Canthium rubrocostatum	F		
Diplorhynchus condylocarpon	SE		
Erythrina abyssinica	BL		
Ficus cyathispula	F		
Ficus urceolaris	F	VS	5
Landolphia owariensis	F	SO	5
Marantochloa leucantha	P	NT	3
Myrianthus holstii	F	SO	5
Parkia filicoidea	F	SP[c]	4
Psychotria peduncularis	F	LS	3
Pterocarpus tinctorius	BL		
Pterocarpus tinctorius	SE		
Smilax kraussiana	LS	NT	3
Sterculia tragacantha	F		
Tinospora caffra	LS		
Trichilia prieuriana	L		

TABLE 1A Continued

Species	Part eaten[a]	Taste	Palatability
Uapaca nitida	F	SA	3
Uvaria angolensis	F		
Other food			
Aframomum spp.	F	LS	4
Costus afer	P	OP	4
Hymenocardia acida	SE	OO	3
Ipomoea rubens	P	BT	3
Marantochloa leucantha	P	NT	3
Parinari curatellifolia	F	SS	5
Rubus pinnatus	F	SO	5
Strychnos innocua	F	SO	4
Syzygium guineense	F	SA	3
Uapaca kirkiana	F	VS	5
Ximenia americana	F	OO	3
Ziziphus mucronata	F	LS	4

[a] P = pith; LS = leaf, shoot; F = fruit; BL = blossom; B = bark; SE = seed
[b] Spinachlike.
[c] Like boiled soybean flour.

TABLE 1B Classification of Taste

VS: Very sweet
SS: Sweet
LS: Lightly sweet
SO: Sweet and sour
SA: Sweet and astringent
SB: Sweet and bitter
SU: Sweet but unpleasant
SP: Sweet and pleasant

OP: Sour and pleasant
OS: Sour and sweet
OO: Sour
VO: Very sour tasty
BT: Bitter but tasty
NT: No taste

TABLE 1C Ratings of Palatability

5. Excellent: I am eager to eat [this food].
4. Good: I like to eat it.
3. Acceptable: I like to eat it when hungry.
2. Tolerable: I usually refuse, but may eat it when I am very hungry.
1. Intolerable: Inedible.

any part of this palm species. Chimpanzees in west Africa eat both the pulp and kernel of the oil palm, although Gombé chimpanzees reject the kernel. The palm tree is a relatively recent introduction to each Africa.

b. Blepharis leaves. The chimpanzees of Mahalé feed on the spiny leaves of *Blepharis buchneri* (Acanthaceae). Since leaves occasionally hurt their mouths and lips, the chimpanzees grimace when chewing them. In contrast, chimpanzees of Gombé have never been seen to feed on them.

c. Dry wood. The chimpanzees of Mahalé sometimes chew or lick the dry wood of a few species of trees. They visit some trees so regularly that huge "caves" are formed in the trunk. When feeding, the chimpanzees creep into these hollows so that only their rumps protrude. This sort of behavior has never been observed at Gombé, even though the same tree species are present and chimpanzees feed on their fruits (Goodall, 1986).

d. Rocks. At Mahalé, chimpanzees sometimes lick rocks lying in streambeds or on the shore of Lake Tanganyika. Curiously, this practice has never been seen at Gombé, although it is common among the baboons there (Goodall, 1986).

e. Ants. Conspicuous differences exist in regard to insect eating. Gombé chimps habitually eat driver ants but eat neither cocktail ants nor carpenter ants. On the other hand, Mahalé chimps eat the latter two habitually but reject the former completely (Nishida and Hiraiwa, 1982; McGrew, 1983).

f. Banana, lemon, mango, and guava. Here I briefly present episodes in the propagation of new food habits in a chimpanzee group.

I introduced bananas to a group of chimpanzees at their feeding site in 1966. The first visitor to come after I had hung up a bunch of bananas was an old female. She came alone, and almost immediately fed on the ripe fruits after a brief gaze. I put up another bundle of bananas. After 4 hours, the same old female came with four adult males. One of the adult males passed by the bananas, apparently not noticing them, but four others stopped on all fours in front of the bananas and began to crowd and push each other, attempting to make a close visual inspection of the fruit. One adult male sat very close to the bananas, and the adult female kept contact with the male, while the two other males almost covered her back, panting in excitement. Meanwhile, the adult female stepped in, and took and ate a few bananas. On seeing this, all the adult males immediately followed suit, reaching and beginning to eat bananas. I got the impression that the adult males would have tasted the bananas even if they had not come with the old female. However, it is true that the sight of the female eating an unfamiliar food facilitated their eating the new item.

Recently, observations similar to those I have just described were made for another group of chimpanzees, which began to eat spontaneously the fruits of cultivated plants, such as mango, guava, and lemon (Takasaki, 1983; Takahata et al., 1986). Mango eating and lemon eating spread very quickly. Mango eating spread among 19 chimps in a few minutes. Lemon feeding spread among all these chimpanzees in a few months. Apparently, the sight of certain individuals eating an unfamiliar food triggered interest in the item and induced other individuals to investigate its potential value. Thus, local enhancement is the likely mechanism of observational learning.

2. Baboons (*Papio cynocephalus*)

a. Dried fish eating. The baboons of the Gombé National Park, Tanzania, have been notorious for stealing small sardinelike fish (*dagaa*) being sun-dried by the local fishermen along the sandy beach of Lake Tanganyika. One troop of baboons ate fish practically every day for as much as 6 months (Ransom, 1981). The baboons of Mahalé, however, neglect *dagaa,* despite their availability. The baboons of Gombé presumably had longer, more peaceful contact with the fishermen because they were protected in the game reserve.

b. Live fish eating. Until recently there has been no single report confirming that wild nonhuman primates eat live fish. This may not be surprising considering that even some people, such as Tasmanians (Hiatt, 1967) and Ugandans, have no habit of eating fish. Recently, however, the eating of live fish was also confirmed in swamp-living cercopithecine monkeys: fecal analysis revealed fish bones and fish eggs. Both Allen's swamp monkey and de Brazza's monkey walk along river beds during the dry season to scoop out fish hatchlings from muddy pools in the evening and predawn hours (Zeeve, 1985). It is still unknown, however, whether this is a local food culture.

Hamilton and Tilson (1985) observed that the chacma baboons in a Namib Desert river canyon feed on both live and dead fish from drying desert pools: dead, or floating live fish were raked by hand from the surface. The baboons also walked into pools and groped about to collect larger dead fish, which sank to the bottom of deeper ponds. In larger pools, some baboons entered the water and seized active fish under boulders.

Baboons covered live fish with sand to immobilize active fish or to facilitate handling. Fish larger than 10 cm were opened with a bite in the anal region. By using the fingers and mouth the flank skin was peeled back toward the head. Only the flesh and gills were eaten.

In the Okavango Swamp, however, during their 7-year-study, Hamilton and Tilson never saw baboons attempt to capture fish in drying pools, nor did they observe situations in which dead fish were available to the baboons. Fish-eating birds there eliminate most fish from pools before they concentrate to the extent noted in the Namib Desert. Thus, live fish capture by this population was made possible by the concentration of fish in small drying pools and the paucity of competing fish-eating birds.

3. Japanese Macaques (*Macaca fuscata*)

a. Elm fruits. The monkeys of Arashiyama feed on the pulp of elm fruits [*Aphananthe aspera* (Thunb.) Planchon] and also on their seeds, by biting open the seed shell. Those of Takasakiyama, however, feed only on the pulp and swallow the seeds intact (Itani, personal communication).

b. Barks. The monkeys of Koshima Island have never been observed to feed on barks of the trees (Watanabe, 1989). In contrast, monkeys of Shimokita and Shiga feed intensively on the barks of trees. Monkeys that live in such snowy areas of temperate deciduous forests depend for their survival in winter on bark feeding (Izawa and Nishida, 1963; Suzuki, 1965). However, differences in the climate and vegetation alone are not enough to explain the lack of bark eating in the monkeys of Koshima, since the monkeys of Takagoyama also feed on the barks of various trees despite living in a similar warm, temperate evergreen forest to the monkeys of Koshima (Nishida, unpublished; Uehara, unpublished).

c. Bulbs. The Japanese monkeys of two troops of Minoo, Osaka, regularly dig and feed on the roots of *Dioscorea* and lilies, while the monkeys of Takasakiyama were never observed to do so, although the plants were available (Kawamura, 1965). It is unknown whether the qualities of each food type are similar or different between the local populations, because the chemical composition of plants sometimes varies within species (Jones et al., 1978; Glander, 1981; van Roosmalen, 1982). A comparative chemical analysis of at least some of the food items is therefore needed.

d. Animal diet of monkeys of Yakushima. Monkeys of Yakushima Island have been observed to prey on frogs (Suzuki, unpublished), lizards (Tsukahara, unpublished), a bird nestling or egg (Takahata, unpublished), and freshwater crabs (Kuroda, unpublished). There has been practically no record of feeding on foods of these types for monkeys elsewhere. However, observation hours in the natural environment have been much longer at Yakushima than at other localities, where observations were mostly made on provisioned troops. It is therefore premature to conclude that the monkeys of Yakushima have a special animal diet.

e. Development of fish eating in the Koshima troop. The Japanese macaques of Koshima Island recently began to eat raw fish and octopus. As recounted by Watanabe (1983, 1989), fish (but not octopus) were either washed ashore by the waves or discarded by fishermen. Formerly, monkeys would neglect fish dried by fishermen on the sandy beach. Fish eating was first observed by a fisherman in 1974. However, the confirmation by a primatologist did not come until 1980, when an old male was observed eating a dried fish. For the next two years observations of fish eating were made only sporadically: a 15-year-old male in 1981 and 4 adult females in 1982. However, a sudden upsurge occurred in 1983, when as many as 32 monkeys ate fish. Another 20 monkeys began to eat fish in the following two years. Thus, almost all the monkeys acquired the fish-eating habit in a few years. The interesting thing about fish eating is that adult males were the first to try the new food.

Monkeys do not appear to like fish very much. In summer, when other food is available, monkeys rarely eat them. Monkeys eat only flesh, avoiding viscera, head, and bones. The onset of fish eating was presumably triggered by starvation, because this habit began after provisioning with wheat was stopped in 1972. Fish eating has not been reported among Japanese macaques of other areas.

B. Acquisition and Processing of Food

Techniques of food manipulation may also be largely acquired through daily observational learning. Chimpanzees and other nonhuman primates can differ in the way they process the same food items.

1. Chimpanzees (*Pan troglodytes*)

a. Strychnos fruit. To open hard-shelled fruits, such as *Strychnos*, the chimpanzees of Gombé bang them against tree trunks or rocks. At Mahalé, the same kind of fruits are always bitten open (Nishida et al., 1983).

b. Hard-shelled nuts. The use of stone and a wooden club as a hammer and anvil for processing hard-shelled nuts provides a remarkable example of local culture. Chimpanzees of west Africa, Sierra Leone, Liberia, Guinea, and the Ivory Coast put hard nuts (e.g., oil palm nuts), as well as fruits of *Panda, Coula,* and *Parinari* on a rock or a tree root, and strike the nuts with a stone hammer or a stout stick. If the materials for the tool are not close at hand, they are sought elsewhere and transported hundreds of meters. In addition, stone anvils are believed to be used repeatedly for many years, and probably from generation to generation.

The hammer-and-anvil technique has never been observed in the chimpanzees of east Africa. The presence or absence of this technique cannot be ascribed simply to the presence or absence of either tool materials or hard nuts. Gombé chimpanzees, for instance, eat the pulp of the oil palm and use stones in social display (Goodall, 1986).

c. Water. Occasionally Gombé chimpanzees crush leaves and use them as sponges to soak up drinking water from holes in the trees. This technique of "leaf sponging" has never been seen in the chimpanzees of Mahalé, although they put their hands into such holes and lick them.

Leaf sponging for water was recently reported for a group of chimpanzees of Bossou, Guinea. However, it is unknown whether this is an established culture. Also, an adult female at Mahalé was observed to use a leafy sponge once in 1988 (Takasaki, unpublished). This is the first observation since the beginning of our study in 1965. It is, therefore, not a culturally established pattern in Mahalé.

d. Driver ants. A Gombé chimpanzee will put a long stick or branch into a swarm of driver ants. Hundreds of ants then stream up it. The chimpanzee watches their progress and when the ants have almost reached its hand, the tool is quickly withdrawn. Immediately, the opposite hand sweeps the length of the tool catching the ants in a mass. These are popped into the open mouth (McGrew, 1974). Chimpanzees of Mahalé do not know this "ant-dipping" technique, but display their own unique technique of ant fishing in trees (Nishida and Hiraiwa,

PRIMATE GASTRONOMY

1982). In both ant fishing and ant dipping, a great amount of trial-and-error learning appears to be needed for tool manipulation, and especially for tactics to counter the ants' antipredator responses.

e. Termites. In catching termites, chimpanzees of Rio Muni and Cameroon, central Africa, appear to use sticks in a different way from that of those in east Africa. The length and thickness of the stick, and the circumstantial evidence surrounding the sticks when they were found, suggest that they were not used for fishing termites but probably for "perforating" the termite mound. Chimpanzees then appear to pick out the termites by hand (Sabater Pi, 1974; Sugiyama, 1985). Such digging sticks have never been used by chimpanzees of east Africa such as in Gombé or Mahalé.

2. Japanese Macaques (*Macaca fuscata*): Seaweed

The Japanese macaques of Kinkazan Island, in northern Japan, feed on a type of seaweed (*Undaria* sp.). When the sea is rough, collecting the seaweed is a very hazardous activity: a monkey enters the water and pulls the seaweed quickly when the high waves are receding; it retreats immediately to dry land before the waves return. Young monkeys do not collect the food because they are afraid of high waves.

The monkeys of the Shimokita Peninsula feed on seaweed washed ashore by waves. However they never enter into the sea (Izawa and Nishida, 1963). The monkeys of Yakushima Island neither enter the sea nor eat anything that is connected with the sea (Furuichi and Takahata, personal communication).

3. Baboons (*Papio cynocephalus*)

Marais (1969) described remarkable local traditions for the acquiring and processing of food or water among chacma baboons. For example, members of a particular troop of baboons dig lily bulbs in a unique way, or break open baobab fruits using a stone. Another troop, however, does not know how to get them. Strum (1975) described how the baboons of Kenya developed a sophisticated way of hunting mammals when competing larger carnivores were removed by people.

4. Capuchin Monkeys (*Cebus apella*)

Brown capuchins of La Macarena National Park, Colombia, employ ingenious techniques of exploiting palm nuts (*Astrocaryum chambira*), according to the ripeness of the palm (Izawa and Mizuno, 1977; Struhsaker and Leland, 1977; Izawa, 1978, 1979). Capuchins also acquire in a unique way frogs, termites, and ants that occur inside of the hollow-stemmed bamboo. Judging from their sophistication, these techniques may very probably be cultural behaviors (Nishida, 1987).

C. Food Versus Drugs

Hamilton et al. (1978) reported a suspected case of drug use by chacma baboons (*Papio cynocephalus ursinus*): these baboons are occasionally observed to feed on traces of poisonous plants (*Datura* spp.). Janzen (1978) listed many anecdotal but suggestive cases of the use of secondary compounds as a medicine by mammals.

Chimpanzees of Mahalé and Gombé display unusual behavioral patterns when feeding on certain plant leaves. They ingest whole leaves of these plants very slowly and without chewing them (Wrangham and Nishida, 1983; Takasaki and Hunt, 1987; Nishida, unpublished). These plant species are used by native Africans as medicinal substances (Watt and Gerdina, 1962). The chemical analysis of one of the plants [*Aspilia mossanbicensis* (Oliv.) Wild] revealed that the leaves included strong antibiotics (Rodriguez et al., 1985). Circumstantial evidence suggests the possibility that *Aspilia* leaves function in expelling intestinal nematodes.

Recently, an adult female chimpanzee who had shown symptoms of a gastrointestinal disorder was observed feeding on stems of *Vernonia* (Compositae), which her companions totally neglected. The female soon recovered from illness (Huffman, 1989).

Thus, chimpanzees show different behaviors when feeding on their usual foods and on druglike foods. I point out that the use of plants as medicine, the "ethnomedicine" of nonhuman primates, is a promising area to seek for cultural diversity.

D. Cooking

Cooking is defined here as any modification of the raw material of foods in a standardized fashion; "modification" includes changing of shape, blending of materials (including "flavoring"), and manipulation of heat (heating, freezing, and smoking).

In nonhuman primates only borderline cases of blending of materials are known.

1. Seasoning

The Japanese macaques of Koshima Island dip sweet potatoes into seawater and eat them. This behavior was initially derived from potato-washing behavior. Later the potatoes were clean when they were provided to the monkeys, because the producers sell only cleaned potatoes. Since, therefore, it is not necessary for the monkeys to clean the potatoes, it seems that dipping them into the sea is a means of getting the salty taste (Kawai, 1965).

2. Wadging

When feeding on meat, eggs, or bananas, chimpanzees pick off green or dry leaves, which they combine with the food (Nishida and Uehara, 1983). Then, the pulpy mass or wadge can be chewed and sucked. It is possible that chimpanzees add some flavor to their favorite food items. An alternative interpretation suggested by Teleki (1973) is that this manner of eating is a means of prolonging the pleasure of masticating flavored foods.

E. Origin of Food Taboo?

The cultural imposition of the "inedible" ban on potentially edible food items appears to be a uniquely human characteristic. However, the origin of such prohibition may have been related to the availability of the items.

Let me give an example. Throughout Africa leopards are rarely, if ever, eaten, and in many places eating them is a taboo. At Mahalé we had an interesting observation concerning a leopard cub (Hiraiwa-Hasegawa et al., 1986). A group of chimpanzees, including adult and immature males and females, found a leopard cub hidden inside a rock cave. One of the adult males dragged it off and killed it. However, none of them would eat the cub. Instead, a juvenile male treated it as a toy and a juvenile female "cared for" it as if it were a doll. Thus, chimpanzees did not regard the leopard as food. This is rather surprising because the chimpanzees of Mahalé eat a variety of mammals including monkeys, bushpigs, antelopes, rodents, and hyraxes, and they occasionally even scavenge dead mammals (Hasegawa et al., 1983). Why do both humans and chimpanzees avoid eating leopards? Both species rarely have the opportunity to have a leopard in hand. This may lead to the aversion to eating leopards, through the principle that Rozin (1977) calls "Strangeness breeds contempt."

IV. DISCUSSION

Information on food may be the most common subject of social transmission in primates and other vertebrates (see Galef, 1976, for review). Monophagous animals such as anteaters might depend a great deal on genetically programmed behavior in their search for prey. On the other hand, polyphagous primates such as chimpanzees, macaques, or capuchins have so varied a food repertoire that they cannot encode the food types and feeding techniques in their genome. Like other animals, these primates are equipped with an inherited gustatory sense to detect nutrients as sweet and tasty, and low-quality and poisonous food as unpleasant or bitter. This gustatory ability, supplemented by attentiveness to the behavior of conspecifics, is a basis for innovative and imitative feeding.

Omnivorous primates are always examining their environment. When a new item becomes available or a particular food item increases in quantity owing to the environmental change, they will sooner or later feed on it experimentally and incorporate it into their food list if it is acceptable. An omnivorous primate frequently peers at his companion's face and sniffs his mouth or foodstuffs, even though the food items are not new. Attentiveness to the natural and social environments is the way of life for primate omnivores.

There may be various explanations for the origin of cultural differences. Most cultural differences in diet selection are more or less related to local differences in habitat environments.

Relative availability limits opportunities to experiment with (e.g., inspect and taste) potential food items. A long history of contact with humans and crop cultivation often provides the animals with more opportunity for experimental feeding. Thus, the chimpanzees of Gombé eat oil palms, but those of Mahalé neglect them.

The presence of food competitors may force animals to give up a certain kind of food item. Fish is exploited by the baboons of the Namib Desert, where fish-eating birds such as storks are rare.

The characteristics of the food type itself may bring out local differences. "Difficult" foods such as driver ants or spiny leaves are least likely to be eaten as an experiment. Moreover, even if they are eaten, they are likely to discourage an animal that tastes them for the first time, and as a result will more likely be ignored by the local group.

Why do the chimpanzees of Mahalé eat carpenter ants but not driver ants, whereas those of Gombé eat driver ants but not carpenter ants? Perhaps the type of ant first eaten in each locality was determined by its relative availability. However, once one kind of ant had been habitually eaten in one place, another species was neglected altogether apparently because the former could be collected more efficiently through the specialization of feeding technique.

Cultural dietary differences in nonhuman primates might originate largely from subtle differences in the environment. As long as the environment influences the selection or rejection of possible food items, the determinant of origins in dietary cultures would seem to be environmental rather than arbitrary.

The hypothesis above can be tested only by the extensive collection of quantitative data on the environment: the absolute and relative abundance of each food item, abundance of food competitors, and chemical analysis of food items.

ACKNOWLEDGMENTS

I thank M. I. Friedman and B. Galef for inviting me to this symposium, and T. Furuichi, M. Kawai, S. Kuroda, S. Suzuki, Y. Takahata, H. Takasaki, T. Tsukahara, S. Uehara, and K. Watanabe for providing me with their unpublished data.

DISCUSSION

Schneider: In the case of the leopard in the tree, wouldn't both humans and chimps know that the leopards were highly predatory and dangerous?

Nishida: I believe, both of them knew that a leopard in the tree could be very dangerous. However, when on the ground, a leopard, even an adult, usually does not pose a grave threat to an adult human or chimpanzee. Moreover, perhaps the male chimpanzee that entered the cave would have had some evidence that there was no mother leopard in the cave.

REFERENCES

Galef, B. G. (1976). Social transmission of acquired behavior: A discussion of tradition and social learning in vertebrates. In *Advances in the Study of Behavior*, Vol. 6, J. S. Rosenblatt, R. A. Hinde, E. Shaw, and C. Beer (Eds.). Academic Press, New York, pp. 77–100.

Glander, K. E. (1981). Feeding patterns in mantled howling monkeys. In *Foraging Behavior: Ecological, Ethological, and Psychological Approaches*, A. Kamil and T. D. Sargent (Eds.). Garland Press, New York, pp. 231–259.

Goodall, J. (1986). *The Chimpanzees of Gombé*. Harvard University Press, Cambridge, MA.

Hamilton, W. J., III, and Tilson, R. L. (1985). Fishing baboons at desert waterholes. *Am. J. Primatol.* **8**:255–257.

Hamilton, W. J., III, Buskirk, R. E., and Buskirk, W. H. (1978). Omnivority and utilization of food resources by chacma baboons, *Papio ursinus. Am. Nat.* **112**:911–924.

Hasegawa, T., Hiraiwa-Hasegawa, M., Nishida, T., and Takasaki, H. (1983). New evidence of scavenging behavior of wild chimpanzees. *Curr. Anthropol.* **24**:231–232.

Hiatt, B. (1967). The food quest and the economy of the Tasmanian aborigines. *Oceania*, **38**:190–219.

Hiraiwa-Hasegawa, M., Byrne, R. W., Takasaki, H., and Byrne, M. E. (1986). Aggression toward large carnivores by wild chimpanzees of Mahalé Mountains National Park, Tanzania. *Folia Primatol.* **47**:8–13.

Huffman, M. A. (1989). Observations on the illness and consumption of a possibly medical plant *Vernonia amygdalina* (Del.), by a wild chimpanzee in the Mahalé Mountains National Park, Tanzania. *Primates*, **30**:51–63.

Izawa, K. (1978). Frog-eating behavior of wild black-capped capuchin (*Cebus apella*). *Primates*, **19**:633–642.

Izawa, K. (1979). Foods and feeding behavior of wild black-capped capuchin (*Cebus apella*). *Primates*, **21**:57–76.

Izawa, K., and Mizuno, A. (1977). Palm-fruit cracking: Behavior of wild black-capped capuchin (*Cebus apella*). *Primates*, **18**:773–792.

Izawa, K., and Nishida, T. (1963). Monkeys living in the northern limits of their distribution. *Primates*, **4**:67–88.

Janzen, D. H. (1978). Complications in interpreting the chemical defenses of trees against tropical arboreal plant-eating vertebrates. In *The Ecology of Arboreal Folivores*, G. G. Montgomery (Ed.). Smithsonian Institution Press, Washington, DC, pp. 73–84.

Jones, D. A., Keymer, R. J., and Ellis, W. M. (1978). Cyanogenesis in plants and animal feeding. In *Biochemical Aspects of Plant and Animal Co-Evolution*, J. B. Harborne (Ed.). Academic Press, London, pp. 21–34.

Kawai, M. (1965). Newly acquired precultural behavior of the natural troop of Japanese monkeys on Koshima Islet. *Primates*, **6**:1–30.

Kawamura, S. (1965). Sub-culture among Japanese macaques. In *Monkeys and Apes: Sociological Studies*, S. Kawamura and J. Itani (Eds.). Chuokoronsha, Tokyo, pp. 237–289.

McGrew, W. C. (1974). Tool use by wild chimpanzees in feeding upon driver ants. *J. Human Evol.* **3**:501–508.

McGrew, W. C. (1983). Animal foods in the diet of wild chimpanzees. Why cross-cultural variation? *J. Ethol.* **1**:46–61.

Marais, E. (1969). *The Soul of the Ape*. Penguin Books, Harmondsworth, pp. 52–59.

Nishida, T. (1974). Ecology of wild chimpanzees. In *Human Ecology*, R. Ohtsuka, J. Tanaka, and T. Nishida (Eds.). Chuokoronsha, Tokyo, pp. 15–60.

Nishida, T., and Hiraiwa, M. (1982). Natural history of a tool-using behavior by wild chimpanzees in feeding upon wood-boring ants. *J. Hum. Evol.*, **11**:73–99.

Nishida, T. (1987). Local traditions and cultural transmission. In *Primate Societies*, B. B. Smuts, D. L. Cheney, R. M. Seyfarth, R. W. Wrangham, and T. T. Struhsaker (Eds.). University of Chicago Press, Chicago, pp. 462–474.

Nishida, T., and Uehara, S. (1983). Natural diet of chimpanzees (*Pan troglodytes schweinfurthii*): Long-term record from the Mahalé Mountains, Tanzania. *Afr. Study Monogr.* **3**:109–130.

Nishida, T., Wrangham, R. W., Goodall, J., and Uehara, S. (1983). Local differences in plant-feeding habits of chimpanzees between the Mahalé Mountains and Gombé National Park, Tanzania. *J. Hum. Evol.* **12**:467–480.

Ransom, T. W. (1981). *Beach Troop of the Gombé*. Bucknell University Press, Lewisburg, PA, pp. 67–72.

Rodriguez, E., Aregullin, M., Nishida, T., Uehara, S., Wrangham, R. W., Abramowski, Z., Finlayson, A., and Towers, G. H. N. (1985). Thiarubrine A, a bioactive constituent of *Aspilia* (Asteraceae) consumed by wild chimpanzees. *Experientia*, **41**:419–420.

Rozin, P. (1977). The significance of learning mechanisms in food selection: Some biology, psychology and sociology of science. In *Learning Mechanisms in Food Selection*, L. M. Barker, M. R. Best, and M. Domjan (Eds.). Baylor University Press, Waco, TX, pp. 557–589.

Sabater Pi, J. (1974). An elementary industry of the chimpanzees in the Okorobiko Mountains, Rio Muni (Republic of Equatorial Guinea). *Primates*, **15**:351–364.

Struhsaker, T. T., and Leland, L. (1977). Palmnut smashing by *Cebus a. apella* in Colombia. *Biotropica*, **9**:124–126.

Strum, S. C. (1975). Primate predation: Interim report on the development of a tradition in a troop of olive baboons. *Science*, **56**:44–68.

Sugiyama, Y. (1985). The brush-stick of chimpanzees found in southwest Cameroon and their cultural characteristics. *Primates*, **26**:361–374.

Suzuki, A. (1965). An ecological study of wild Japanese monkeys in snowy areas—Focused on their food habits. *Primates*, **6**:31–72.

Takahata, Y., Hiraiwa-Hasegawa, M., Takasaki, H., and Nyundo, R. (1986). Newly acquired feeding habits among the chimpanzees of the Mahalé Mountains National Park, Tanzania. *Hum. Evol.* **1**:277–284.

Takasaki, H. (1983). Mahalé chimpanzees taste mangoes—Toward acquisition of a new food item? *Primates*, **24**:273–275.

Takasaki, H., and Hunt, K. (1987). Further medical plant consumption in wild chimpanzees? *Afr. Study Monogr.* **8**:125–128.

Teleki, G. (1973). *The Predatory Behavior of Wild Chimpanzees.* Bucknell University Press, Lewisburg, PA.

van Roosmalen, M. G. M. (1982). Habitat preferences, diet, feeding behavior and social organization of the black spider monkey, *Ateles paniscus paniscus,* in Surinam. Ph.D. thesis, University of Waageningen, Holland.

Watanabe, K. (1983). Monkeys eating fish. *Monkey,* **27**(5, 6):24.

Watanabe, K. (1989). Fish: A new addition to the diet of Japanese macaques on Koshima Island. *Folia Primatol.* **52**:124–131.

Watt, J. M., and Gerdina, M. (1962). *The Medicinal and Poisonous Plants of Southern and Eastern Africa,* 2nd ed. E. & S. Livingstone, Edinburgh and London.

Wrangham, R. W., and Nishida, T. (1983). *Aspilia* spp. leaves: A puzzle in feeding behaviour of wild chimpanzees. *Primates,* **24**:276–282.

Zeeve, S. R. (1985). Swamp monkeys of the Lomako Forest, Central Zaire. *Primate Conserv.* No. 5, pp. 32–33.

11
Innate, Learned, and Evolutionary Factors in the Hunger for Salt

Jay Schulkin

University of Pennsylvania
Philadelphia, Pennsylvania

I. INTRODUCTION

Curt Richter showed that the adrenalectomized rat chronically loses sodium, is in a fatal situation, and ingests salt to survive (Richter, 1936; Fig. 1). Moreover, when offered choices among solutes, these salt-hungry rats tend to ingest the sodium salts (see, e.g., Clark and Clausen, 1942; Richter and Eckert, 1938).

Richter hypothesized that the appetite for salt that results from body sodium deficiency is innate (1939, 1956). Others concurred (Bare, 1949). One study, for example, demonstrated that salt-hungry rats ingest salt within an hour upon first exposure to it (Epstein and Stellar, 1955). Later studies provided persuasive evidence that when salt hungry for the first time, rats and sheep would ingest salt by 20 minutes (Falk and Herman, 1961) or 10 minutes (Nachman, 1962; Denton and Sabine, 1963) upon being exposed to it. This response was also demonstrated within seconds (Handal, 1965; Wolf, 1969; Fig. 2), and it occurs in the 12-day-old suckling rat that has never tasted NaCl and has never been sodium deficient before (Moe, 1986). It is believed that this innate response, at least in the adult, is quite specific for sodium salts, because other salts tend not to be immediately ingested by the salt-hungry rat (see, e.g., Nachman, 1962; Handal, 1965; Fig. 2; see also Denton, 1982).

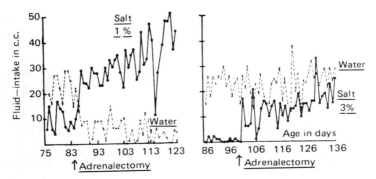

FIGURE 1 Twenty-four hour salt and water intake before and following adrenalectomy. (From Richter, 1936.)

Salt-hungry rats also ingest salty food upon immediate exposure (Bolles, et al., 1964). Unlike their behavior with other mineral or vitamin deficiencies (Rozin, 1976), when sodium-deprived rats are offered a choice of diet with added Na, or a novel diet but without the needed salt, they still prefer the old but now salted food (Rodgers, 1967).

It had been suggested for some time that the taste of salt is more rewarding when the rat is in need of sodium (Richter, 1956; Nachman, 1962; Morrison and Young, 1972; Denton, 1982). This phenomenon was couched under the "avidity theory" (Katz, 1937; Young, 1952; or for an analogous concept, see Cabanac, 1979—alliesthesia), which holds that salt tastes better to the salt-hungry animal. There is evidence now to support this hypothesis. By using the taste–reactivity methodology (Grill and Norgren, 1978) to measure the oral–facial profile to intraorally infused solutes, it was shown that rats switch from their mixed ingestion–rejection response to hypertonic NaCl when they are sodium replete to a wholly ingestive sequence when they are sodium deplete (Berridge et al., 1984; Schulkin et al., 1985; Grill and Bernstein, 1988; Berridge and Schulkin, 1989; Fig. 3). They did not change their ingestive response to HCl at a concentration that provokes equally mixed ingestive–aversive responses. Importantly, the rats displayed this shift in their oral–facial profile the first time they were salt hungry and exposed to the NaCl. This positive facial display to infusions of hypertonic NaCl may be mediated by gustatory fibers in the seventh cranial nerve that decrease their firing patterns to the NaCl when the rat is salt hungry (Contreras, 1977), and the gustatory region of the solitary nucleus, which also changes its firing pattern when the animal is salt hungry (Jacobs et al., 1989). When sodium deprived, the "sodium best" neurons in the solitary nucleus are activated much less to infusions of NaCl into the oral cavity, while "sucrose best" neurons are activated much more; other taste modalities tend not to change.

FACTORS IN THE HUNGER FOR SALT

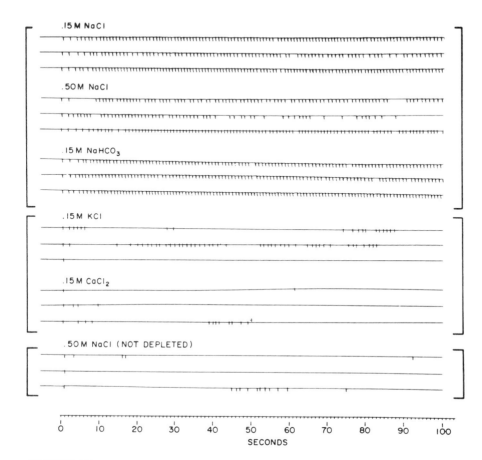

FIGURE 2 Drinkometer records of individual rats. Every fifth lick is represented by a vertical mark. Records within the upper brackets are from 9 sodium-depleted rats drinking solutions of sodium salts. Records within the middle brackets are from 6 sodium-depleted rats drinking solutions of nonsodium salts. Records within the lower brackets are from 3 nondepleted rats drinking a sodium salt solution. (From Handal, 1965; Wolf, 1969.)

II. LATENT LEARNING STUDIES

Krieckhaus and Wolf (1968) and Krieckhaus (1970) provided evidence that thirsty rats taught to press a bar that delivered isotonic saline, or run down an alleyway for it, learn something about the availability of saline. When rendered salt hungry and no longer thirsty immediately after this experience, they bar pressed or ran down the alley for the salt under extinction conditions even though

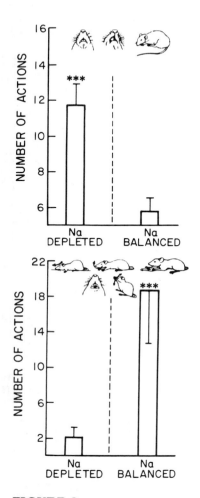

FIGURE 3 Taste reactivity profile to 0.5 M NaCl while either sodium balanced or sodium depleted. Top: Combined mean (SEM) number of ingestive actions (rhythmic tongue protrusions, nonrhythmic lateral tongue protrusions, and paw licks). Bottom: Combined mean (SEM) number of aversive actions (chin rubs, head shakes, paw treads, gapes, face washes, and forelimb flails).

there was now no salt available (Fig. 4). They did not bar press (or run significantly) for other solutes (Fig. 5). However, after an 8-hour delay they bar pressed about the same (having been trained with either NaCl or KCl) under these extinction conditions (Dickinson and Nicholas, 1983b). One hypothesis is that what they remember over time is ''saltiness,'' perhaps because the other attributes of KCl fade. That is, the preferred salient taste that is tied to memory is its being

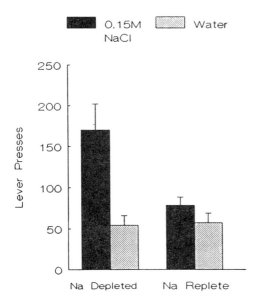

FIGURE 4 Mean (SEM) number of lever presses of rats during a 1-hour extinction test. Rats tasted either NaCl or water during the training, and are now sodium depleted or sodium replete. (From Krieckhaus and Wolf, 1968.)

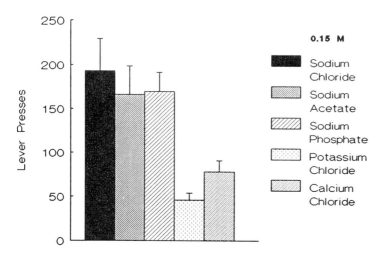

FIGURE 5 Mean (SEM) number of lever presses of rats during a 1-hour extinction test. All rats are sodium depleted after having been exposed to one of the five salt solutions. (From Krieckhaus and Wolf, 1968.)

salty—Shakespeare's "saltiness of time." Potassium chloride tastes somewhat salty to humans (see, e.g., Schulkin, 1982), and therefore it is conceivable that the other taste properties of the KCl solution drop out, and the salt-hungry rat bar presses for the salty taste.

The finding of Krieckhaus and Wolf has been replicated a number of times (see, e.g., Dickinson and Nicholas, 1983a,b; Weisinger et al., 1970), and other drives such as hunger or thirst do not elicit this behavior for salt (Kharavi and Eisman, 1971; Dickinson and Nicholas, 1983; Dickinson, 1986). Moreover, just a brief exposure to the saline is sufficient to elicit this effect; that is, thirsty rats bar pressing for water and then given 2 to 5 licks (of saline) over 2–4 minutes, just prior to being rendered salt hungry, demonstrate the phenomenon (Wirsig and Grill, 1982; Bregar et al., 1983; Fig. 6). In addition, rats with central gustatory damage, disconnecting afferents to the ventral forebrain, which normally do not ingest salt (Wolf, 1968), will do so if they taste salt preoperatively (Ahern et al., 1978; Paulus et al., 1984). Just 30 seconds of exposure to the saline is sufficient to elicit this protective effect (Hartzell et al., 1985). If the salt is moved to a new position postoperatively (Fig. 7), they do not ingest the salt (Paulus et al., 1984; Fig. 8). Place learning for salt, therefore, figured importantly in these animals.

Salt-hungry rats not only overcome any aversion they may have acquired to salt

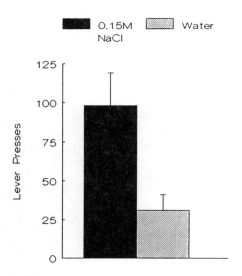

FIGURE 6 Mean (SEM) number of lever presses of rats during a 1-hour extinction test. Sodium-depleted rats tasted either NaCl or water during the training.

FACTORS IN THE HUNGER FOR SALT

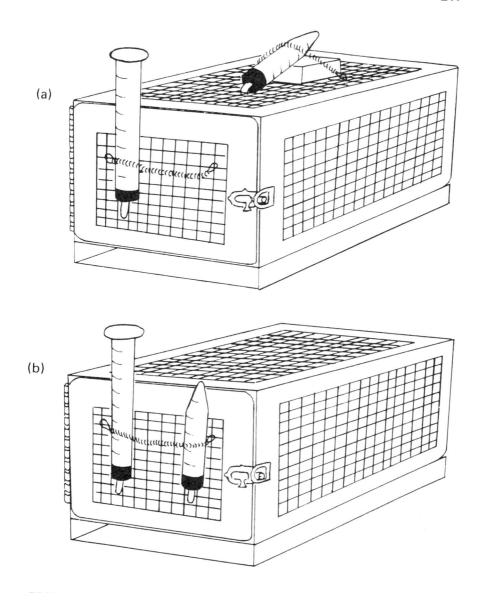

FIGURE 7 (a) Position of 0.5 M NaCl and water tubes during preoperative experience for the same-place group. (This is also the position of two water tubes during preoperative experience for the water control group and the position of salt and water tubes during postoperative salt appetite test for all groups.) (b) Position of salt and water tubes during preoperative experience for the different-place groups.

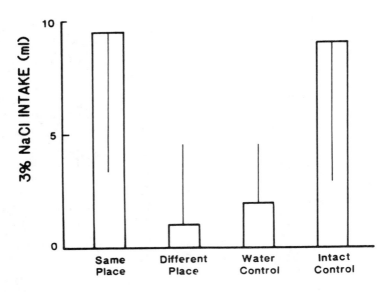

FIGURE 8 Median increases of salt intake over baseline intake, with the interquartile range indicated, for the experimental setup of Figure 7.

when sodium replete (see, e.g., Stricker and Wilson, 1970), they also learn to associate another taste with sodium when the two are given together as a compound stimulus. For example, thirsty rats offered a mixture of sodium and quinine, and then rendered salt hungry but no longer thirsty, will ingest the quinine; those not exposed to the salt–quinine mixture will not (Rescorla and Freberg, 1978). Similar effects have been found with other tastes associated with sodium in rats that were not water-deprived during the learning period (Fudim, 1978; Berridge and Schulkin, 1981; Fig. 9). Moreover, salt-hungry rats demonstrate a change in their oral–facial profile to the taste associated with sodium (when they were sodium replete) when these rats are salt hungry for the first time (Berridge and Schulkin, 1989; Fig. 10).

The idea of the above-mentioned latent learning studies was to demonstrate that animals could learn about the significance of salt and how to obtain it or what it is associated with at a time in which they were not in need of sodium. But water deprivation, which was used in many of the studies, results in a sodium loss of about 2 milliequivalents a day and a salt appetite (Weisinger et al., 1983). In other words, the water-deprived rats have a salt appetite. The validity of the latent learning studies could be questioned on that account. However, three of the studies did not use water deprivation. Two of these used sensory preconditioning (Fudim, 1978; Berridge and Schulkin, 1989), and the third used place learning (Paulus et al., 1984). All these studies demonstrated that rats learned where salt

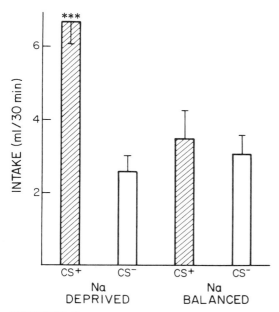

FIGURE 9 Mean (SEM) amounts consumed of the conditioned stimulus (CS$^+$-quinine or citric acid) and CS$^-$ solutions in a 30-minute intake test while either sodium balanced or sodium depleted.

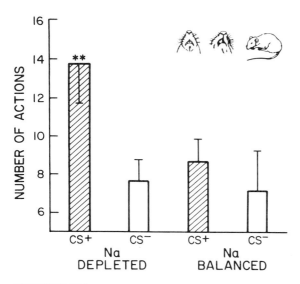

FIGURE 10 Ingestive taste reactivity to conditioned tastes. Mean (SEM) number of combined ingestive actions emitted to the CS$^+$ (quinine or citric acid) and CS$^-$ while either sodium depleted or sodium replete; for oral–facial profiles, see legend to Figure 3.

was located or what it was associated with when they were sodium replete and then used this knowledge when subsequently salt hungry.

III. SALTY TASTE INGESTION

I have suggested that the salt-hungry rat is searching for a salty taste (Schulkin, 1982). The evidence for this claim stems from the fact that lithium chloride (LiCl) is also ingested by salt-hungry rats (see, e.g., Nachman, 1963; Schulkin, 1982; Fig. 11). In fact, salt-hungry rats will run down a runway faster for LiCl than for some sodium salts (Schulkin et al., 1985a). In rats, lithium and sodium are known to activate the same gustatory sodium transport system, whereas potassium does not activate this system (see, e.g., Boudreau et al., 1983). In addition, KCl to some extent is also ingested by the salt-hungry rat (Falk, 1965; Jalowiec et al., 1966; Schulkin, 1982; Fig. 12). In some instances, rats will ingest more KCl than NaCl solution (Falk, 1965), though this phenomenon is rather rare. In humans, LiCl is reported to taste very salty, whereas KCl tastes somewhat salty but also

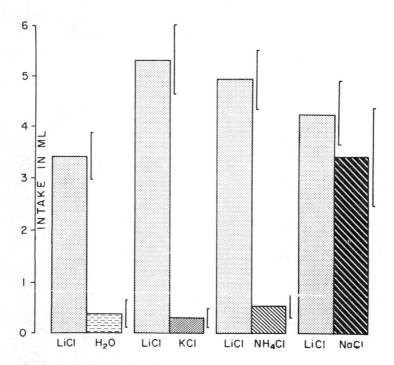

FIGURE 11 Mean intake of solutions in 15 minutes by adrenalectomized rats in two-bottle preference tests. The salts are all 0.12 M solutions, standard error of each mean is marked off by vertical lines. (From Nachman, 1963.)

FACTORS IN THE HUNGER FOR SALT

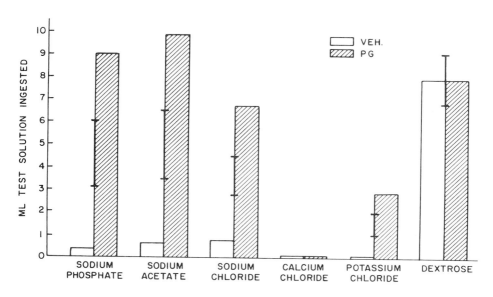

FIGURE 12 Ingestion of test solutions during a 15-minute period when offered one of the test solutions in addition to water. (From Jalowiec et al., 1966.)

bitter and chalky (see, e.g., Bartoshuk, 1980; Schulkin, 1982). Sodium deficiency is also known to reduce the firing rate of Na-sensitive fibers in the chorda tympani or seventh cranial nerve when sodium and lithium, and to a lesser extent KCl, are delivered to the oral cavity (Contreras et al., 1984). Furthermore, sodium-deprived humans in New Guinea grind up and eat potassium (Denton, 1982), perhaps because it tastes salty. This is of course speculative. What is not, however, is that when Na-deprived rats are offered a choice based on human psychophysical judgments of saltiness, they tend to choose the saltier solutions (Schulkin, 1982). This holds for both sodium and nonsodium salts. Recall that NaCl is a prototypical salty taste. All other salts have additional taste properties to saltiness (Bartoshuk, 1980; Schiffman, 1980). This may be why it is preferred to other salt solutions (Schulkin, 1982).

Two types of "salt best" fiber in the chorda tympani nerve have been hypothesized, a specialist that responds to sodium and lithium and a generalist that is largely responsive to other or mixed salty tastes (Frank, 1985). The specialist Na taste fiber picks out the NaCl or LiCl within seconds, but is not responsive to lesser salty tasting solutions like KCl. One hypothesis is that KCl is tasted over time, possibly by the activation of the generalist taste system.

But the specialist salty taste fibers and the generalist salty taste fibers should be activated by the ingestion from salt licks in nature. After all, in nature sodium-deficient animals are not offered a pure sodium stimulus. They are not provided pure NaCl (or Na_2CO_3 as is used in the sheep studies) to drink. The ingesta are mixed with a variety of minerals, and the salty taste is a mixed one. The debate about the specificity of the appetite is therefore to some extent contrived. It isn't as if salt-hungry rats or sheep do not generally ingest sodium ions preferentially under laboratory conditions. They certainly do. But the fact that they ingest nonsodium salts like potassium suggests that the nonspecific "salt best" gustatory fibers may be playing a role in the selection process. And the mixed taste of the salt licks is bound to activate both senses of salt gustation.

IV. DETERMINING THE CONSEQUENCES OF SALT INGESTION

Salt-hungry rats or sheep with gastric (Denton, 1982; Tordoff et al., 1987) or esophageal fistulas (Mook, 1969) ingest at least two times more salt with the fistula open than with it closed (Fig. 13). With the fistula closed, salt could reach tissues in the alimentary tract that are monitoring sodium levels. In fact there is a large reservoir of sodium within the alimentary tract, especially in ruminants (Michell, 1986; see also Falk, 1965b; and Wolf and Stricker, 1967, for a discussion of sodium reservoir in bone).

Organs in the alimentary tract determine the consequences of salt ingestion. Increases or decreases of NaCl ingestion result from sodium delivered by gavage (McCleary, 1953; Stellar et al., 1954; Epstein and Stellar, 1955; Mook, 1963). Moreover, the ingestion of salt is not necessary to reduce the hunger for it (see, e.g., Kissileff and Hoeffer, 1975; Wolf et al., 1984; Tordoff et al., 1987). In other words, one can bypass the oral cavity by sodium gavage or infusions into the hepatic portal vein and reduce the ingestion of salt (see, e.g., Wolf et al., 1984; Tordoff et al., 1986, 1987). It is possible that these postoral sites contribute to the learning about the consequences of salt ingestion (see, e.g., Smith, 1972).

One postabsorptive site that monitors sodium levels from salt ingestion and then signals the brain as to its consequences is the liver (see, e.g., Lin and Blake, 1971). Learned preferences have been demonstrated by direct infusions into the hepatic portal circulation (Tordoff and Friedman, 1986). The fact that hepatic signals about sodium and osmolarity in the hepatic portal circulation (see, e.g., Novin et al., 1981) activate brain stem gustatory sites suggests that this activation may be influencing salt ingestion. Damage to the hepatic vagus (Contreras and Kosten, 1981; Tordoff et al., 1986) or these brain stem gustatory sites reduces or abolishes the salt ingestion to body sodium depletion (Hill and Almli, 1983; Schulkin et al., 1985b).

V. MINERAL DEFICIENCIES

It appears that other mineral deficiencies trigger the ingestion of sodium. Why should that be? One hypothesis is that sodium is ingested by mineral-deficient

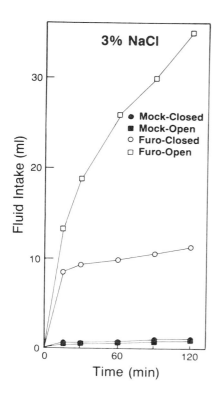

FIGURE 13 Ingestion of salt with the fistula open and closed.

animals because in nature they search for salt licks (see below) in which a composite of salts is found (Jones and Hanson, 1985). Therefore by having this one innate hunger for sodium or salty-tasting commodities, they are able to solve their mineral-related requirements. Consider some of the evidence for this hypothesis.

1. Young rats raised on a potassium-deficient diet for a week or more will ingest KCl when they are exposed to it (Milner and Zucker, 1965; Zucker, 1965). This ingestion is expressed within minutes (Zucker, 1965). However, when potassium-deficient rats are offered a choice between KCl and a variety of concentrations of NaCl solutions, they choose the NaCl solutions (Zucker, 1965; Blake and Jurf, 1968; Adam and Dawborn, 1972; Schulkin, 1986, Fig. 14). They do the same when offered a choice between NaCl and HCl—a salty and a sour taste. This increase in NaCl ingestion is independent of changes in body sodium (Blake and Jurf, 1968) and is not reversed by intragastric intubation of NaCl either immediately or an hour before (Adam and Dawborn, 1972; Schulkin, unpublished observations).

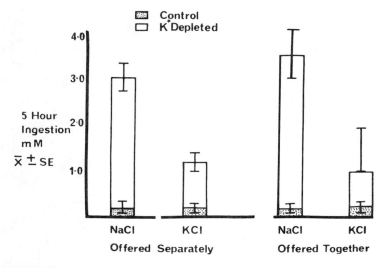

FIGURE 14 The 5-hour ingestion (means and SEM) of 0.67 M NaCl and 0.4 M KCl when offered separately or together in potassium-depleted and control rats. (From Adam and Dawborn, 1972.)

2. Although calcium-deficient rats increase their calcium ingestion (see, e.g., Richter and Eckert, 1937; Scott et al., 1950), like potassium-deficient rats, they tend to ingest NaCl. Both eventually ingest the CaCl or KCl, which they need (Fig. 15). But they at first go after the sodium, and continue to ingest it throughout the experiment. Thiamine-deficient rats tend not to ingest the salts (Fig. 15). Moreover, not only do calcium-deficient rats ingest NaCl solutions, they bar press for NaCl solutions (Lewis, 1968). Other mineral deficiencies, such as zinc deficiency, also increased NaCl ingestion (see, e.g., Catalanotto, 1978).

VI. SALT LICK

Periodically the rain depletes the grazing sources of sodium for land-dwelling herbivorous animals far from seawater, and this sodium deprivation may have resulted in their evolving a hunger for salt (Denton, 1982). A set of elegant naturalistic experiments carried out in Australia demonstrated the phenomenon of salt hunger (Blair-West et al., 1968). During a time when vegetation was low in sodium, and rabbits had elevated mineralocorticoid levels, these animals were photographed licking sodium sticks that had been placed by the experimenters (Fig. 16).

FACTORS IN THE HUNGER FOR SALT

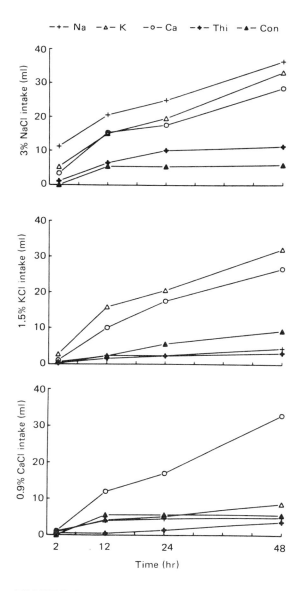

FIGURE 15 Ingestion of various minerals in young rats ($\approx$ 45 days of age) placed on either a sodium-deficient, potassium-deficient, calcium-deficient, thiamine-deficient, or control diet. Rats were generally on the diet for 2 weeks, then were given access to the salt solutions. Intakes were monitored 2, 12, 24, and 48 hours.

FIGURE 16 Telescopic lens photograph of sodium-deficient alpine wild rabbits attacking pegs impregnated with $NaHCO_3$ and NaCl. (From Blair-West et al., 1968.)

We have observed that salt-hungry rats would run in a runway for pure salt, as well as salt mixed with water (Schulkin et al., 1985a). We have also observed that salt-hungry rats regulate their salt intake via a salt lick (Schulkin et al., unpublished observations). This was demonstrated in three contexts: adrenalectomy, acute sodium deprivation, and mineralocorticoid treatment (Schulkin et al., unpublished observations; Fig. 17). We also found that salt-hungry rats regulate their deficiency by ingesting from a mineral lick.

There are many instances in the wild of mammals searching for deposits of salt in salt licks. Moose, white-tailed deer, reindeer, caribou, goats, sheep, elephants, and even foxes are known to search for vegetation rich in sodium, or to lick at mineral deposits in which sodium is only one of several mineral elements (see, e.g., Cowan and Brink, 1949; Herbert and Cowan, 1971; Botkin et al., 1973;

FACTORS IN THE HUNGER FOR SALT

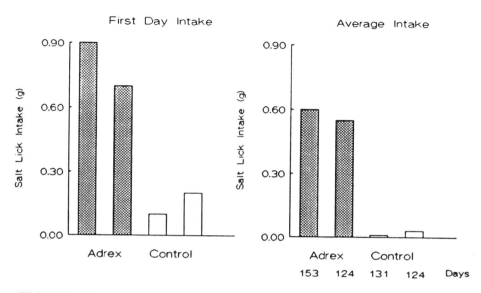

FIGURE 17 Salt lick intake of adrenalectomized (adrex) and control rats.

Belovsly, 1978; Weeks and Kirkpatrick, 1976; and see Denton, 1982, for review). Closer to home, the gorilla (Schaller, 1963) and the chimpanzee (Goodall, 1986) have been observed eating deposits of an assortment of minerals in dirt or rock or mineral springs. No doubt social influences contribute in the genesis of this ingestive behavior (Galef, 1986).

In addition to mammals, a variety of birds (e.g., crossbill, gold finch, and vulture: see Aldrich, 1939; Peterson, 1942; Coleman and Fraser, 1985) are known to migrate to salt licks. This is particularly observable during the spring, the mating season, when the female is known to be at the salt lick to a greater extent than the male (Dixon, 1958). In fact, females are known to ingest a variety of minerals during pregnancy and lactation (Richter, 1956; Denton, 1982; Fig. 18). This also occurs when they are given the hormones of sexual reproduction (oxytocin, estrogen, and prolactin: Shulkes et al., 1972; Denton, 1982).

VII. EVOLUTION OF SALT HUNGER

A number of herbivorous vertebrates increase their salt intake in response to experimental manipulations (see Denton, 1982, for review). But while there is selective pressure on herbivores, as indicated already, a variety of omnivores, of which the rat is a member par excellence, express a salt hunger (e.g., the rhesus monkey, see Schulkin et al., 1984). There are several instances of carnivores

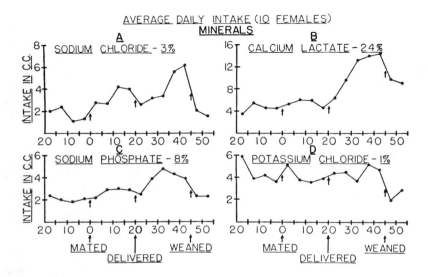

FIGURE 18 Curves showing average daily intake of the various mineral solutions in 5-day periods for 20 days before mating, for 20 days of pregnancy, for 25 days of lactation, and for 10 days after weaning. (From Richter and Barelare, 1938.)

ingesting salt (Fitzsimons and Moore-Gillon, 1980; Ramsay and Reid, 1981). Omnivorous birds are also known to increase their salt intake to natriorexigenic treatments (Epstein and Massi, 1987). Thus while there are a variety of species and strain differences with regard to the ingestion of salt to experimental manipulations (see Rowland and Fregly, 1988). The phenomenon of salt hunger is expressed in a variety of vertebrates. Interestingly, it has not been demonstrated with invertebrates (Dethier, personal communication).

Salt hunger, and more generally the behavioral regulation of body fluid homeostasis, probably evolved later than the regulation of food intake, because it was not until vertebrate animals inhabited land-dwelling terrains that finding and ingesting fluids was an issue at all. This is represented neurally by the fact that decerebrated rats do not increase their salt ingestion when they are depleted of body sodium (Grill et al., 1986), or to body fluid challenges more generally (Grill and Miselis, 1979); they do increase their food intake to hunger-related signals (Flynn and Grill, 1983; Norgren and Grill, 1984). Therefore, the forebrain is essential in the genesis of salt ingestion.

Two factors are important in the evolution of salt hunger. One is the regulation of extracellular fluids. Animals must drink both water and sodium to regulate extracellular fluid volume (Wolf and Stricker, 1967; Stricker, 1973; Fitzsimons, 1979). The hormones angiotensin and aldosterone figure heavily in its regulation (see, e.g., Fluharty and Epstein, 1983; Sakai, 1986; see review by Fregly and Rowland, 1985). The second factor is the enhanced avidity for salt in females, who typically ingest more salt than males (see, e.g., Krecek et al., 1972; Wolf, 1982). This behavior has its roots in reproduction (Richter, 1956; Denton, 1982), where there are greater extracellular fluid demands.

VIII. CONCLUSIONS

I have outlined some of the interesting observations of the innate expression of salt hunger and its interactions with several kinds of learning. In addition, I have suggested that the hunger for salty-tasting substances is operative when mammals and birds are deficient in minerals of several kinds. The innate hunger for salt may serve in fulfilling requirements for several different minerals. Although the gustatory system is important in identifying salt in the environment, organs within the alimentary tract assess the consequences of the salt ingestion, and forebrain sites are responsible for the genesis of the hunger for salt.

ACKNOWLEDGMENTS

In this brief review of the literature I have focused almost exclusively on experiments with rats. The figures reflect this fact. But there is also a large body of interesting and generally supportive gustatory–experimental work on sheep by the investigators at Howard Florey Institute at the University of Melbourne.

I thank Alan Epstein, John Sabini, and Eliot Stellar for their suggestions on the manuscript. I also thank three colleagues at Monell who I have enjoyed very much over the years: Mark Friedman, Michael Tordoff, and Gary Beauchamp. This research is supported by grant 00678 from the National Institute of Mental Health.

DISCUSSION

Grill: While I would agree with you that aqueous solutions of individual compounds are rarely found in the wild, your assertion that the specificity of animal behavior toward compounds that are chemically categorized as salt is an artifact of laboratory testing. . . .

Schulkin: I said "contrived." I didn't say "artifact." "Artifact" is a stronger term than I'd use.

Grill: Well, I'll accept your term "contrived." I think that it is quite clear that in particular sets of species, and not all species, the taste system specifies the animal's capacity to differentiate sodium and lithium ions from all other ions. The very data that you present showing the ability of the wild rabbits to select sodium compounds from other salts demonstrates quite clearly—and there are many demonstrations in the literature—this specificity of sodium and lithium ions in comparison to all other cations. So while you suggest that the system is potentially sloppy, I suggest that the system is highly tuned to see particular cations. This is not true in all species, it's only true in some.

Schulkin: I think if you take the rat or the sheep, which are the two animals that have really been studied, and you put them in a laboratory, there's no issue about it.

Grill: In the real world, there's no issue about it either.

Schulkin: Well, you mean the fact that they have the salt pegs as opposed to the salt composites, which it normally finds. That's a laboratory version in the wild of what we've produced, and I don't think that's a good argument.

Grill: It's the very issue of sodium appetite as an appetite. The slide that you show that an animal within a few licks, literally within a few fractions of a second, can identify the presence of these cations—they can do that in mixtures as well. The fact that the animal is encountering salt in mixtures does not say that the animal is sloppy. It further shows the animal's capacity to differentiate among cations in mixtures.

Schulkin: There's probably four or five studies that also show animals ingesting potassium from time to time. Never to the same extent as sodium. That's a rare phenomenon, but it's often the case that you can see animals ingesting a fair amount of potassium as well. Potassium tastes somewhat salty, doesn't it?

Grill: There is a confound in your slide showing intake of potassium and calcium and sodium. If you notice, the concentrations that were presented were quite different. Sodium concentration was highest, and the concentration of potassium and calcium were much lower. That very fact itself could bias intake of those compounds.

Schulkin: You get the same effect if you keep the molarity the same.

Grill: . . . I'm not convinced.

Schulkin: OK.

Thrasher: I have a two-part question. The demonstration that the rat is hard-wired for recognition of sodium I think is beyond doubt. Are the data as good for sheep or ruminants? That is, are they hard-wired?

Schulkin: As I read Denton, it looks that way. It looks very much that way. They use sodium bicarbonate; that's their mixture. The sheep is much more avid for that. Of course, that's what they're losing with the parotid fistula. But it looks very tight.

Thrasher: The second part is, the last I heard there was a raging controversy about the mechanism for salt appetite in the rat versus the ruminant model.

Schulkin: Still true.

Thrasher: What are the possibilities that this phenomenon arose twice in evolution rather than being a single event with common background mechanisms?

Schulkin: There are very clear functional differences; for what holds in the rat clearly doesn't hold in the sheep. Moreover, it doesn't hold in the dog. You don't think a dog has salt appetite anyway. But there are very clear species differences. Where brain sodium plays a prominent role in the regulation in the sheep, it seems to play very little or no role in the rat. The hormones of sodium homeostasis play a role in the genesis of salt hunger in the rat. You can get angiotensin-induced drinking in the sheep, but they think it's due to natriuresis. So there are clear functional differences in terms of physiology but a lot of the behavior is the same, which suggests a philosophical thesis: that one can have very different physical substrates and still get the same behavior.

REFERENCES

Adam, W. R., and Dawborn, J. K. (1972). Effect of potassium depletion on mineral appetite in the rat. *J. Comp. Physiol. Psychol.* **418**:51–58.

Ahern, G. L., Landin, M. L., and Wolf, G. (1978). Escape from deficits in sodium intake after thalamic lesions as a function of preoperative experience. *J. Comp. Physiol. Psychol.* **92**:544–554.

Aldrich, E. C. (1939). Notes on the salt-feeding habits of the red crossbill. *Condor*, **66**:30.

Bare, J. K. (1949). The specific hunger for sodium chloride in normal and adrenalectomized white rats. *J. Comp. Physiol. Psychol.* **42**:242–253.

Bartoshuk, L. M. (1980). Sensory analysis of the taste of NaCl. In *Biological and*

Behavioral Aspects of Salt Intake, M. Kare, M. Fregly, and R. Bernard (Eds.). Academic Press, New York.

Belovsky, G. E. (1978). Diet optimization in a generalist herbivore: The moose. *Theor. Popul. Biol.* **14**:105–134.

Berridge, K. C., and Schulkin, J. (1989). Palatability shift of a salt-associated incentive drive during sodium depletion. *Q. J. Exp. Psychol.* **4**:121–138.

Berridge, K. C., Flynn, F. W., Schulkin, J., and Grill, H. J. (1984). Sodium depletion enhances salt palatability in rats. *Behav. Neurosci.* **98**:652–660.

Blair-West, J. R., Coghlan, J. P., Denton, D. A., Nelson, J. F., Orchard, E., Scoggins, B. A., Wright, R. D., Myers, K., and Junqueira, C. L. (1968). Physiological, morphological and behavioural adaptation to a sodium deficient environment by wild native Australian and introduced species of animals. *Nature*, **217**:922–928.

Blake, W. D., and Jurf, A. N. (1968). Increased voluntary Na intake in K deprived rats. *Behav. Biol.* **1**:1–7.

Bolles, R. C., Sulzbacher, S. I., and Arant, H. (1964). Innateness of the adrenalectomized rat's acceptance of salt. *Psychonomic Sci.* **1**:21–22.

Botkin, D. B., Jordan, P. A., Dominski, A. S., Lowendorf, H. S., and Hutchinson, G. E. (1973). Sodium dynamics in a northern ecosystem. *Proc. Natl. Acad. Sci. USA*, **70**:2745–2748.

Boudreau, J. C., Hoang, N. K., Oravec, J., and Do, L. T. (1983). Rat neurophysiological taste responses to salt solutions. *Chem. Senses*, **8**:131–150.

Bregar, R. E., Strombakis, N., Allan, R. W., and Schulkin, J. (1983). Brief exposure to a saline stimulus promotes latent learning in the salt hunger system. *Neurosci. Abstr.* **9**.

Cabanac, M. (1979). Sensory pleasure. *Q. Rev. Biol.* **54**(1):1–29.

Catalanotto, F. A. (1978). Effects of dietary methionine supplementation on preferences for NaCl solutions. *Behav. Biol.* **24**:457–466.

Clark, W. G., and Clausen, D. F. (1942). Dietary "self-selection" and appetites of untreated and treated adrenalectomized rats. *Am. J. Physiol.* **133**:70–76.

Coleman, J. S., Fraser, J. D., and Pringle, C. A. (1985). *Condor*, **87**:291–292.

Contreras, R. J. (1977). Changes in gustatory nerve discharges with sodium deficiency: A single unit analysis. *Brain Res.* **121**:373–378.

Contreras, R. J., and Kosten, T. (1981). Changes in salt intake after abdominal vagotomy: Evidence for hepatic sodium receptors. *Physiol. Behav.* **26**:575–582.

Contreras, R. J., Kosten, T., and Frank, M. E. (1984). Activity in salt taste fibers: Peripheral mechanism for mediating changes in salt intake. *Chem. Senses*, **8**(3):275–288.

Cowan, I., and Brink, V. C. (1949). Natural game licks in the Rocky Mountain national parks of Canada. *J. Mammal.* **30**:379–387.

Denton, D. A. (1982). *The Hunger for Salt*. Springer-Verlag, New York.

Denton, D. A., and Sabine, J. R. (1963). The behaviour of Na deficient sheep. *Behaviour*, **4**:364–376.

Dickinson, A. (1986). Re-examination of the role of the instrumental contingency in the sodium-appetite irrelevant incentive effect. *Q. J. Exp. Psychol.* **38B**:161–172.

Dickinson, A., and Nicholas, D. J. (1983a). Irrelevant incentive learning during training on ratio and interval schedules. *Q. J. Exp. Psychol.* **35B**:235–247.

Dickinson, A., and Nicholas, D. J. (1983b). Irrelevant incentive learning during instrumental conditioning: The role of the drive–reinforcer and response–reinforcer relationships. *Q. J. Exp. Psychol.* **35B**:249–263.

Dixon, J. S. (1958). Some biochemical aspects of deer licks. *J. Mammal.* **20**:109.

Epstein, A. N., and Massi, M. (1987). Salt appetite in the pigeon in response to pharmacological treatments. *J. Physiol.* **393**:555–568.

Epstein, A. N., and Stellar, E. (1955). The control of salt preference in the adrenalectomized rat. *J. Comp. Physiol. Psychol.* **48**:167–172.

Falk, J. L. (1965a). Limitations to the specificity of NaCl appetite in sodium-depleted rats. *J. Comp. Physiol. Psychol.* **60**:393–396.

Falk, J. L., and Herman, T. S. (1961). Specific appetite for NaCl without postingestional repletion. *J. Comp. Physiol. Psychol.* **54**:405–408.

Falk, J. L., and Lipton, J. M. (1967). Temporal factors in the genesis of NaCl appetite by intraperitoneal dialysis. *J. Comp. Physiol. Psychol.* **63**:247–251.

Fitzsimons, J. T. (1979). *The Physiology of Thirst and Sodium Appetite.* Cambridge University Press, Cambridge.

Fitzsimons, J. T., and Moore-Gillon, M. J. (1980). Drinking and antidiuresis in response to reductions in venous return in the dog: Neural and endocrine mechanisms. *J. Physiol.* **308**:403–416.

Fluharty, S. J., and Epstein, A. N. (1983). Sodium appetite elicited by intracerebroventricular infusion of angiotensin II in the rat: II. Synergistic interactions with systemic mineralocorticoids. *Behav. Neurosci.* **97**:746–758.

Flynn, F. W., and Grill, H. J. (1983). Insulin elicits ingestion in decerebrate rats. *Science,* **221**:188–190.

Frank, M. E. (1985). On the neural code for sweet and salty tastes. In *Taste, Olfaction and the CNS,* D. Pfaff (Ed.). Rockefeller University Press, New York, pp. 107–128.

Fregly, M. J., and Rowland, N. E. (1985). Role of renin–angiotensin–aldosterone system in NaCl appetite of rats. *Am. J. Physiol.* **248**:R1–R11.

Fudim, O. W. (1978). Sensory preconditioning of flavors with a formalin-produced sodium need. *J. Exp. Psychol.: Anim. Behav. Processes,* **4**:276–285.

Galef, B. G., Jr. (1986). Social interaction modifies learned aversions, sodium appetite, and both palatability and handling-time-induced dietary preference in rats. *J. Comp. Psychol.* **100**:432–439.

Goodall, J. (1986). *The Chimpanzees of Gombe.* Belknap Press of Harvard University Press, Cambridge, MA.

Grill, H. J., and Bernstein, I. (1988). Strain differences in taste reactivity to NaCl. *Am. J. Physiol.* **10**:R424–R430.

Grill, H. J., and Miselis, R. R. (1979). Lack of ingestive compensation to dehydrational stimuli in decerebrates. *Am. J. Physiol.* **240**:R81–R86.

Grill, H. J., and Norgren, R. (1978). The taste reactivity test: 1. Mimetic response to gustatory stimuli in neurologically normal rats. *Brain Res.* **143**:263–279.

Grill, H. J., Schulkin, J., and Flynn, F. W. (1986). Sodium homeostasis in chronic decerebrate rats. *Behav. Neurosci.* **100**:536–543.

Handal, P. J. (1965). Immediate acceptance of sodium salts by sodium deficient rats. *Psychonomic Sci.* **3**:315–316.

Hartzell, A. K., Paulus, R. A. and Schulkin, J. (1985). Brief preoperative exposure to saline protects rats against behavioral impairments in salt appetite following central gustatory damage. *Behav. Brain Res.* **15**:9–13.

Hebert, D., and McTaggart Cowan, I. (1971). Natural salt licks as a part of the ecology of the mountain goat. *Can. J. Zool.* **49**:605–610.

Hermann, G. E., Kohlerman, N. J., and Rogers, R. C. (1983). Hepatic–vagal and gustatory afferent interactions in the brainstem of the rat. *J. Auton. Nerv. Syst.* **9**:477–495.

Hill, D. L., and Almli, C. R. (1983). Parabrachial nuclei damage in infant rats produces residual deficits in gustatory preferences/aversions and sodium appetite. *Dev. Psychobiol.* **16**:519–533.

Jacobs, K. M., Mark, G. P. and Scott, T. R. (1988). Taste responses in the nucleus tractus solitarius of sodium-deprived rats. *J. Physi.* **406**:393–410.

Jalowiec, J. E., Crapanzano, J. E., and Stricker, E. M. (1966). Specificity of salt appetite elicited by hypovolemia. *Psychonomic Sci.* **6**(7):331–332.

Jones, R. L., and Hanson, H. C. (1985). *Mineral Licks, Geography, and Biogeochemistry of North American Ungulates.* Iowa State University Press, Ames.

Katz, D. (1937). *Animals and Men.* Longmans & Green, New York.

Khavari, K. A., and Eisman, E. H. (1971). Some parameters of latent learning and generalized drives. **77**:463–469.

Kissileff, H. R., and Hoeffer, R. (1975). Reduction of saline intake in adrenalectomized rats during chronic intragastric infusions of saline. In *Control Mechanisms of Drinking*, G. Peter, J. T. Fitzsimons, and L. Peters-Haefeli (Eds.). Springer-Verlag, Heidelberg.

Krecek, J., Novakova, V., and Stribral, K. (1972). Sex differences in the taste preference for a salt solution in the rat. *Physiol. Behav.* **8**:183–188.

Krieckhaus, E. E. (1970). "Innate recognition" aids rats in sodium regulation. *J. Comp. Physiol. Psychol.* **73**:117–122.

Krieckhaus, E. E., and Wolf, G. (1968). Acquisition of sodium by rats: Interaction of innate mechanisms and latent learning. *J. Comp. Physiol. Psychol.* **65**:197–201.

Lewis, M. (1968). Discrimination between drives for sodium chloride and calcium. *J. Comp. Physiol. Psychol.* **65**:208–212.

Lin, K. K., and Blake, W. D. (1971). Hepatic sodium receptor in control of saline drinking behavior. *Commun. Behav. Biol.* **5**:359–363.

McCleary, R. A. (1953). Taste and postingestion factors in specific-hunger behavior. *J. Comp. Physiol. Psychol.* **46**:411–421.

Michell, A. R. (1986). The gut: The unobtrusive regulator of sodium balance. *Perspect. Biol. Med.* **29**:203–213.

Milner, P., and Zucker, P. (1965). Specific hunger for potassium in the rat. *Psychonomic Sci.* **2**:17–18.

Moe, K. (1986). The ontogeny of salt intake in rats. In *The Physiology of Thirst and Sodium Appetite*, G. deCaro, A. N. Epstein, and M. Massi (Eds.). Plenum Press, New York.

Mook, D. (1963). Oral and postingestional determinants of the intake of various solutions in rats with esophageal fistulas. *J. Comp. Physiol. Psychol.* **56**:645–659.

Mook, D. (1969). Some determinants of preference and aversion in the rat. *Ann. N.Y. Acad. Sci.* **157**:1158–1170.

Morrison, G. R., and Young, J. C. (1972). Taste control over sodium intake in sodium deficient rats. *Physiol. Behav.* **3**:29–32.

Nachman, M. (1962). Taste preferences for sodium salts by adrenalectomized rats. *J. Comp. Physiol. Psychol.* **55**:1124–1129.

Nachman, M. (1963). Taste preferences for lithium chloride by adrenalectomized rats. *Am. J. Physiol.* **205**:219–221.

Norgren, R., and Grill, H. J. (1982). Brain-stem control of ingestive behavior. In *The Physiology of Motivation*, D. Pfaff (Ed.). Springer-Verlag, New York.

Novin, D., Rogers, R. C., and Hermann, G. (1981). Visceral afferent and efferent connections in the brain. *Diabetologia*, **20**:331–336.

Paulus, R. A., Eng, R., Schulkin, J. (1984). Preoperative latent place learning preserves salt appetite following damage to the central gustatory system. *Behav. Neurosci.* **98**:146–151.

Peterson, J. G. (1942). Salt feeding habits of the house finch. *Condor*, **44**:73.

Ramsay, D. J., and Reid, I. A. (1981). Salt appetite in dogs. *Neurosci. Abstr.* **7**.

Rescorla, R. A., and Freberg, L. (1978). The extinction of within-compound flavor associations. *Learn. Motiv.* **9**:411–427.

Richter, C. P. (1936). Increased salt appetite in adrenalectomized rats. *Am. J. Physiol.* **115**:151–161.

Richter, C. P. (1939). Salt taste thresholds of normal and adrenalectomized rats. *Am. J. Physiol.* **118**:367–371.

Richter, C. P. (1956). Salt appetite in mammals—Its dependence on instinct and metabolism. In *L'Instinct*, M. Autouri (Ed.). Masson, Paris.

Richter, C. P., and Eckert, J. F. (1937). Increased calcium appetite of parathyroidectomized rats. *Endocrinology*, **21**:50–54.

Richter, C. P., and Eckert, J. F. (1938). Mineral metabolism of adrenalectomized rats studied by the appetite method. *Endocrinology*, **22**:214–224.

Rodgers, W. L. (1967). Specificity of specific hungers. *J. Comp. Physiol. Psychol.* **64**:49–58.

Rowland, N. E., and Fregly, M. J. (1988). Sodium appetite: Species and strain differences and role of renin–angiotensin–aldosterone system. *Appetite*, **11**:143–178.

Rozin, P. (1976). The selection of foods by rats, humans and other animals. In *The Study of Behavior*, J. B. Rosenblatt, R. A. Hinde, E. Shaw, and C. Beer (Eds.). Academic Press, New York.

Sakai, R. R. (1984). The hormones of renal sodium conservation act synergistically to arouse sodium appetite in the rat. In *The Physiology of Thirst and Sodium Appetite*, G. deCaro, A. N. Epstein, and M. Massi. (Eds.). Plenum Press, New York, pp. 425–430.

Schaller, G. B. (1963). *The Mountain Gorilla*. University of Chicago Press, Chicago.

Schiffman, S. S. (1980). Contribution of the anion to the taste quality of sodium salt. In *Biological and Behavioral Aspects of Salt Intake*, M. Kare, M. Fregly, and B. Bernard (Eds.). Academic Press, New York.

Schulkin, J. (1982). Behavior of sodium deficient rats: The research for a salt taste. *J. Comp. Physiol. Psychol.* **96**:628–634.

Schulkin, J. (1986). The evolution and expression of salt appetite. In *The Physiology of Thirst and Sodium Appetite*, G. deCaro, A. N. Epstein, and M. Massi (Eds.). Plenum Press, New York.

Schulkin, J., Leibman, D., Ehrman, R. N., Norton, N. W., and Ternes, J. W. (1984). Salt hunger in the rhesus monkey. *Behav. Neurosci.* **98**(4):753–756.

Schulkin, J., Arnell, P., and Stellar, E. (1985a). Running to the taste of salt in mineralocorticoid-treated rats. *Horm. Behav.* **19**:413–425.

Schulkin, J., Flynn, H. J., Grill, H. J., and Norgren, R. (1985b). Central gustatory lesions: Effects on salt appetite and taste aversion learning. *Neurosci. Abstr.* **11**:1259.

Schulkin, J., Arnell, P., and Stellar, E. The ingestion of salt licks in salt-hungry rats. Unpublished observations.

Scott, E. M., Verney, E. L., and Morissey, P. D. (1950). Self-selection of diet: XI. Appetites for calcium, magnesium and calcium. *J. Nutr.* **41**:187–202.

Shulkes, A. A., Covelli, M. D., Denton, D. A., and Nelson, J. F. (1972). Hormonal factors influencing salt appetite in lactation. *Aust. J. Exp. Biol. Med. Sci.* **L(7)**:819–826.

Smith, M. H. (1972). Evidence for a learning component of sodium hunger in rats. *J. Comp. Physiol. Psychol.* **78**:242–247.

Stellar, E., Hyman, R., and Samet, S. (1954). Gastric factors controlling water- and salt-solution-drinking. *J. Comp. Physiol. Psychol.* **47**:220–226.

Stricker, E. M. (1973). Thirst, sodium appetite, and complementary physiological contributions to the regulation of intravascular fluid volume. In *The Neurophysiology of Thirst*, A. N. Epstein, H. R. Kissileff, and E. Stellar (Eds.). V. H. Winston & Sons, Washington, DC.

Stricker, E. M., and Wilson, N. E. (1970). Salt-seeking behavior in rats following acute sodium deficiency. *J. Comp. Physiol. Psychol.* **72**:416–420.

Tordoff, M. G., and Friedman, M. I. (1986). Hepatic portal glucose infusions decrease food intake and increase food preference. *Am. J. Physiol.* **251**:R192–R196.

Tordoff, M. G., Schulkin, J., and Friedman, M. I. (1986). Hepatic contribution to the satiation of salt appetite. *Am. J. Physiol.* **251**:1095–1102.

Tordoff, M. G., Schulkin, J., and Friedman, M. I. (1987). Further evidence for a hepatic control of salt intake. *Am. J. Physiol.* **22**:444–449.

Weeks, H. P., Jr., and Kirkpatrick, C. M. (1976). Adaptations of white-tailed deer to naturally occurring sodium deficiencies. *J. Wildl. Manage.* **40**:610–625.

Weisinger, R. S., Woods, S. C., and Skorupski, J. D. (1970). Sodium deficiency and latent learning. *Psychonomic Sci.* **19**:307–308.

Weisinger, R. S., Denton, D. A., McKinley, M. J., and Nelson, J. F. (1985). Dehydration-induced sodium appetite in rats. *Physiol. Behav.* **34**:45–50.

Wirsig, C. R., and Grill, H. J. (1982). Contribution of the rat's neocortex to ingestive control: 1. Latent learning for the taste of sodium chloride. *J. Comp. Physiol. Psychol.* **96**:615–627.

Wolf, G. (1968). Thalamic and tegmental mechanisms for sodium intake: Anatomical and functional relations to lateral hypothalamus. *Physiol. Behav.* **3**:997–1007.

Wolf, G. (1969). Innate mechanisms for regulation of sodium intake. In *Olfaction and Taste*, Pfaffman (Ed.). Rockefeller University, New York.

Wolf, G. (1982). Refined salt appetite methodology for rats demonstrated by assessing sex differences. *J. Comp. Physiol. Psychol.* **96**:1016–1021.

Wolf, G., and Stricker, B. M. (1967). Sodium appetite elicited by hypovolemia in adrenalectomized rats: Reevaluation of the "reservoir" hypothesis. *J. Comp. Physiol. Psychol.* **63**:252–257.

Wolf, G., Schulkin, J., and Simpson, P. E. (1984). Multiple factors in the satiation of salt appetite. *Behav. Neurosci.* **98**:661–673.

Young, P. T. (1952). The role of hedonic processes in the organization of behavior. *Psychol. Rev.* **59**:249–262.

Zucker, I. (1965). Short-term salt preference of potassium-deprived rats. *Am. J. Physiol.* **208**:1071–1074.

12
Metabolic Basis of Learned Food Preferences

Michael G. Tordoff

Monell Chemical Senses Center
Philadelphia, Pennsylvania

I. INTRODUCTION

Although social interaction influences what is considered to be edible under many circumstances (see reviews in this volume by Birch, Chapter 15; Galef, Chapter 9), ultimately an animal must use its own judgment to assess a food's value. Individuals use both innate and learned cues to recognize food. However, the only well-accepted innate cue is sweetness (Steiner, 1973; Grill et al., 1984; see review by Beauchamp and Cowart, 1987). There may be other innate cues that identify food—for example, starch (Sclafani, 1987) and umami (Kawamura and Kare, 1987)—and some that identify an item as inedible (e.g., bitter). However, the vast majority of flavors used that signify food do so because of prior learning.

Learning can be supported by several properties of food. The potential of food to induce or reduce malaise has been studied in detail (i.e., conditioned taste aversion or "medicine" effects in response to a deficiency; see reviews by Rozin, 1967; Riley and Clarke, 1977). However, it is unlikely that this is germane to the majority of food choices outside the laboratory because malaise is relatively rare. A less studied but more obvious source of reinforcement is the food's nutrient content. After experience with flavored foods and their physiological effects, the animal can discriminate among foods with different nutrient values by flavor alone. This implies that information concerning "nutrient value" is detected and processed. Where and how this might be done is the subject of this review.

Learning plays a role in when, what, and how much is eaten (i.e., conditioned initiation of eating, conditioned flavor preference, and conditioned satiation or conditioned desatiation, respectively). These different facets of feeding behavior may involve common mechanisms. In particular, what and how much is eaten appear to be closely related in that they may share the same conditioned and unconditioned stimuli. Indeed, some investigators (e.g., Booth, 1972; Bolles et al., 1981) imply that conditioned satiation (reduction of meal size in a single flavor source test) and conditioned preference (choice between flavors in a two-source test) are one and the same. This is based on studies in which flavors associated with a high-calorie meal were eaten in small quantities when presented as the only source of food but were eaten in relatively large quantities (i.e., preferred) in a choice test. The idea that the most satiating foods are the most palatable is conceptually attractive but incorrect. There are examples of a preferred flavor in two-flavor choice tests being consumed during training (one-flavor tests) in greater, the same, or smaller quantity than the unpreferred flavor (G. L. Holman, 1968; Tordoff and Friedman, 1986; Tordoff et al., 1987, 1989, 1990; Elizalde and Sclafani, 1988; Sclafani and Nissenbaum, 1988). Moreover, rats form flavor preferences even if they have never experienced satiety in association with the flavor (Deutsch and Wang, 1977; Van Vort and Smith, 1983).

II. PHENOMENOLOGICAL STUDIES

Booth (1972) was the first to attempt to produce a conditioned flavor preference by manipulating the nutrient composition of flavored food. He fed rats flavored meals differing in starch content (4.1 versus. 1.7 kcal/g). When, after 7–8 pairs of training trials, the rats were given a choice between the two flavors presented in a midcalorie (2.9 kcal/g) mixture of the diets, they preferred the flavor that had been previously incorporated into the high-calorie diet. This basic paradigm has been repeated and extended in a variety of ways. Most impressively, when high- and low-calorie flavored foods are given together for several days, robust and long-lasting preferences develop for the flavor of the high-calorie food (Bolles et al., 1981). Booth et al. (1972) found that preferences for a sweet solution could be increased by pairing the solution with another of higher caloric value, showing that even innate preferences can be modified by learning. The results found in rats probably also apply to humans: flavor choice and/or preference ratings have been conditioned in humans given flavors incorporated in foods of varying starch content (Booth et al., 1976, 1982) or in sugar solutions (Zellner et al., 1983).

Although studies with humans suggest that consumption of nutrients can alter the palatability of foods, learning affects more than simply hedonic rating. Fedorchak and Bolles (1989) showed that rats' prior history determined their intake of a flavor ingested after cholecystokinin (CCK) administration. CCK reduced intake of a flavor previously paired with calories (an ethanol solution), but the hormone was ineffective in suppressing intake of the same flavor if it had previously been paired with water. Thus, the information that becomes associated with a nutrient-paired flavor is used as a guide to the expected nutritive value of the food.

A. Dissociation of Conditioned and Unconditioned Stimuli

The studies mentioned above have merit in that they mimic the sort of choices animals could potentially make in the wild, but from an analytical viewpoint most are beset with problems. This is because during training, the conditioned stimulus (CS; flavor) is compounded with the unconditioned stimulus (UCS; nutrients). The nonnutritive diluents (cellulose, kaolin, paraffin, mineral oil, and/or chalk) or carbohydrates (sugar or starch) that are added to diets or solutions to alter caloric content also alter the taste, flavor, and texture of the food, so that the subject can (potentially) discriminate between high- and low-calorie diets by oral cues alone. Thus, the results of these experiments could be due to learned discriminations based on either the food's physiological effects or its orosensory ones.

One solution to this problem is to dissociate CS and UCS presentation in time. However, some authors find that learning does not occur if a delay of more than a few minutes is imposed between ingestion of a flavor and a caloric load (see, e.g., Simbayi et al., 1986; see also E. W. Holman, 1975; Boakes et al., 1987; Capaldi et al., 1987a; Boakes and Lubart, 1988). The recent work of Elizalde and Sclafani (1988) has helped to clarify this. These authors found that flavor preference learning did not occur if a 20-minute delay was imposed between the presentation of a flavor and a Polycose solution. However, if rats were given prior exposure to Polycose containing Acarbose, which prevents the absorption of Polycose, preferences were obtained with delays of 10, 30, or 60 minutes. Apparently, prior exposure to a Polycose–Acarbose solution "teaches" the rat that the taste of the Polycose is not associated with calories. Thus, when later trained with a flavor and Polycose, the animal attributes the caloric effects to the novel flavor. It appears that the rat can analyze and identify the nutrient value of foods ingested just a few minutes apart. Since the postingestive effects of foods can last for hours, it remains a mystery how the rat can discriminate between the postingestive effects of two or more flavors consumed in close succession (or even simultaneously: Puerto et al., 1976; Deutsch and Wang, 1977; Bolles et al., 1981).

B. Deprivation State

Hartley and Perelman (1963) found that subjects who tasted five unusual foods after an overnight fast later rated them as more pleasant than did nonfasted controls. Similarly, Booth et al. (1982) found that a high-calorie soup conditioned a more positive taste preference in subjects who were hungry than in those who were not. It appears that the deprivation state of subjects during training influences their learned response to the food flavor.

This deprivation-dependent effect of calories on learning seems to be intuitively logical, whether it is interpreted within a psychological (i.e., different reductions in drive), or physiological (i.e., different metabolic fate of food) context. A number of studies have attempted to determine whether hunger state during training influences the acquisition of conditioned flavor preferences in rats, but the results are inconsistent (see, e.g., Revusky, 1967; Capaldi and Myers, 1982; Capaldi et al., 1983; Campbell et al., 1987; Fedorchak and Bolles, 1987). One thing is clear, however: rats with ad libitum access to food develop conditioned taste preferences (see, e.g., Bolles et al., 1981; Sclafani and Nissenbaum, 1988; Tordoff and Friedman, 1986). Assuming that rats with continuous access to food are never hungry, it would appear that satiation of the drive for food is not essential for learning to occur (see also Van Vort and Smith, 1983).

To make the matter more complicated, Booth (e.g., Booth, 1977) has suggested that preferences for flavors paired with calories depend on deprivation state during testing, as well as deprivation state during training. Thus, a flavor that has previously been paired with a high-calorie food will be preferred when the subject is hungry and avoided when the subject is fed. This implies that if given a choice between flavors associated with low- and high-calorie foods during a meal, the subject should initially choose the high-calorie paired flavor and then, as hunger is satisfied, switch to the low-calorie flavor. This is an intriguing idea, but the evidence for it is weak. Booth cites data from an unpublished manuscript in various reviews (e.g., ref. 12 in Booth, 1977). However, what few data are given cannot be evaluated critically because they are transformed into differences in preference ratios. Moreover, the shift from preference to aversion occurs only with flavors paired with 40% versus 10% starch loads and not 25% versus 0% starch loads. Given that no other investigators have seen reversals of calorie-based flavor preferences while rats refeed (e.g., Bolles et al., 1981; Tordoff et al., 1987, 1990), some skepticism is warranted. One also wonders about the purpose of this strategy. Booth (1977) argues that at the end of the meal, the rat must be cautious in "taking absorptive energy flow up to the optimum value" (p. 321). This assumes that there is an optimum value, and avoids the question of why the rat fills itself with the food it believes is largely nonnutritive rather than eating a few bites less of the food it believes is of high caloric density.

III. THE SITE OF UNCONDITIONED STIMULUS TRANSDUCTION

The studies described above indicate that experience with a food's nutrient value can influence how the food is later perceived. Some studies also delineate the associative mechanisms involved, but they do little to delineate physiological mechanisms. A physiological analysis requires identification of the site(s) of UCS transduction, the method of transduction, and the method of neural or humoral communication of the transduced information to the brain. Below, I first summarize work that helps identify the site of UCS transduction. Most of the studies use nutrient infusions. In addition to helping localize the site of UCS transduction, infusions have the advantage of bypassing the oral cavity and thus allowing discrete presentation of a CS (flavor) and UCS (nutrients) without the UCS interfering with the orosensory perception of the CS. They have the potential disadvantage of allowing the investigator to administer nutrients in a manner that is of no physiological significance to the animal. It is easy to induce malaise with inappropriate infusions of nutrients (see, e.g., Deutsch et al., 1976); and even with appropriate ones, oral stimulation influences the metabolic fate of the administered nutrients (see, e.g., Ramirez, 1985). It should also be recognized that the site of administration of a nutrient load is not necessarily its site of action.

A. Oral

The possibility that both CS and UCS originate from the orosensory aspects of the food is based on findings that rats come to prefer an arbitrary flavor that has been ingested as part of a mixture with saccharin solution (see, E. W. Holman, 1975, 1980; Faneslow and Birk, 1982). Because neither the flavor (CS) nor the saccharin (UCS) provides nutrients, it has been assumed that incentive motivation or "hedonics" must be involved. Recently, however, this interpretation has been questioned: Simple "flavor-flavor" conditioning cannot account for findings that flavored solutions do not become preferred if, during training, they are presented before the saccharin solution, rather than mixed with it (Capaldi et al., 1987a,b; Tordoff and Friedman, 1989c). Work using a sensory preconditioning technique suggests that a delay of more than a few seconds between flavor and saccharin solution is sufficient to prevent learning (Lavin, 1976). On the other hand, a flavor preference develops if the saccharin is presented before the flavor.

The most coherent explanation of these results is that the reinforcing effects of saccharin outlast the period of orosensory stimulation. That the reinforcement involves more than simply hedonic effects of the "taste" of saccharin is suggested by findings that a flavor preference develops when rats ingest saccharin along with a generous supply of flavored food, but an aversion develops if rats ingest the same volume of saccharin but have only a limited ration of flavored food (Tordoff and Friedman, 1989c). Moreover, rats given saccharin to drink increase their short-term food intake (Tordoff and Friedman, 1989a,b,c,d). It appears that consumption of this "extra" food is necessary to reinforce the food preference (Tordoff and Friedman, 1989c,d).

There is little evidence for a direct postingestive effect of saccharin on metabolism (cf. Rogers and Blundell, 1989, for a review). It seems more likely that saccharin influences metabolism by virtue of its sweet taste. Recent experiments suggest that a hepatic cephalic phase reflex is the primary mediator of saccharin's reinforcing effects (Tordoff and Friedman, 1989d). The most direct evidence was that, unlike intact rats, rats with hepatic vagotomy did not develop a preference for flavored food ingested with a saccharin solution. This and other findings led to the conclusion that neural reflexes initiated by saccharin alter liver metabolism in a manner that diverts fuel into storage, away from oxidation. The resulting reduction in fuel oxidation induces hunger. Food that ameliorates this hunger is reinforcing and its flavor therefore becomes preferred (Tordoff, 1988).

If the reinforcing effects of a prototypical "hedonic" stimulus such as saccharin can be accounted for by its indirect influence on liver metabolism, does "hedonic" or flavor–flavor conditioning occur at all? This is difficult to answer because it is rarely possible to be certain that the flavors used as UCSs have only hedonic effects. In any case, the question is largely moot, since flavors are rarely consumed without calories. In the face of the powerful effects of flavor–nutrient conditioning, any subtle association of two or more nonnutritive flavors is likely to be overwhelmed.

1. Sham Feeding

Dissociation of orosensory and caloric effects can be achieved with sham feeding. The first attempt to do this (Hull et al., 1951) used only one subject and some of the data were mislaid, but it warrants attention as the earliest effort to localize the rewarding effect of food. A dog with an esophageal fistula ate 50 ml of gruel in one arm of a Y-maze and sham-ate the same volume in the other arm. After a few days of experience, the dog always entered the arm where it could eat normally. Later, the dog was given a choice between sham-fed gruel and nothing, and it learned to approach the side of the maze where it could sham-feed gruel. Moreover, if given an esophageal load of gruel in one arm of the maze and a mock infusion (no load) in the other, the dog preferred the side of the maze where it received the esophageal load. Thus, it appears that both the orosensory effects of gruel (or the act of sham feeding) and its postesophageal effects were positively reinforcing.

A different conclusion was reached by Van Vort and Smith (1983). These investigators gave 17-hour food-deprived, gastric cannulated rats training trials with flavored milk to drink (until satiety) in one location on one day and an alternative flavor of milk to sham drink in another location on the next. Every few days, a 4-minute choice between the two flavors was given. Rats allowed to sham feed for about 10–30 minutes during training neither preferred nor avoided the sham-fed flavor. The authors interpret this as evidence that satiety (and the postingestive effect of food) has no positive reinforcing effect. On the other hand, a 4-minute choice test in 17-hour food-deprived rats is not very sensitive, and in 9 of 11 tests, the rats tended to drink more of the flavor associated with "real" feeding. Moreover, rats that sham fed for 60 minutes during training preferred the flavor of the food associated with "real" feeding. Why extended sham feeding should be more aversive than brief sham feeding is unclear, particularly since some of the nutrients sham-fed by rats with gastric cannulas are absorbed (Sclafani and Nissenbaum, 1985, 1987; Reed et al., 1990). Perhaps the proportion of a sham-fed load that escapes the stomach into the intestines decreases after the beginning of sham feeding, so that intake late in the sham-feeding period realizes fewer calories than intake early on. Alternatively, extended sham feeding may lead to a reduction in flavor preference due to sensory-specific satiety (see Rolls, 1986) or simply fatigue.

B. Gastric

Miller and Kessen (1952) gave rats an intragastric (IG) infusion of milk or 0.9% NaCl, depending on which arm of a T-maze they entered. Over 40 days, the rats developed a preference for the side associated with milk infusion. Other groups that drank the milk learned the discrimination much faster, even if the presentation of the milk was delayed by 7.5 minutes. This suggested that oropharyngeal stimulation is not required for, but can facilitate, the development of a side preference (see also Hull et al., 1951). Similarly, G. L. Holman (1968) trained rats to bar press in order to receive a drink of saccharin, an IG infusion of liquid diet, or the combination. Rats pressed the bar during extinction only after experience with the combination. The same result was found if rats were trained with cold IG infusions (which provided nasopharyngeal sensation because the IG catheter was implanted by an intranasal route) but not with body temperature IG infusions. This has been interpreted as evidence that IG nutrients enhance the reward value of a "neutral" oral stimulus; however, because oral stimulation modifies nutrient digestion (Ramirez, 1985), the results of this and other studies comparing IG and oral nutrients may reflect differences in postabsorptive handling of the IG-infused diet caused by the oral stimulation.

In a subsidiary experiment, G. L. Holman (1968) gave hungry rats three pairs of training trials, consisting of 5-minutes of access to a flavored drink followed immediately by an IG injection of liquid diet or water. In a two-bottle preference test conducted the day after the last training trial, 17 of 18 rats initially preferred the diet-paired solution. This was the first demonstration that the postingestive effects of nutrients can increase food preference.

There have been several other studies showing that rats develop a preference for flavored food or water paired with IG administration of nutrients (e.g., Mather et al., 1978; Tordoff et al., 1990). Of particular interest, Sclafani and Nissenbaum (1988) gave rats grape- or cherry-flavored Kool-Aid to drink for 2 days each. When the rats drank one flavor they received an IG infusion of starch (Polycose), and when they drank the other flavor they received IG water. During a subsequent 4-day preference test, the rats strongly and persistently preferred the IG starch-paired flavor. This was not due to an aversive effect of IG water infusion—in subsequent choice tests between each flavor and unflavored water, the rats showed indifference to the flavor paired with IG water and preferred the flavor paired with IG starch.

In these studies, however, the infused nutrients were not confined to the stomach and could easily drain into the intestine and be absorbed. Thus, they say little about where the UCS is detected, except that the site is postoral. Deutsch and colleagues have attempted to localize the effects of IG manipulations to the stomach. In one study (Puerto et al., 1976), fasted rats were given simultaneous access to almond- and banana-flavored water. Ingestion of one flavor initiated an IG infusion of predigested whole milk, whereas ingestion of the other flavor either had no effect or initiated an IG 0.9% NaCl infusion. After either two 10-minute tests (milk versus no infusion) or on the first test (milk versus 0.9% NaCl), the rats preferred the nutrient-paired flavor. The authors infer from this very rapid (i.e., < 10 min) development of a discrimination that a signal from the upper gastrointestinal tract must be involved.

In another study, the same method was used except injected nutrients were restricted to the stomach by inflation of a pyloric cuff (Deutsch and Wang, 1977). These animals showed a slight preference for the milk-paired flavor during the first 10-minute test and gradually increased preference with repeated testing. Unfortunately, these findings cannot be attributed to gastric effects because the pyloric cuffs were deflated (allowing drainage) immediately after the training period (see Baker and Booth, 1989). This means that an association could form between the ingested flavors and the delayed, postgastric consequences of nutrients. It is noteworthy that Deutsch and colleagues found some evidence of discrimination on the first training trial (i.e., before delayed postingestive effects could be involved), but only when the UCS was an infusion of predigested milk, and not undigested milk or glucose. The authors suggest this implies that a breakdown product of the milk is sensed in the stomach. However, an alternative explanation is that milk extracted from the stomach of donor rats could contain carbon disulfide or related compounds that communicate food safety and enhance preference (Galef et al., 1988; see review by Galef, this volume, Chapter 9). The rats may initially prefer the milk-paired flavor because they smell the mixture of ingested flavor and carbon disulfide.

It is possible that a CS rather than, or as well as, a UCS may be provided by gastrointestinal effects of the food (perhaps by stretch or nutrient receptors). This could be part of a mechanism controlling meal size that is sensitive to gastric fill (i.e., a particular level of gastric fill could cause satiety as a result of its prior association with a postabsorptive effect). Although some of the work of Deutsch and colleagues (see review by Deutsch, 1983) alludes to this possibility, there has been no serious consideration of it. Booth (1989) provides an excellent review of this and other possible conditioned stimuli that have been, or could plausibly be, used.

C. Intestinal

The small intestine contains receptors sensitive to macronutrients (glucose, protein, and fat), and osmotic pressure, making it a candidate site for detection of the UCS. However, to date there is no evidence for this possibility, and some against.

Here, the problem of localization is extremely difficult because there is no easy way to prevent nutrient absorption, and thus confine the nutrient to the intestine. The work of Elizalde and Sclafani (1988) bears on this issue. They trained two groups of rats to drink either grape- or cherry-flavored sweetened Kool-Aid. For both groups, one drink contained Polycose and the other did not; however, for one group the Polycose drink also contained Acarbose, an inhibitor of starch hydrolysis. The control group acquired a strong preference for the flavor paired with Polycose, but the group given Polycose with Acarbose did not. Because Acarbose prevents the hydrolysis of glucose from Polycose in the duodenum, this finding suggests that duodenal or postabsorptive glucose supports the development of taste preferences. It also argues against the involvement of preabsorptive Polycose receptors.

A recent paper has emphasized the importance of intestinal absorption for food preferences from an ecological perspective (Martinez del Rio and Stevens, 1989). Starlings drink solutions of glucose and fructose but not sucrose. These birds lack intestinal membrane disaccharidases that convert sucrose to glucose and fructose, so they derive no caloric benefit from sucrose consumption. Perhaps the birds do not drink sucrose because of this.

D. Hepatic

Hepatic involvement in the conditioned control of food intake was first proposed by Russek. He found that dogs given hepatic–portal glucose infusions were anorectic when fed in the apparatus where they were infused but ate normally if fed in their home cages. Two dogs administered glucose in the presence of a bright light were later anorectic when the light was on and no infusion was given (Russek, 1970). These results led Russek to propose that "preabsorptive satiation" was the result of sensory influences on hepatic glucose production, mediated by the autonomic nervous system (see Russek, 1981).

Although the role of the liver in unconditioned feeding behavior has been studied for many years (see Russek, 1981; Tordoff and Friedman, 1986; Tordoff et al., 1990, for reviews), the conditioned aspects were largely neglected until recently. Tordoff and Friedman (1986) gave rats flavored foods in association with hepatic–portal infusions of either saline or glucose. When allowed to choose between the glucose- and saline-paired flavored foods, the rats preferred the one given previously with glucose. Rats tested concurrently under identical conditions but with jugular infusions did not alter food preference, indicating that the primary metabolic effects of the hepatic–portal glucose infusions were confined to the liver.

Another approach to the issue of localization in the liver involves taking advantage of the differential metabolic action of glucose and fructose. Unlike glucose, which is metabolized by nearly all tissues, fructose does not enter the brain, is metabolized poorly by most peripheral tissues, but is metabolized well by the liver (Van den Berghe, 1978). Thus, gram for gram, fructose provides more fuel for the liver than does glucose. A critical experiment was to pair one flavored food with a drink of fructose and another flavored food with an equicaloric drink of glucose. After training, the rats preferred to eat the flavored food paired with fructose rather than the flavored food paired with glucose. The same result was obtained in a second experiment when the fructose and glucose solution were gavaged (rather than ingested voluntarily; Tordoff et al., 1990).

E. Parenteral and Systemic

The earliest attempt to determine whether intravenous injections are reinforcing was conducted by Coppock and Chambers (1954). Restrained rats received tail vein infusions of either glucose or 0.9% NaCl when their head was turned away from the side they spontaneously preferred. There was a nonsignificant tendency for rats infused with glucose to keep their heads on the reinforced side longer than did those infused with saline.

Le Magnen (1955; reviewed in English by Le Magnen, 1967, 1969) gave rats flavored food during two 1-hour periods daily for 12 days. Immediately after eating one of the flavored foods, the rats received 25% more calories (20% glucose) by parenteral injection; after eating the other flavor, they received an injection of 0.9% NaCl. When given a choice between the two flavored foods, the rats avoided the food paired with the glucose injection. Le Magnen interpreted this as evidence that the rats learn that the glucose-paired flavored food is (from the rat's point of view) of increased caloric density, and therefore should be eaten in smaller quantities. It seems more likely that the avoidance of the glucose-paired flavor was due to malaise produced by the large dose of hypertonic glucose given to already satiated rats (see, e.g., Booth et al., 1972; Deutsch et al., 1976).

Revusky et al. (1971) gave hungry rats five training trials consisting of 30-minute access to a coffee-flavored saccharin solution followed 10–20 minutes later by a 53-minute infusion (5 or 10 ml of 10% glucose) into the jugular vein. Control groups were infused with an identical volume of 0.9% NaCl or nothing. After training, each group received unflavored saccharin during a 6–8 hour feeding period. The following day, when the animals were again hungry, a preference text between unflavored and coffee-flavored saccharin was conducted. Rats previously given NaCl infusions after drinking coffee avoided this flavor relative to those given glucose or no infusion. Rats given the coffee–glucose pairing showed a very slightly greater (nonsignificant) preference for the coffee flavor than did those given no infusion. The apparently aversive effect of 0.9% NaCl infusion is surprising. The investigators had difficulty maintaining the patency of the rats' catheters; it may be that the NaCl solution was not sterile, causing infections that could lead to both an aversive, pyrogenic response and catheter blockage.

The only clear demonstration of a reinforcing effect of systemic glucose was provided by Mather et al. (1978). These authors gave free-feeding rats meal-contingent jugular infusions. Commencing 5 minutes after the onset of each feeding bout, 30% glucose or 0.9% NaCl was infused at 3 ml/h for an unspecified period (probably until the end of the meal). In one group, glucose infusions were given on days when either citral- or eucalyptol-flavored food was available, and NaCl infused on days when the other flavor was available. When given a choice, the rats strongly preferred the flavor associated with glucose infusion to that associated with saline infusion.

The reinforcing effect of systemic glucose in all these experiments may simply reflect a portion of the infused glucose being used by the liver. In a direct comparison of the potency of systemic and hepatic–portal glucose infusions as reinforcers of food preference, Tordoff and Friedman (1986) gave free-feeding rats access to chocolate- or chicken-flavored chow during the first 2 hours of the dark period, and at the same time either jugular or hepatic–portal infusions of NaCl or glucose. Although hepatic–portal glucose infusions induced a preference for the concurrently eaten flavored food (see above), preference for the jugular NaCl- and glucose-paired flavored foods was equal. In other work (Tordoff and Friedman, 1988), it was found that infusion of this quantity of glucose into the jugular vein had no observable influence on liver metabolism. However, it is unclear whether the same can be said for the infusions used by other investigators.

F. Cranial

There have been no attempts to examine whether nutrients infused directly into the brain support a conditioned flavor preference. However, flavor preferences have been produced by electrical stimulation of "rewarding" regions of the brain (see, e.g., Ettenberg and White, 1978). It is apparently not the stimulation itself that acts as the UCS for flavor preference learning, because rats develop a preference for flavored water ingested after a period of rewarding stimulation (Tordoff et al., 1982). Perhaps the UCS involves a change in brain neurochemistry that outlasts the electrical stimulation. Alternatively, stimulating the brain may simply be a way of activating a hepatic mechanism, because such stimulation alters liver metabolism (see, e.g., Shimazu, 1983). It is interesting in this regard that total subdiaphragmatic vagotomy attenuates rates of intracranial electrical self-stimulation (Ball, 1974).

There are also studies showing that pharmacological treatments with substances that act on the CNS can support conditioned flavor or place preferences (see review by Blass, this volume, Chapter 14). However, these agents have generally been administered into the periphery, and there has been no attempt to rule out the possibility that they exert their rewarding effects by interfering with metabolism; they could act directly on the autonomic nervous system or indirectly on CNS autonomic efferents to the periphery. Thus, a role for the brain as a site of UCS transduction for flavor preference learning has not been demonstrated. This is not to say that the brain does not participate in the acquisition of taste preferences. There must be neural integration of the CS and UCS (see discussions in this volume by Powley and Berthoud, Chapter 19, and Scott, Chapter 22). Perhaps this circuitry involves an opiate-mediated component, which could account for the changes in preference and other behaviors that are produced by opiate blockade (see review by Blass, Chapter 14).

G. What Is Transduced?

The UCS for nutrient-based learning must, by definition, provide the animal with information concerning the nutritional benefit of the food. Unfortunately, it is not clear what "nutritional benefit" is. An early report suggested that the increases in temperature or changes in respiration produced by nutrients might act as the reinforcer (Chambers, 1956), although this now seems unlikely.

The answer to the question "Where is the UCS transduced?" provides insight into what is transduced. The conditioned flavor preferences produced by hepatic–portal infusion of glucose, and by drinking or gavage of fructose relative to drinking or gavage of glucose, argue strongly that the liver is a site of UCS transduction. It therefore seems reasonable to surmise that a metabolic signal is involved. Studies on the unconditioned control of food intake suggest three possibilities. First, nutrients arriving in the liver could activate chemospecific neural receptors (Russek, 1970). One problem with this is that whereas fructose is more potent than glucose as a UCS (Tordoff et al., 1990), electrical recordings from the vagus nerve support the existence of fibers that are sensitive to glucose but not to fructose (Niijima, 1983). A second possibility is that the UCS is generated by a metabolite such as lactate or pyruvate (Russek, 1981), which is common to glucose and fructose metabolism. This hypothesis is consistent with findings that hepatic pyruvate is elevated higher and more rapidly by administration of fructose than by glucose (Burch et al., 1970).

The problem with locating the origin of the neural representation of the UCS in a process related to a glycolytic intermediary is that this cannot easily account for the preference produced by pairing flavors with administration of corn oil or alcohol, which are not metabolized by this pathway. We have speculated that the UCS is generated by a signal within mitochondria that is common to the metabolism of carbohydrate and fat, perhaps by their biological oxidation (Tordoff et al., 1987). This is based in large part on work examining the unconditioned metabolic control of food intake (see review by Friedman, this volume, Chapter 2). With respect to conditioned food preferences, confirmatory evidence comes from studies comparing the acquisition of flavor preferences in normal and diabetic rats given fixed rations (13.3 kcal) of glucose solution or corn oil emulsion to drink in association with a flavored food. The controls equally preferred flavored food paired with glucose and with oil, whereas the diabetics preferred the flavor paired with fat and avoided the flavor paired with glucose (Tordoff et al., 1987). This is consistent with the limited ability of diabetic rats to oxidize carbohydrate. The

LEARNED FOOD PREFERENCES

"equipotentiality" of calories from different fuels as a source of the UCS is also seen with respect to carbohydrates and alcohol. Meheil and Bolles (1984) found that rats equally preferred grape- and orange-flavored Kool-Aid after one flavor was paired with ethanol solution and the other with a sucrose solution of equal caloric density.

H. Information Transfer from Transduction Site to Brain

Given that a food's nutrients have been detected, how is this information transferred to the brain? Three possibilities have been suggested, each depending on a different site of transduction.

1. Intestine–Cholecystokinin

Cholecystokinin (CCK) is an intestinal factor of interest as a putative control of unconditioned feeding behavior. It has also been suggested to provide the (internalized) UCS for learned flavor preferences (Meheil and Bolles, 1988a,b). However, the evidence for this is weak at best: one unpublished study shows that exogenous CCK administration can produce a flavor preference (Meheil, 1989); but a number show either no effect or a pronounced aversion (e.g., Gibbs et al., 1973; Holt et al., 1974; Deutsch and Hardy, 1977).

2. Pancreas–Insulin

Insulin has been implicated in conditioned taste preference learning by the finding that rats allowed to sham feed one flavor of liquid diet after saline injection and another flavor after insulin injection preferred the flavor paired with insulin (Oetting and VanderWeele, 1985). Moreover, diabetic rats, which have negligible insulin, developed an aversion to flavored food paired with a drink of glucose (Tordoff et al., 1987). These findings raise the possibility that insulin may be involved in the generation of the UCS.

The possibility that insulin per se conveys information concerning the UCS appears unlikely: the finding that rats develop a preference for the flavored food they sham drink after insulin injection may well be an artifact of insulin-induced gastric emptying, and thus does not generalize to normal ingestion (VanderWeele and Deems, 1989). There are several examples of flavor aversions produced by exogenous insulin injection (e.g., Le Magnen, 1967; Lovett et al., 1968). Moreover, despite their lack of insulin, diabetic rats form taste preferences for food flavors paired with fat (Tordoff et al., 1987), and intact rats prefer flavored food that has been paired with fructose to one paired with glucose (Tordoff et al., 1990), even though fructose is the less insulinogenic sugar. Thus, it is difficult to believe that insulin carries information concerning the nutrient content (rewarding value) of food. It seems more likely that it is only indirectly involved, through its actions affecting fuel supply and disposition in the liver (see Tordoff et al., 1989) and other tissues (see review by Newsholme and Start, 1973).

3. Liver–Vagus

The hepatic vagus nerve has been implicated in flavor preference conditioning by the finding that although intact rats preferred flavored food eaten after intubation of fructose to another flavored food eaten after intubation of glucose, rats with hepatic vagotomy did not. More impressively, whereas intact rats showed a strong preference for flavored food paired with a drink of fructose relative to flavored food paired with only water to drink, rats with hepatic vagotomy preferred the two flavored foods equally (Tordoff et al., 1990). It appears as if hepatic vagotimized rats are insensitive to the caloric benefit of the sugar they ingest.

These results with hepatic vagotomy provide additional evidence for localization of the UCS in the liver. It is tempting to speculate that a signal transduced in the liver is conveyed to the brain via the hepatic vagus nerve. However, there is also the possibility that vagal innervation interferes with the metabolism of fructose and thus prevents the genesis of the neural signal. We have recently found that hepatic vagotomy eliminates the rise in hepatic pyruvate concentration after gavage of fructose or glucose (Sandler and Tordoff, 1989). This implies that rats with hepatic vagotomy have a reduced glycolytic capability. If the UCS is generated by hepatic pyruvate concentration (Russek, 1981) or by postglycolytic metabolites (or processes), then the signal will be absent or distorted in rats with hepatic vagotomy. It is also possible that the hepatic vagus both modulates the metabolic-to-neural transduction of the UCS and carries this signal from the liver to the brain.

IV. CONCLUSIONS

Individuals can associate a food's flavor with its nutrient value. Although there is still much work to be done, the best available evidence suggests that information about "nutrient value" is derived from the oxidation of fuel in the liver. This information could travel to the brain in the hepatic vagus nerve, although other neural routes and humoral signals have not been excluded.

The possibility that a signal related to hepatic fuel oxidation provides the UCS for flavor preference learning has several implications. For example, consideration of fuel supply to the liver, rather than simply fuel intake, can explain why satiety is not necessarily rewarding. It may be that oral and gastrointestinal factors causing satiety, such as gut distension, osmotic pressure, and CCK, have a neutral or aversive hedonic influence. Thus, the cessation of eating caused by activation of these controls is not rewarding. A signal based on hepatic fuel oxidation is consistent with findings that conditioned flavor preference learning is influenced by food deprivation. Fasting differentially affects the metabolic disposition of carbohydrate and fat. This implies that a foods' rewarding value depends on an interaction between the macronutrient composition of the food and the deprivation (metabolic) state of the subject.

It would be possible but premature to conclude that the liver is the only source of UCS transduction. Experiments that localize nutrient action to a particular component of the gastrointestinal tract have not been done. Even if the liver is the only source of UCS transduction, it is unlikely to act alone. There are suggestions (Bolles et al., 1981; Hayward, 1983) that the effective UCS for flavor preference learning is related to nutrient density (i.e., kcal/g), not simply total nutrient content. It is possible that differences in fuel disposition related to nutrient density could account for such findings. However, it is also consistent with the integration of a metabolic signal with a signal related to food size or bulk, which could originate from an oral meter or gastric fill detector.

On the other hand, it would be a mistake to assume that all controls of unconditioned feeding behavior also participate in the generation of learned behaviors. Presumably, the acquired oral control of food intake and preference is a strategy, honed by evolution, to optimize intake of foods that vary in macronutrient and caloric density. There seems little doubt that animals use a signal derived from fuel oxidation to make unconditioned adjustments to variations in diet composition (see review by Friedman, Chapter 2). It would make sense for the animal to rely on this information to generate the unconditioned stimulus needed to acquire an oral control of food intake and preference.

ACKNOWLEDGMENTS

Much of the work performed at Monell and reported in this chapter was supported by National Institutes of Health grant DK-36339.

DISCUSSION

Weingarten: Let me ask you about the aversion that you see when the animal isn't satisfied after ingesting saccharin with a limited food ration. Because the animal is eating something, would there not be increased fuel oxidation under those circumstances?

Tordoff: Yes, I think that there probably would be an increase in fuel oxidation, but I don't think it's sufficient to counteract the effects of the sweetener. The sweetener is making the animal hungrier, but the fuel that we're giving them is not sufficient to increase oxidation above control levels, and thus completely eliminate this hunger. I think that you could perhaps titrate the hunger caused by the sweetness and the reduction of hunger caused by the food.

Weingarten: If it's not sufficient, then the animal must have some expectation of what it should be getting. How does that work? I mean if the animal detects that it's not sufficient—how much he should have gotten—he must have known how much he should have gotten to do that comparison. How does the animal have that information?

Tordoff: I'm not sure that he does need to. He just takes it as a moment-by-moment situation. He knows whether he has enough fuel available by the mechanism in the liver which is measuring fuel oxidation.

Weingarten: I guess my only comment is, fuel oxidation goes up in both circumstances, yet one leads to a preference and one leads to aversion. So you need some other construct other than fuel.

Tordoff: It's not just the direction. What we're saying is that it's a particular amount of fuel oxidation that's required. In one situation, the effect is a decrease because of the sweetness and then a slight increase because of the food; in the other situation it's going beyond the initial value.

VanItallie: I'm a little confused by the way you use the term "reward." In Bart Hoebel's study, where the animals self-stimulate for reward, the self-stimulation is depressed when calories are utilized. So, in a sense, calorie utilization is a reward reduction phenomenon in that paradigm. And yet it seemed to me that you were saying that utilization of calories was rewarding. Now that confuses me. Could you clarify that?

Tordoff: I'm not sure I could. One of the interesting things about the self-stimulation literature, of course, is that vagotomy attenuates self-stimulation, which is interesting when you think of this as just a central mechanism. It's not surprising that with different levels of hunger there should be different levels of self-stimulation because there may be some control of the gain in a situation. But the answer is, I really don't know.

Cabanac: It's somewhat the same question. In your basic experiment, is not the sequence of presentation of stimuli important? Wouldn't you expect the hepatic load coming after the oral stimulus to be rewarding? I see a problem there, although it works. You have perfectly demonstrated that it works. But would you not expect a bigger effect if you had some delay before the hepatic load and presentation of the stimulus, rather than simultaneous or even the reverse sequence?

Tordoff: Although, in the portal infusion experiment, we gave the food and the infusion at the same time, the effects of the infusion are not felt for some time, and the animals, which are being tested in the dark period, eat straight away. So we do have a good CS-UCS relationship in that situation. But I agree that those experiments need to be done—to look at the contingencies involved and the delay that's available. For example, it would be very interesting to know whether we can get the long delay learning that you get with the conditioned taste aversion paradigm.

LEARNED FOOD PREFERENCES

Campfield: These are very interesting studies, Mike. I have two questions. The first set of studies on the basic flavor preference—preference for the food paired with the glucose infusion: I would imagine from your hypothesis that you could titrate that by the amount of metabolic fuel you apply to the liver. I'd like to know if that is the case, if you could get sort of a dose–response of preference either in time course or magnitude. The second question relates to the comparison between glucose and fructose. We and many other people have looked at the blood glucose effects of oral ingestion of these two hexoses and they're very different, as you know. Glucose causes large changes in plasma glucose, and fructose does essentially very little in the light phase. A paper now in press shows that either oral ingestion or infusion of fructose can cause transient decreases in blood glucose that were within the magnitude that the CNS can clearly see. So I think perhaps we need to be careful about no direct effect of fructose.

Tordoff: In response to your second question, that's true, but then why does hepatic vagotomy remove the effect? Unless you're going to say that the effects of hepatic vagotomy are to alter the dynamics of the fructose or glucose, then I'm afraid I can't accept that.

Campfield: But there's also a very different peripheral effect in blood glucose in animals drinking glucose or fructose that I think you need to account for in explaining the behavior.

Tordoff: We can show that it is not a glucose effect. In the infusion experiments, we've shown over and over again that you can infuse glucose into liver and get no change in peripheral glucose levels. So we can do specific manipulations of the liver without altering systemic glucose.

Greenberg: Also with your glucose and fructose experiment, if I understood correctly, you said you were at times intubating the glucose. As you know, Deutsch has shown that intubation alone can be the stimulus for a conditioned taste aversion. Therefore, what you could be showing in that case is a reduction of preference for glucose rather than a preference for fructose.

Tordoff: Yes, that's certainly one problem, and that's why we've done it in both the intubated animals and in animals that drink it themselves. The results are identical. I showed you only one today just to save time. Also, we can get the effect if we intubate fructose versus nothing; we still get the preference for the fructose. So that doesn't seem to be a problem, at least in our hands.

VanItallie: Since fructose has been given a lot of attention, I wonder whether you considered what happens to it when it enters the liver. I've always heard that it was a very good glycogen substrate. The question is, How much of it is forming glycogen and how much is actually being oxidized?

Tordoff: Well, that's a question for my next grant proposal! There is data to suggest that is the case, but one thing that we do know is that, although some of it goes to glycogen, much of it is pushed through glycolysis and is rapidly oxidized. It's much more rapidly sent through glycolysis than is glucose.

VanItallie: You get a rise of lactic acid?

Tordoff: Yes, and pyruvic acid.

REFERENCES

Baker, B. J., and Booth, D. A. (1989). Preference conditioning by concurrent diets with delayed proportional reinforcement. *Physiol Behav.* **46**:585–590.

Ball, G. G. (1974). Vagotomy: Effect on electrically elicited eating and self stimulation in the lateral hypothalamus. *Science,* **184**:484–485.

Beauchamp, G. K., and Cowart, B. J. (1987). Development of sweet taste. In *Sweetness,* J. Dobbing (Ed.). Springer-Verlag, London, pp. 127–138.

Boakes, R. A., and Lubart, T. (1988). Enhanced preference for a flavour following reversed flavour–glucose pairing. *Q. J. Exp. Psychol.* **40B**:49–62.

Boakes, R. A., Rossi-Arnaud, C., and Garcia-Hoz, V. (1987). Early experience and reinforcer quality in delayed flavour-food learning in the rat. *Appetite,* **9**:191–206.

Bolles, R., Hayward, L., and Crandall, C. (1981). Conditioned taste preferences based on caloric density. *J. Exp. Psychol.: Animal Behav. Processes,* **7**:59–69.

Booth, D. A. (1972). Satiety and behavioral caloric compensation following intragastric glucose loads in the rat. *J. Comp. Physiol. Psychol.* **73**:412–432.

Booth, D. A. (1977). Appetite and satiety as metabolic expectancies. In *Food Intake and Chemical Senses,* Y. Katsuki, M. Sato, S. F. Takagi, and Y. Oomura (Eds.). University of Tokyo Press, Tokyo, pp. 317–330.

Booth, D. A. (1989). How to measure learned control of food or water intake. In *Feeding and Drinking,* F. M. Toates and N. Rowland (Eds.). Elsevier Applied Science, London, pp. 111–149.

Booth, D. A., Lovett, D., and McSherry, G. M. (1972). Postingestive modulation of the sweetness preference gradient in the rat. *J. Comp. Physiol. Psychol.* **78**:485–512.

Booth, D. A., Lee, M., and McAleavy, C. (1976). Acquired sensory control of satiation in man. *Br. J. Psychol.* **67**:137–147.

Booth, D. A., Mather, P., and Fuller, J. (1982). Starch content of ordinary foods associatively conditions human appetite and satiation, indexed by intake and eating pleasantness of starch-paired flavours. *Appetite,* **3**:163–184.

Burch, H. B., Lowry, O. H., Meinhardt, L., Max, P., Jr., and Chyu, K.-J. (1970). Effect of fructose, dihydroxyacetone, glycerol, and glucose on metabolites and related compounds in liver and kidney. *J. Biol. Chem.* **245**:2092–2102.

Campbell, D. H., Capaldi, E. D., and Myers, D. E. (1987). Conditioned flavor preferences as a function of deprivation level: Preferences or aversions? *Animal Learn. Behav.* **15**:193–200.

Capaldi, E. D., and Myers, D. E. (1982). Taste preference as a function of food deprivation during original taste exposure. *Animal Learn. Behav.* **10**:211–219.

Capaldi, E. D., Myers, D. E., Campbell, D. H., and Sheffer, J. D. (1983). Conditioned flavor preferences based on hunger level during original flavor exposure. *Animal Learn. Behav.* **11**:107–115.

Capaldi, E. D., Campbell, D. H., Sheffer, J. D., and Bradford, J. P. (1987a). Conditioned flavor preferences based on delayed caloric consequences. *J. Exp. Psychol.: Animal Behav. Processes,* **13**:150–155.

Capaldi, E. D., Campbell, D. H., Sheffer, J. N., and Bradford, J. P. (1987b). Non-reinforcing effects of giving "dessert" in rats. *Appetite,* **9**:99–112.

Chambers, R. M. (1956). Some physiological bases for reinforcing properties of reward injections. *J. Comp. Physiol. Psychol.* **49**:565–568.

Coppock, H. W., and Chambers, R. M. (1954). Reinforcement of position preference by automatic intravenous injections of glucose. *J. Comp. Physiol. Psychol.* **47**:355–357.

Deutsch, J. A. (1983). Dietary control and the stomach. *Prog. Neurobiol.* **20**:313–332.

Deutsch, J. A., and Hardy, W. T. (1977). Cholecystokinin produces bait shyness in rats. *Nature,* **266**:196.

Deutsch, J. A., and Wang, M.-L. (1977). The stomach as a site for rapid nutrient reinforcement sensors. *Science,* **195**:89–90.

Deutsch, J. A., Mólina, F., and Puerto, A. (1976). Conditioned taste aversion caused by palatable nontoxic nutrients. *Behav. Biol.* **16**:161–174.

Elizalde, G., and Sclafani, A. (1988). Starch-based conditioned flavor preferences in rats: Influence of taste, calories and CS–US delay. *Appetite,* **11**:179–200.

Ettenberg, A., and White, N. (1978). Conditioned taste preferences in the rat induced by self-stimulation. *Physiol. Behav.* **21**:363–368.

Faneslow, M. S., and Birk, J. (1982). Flavor-flavor associations induce hedonic shifts in taste preference. *Animal Learn. Behav.* **10**:223–228.

Fedorchak, P. M., and Bolles, R. C. (1987). Hunger enhances the expression of calorie- but not taste-mediated conditioned flavor preferences. *J. Exp. Psychol.: Animal Behav. Processes,* **13**:73–79.

Fedorchak, P. M., and Bolles, R. C. (1989). Nutritive expectancies mediate cholecystokinin's suppression-of-intake effect. *Behav. Neurosci.* **102**:451–455.

Friedman, M. I., and Stricker, E. M. (1976). The physiological psychology of hunger: A physiological perspective. *Psychol. Rev.* **83**:409–431.

Galef, B. G., Jr., Mason, J. R., Preti, G., and Bean, N. J. (1988). Carbon disulfide: A semiochemical mediating socially-induced diet choice in rats. *Physiol. Behav.* **42**:119–124.

Gibbs, J., Young, R. C., and Smith, G. P. (1973). Cholecystokinin decreases food intake in rats. *J. Comp. Physiol. Psychol.* **87**:488–495.

Grill, H. J., Berridge, K. C., and Ganster, D. J. (1984). Oral glucose is the prime elicitor of preabsorptive insulin secretion. *Am. J. Physiol.* **246**:R88–R95.

Hartley, E. L., and Perelman, M. A. (1963). Deprivation and the canalization of responses to food. *Psych. Rep.* **13**:647–650.

Hayward, L. (1983). The role of oral and postingestional cues in the conditioning of taste preferences based on differing caloric density and caloric outcome in weanling and mature rats. *Animal Learn. Behav.* **11**:325–331.

Holman, E. W. (1975). Immediate and delayed reinforcers for flavor preferences in rats. *Learn. Motiv.* **6**:91–100.

Holman, E. W. (1980). Irrelevant-incentive learning with flavors in rats. *J. Exp. Psychol. Animal Behav. Processes,* **6**:126–136.

Holman, G. L. (1968). Intragastric reinforcement effect. *J. Comp. Physiol. Psychol.* **69**:432–441.

Holt, J., Antin, J., Gibbs, J., Young, R. C., and Smith, G. P. (1974). Cholecystokinin does not produce bait shyness in rats. *Physiol. Behav.* **12**:497–498.

Hull, C. L., Livingston, J. R., Rouse, R. O., and Barker, J. R. (1951). True, sham, and esophageal feeding as reinforcements. *J. Comp. Physiol. Psychol.* **44**:236–245.

Kawamura, Y., and Kare, M. R. (1987). *Umami: A Basic Taste.* Dekker, New York.

Lavin, M. J. (1976). The establishment of flavour–flavour associations using sensory preconditioning training procedure. *Learn. Motiv.* **7**:173–176.

Le Magnen, J. (1955). La satiété induite par les stimuli sucrés chez le rat blanc. *C. R. Séances Soc. Biol.* **149**:1339–1342.

Le Magnen, J. (1967). Food habits. In *Handbook of Physiology: Alimentary Canal.* Volume 1 Section 6 American Physiological Society, Washington, DC, pp. 11–30.

Le Magnen, J. (1969). Peripheral and systemic actions of food in the caloric regulation of intake. *Ann. N.Y. Acad. Sci.* **57**:1126–1156.

Lovett, D., Goodchild, P., and Booth, D. A. (1968). Depression of intake of nutrient by association of its odor with effects of insulin. *Psychonomic Sci.* **11**:27–28.

Martinez del Rio, C., and Stevens, B. R. (1989). Physiological constraint on feeding behavior: Intestinal membrane disaccharidases of the starling. *Science,* **243**:794–796.

Mather, P., Nicolaidis, S., and Booth, D. A. (1978). Compensatory and conditioned feeding responses to scheduled glucose infusions in the rat. *Nature,* **273**:461–463.

Meheil, R. (1989). Rats learn to like flavors paired with CCK. Paper presented at the Annual Meeting of the Eastern Psychological Association.

Meheil, R., and Bolles, R. C. (1984). Learned flavor preferences based on caloric outcome. *Animal Learn. Behav.* **12**:421–427.

Meheil, R., and Bolles, R. C. (1988a). Learned flavor preferences based on calories are independent of initial hedonic value. *Animal Learn. Behav.* **16**:383–387.

Meheil, R., and Bolles, R. C. (1988b). Hedonic shift learning based on calories. *Bull. Psychonomic Soc.* **26**:459–462.

Miller, N. E., and Kessen, M. L. (1952). Reward effects of food via stomach fistula compared with those of food via mouth. *J. Comp. Physiol. Psychol.* **45**:555–564.

Newsholme, E. A., and Start, C. (1973). *Regulation in Metabolism.* Wiley, London.

Niijima, A. (1983). Glucose-sensitive afferent nerve fibers in the liver and their role in food intake and blood glucose regulation. *J. Auton. Nerv. Syst.* **9**:207–220.

Oetting, R. L., and VanderWeele, D. A. (1985). Insulin suppresses intake without inducing illness in sham feeding rats. *Physiol. Behav.* **34**:557–562.

Puerto, A., Deutsch, J. A., Mólina, F., and Roll, P. L. (1976). Rapid discrimination of rewarding nutrient by the upper gastrointestinal tract. *Science,* **192**:485–487.

Ramirez, I. (1985). Oral stimulation alters digestion of intragastric meals in rats. *Am. J. Physiol.* **248**:R459–R463.

Reed, D. R., Tordoff, M. G., and Friedman, M. I. (1990). Sham feeding of corn oil by rats: sensory and postingestive factors. *Physiol. Behav.* **47**:779–781.

Revusky, S. H. (1967). Hunger level during food consumption: Effects on subsequent preference. *Psychonomic Sci.* **7**:109–110.

Revusky, S. H., Smith, M. H., Jr., and Chalmers, D. V. (1971). Flavor preference: Effects of ingestion-contingent intravenous saline or glucose. *Physiol. Behav.* **6**:341–343.

Riley, A. L., and Clarke, C. M. (1977). Conditioned taste aversions: A bibliography. In *Learning Mechanisms in Food Selection*, L. M. Barker, M. R. Best, and M. Domjan (Eds.). Baylor University Press, Waco, TX, pp. 593–615.

Rogers, P. J., and Blundell, J. E. (1989). Separating the actions of sweetness and calories: Effects of saccharin and carbohydrates on hunger and food intake in human subjects. *Physiol. Behav.* **45**:1093–1099.

Rolls, B. J. (1986). Sensory specific satiety. *Nutr. Rev.* **44**:93–101.

Rozin, P. (1967). Thiamine specific hunger. In *Handbook of Physiology: Alimentary Canal*. Volume 1 Section 6 American Physiological Society, Washington, DC, pp. 411–431.

Russek, M. (1970). Demonstration of the influence of an hepatic glucosensitive mechanism on food intake. *Physiol. Behav.* **5**:1207–1209.

Russek, M. (1981). Current status of the hepatostatic theory of food intake control. *Appetite*, **2**:137–143.

Sandler, F., and Tordoff, M. G. (1989). Hepatic vagotomy prevents the conversion of fructose to pyruvate and lactate. Paper presented at the Annual Meeting of the Society for the Study of Ingestive Behavior.

Sclafani, A. (1987). Carbohydrate taste, appetite and obesity: An overview. *Neurosci. Biobehav. Rev.* **11**:131–153.

Sclafani, A., and Nissenbaum, J. W. (1985). Is gastric sham feeding really sham feeding? *Am. J. Physiol.* **248**:R387–R390.

Sclafani, A., and Nissenbaum, J. W. (1987). Oral versus postingestive origin of polysaccharide appetite in the rat. *Neurosci. Biobehav. Rev.* **11**:169–172.

Sclafani, A., and Nissenbaum, J. W. (1988). Robust conditioned flavor preference produced by intragastric starch infusions in rats. *Am. J. Physiol.* **255**:R672–R675.

Shimazu, T. (1983). Reciprocal innervation of the liver: Its significance in metabolic control. *Adv. Metab. Disorders*, **10**:355–384.

Simbayi, L. C., Boakes, R. A., and Burton, M. J. (1986). Can rats learn to associate a flavour with the delayed delivery of food? *Appetite*, **7**:41–53.

Steiner, J. E. (1973). The gusto-facial response: Observation on normal and anencephalic newborn infants. In *Fourth Symposium on Oral Sensation and Perception*, J. F. Bosmas (Ed.). Government Printing Office, Washington, DC.

Tordoff, M. G. (1988). How do nonnutritive sweeteners increase food intake? *Appetite*, **11**(Suppl.):5–11.

Tordoff, M. G., and Friedman, M. I. (1986). Hepatic–portal glucose infusions decrease food intake and increase food preference. *Am. J. Physiol.* **251**:R192–R195.

Tordoff, M. G., and Friedman, M. I. (1988). Hepatic control of feeding: Effect of glucose, fructose and mannitol infusion. *Am. J. Physiol.* **254**:R969–R976.

Tordoff, M. G., and Friedman, M. I. (1989a). Drinking saccharin increases food intake and preference: I. Comparison with other drinks. *Appetite*, **12**:1–10.

Tordoff, M. G., and Friedman, M. I. (1989b). Drinking saccharin increases food intake and preference: II. Hydrational factors. *Appetite*, **12**:11–21.

Tordoff, M. G., and Friedman, M. I. (1989c). Drinking saccharin increases food intake and preference: III. Sensory and associative factors. *Appetite*, **12**:23–35.

Tordoff, M. G., and Friedman, M. I. (1989d). Drinking saccharin increases food intake and preference: IV. Cephalic phase and metabolic factors. *Appetite*, **12**:37–56.

Tordoff, M. G., Jelinek, M., and Holman, E. W. (1982). Backward conditioning of taste preferences using electrical stimulation of the brain as a reinforcer. In *The Neural Basis of Feeding and Reward,* B. G. Hoebel and D. Novin (Eds.). Haer Institute, Brunswick, MA, pp. 123–127.

Tordoff, M. G., Tepper, B. J., and Friedman, M. I. (1987). Food flavor preferences produced by drinking glucose and oil in normal and diabetic rats: Evidence for conditioning based on fuel oxidation. *Physiol. Behav.* **41**:481–487.

Tordoff, M. G., Tluczek, J. P., and Friedman, M. I. (1989). Effect of hepatic–portal glucose concentration on food intake and metabolism. *Am. J. Physiol.* **257**:R1474–R1480.

Tordoff, M. G., Ulrich, P. M., and Sandler, F. (1990). Flavor preferences and fructose: Evidence that the liver detects the unconditioned stimulus for calorie-based learning. *Appetite,* **14**:29–44.

Van den Berghe, G. (1978). Metabolic effects of fructose in the liver. In *Current Topics in Cellular Regulation,* B. L. Horecker and E. R. Stadtman (Eds.). Academic Press, New York, pp. 98–135.

VanderWeele, D. A., and Deems, R. O. (1989). Insulin-paired milk flavors avoided in real-feeding rats. Paper presented at Annual Meeting of the Eastern Psychological Association.

Van Vort, W., and Smith, G. P. (1983). The relationships between the positive reinforcing and satiating effects of a meal in the rat. *Physiol. Behav.* **30**:279–284.

Zellner, D. A., Rozin, P., Aron, M., and Kulish, C. (1983). Conditioned enhancement of human's liking for flavor by pairing with sweetness. *Learn. Motiv.* **14**:338–350.

13
Protein- and Carbohydrate-Specific Cravings: Neuroscience and Sociology

David A. Booth

University of Birmingham
Birmingham, England

I. SOCIOLOGY OF APPETITE SCIENCE

Measurements of dietary uptake have been widely used in sciences that are not primarily concerned with the determinants of dietary behavior. Measurements of food intake by themselves, however, do not advance our understanding of appetite (Booth, 1987b). Yet such behaviorally uninterpretable data dominate the research literature on "feeding."

A scientific symposium on appetite should therefore be selective. The research presented at this conference uses many tools from modern biology and chemistry. Yet the experiments also address issues about the psychological and physiological processes involved in eating behavior, or at least they yield phenomena that are susceptible to an analysis of the mechanisms of appetite.

A. Biological Reductionism

The diversion of research away from the behavioral causation within appetite reflects in part a sociological phenomenon in the sciences generally, especially in the United States. Behavioral and social scientists have been neglecting, or even deserting, their home disciplines for biological areas such as neuroscience, pharmacology, metabolism, or genetics. Discoveries about a behavioral phenomenon or a social problem seldom get much of a scientific audience unless they invoke a place in the brain, a neurotransmitter receptor, a sequenced gene, or something of that ilk.

Such biological or even chemical reductionism is strangely blind to a glaring deficiency in its own logic. This flaw is exposed by the paradox that biologists resist reductionist claims from physics. It seems so clear to them that organisms and molecules really exist and work according to their own rules, without casting in question the fundamental physical forces and structures (whatever they turn out to be). Yet, if life exists, then assuredly mental processes exist also. Beyond the minds of individual organisms, furthermore, social organizations have a life of their own, while being no less comprised of debris from the Big Bang.

Thus, neither the brain nor any other part of human biology will be understood unless research also takes on board the realities of both the mental processes organizing the observed neural and bodily functioning and the cultural processes to which human brains and bodies become socialized as the individual person develops (Booth, 1987a, 1988; Hatfield, 1988).

B. Measurement of Mechanisms of Appetite

The scientific understanding of appetite cannot advance without sociological, psychological, and physiological analyses of the processes organizing ingestive behavior.

Despite this fact, little research on "feeding" and its neuroscience—or, for that matter, applied human nutrition or food marketing—measures the moment-to-moment somatic, sensory, and social influences on the behaviors that together generate the observed intakes. Instead, the disappearance of food from stock is measured more and more reliably over periods of an hour or even a day. Intake is subjected to finer and finer temporal analysis. This does not assess or control the inputs that are the determinants of such output. Therefore, it must miss the scientific issues about appetite.

1. Dietary Selection

Nutritional science cannot advance without including behavioral data. To analyze behavior, we must measure and manipulate a great deal more than the nutrient contents of the diet and its effects on tissues.

Nutrient preparations affect ingestive behavior directly through their sensory attributes, their effects in the gut, and effects in the tissues that are reached during eating. It is therefore quite unscientific to give the name "nutrient selection" to the relative intakes of diets differing in nutrient composition. It is misleading even to report the observed intakes in terms of the generic nutrients (i.e., carbohydrate, protein, fat), rather than as the specific preparations used (e.g., dextrin, casein, corn oil, with details of supplier and batch). This is because the key issues always center round the sensory characteristics of the diets. Neither the choice between diets nor their relative intakes needs to be caused by the nutrients as such. The immediately controlling factors are liable to be flavors or textures that have no reliable connection with the nutritional effects of the nutrient preparations used, whether in the species' history or in the eater's own past experience.

It is now becoming more widely appreciated that adaptation to a diet often teaches the organism what sensory characteristics are associated with what nutritional effects. Verbal acknowledgment of the role of learning does not, however, do anything to avoid the fallacy of confusing intakes of particular preparations with nutrient selection. The effect of a drug or a dietary preload on dietary selection does not have to be on the acquired sensory mediation of nutritional control. It can arise from any number of other influences on intake.

Thus, we shall build a sound neuroscience of human nutrition only if we take full empirical account of the cognitive and behavioral processes involved in food choices, and indeed of the sociology of eating (Booth, 1987a).

2. Satiety Values

Equally unproductive and even misleading is a so-called psychobiology of food intake that ignores both the biological processes and the psychological processes by which foods control eating. This is illustrated by current interest in the "satiety values" of foods and food constituents. It is not bits of the diet that have quantitative effects on intake. The suppressant or indeed excitant effect of eating a food on subsequent intake is an interaction between the mechanisms that influence the eater's appetite and that person's subsequent occasions for eating (Booth, 1989a).

Sweetness, for example, is liable to provoke both visceral and also cognitive processes that modify the control of ingestion. Without any attempt to measure such processes, experiments on the effects of sweeteners on food intake, or on ratings predictive of intake, cannot yield interpretable results (Booth, 1987c) or even be designed effectively (Booth, 1989a). Furthermore, different sweeteners cannot be shown to differ in their effects until significantly different results are demonstrated within the same experiment, accompanied by data showing that the sweetness levels were not discriminable.

3. Data for Other Sciences

Mere intake data can of course be useful to enterprises—from biochemical nutrition to experimental neuropsychology—in which analysis of ingestive behavior is beside the point. Pharmacologists and geneticists, for example, can advance their own sciences by varying the diet in any consistent fashion and recording the effects of drugs or animal strains on intake. But differences between drugs or genes in their effects in a food intake test can tell us only about the drug receptors or the genetic differences. Such work does nothing to advance the understanding of behavior, even if it professes to be psychopharmacology or psychogenetics. Drugs or genes can be useful tools in the study of ingestive behavior only if the ongoing causal influences on the observed intakes are measured at the same time (Booth, 1989b).

Much pharmacological and nutritional research on dietary intakes thus has no clear prospect of connecting with behavioral and physiological knowledge. By careful comparisons among the chapters in this book, the reader will see how much more rapidly scientific understanding can advance in a multidisciplinary field such as nutrition when adequate attention is paid to the sensory and learned aspects of behavior and their relationships to physiological processes in the gut, liver and brain.

II. SOCIAL, SENSORY, AND SOMATIC FACTORS IN APPETITE

Appetite, the disposition to eat and drink (Booth, 1976; Bolles, 1980) is subject to a myriad of influences from the external and internal environments.

The tradition has been to divide these influences into two groups. On the one hand, it is supposed, there is the palatability of a foodstuff; other terms for this include reward, pleasure ("hedonics"), craving, incentive, or even—confusingly—appetite. On the other hand, the assumption goes, there are the somatic factors facilitating ingestion and foods' postingestional inhibitory influences; often, no less confusedly, these are labeled hunger or satiety, respectively.

This categorization remains highly influential. Yet the summation of fixed and independent sensory and somatic influences fails to account for most of the facts of ordinary eating.

A. Social Influences on Eating

For one thing, a theory considering only physiological signals and the sensory characteristics of the diet ignores the sociology of food. Eating is subject to many social and physical influences from the external environment, in addition to those from foodstuffs themselves. A dinner bell or the clock moving toward a habitual mealtime augments or even creates the desire to eat. The more enthusiastic the eating around the table, the more someone in that company is liable to eat. These are not peculiarly human attributes either. The rat, like Pavlov's dogs, can be trained to eat when a bell rings (Valle, 1968; Weingarten, 1984). Galef (this volume, Chapter 9) has provided many examples of social facilitation of eating in rats.

B. Multiple Appetites and Satieties

The notion that satiety gates out palatability is not just incomplete. It is comprehensively refuted by the evidence for particular appetites and satieties, summarized below. In other words, palatability is not stable but often highly contingent on somatic and social context.

An appetite or a satiety is ingestive behavior under the control of both a distinctive set of sensory characteristics and a contextual factor such as a particular bodily state. That is, the behavior is specific to a food and 100% under "orosensory" control. Yet, at the same time, it is specific to an internal state and 100% controlled by signals of incipient deficit or repletion.

The most familiar examples are water-deficit-induced drinking within the broader phenomenon of thirst, the appetite for water, and food-deprivation-facilitated eating within the phenomenon of hunger, presumed to be the appetite for energy. More clearly and conclusively analyzed examples, however, are the innate hunger for sodium (Schulkin, this volume, Chapter 11) and carbohydrate-conditioned satiety and hunger for arbitrary odors, tastes, and textures (Booth, 1972b; Booth and Davis, 1973). In such cases, it is logically impossible to quantitate the sensory effect separately from quantitating the somatic effect and then to add the two quantities to give the observed behavior. Rather, the behavior is the result of presenting a unique combination of particular external and internal stimuli.

No less a refutation of the traditional view is the way that the relative and absolute attractiveness of a food often interact with social circumstance. For many people, some conventional breakfast foods are not at all palatable except at the first meal of the day (Birch et al., 1984). Another example is the parent lucky enough to have an adaptable palate and stomach, who is able to enjoy both eating with the children and also the exotic fare at a banquet.

III. A THEORY OF APPETITE

We proposed nearly 20 years ago a physiological and cognitive theory of appetite (Booth, 1969, 1972a,b; Booth et al., 1972) that provided the basis for integrating quantitative data on the psychobiological and psychosocial phenomena into a psychobiosocial system of causal processes (Booth and Toates, 1974; Booth, 1978, 1988; Booth and Mather, 1978).

A. Framework for Appetite Neuroscience

This analytical and synthetic approach was intended to replace (Booth, 1967, 1968) the strategy hitherto used of merely invoking places in the brain to organize appetite, such as dual centers in the hypothalamus. Instead, the neuroscience of feeding could be built on integration of visceral, dietary, and wider environmental inputs over multiple pathways through the brain to the skeletomotor control systems (Booth, 1976, 1978). The specification also of visceral functions in appetite could be used to analyze autonomic efferent–afferent loops and endocrine modulation of metabolism and gastrointestinal processing (Booth et al., 1976b; Booth 1980a).

Neural pathways research is presented in Part IV of this book. The quantitative model and the qualitative theory have been highly successful in guiding our own research both on learned controls of eating and also on some well-known neurobiological phenomena, such as obesity induced by ventromedial hypothalamic lesions and its basis in rapid gastric emptying (Duggan and Booth, 1986), visceral and brain stem actions of appetite-suppressant drugs and hormones (Booth et al., 1986a), noradrenergic gating of learned satiety (Booth et al., 1986b; Gibson and Booth, 1989), and the macronutrient-specific appetites to be detailed here (Baker et al., 1987; Gibson and Booth, 1988). Such behavioral and physiological evidence on brain input–output relationships has not usually been allowed for by those tracing sensory and autonomic pathways, and so such work has not been as productive as it might have been.

B. Regulation and Function in Appetite

This attention to the causal processes in appetite exposed the fallacy in dividing ingestion or the influences on it into regulatory and nonregulatory categories (Booth et al., 1976b). Some influences on intake arise directly or indirectly from deficits or supplies of water, energy, or other nutrients and so are capable of contributing to the immediate regulation of those nutrient balances. Other influences might have an immediate dysregulatory impact in certain circumstances. Yet such influences may have had life-preserving or reproduction-promoting functions in our ancestors' ecology. The measurement of regulatory responses to disturbances is uninformative unless the mechanisms by which the organism responds to the challenge are examined.

Predation risk and the work cost of foraging must operate on the food intake pattern through cuing mechanisms, just as depletion and repletion must do. If cues from one rat's sickness do not condition aversion in another rat to cues from dietary residues on the first rat, then we need to specify in what other way that exchange of information may occur. Functional analysis is not a rival to mechanistic analysis. The alternative to mechanism is magic.

C. Development of Appetite

The cognitive theory was that all appetite in familiar situations is a learned reaction to combinations of cues (Booth, 1972b; Booth et al., 1972). It follows that sufficiently salient and recurrent conjunctions of features of the diet, the body, and the culture could become incorporated in a conscious and linguistic superstructure in human appetite (Booth, 1987a). This integrative learning begins from the earliest acquaintances with foods and drinks (Booth et al., 1974; Harris and Booth, 1987) and continues throughout adulthood (Booth, 1972a; Booth et al., 1986b). Therefore, moderate differences in sensitivity between people or with aging, for example, are unlikely to account for differences in food preferences, satiability, or cultural practices.

IV. IMMEDIATE SENSORY CONTROL OF INGESTION

It is part of appetite if the sensed characteristics of materials affect any essential component of the approach to and ingestion of materials or their rejection or egestion. Such aspects of appetite are variously known as sensory preferences and aversions (relative or absolute), (un)palatability, or food hedonics, reward, incentive, etc. A diet's sensory influence on ingestion is sometimes talked about as a qualitative effect (e.g., a preference for sweetness or crispness). Obviously, though, the stimulus level is crucial.

Dietary stimuli influence ingestion with various degrees of sophistication. One dimension of behavioral sophistication is the specificity and complexity of the dietary stimuli controlling appetite. Another dimension is the degree to which the dietary control of appetite is tied to context. Both forms of complexity in appetite follow directly from learning. Habituated stimuli, classically conditioned cues, and chunks of verbal information are all highly specific and often multidimensional and contextualized. Nevertheless, there are a few examples of innately organized sensory control of ingestive behavior.

A. Innate Preferences and Aversions

Some dietary stimuli affect the organism's disposition to ingest, regardless of bodily state or social context. These preferences or aversions are usually independent of experience.

In the case of the congenital ingestive response to sweetness or the gaping reflex to bitterness, the vigor of the response is monotonically (or at any rate asymptotically) related to the strength of the taste. This is presumably because the function of such innate responses is more important when the stimulus is more concentrated.

It may prove relevant to protein preference to note that, contrary to the almost universal assumption, sweetness cannot mean calories. The hungry primate needs no encouragement to eat fruit that has become softer and less acid. Rather, we have suggested, the sweet preference is a protein–peptide–amino acid preference (Booth et al., 1987). Receptors sensitive to the hydroxyl groups in amino acids, connected to an ingestive reflex, would prevent novel dietary sources of protein from being spat out by countering stimulation by amino acids of the aversion to organic nitrogen groups, which is necessary to avoid plant toxins. It is absolutely vital to protect neonatal suckling from disruption by the nitrogen aversion (bitterness). On this theory, there would be no selection pressure on a protein-preference receptor to detect sugars. So, receptors for aliphatic hydroxyls in omnivorous mammals would have the side effect of a sugar preference for the plant kingdom to exploit.

B. Learned Preferences

Learned responses are quite different. They do not increase indefinitely with increases in stimulus strength. Any habituated or conditioned stimulus is highly specific: the learned response is weaker when the test stimulus is weaker or stronger than the trained stimulus. Indeed, the more different the stimulus level is from the learned level, the greater is this "generalization decrement."

1. Food Sweetness Preferences

This two-sided gradient in learned responses applies even to sweet tastes. In what was the first demonstration of calorically conditioned food preferences (Booth et al., 1972), postingestional effects of carbohydrate preparations that did not have aversive osmotic effects were shown to condition increased selection and intake of a particular level of a taste (refuting the assumption that only aversions could be strongly conditioned to tastes).

The taste was sweetness, furthermore, at any level of a sugar or saccharin. Thus, the innate preference for stronger sweetness could be reversed by learning that a lower sweetness was associated with greater caloric effects. In that case, not only a weaker sweetness but also a sweetness stronger than the conditioned level was less liked.

Similarly, adult human preference for the sweetness of a familiar food or drink shows a peak at a particular level. That is, there is a stimulus generalization decrement around the most preferred level of sweetness. This ideal point varies among foods and among people, presumably as a result of past experience (Conner et al., 1988a). Indeed, there are two sorts of "sweet tooth," which can only have been induced by differences in eating habits. Sweetness preferences for snack foods and drinks and desserts tend to intercorrelate, while sweetness preferences for vegetables and fruit tend to group together separately (Conner and Booth, 1988; Conner et al., 1988a). As argued below, the "snacking sweet tooth" is a more plausible concept than "carbohydrate craving."

2. Multidimensional Preferences

When the psychophysical function for a sensory constituent is measured without biasing performance, the asymmetrical inverted U of the traditional hedonic curve is not obtained. The true preference peak is an isosceles triangle, with equal and opposite slopes on either side of the peak. This applies to saltiness (Booth et al., 1983; Conner et al., 1988b) as well as to sweetness, and to every other salient attribute that has been tested so far.

A further major principle is that ingestive responses can become attached to combinations of food attributes, within the taste modality or other modalities and between sensory modalities (Booth, 1987d; Booth and Blair, 1988). The most preferred hot drink may have a set of particular levels of volume, temperature, brown color and bitterness, sweetness, thickness, and oiliness (coffee with sugar and cream).

This complexity in the stimulus is, however, partly a matter of level of analysis: a perceptually unitary smell may be comprised of many volatile compounds; a combination of olfactory, gustatory, textural, and color information may make up the full flavor of a fruit; a one-termed multidimensional formula could describe a person's ideal cup of coffee (Booth, 1987d).

C. Conditioned Appetites and Satieties

It is no great conceptual step from the learned preferences and aversions for unique combinations of features of foods to the learned appetites and satieties, that is, the control of ingestion by combinations that include features of bodily state or of other contexts. If such context does not predict a particular nutritional effect, then the resulting appetite or satiety is not nutrient specific, even though a particular nutrient might have conditioned it.

1. Conditioned Desatiation

The first demonstrated internal-state-dependent sensory preference or aversion was carbohydrate-conditioned desatiation (Booth, 1972b; Booth and Davis, 1973). This learned loss of satiation was also the caloric conditioning of flavor preference first shown with sweet tastes by Booth et al. (1972), but the acquired control of ingestion was demonstrated to be more refined than just by the sensory attributes of the diet. The preference for the richer flavor was confined to choices made early in the meal or while mildly hungry (Booth and Davis, 1973), as has subsequently been confirmed many times in rats (Booth, 1977, 1980b; Van Vort and Smith, 1988; Gibson and Booth, 1989) and people (Booth et al., 1976a, 1982; Booth and Toase, 1983).

Thus, a dramatic increase in meal size (Booth, 1972b) was caused by a conditioning of preference to the low-carbohydrate flavor specifically at the end of the meal. That is, internal repletion cues had gained control of ingestion as well as the dietary cues.

Gibson and Booth (1989) have shown that distension of the gastrointestinal tract is sufficient to serve as the internal contextual cue for carbohydrate-conditioned preference; nothing chemically specific is required. This fits the universal assumption that gastric distension is a satiety signal. More importantly, it supports a suggestion that has not been taken seriously enough even by the advocates of gastric satiety, that the satiating effect of normal moderate distension is entirely learned. The evidence is that gastric satiety gains chemical specificity by being food specific, hence also that the sensory control of satiation is very strong.

Even though distension can suffice as an internal cue for a carbohydrate-conditioned preference, the flavor liking might also be contextualized to carbohydrate need and thus become a learned carbohydrate-specific appetite. We have preliminary evidence that this can happen.

2. Conditioned Satiety and Noradrenergic Feeding

None of the foregoing data show the conditioning of satiety in an absolute sense. That would be the acquisition of a genuine aversion in the replete state (Booth, 1972a, 1980b). A large dose of sufficiently concentrated maltodextrin has recently been demonstrated to do this (Booth et al., 1986b).

It remains to be shown whether this learned food-specific satiety depends on a carbohydrate excess. However, Leibowitz et al. (1985) claim that the rat prefers to eat a familiar diet that is rich specifically in carbohydrate when a meal is elicited by injection of norepinephrine near the paraventricular nucleus. Such an injection specifically disrupts conditioned satiation (Matthews et al., 1985; Booth et al., 1986b). Thus, it would be worth ascertaining whether carbohydrate-conditioned satiety can be dependent on glucoreceptors in the intestine wall or parenterally. On present evidence, this is the only mechanism that would make scientific sense of the idea that norepinephrine elicits eating by inhibiting a carbohydrate-selective satiety, releasing a relative appetite for carbohydrate.

D. Nutrient-Specific Sensory Preferences

For the case of carbohydrate, this brings us to the issue of whether there are any nutrient-specific appetites, as so often is claimed. The criterion of nutrient specificity, whether learned or innate, must be a facilitation of ingestion that comes both from a sensory predictor of supply of the nutrient, independently of its form in a particular diet, and also at the same moment from a somatic predictor of a deficit of the same particular nutrient.

1. Craving for Carbohydrate or for Conventional Snack Foods?

As just explained, this crucial evidence for a carbohydrate-specific appetite has yet to be established in the rat (or other animals). Notwithstanding much talk about "carbohydrate cravings," such an appetite has yet to be found in human subjects.

The existence of people wanting or eating foods that happen to be high in carbohydrate (or fat) is no evidence whatsoever for selection of carbohydrate (or fat). Nor is preference for high-protein foods evidence of protein selection. Our culture provides low- and high-protein foods and drinks for use in different circumstances and attaches different attributions to them. For technological reasons (low water activity and so good storage), convenience foods are high in lipids, sugar, salt, and/or dried starch. They are thought of as foods to eat away from meals and also as nutritionally poor. Choice of such foods must therefore be assumed, until proved otherwise, to be merely a craving for conventional snack foods or even for supposedly "junk" food, as opposed to behavior under the control of any specific nutrient's action in the body.

In the case of "protein craving," there has recently been some progress toward identifying nutrient specificity.

V. MECHANISMS OF PROTEIN-SPECIFIC APPETITE

A. Innate Protein Appetite

There is no firm evidence for an innate protein appetite. It is difficult to specify a sensory cue that would reliably indicate the presence, amount, or quality of all the sorts of protein in natural materials. Bacterial decomposition could produce strong odors from certain amino acids, such as isobutyric acid from the branched-chain amino acids and trimethylamine from others (Amoore, 1975).

Prompt preferences for some proteins in protein-deprived rats have been interpreted as evidence for an innate protein appetite (Heinrichs et al., 1988). Generalization from protein in chow was not excluded, however. Furthermore, just as with glucose from starch (Pilcher et al., 1974), within a minute of ingestion of protein, amino acids will begin to be released to duodenal receptors and absorption, and so such data do not exclude postingestional reinforcement.

B. Learned Protein Appetite

Any particular source of good-quality protein would have manifold sensory characteristics, and indeed, for human beings at least, a conceptual character such as "meat" or "beans." Each major protein source could thus become preferred as a result of postingestional reinforcement—for example, from relief of limiting levels of an essential amino acid in the detector site in prepyriform cortex (Leung and Rogers, 1971; Booth and Stribling, 1978; Gibson et al., 1987).

Furthermore, the reinforcing action of the essential amino acids could itself serve to identify an internal cue to the specific depletion that is being repaired. This need signal could be configured with the cues distinguishing the dietary protein source. Then the protein-conditioned sensory preference would be dependent also on the presence of a need state. Just such a protein-specific appetite is learned in one or two experiences by mildly deprived rats, using either odor cues (Gibson and Booth, 1986; Baker et al., 1987) or textural cues (Booth and Baker, 1988).

Rats with trigeminal lesions lose the ability to avoid insufficient intake of protein when given the usual laboratory choice between high casein and high dextrin diets (Miller and Teates, 1985). Thus, some textural characteristics of the casein are learned, to cue the effects of protein. Drugs that act on the brain stem pattern generators for ingestion, as both serotoninergic and dopaminergic agents do, will modify the effects of texture on ingestion regardless of the textural cue's learned nutritional significance. Thus the basis for supposing that the serotonin is involved in selection of carbohydrate over protein has proved to be confounded (Booth and Baker, 1988; Gibson and Booth, 1988). Some other explanation must therefore also be considered for the reported suppression of intake of high-carbohydrate foods after and between meals by serotonergic drugs in human subjects (Wurtman et al., 1981; Wurtman and Wurtman, 1986).

In fact, neither fenfluramine nor amphetamine in doses that suppress intake has been found to alter the preference for a sensory cue to either protein or carbohydrate in rats (Gibson and Booth, 1988). This is true even though the protein-need state on which the protein-conditioned preference depends involves a halving of brain levels of amino acid precursors for both serotonin and the catecholamines (Gibson et al., 1987).

There is thus no reason to invoke the serotonin or norepinephrine "theories" of nutrient selection if people's food preferences are changed by drugs or dietary preloads toward or away from protein-rich foods. The first mechanisms to consider are either sensory interactions (e.g., food-specific habituation from the preload) or cognitive mechanisms (e.g., awareness that an appetite-suppressant drug is limiting intake and a resulting focus of eating choices, therefore, a deliberate focus of eating or foods supposed to be nutritious, which are liable in fact to be higher in protein content).

VI. THE RESEARCH NEED: APPETITE MECHANISMS, NOT INTAKE EFFECTS

These results should make it very clear that substantial advances in our understanding of appetite depend on investigations that assess specific mental, physiological, and social processes. This cannot be achieved by the usual procedures of measuring food intake or collecting ratings of appetite without also unconfoundedly controlling or independently measuring the influences on such intakes or verbal data. Such unanalytical research at best raises questions; at worst, it obfuscates the scientific issues and fails to help or even harms the general public.

Fundamental discovery and applicable conclusions depend on identifying and measuring the causal processes operative in the individual. Causation can be identified only by double dissociation. The strength of each causal process can be measured only as the sensitivity of its physicopsychometric (dose–response) function. These elementary and scientifically universal methodological requirements still apply if the risk must be taken of assuming that everybody operates in qualitatively identical causal networks.

The particular appetite mechanism of protein- or carbohydrate-specific dietary selection is demonstrated to exist insofar as approach behavior and ingestion are shown to be activated by cues only to foods' contents of the nutrient in question, rather than activated by other features of the foods. No such demonstration has been forthcoming for the basic phenomenon called "carbohydrate craving," let alone its dependence on serotoninergic transmission.

All the analytical evidence is that the differences in dietary selection that have been observed in rats and people after pharmacological or dietary manipulations are not nutritionally controlled behavior, mediated specifically by any monoamine neurotransmitter. In people, what obviously could be happening is that perceptions of the experimental manipulations are interacting with snacking habits and other food choices, to accord with cultural stereotypes of healthy eating and weight control practices (Blair et al., 1989).

These and other causal hypotheses must be tested before any interpretation can responsibly be laid before the public, let alone claims made that effective means to reduce obesity or emotional distress have been identified. The alleged craving for carbohydrate rather than protein is not only a clear case of unrecognized interactions between neuroscience and sociology, and at more than one level. It is also a major example of threats to health arising more widely from the neglect of behavioral science within nutrition (Booth, 1989c).

What needs addressing is the conventional practice of consuming energy-containing drink and food items between and after meals. Social psychology and food technology should have far more to say about that than biochemical nutrition or neuroscience.

DISCUSSION

Beauchamp: I have one question. You made an analogy, if I understood you right, between salt and protein hunger. Salt hunger has an innate sensory component to it. Do you see any innate sensory component to protein hunger—for example, amino acid flavors or something of that sort?

Booth: The analogy I drew was a mechanistic one saying that the hunger in this sense has to have both the sensory information and a bodily source of information. The innate sodium appetite is, I believe, a reduction in the aversiveness of strong salt taste, which by inheritance is peculiar to salt deficiency in the body. Taking that as a structure, whether it's right or wrong, the analogy I'm drawing is that as a result of learning, then one goes to whatever flavor signals a protein supply in the diet, when and only when the body is telling one that one needs protein. So the analogy is between the dual source of information that is controlling the behavior. The difference, the important difference as I see it, is that one is programmed by the genes, the other is programmed by personal experience.

Beauchamp: The answer is "no."

Booth: The mechanism is the same; how it gets there is quite different.

REFERENCES

Amoore, J. (1975). Four primary odor modalities of man: Experimental evidence and possible significance. In *Olfaction and Taste V*, D. A. Denton and J. P. Coghlan (Eds.). Academic Press, New York, pp. 283–289.
Baker, B. J., Booth, D. A., Duggan, J. P., and Gibson, E. L. (1987). Protein appetite demonstrated: Learned specificity of protein-cue preference to protein need in adult rats. *Nutr. Res.* **7**:481–487.
Birch, L. L., Billman, J., and Richards, S. S. (1984). Time of day influences food acceptability. *Appetite*, **5**:109–116.
Blair, A. J., Lewis, V. J., and Booth, D. A. (1989). Official and informal weight reduction techniques: successful and unsuccessful. *Psychol. Health*, **3**:195–206.
Bolles, R. C. (1980). Historical note on the term "appetite." *Appetite*, **1**:3–6.
Booth, D. A. (1967). Localization of the adrenergic feeding system in the rat diencephalon. *Science*, **158**:515–517.

Booth, D. A. (1968). Effects of intrahypothalamic glucose injection on eating and drinking elicited by insulin. *J. Comp. Physiol. Psychol.* **65**:13–16.
Booth, D. A. (1969). Food preferences and nutritional control in rats and people. *British Association for Advancement of Science, Exeter Meeting Abstracts*, p. 90.
Booth, D. A. (1972a). Satiety and behavioral caloric compensation following intragastric glucose loads in the rat. *J. Comp. Physiol. Psychol.* **78**:412–432.
Booth, D. A. (1972b). Conditioned satiety in the rat. *J. Comp. Physiol. Psychol.* **81**:457–471.
Booth, D. A. (1976). Approaches to feeding control. In *Appetite and Food Intake*, T. Silverstone (Ed.). Abakon/Dahlem Konferenzen, Berlin, pp. 417–478.
Booth, D. A. (1977). Appetite and satiety as metabolic expectancies. In *Food Intake and Chemical Senses*, Y. Katsuki, M. Sato, S. F. Takagi, and Y. Oomura (Eds.). University of Tokyo Press, Tokyo, pp. 317–330.
Booth, D. A. (1978). Prediction of feeding behaviour from energy flows in the rat. In *Hunger Models*, D. A. Booth (Ed.). Academic Press, London, pp. 227–278.
Booth, D. A. (1980a). Acquired behavior controlling energy intake and output. In *Obesity*, A. J. Stunkard (Ed.). Saunders, Philadelphia, pp. 101–143.
Booth, D. A. (1980b). Conditioned reactions in motivation. In *Analysis of Motivational Processes*, F. M. Toates and T. R. Halliday (Eds.). Academic Press, London, pp. 77–102.
Booth, D. A. (1987a). Cognitive experimental psychology of appetite. In *Eating Habits*, R. A. Boakes, M. J. Burton, and D. A. Popplewell (Eds.). Wiley, Chichester, pp. 175–209.
Booth, D. A. (1987b). How to measure learned control of food or water intake. In *Feeding and Drinking*, F. M. Toates and N. E. Rowland (Eds.). Elsevier, Amsterdam, pp. 111–149.
Booth, D. A. (1987c). Evaluation of the usefulness of low-calorie sweeteners in weight control. In *Developments in Sweeteners*, Vol. 3, T. H. Grenby (Ed.). Elsevier Applied Science, London, pp. 287–316.
Booth, D. A. (1987d). Objective measurement of determinants of food acceptance: Sensory, physiological and psychosocial. In *Food Acceptance and Nutrition*, J. Solms, D. A. Booth, R. M. Pangborn, and O. Raunhardt (Eds.). Academic Press, London, pp. 1–27.
Booth, D. A. (1988). A simulation model of psychobiosocial theory of human food-intake controls. *Int. J. Vitam. Nutr. Res.* **58**:55–69.
Booth, D. A. (1989a). The effect of dietary starches and sugars on satiety and on mental state and performance. In *Dietary Starches and Sugars in Man*, J. Dobbing (Ed.). Springer-Verlag, London, pp. 225–249.
Booth, D. A. (1989b). Nutrient- or mood-specific appetites: Physiological and cultural bases for eating disorders. *Ann. N.Y. Acad. Sci.*, **575**:122–135.
Booth, D. A. (1989c). Effective habit messages and food marketing to prevent obesity. In *Promotion of Healthy Nutrition as Part of Lifestyles Conducive to Health*, P. Voss (Ed.). WHO European Office, Copenhagen, pp. 81–87.
Booth, D. A., and Baker, B. J. (1988). Main serotoninergic effects of fenfluramine on food intake may not involve central appetite pathways. *Soc. Neurosci. Abstr.* **14**:613.
Booth, D. A., and Blair, A. J. (1988). Objective factors in the appeal of a brand during use by the individual consumer. In *Food Acceptability*. D. M. H. Thomson (Ed.). Elsevier Applied Science, London, pp. 329–346.

Booth, D. A., and Davis, J. D. (1973). Gastrointestinal factors in the acquisition of oral sensory control of satiation. *Physiol. Behav.* **11**:23–29.

Booth, D. A., and Mather, P. (1978). Prototype model of human feeding, growth and obesity. In *Hunger Models,* D. A. Booth (Ed.). Academic Press, London, pp. 279–322.

Booth, D. A., and Stribling, D. (1978). Neurochemistry of appetite mechanisms. *Proc. Nutr. Soc., London,* **37**:181–191.

Booth, D. A., and Toase, A. M. (1983). Conditioning of hunger/satiety signals as well as flavour cues in dieters. *Appetite,* **4**:235–236.

Booth, D. A., and Toates, F. M. (1974). A physiological control theory of food intake in the rat: Mark 1. *Bull. Psychonomic Soc.* **3**:442–444.

Booth, D. A., Lovett, D., and McSherry, G. M. (1972). Postingestive modulation of the sweetness preference gradient in the rat. *J. Comp. Physiol. Psychol.* **78**:485–512.

Booth, D. A., Stoloff, R., and Nicholls, J. (1974). Dietary flavor acceptance in infant rats established by association with effects of nutrient composition. *Physiol. Psychol.* **2**:313–319.

Booth, D. A., Lee, M., and McAleavey, C. (1976a). Acquired sensory control of satiation in man. *Br. J. Psychol.* **67**:137–147.

Booth, D. A., Toates, F. M., and Platt, S. V. (1976b). Control system for hunger and its implications in animals and man. In *Hunger,* D. Novin, W. Wyrwicka, and G. A. Bray (Eds.). Raven Press, New York, pp. 127–142.

Booth, D. A., Mather, P., and Fuller, J. (1982). Starch content of ordinary foods associatively conditions human appetite and satiation, indexed by intake and eating pleasantness of starch-paired flavours. *Appetite,* **3**:163–184.

Booth, D. A., Thompson, A. L., and Shahedian, B. (1983). A robust, brief measure of an individual's most preferred level of salt in an ordinary foodstuff. *Appetite,* **4**:301–312.

Booth, D. A., Gibson, E. L., and Baker, B. J. (1986a). Gastromotor mechanism of fenfluramine anorexia. *Appetite,* **10**(Suppl.):57–69.

Booth, D. A., Gibson, E. L., and Baker, B. J. (1986b). Behavioral dissection of the intake and dietary selection effects of injection of fenfluramine, amphetamine or PVN norepinephrine. *Soc. Neurosci. Abstr.* **15**:593.

Booth, D. A., Conner, M. T., and Marie, S. (1987). Sweetness and food selection: Measurement of sweeteners' effects on acceptance. In *Sweetness,* J. Dobbing (Ed.). Springer-Verlag, London, pp. 143–160.

Conner, M. T., and Booth, D. A. (1988). Preferred sweetness of a lime drink and preference for sweet over non-sweet foods, related to sex and reported age and body weight. *Appetite,* **10**:25–35.

Conner, M. T., Haddon, A. V., Pickering, E. S., and Booth, D. A. (1988a). Sweet tooth demonstrated: Individual differences in preference for both sweet foods and foods highly sweetened. *J. Appl. Psychol.* **73**:275–280.

Conner, M. T., Booth, D. A., Clifton, V. J., and Griffiths, R. P. (1988b). Individualized optimization of the salt content of white bread for acceptability. *J. Food Sci.* **53**:549–554.

Duggan, J. P., and Booth, D. A. (1986). Obesity, overeating and rapid gastric emptying in rats with ventromedial hypothalamic lesions. *Science,* **231**:609–611.

Gibson, E. L., and Booth, D. A. (1986). Acquired protein appetite in rats: Dependence on a protein-specific need state. *Experientia,* **42**:1003–1004.

Gibson, E. L., and Booth, D. A. (1988). Fenfluramine and amphetamine suppress dietary intake without affecting learned preferences for protein or carbohydrate cues. *Behav. Brain Res.* **30**:25–29.

Gibson, E. L., and Booth, D. A. (1989). Dependence of carbohydrate-conditioned flavor preference on internal state in rats. *Learn. Motiv.* **20**:36–47.

Gibson, E. L., Barber, D. J., and Booth, D. A. (1987). 5HT and CA precursor levels during protein selection. *Soc. Neurosci. Abstr.* **13**:15.

Harris, G., and Booth, D. A. (1987). Infants' preference for salt in food: Its dependence upon recent dietary experience. *J. Reprod. Infant. Psychol.* **5**:97–104.

Hatfield, G. (1988). Neuro-philosophy meets psychology: Reduction, autonomy, and physiological constraints. *Cognitive Neuropsychol.* **5**:723–746.

Heinrichs, S. C., Moore, B. O., and Deutsch, J. A. (1988). A role for both neophilia and postingestive feedback in diet selection by protein deficient rats. *Soc. Neurosci. Abstr.* **14**.

Leibowitz, S. F., Weiss, G. F., Yee, F., and Tretter, J. B. (1985). Noradrenergic innervation of the paraventricular nucleus: Specific role in control of carbohydrate ingestion. *Brain Res. Bull.* **14**:561–567.

Leung, P. M. B., and Rogers, Q. R. (1971). Importance of prepyriform cortex in food intake response of rats to amino acids. *Am. J. Physiol.* **221**:929–935.

Matthews, J. W., Gibson, E. L., and Booth, D. A. (1985). Norepinephrine-facilitated eating: Reduction in saccharin preference and conditioned flavor preferences with increase in quinine aversion. *Pharmacol. Biochem. Behav.* **22**:1045–1052.

Miller, M. G., and Teates, J. F. (1985). Acquisition of dietary self-selection in rats with normal and impaired oral sensation. *Physiol. Behav.* **34**:401–408.

Pilcher, C. W. T., Jarman, S. P., and Booth, D. A. (1974). The route of glucose to the brain from food in the mouth of the rat. *J. Comp. Physiol. Psychol.* **87**:56–61.

Valle, F. P. (1968). Effect of exposure to feeding-related stimuli on food consumption in rats. *J. Comp. Physiol. Psychol.* **66**:773–776.

Van Vort, W., and Smith, G. P. (1988). Sham feeding experience produces a conditioned increase in meal size. *Appetite*, **9**:21–29.

Weingarten, H. P. (1984). Meal initiation controlled by learned cues: Basic behavioral properties. *Appetite*, **5**:147–158.

Wurtman, R. J., and Wurtman, J. J. (1986). Carbohydrate craving, obesity and brain serotonin. *Appetite*, **7**(Suppl.):99–103.

Wurtman, J. J., Wurtman, R. J., Gowdon, J. H., Henry, P., Lipscomb, A., and Zeisel, S. (1981). Carbohydrate craving in obese people: Suppression by treatments affecting serotonergic transmission. *Int. J. Eating Disorders*, **1**:2–15.

Part III Discussion

Friedman: I have a question for Jeff Galef. You describe very elegantly how pairing of the social experience with the flavors can create these preferences and can even overcome learned taste aversions. Have you or anyone looked at the opposite situation, namely, when an animal communicates to another animal that food is safe, but communicates inaccurately? Would one animal form an aversion to another animal that lied to him?

Galef: The answer is that neither I nor anyone else has done that particular experiment. There is some work now on deceit, and Marler's group is looking at the question of deceitful communications with respect to feeding. I don't know whether they've looked to see whether an individual learns an aversion to deceitful others. My feeling is you could probably teach an animal such a thing. We've done experiments in which one rat serves as a signal to another as to where to go to look for food, and it's our experience that over trials, if the informant is a liar, certainly the recipient of the information learns to ignore that information. Whether or not it's particular to individuals though, I couldn't tell you.

Spector: I have a question for Michael Tordoff. In the experiment where you were pairing one flavor with fructose and the other flavor with glucose, the logic was that since the brain can't use fructose, you might expect a differential preference. And in fact, you found that the rats preferred flavors paired with fructose over glucose. Wouldn't you expect the same findings in the experiment where you were infusing glucose? It seemed like they didn't show a preference one way or the other.

Tordoff: I think that the question is not what the brain can use, but what the liver can use. When you give a load of fructose, more of that is going to be used by the liver than is an equivalent load of glucose. The same may be true for fat. The fat is not going to be used very well at all by the brain, and neither is the fructose. It's the one that the liver is using and not what the brain is using.

Cabanac: Also a question to the same speaker. How do your results relate with Russek's hypothesis putting the main input for satiety in the liver?

Tordoff: Well I think Russek is very happy with my results because one of the things that they show is that the infusions he's using may not be causing a taste aversion. The way that we did the infusions, we get a taste preference. I don't see any problems with Russek's theory except that he says that the signal originates from lactate or pyruvate. I would disagree with the precise nature of the signal, but not with the general ideas of what Russek is saying.

Blundell: This is a question for Mike Tordoff about the model that included the effects of sweetness on fuel storage and oxidation. Most of the experiments, if not all those you did, use saccharin as the sweet stimulus. But, as you've shown with the hepatic vagotomy data, and as Michael Naim showed yesterday with cyclic AMP, and as there is data in the literature showing that saccharin blocks glucose-6-phosphatase, saccharin is having quite a profound metabolic effect. The question is whether the effects of sweeteners that you indicated were pertinent in the model are really saccharin-specific. Can you disassociate the sweetness effect per se from the metabolic consequences of saccharin ingestion, which appear to be fairly profound?

Tordoff: I think that it's the metabolic consequences of sweetness that are important, not the saccharin itself. We don't think that it's the pharmacological effects of saccharin that are important in this particular model because Dani Reed and I have shown, for example, that a rat that sham feeds sucrose will increase its food intake.

Blundell: Well you've done the experiment to disengage sweetness per se from its metabolic consequences.

Booth: As far as Galef's evidence goes, the social reinforcer of the information that the demonstrator is providing is, so to speak, "good breath"; they are all particular odorants in the animal's breath. In that sense, you have a molecule, or a few molecules, that actually mediate this social information. Do you have evi-

PART III DISCUSSION

dence or reason to think that there are much more complicated mechanisms actually inducing the learning?

Galef: I'm simply going to speculate because I really have no evidence. I would be very surprised if the carbon disulfide is in some sense a reinforcer. I suspect it's more of a signal for a much more complex series of events that gets foods that are experienced in contiguity with that signal treated in a very different way from foods that are simply experienced without that signal. I guess I would agree with the latter rather than the former model that you proposed.

Part IV
Appetite During Life Stages

14
Suckling: Opioid and Nonopioid Processes in Mother–Infant Bonding

Elliott M. Blass

*Cornell University
Ithaca, New York*

I. INTRODUCTION

Almost all studies of ingestive behavior focus on factors determining volume or rate of food eaten or liquid drunk. This focus has been on such internal factors as gastric fill and emptying rate, on sensory characteristics of the food as they affect olfactory and gustatory sensation, on social and cognitive determinants and, of course, on the neurology that manages feeding adjustments to each of these factors.

The approach we have taken during the past few years in studying the development of ingestive behavior evaluates how early ingestion influences immediate state and long-term behavior, not the traditional converse of how immediate state and past history affect intake. The rationale is that suckling and feeding take place within a complex social nexus that is changed through the dynamics of mother–infant contact and ingestion. This rationale gives rise to a number of testable hypotheses concerning the consequences of suckling and its various components of maternal contact, milk ingestion and milk digestion.

In nested rats, the nursing–suckling sequence consists of the female returning to her young after a period of separation that reflects the litter's age. She arouses them by rough handling, licks their anogenital region, and then, settling over the awake litter, she makes her 12 nipples available for suckling. She does not aid in nipple location or attachment; indeed, she will not return the occasional pup that has fallen from the nest. Attachment eventually occurs, and milk letdown ensues once every 4–12 minutes. Pups sleep in between milk letdowns, become excited at the time of letdown, withdraw milk in a stereotyped fashion (see Blass and Teicher, 1980, for a photographic sequence of this striking act), and in the normal course of events, release the nipple to take up a second, or even a third per milk letdown. Eventually, depending on litter mass and metabolic factors, ambient temperature, and the dam's ability to dissipate a heat load, the mother, for thermal reasons, leaves the nest, staying away for varying periods of time until the sequence begins again.

Our earlier research efforts were aimed at understanding the basis of nipple attachment and milk intake. They grew out of the traditional paradigms for the analysis of ingestive behavior. The current studies ask about the consequences of the complex mother–infant interactions described above on the infant's state, on the infant's adjustments to repeated absences it must experience, and on the consequences of maternal return. It focuses on the three facets of the maternal stimulus array—the dam's interactions with her infants, her physical properties (especially thermotactile), and the effects of her milk and its components. The goal of this enterprise is to understand the reciprocity between ingestion and subsequent behavior during ontogeny as it affects immediate state, influences formation of the mother–infant bond, and casts long-term influences on what and with whom an animal or human will interact beyond the weaning period when its decisions are no longer influenced proximately by its mother.

There are two broad classes of approaches to gain insight into suckling dynamics. One has it that suckling and milk ingestion are rewarding. This approach seeks to understand these consequences and mechanisms of action. This approach, of course, is steeped in everyday parental experience and has been repeatedly verified experimentally in human and infrahuman mammals. The second class of hypothesis states that suckling and its component parts reduce tension and perhaps even pain. This too is verified informally every time a crying infant is picked up and held or when it suckles nutritively or nonnutritively. In fact, an entire industry of pacifiers is founded on this assumption. Obviously the two are not mutually exclusive, conceptually or pragmatically.

The rationale for studying these functions is that very little is known about their underlying mechanisms as to both immediate and long-term effects on motivation in general and on subsequent ingestive behavior in particular. Because initial feeding choices and locations are influenced by mother's diet and behavior and because of the great diversity in human and other primate feeding patterns (see Nishida, Chapter 10, this volume), the effects of food on subsequent behavior were determined in rat and human infants.

Parallel findings emerged from both sets of studies and can be summarized as follows.

1. The suckling act and its consequences are calming, strongly antinociceptive, and rewarding to both rat and human infants.
2. These effects are achieved through simultaneous, additive mediation by systems that utilize opioid pathways and those that do not.
3. Thermotactile contact between mother and infant, whether general, as in whole-body contact, or local, as in oropharyngeal contact with the nipple, are mediated by mechanisms that do not appear to have an opioid component.
4. The effects of milk, at least two sugars, fat, and polysaccharides are mediated through mechanisms that have an opioid component.
5. There are multiple loci in the gastrointestinal tract for the effectiveness of some of the substances identified in item 4.
6. The characteristics of opioid calming and analgesia differ qualitatively and quantitatively from those mediated by nonopioid systems.

The remainder of this chapter will provide empirical flesh to these statements. Data presentation will be shaped by system characteristics and will cut across species. To our knowledge, these represent the only data that deal experimentally with the mechanisms underlying mother–infant interactions that surround ingestion. There is reason to believe that the phenomena are widespread, and we encourage more detailed and comparative study. The basis for this optimism is that morphine and other exogenous opioids significantly affect behavior in a variety of mammals, and we are demonstrating opioid mediation of calming and analgesia by natural food substances. In a series of important developmental studies, Panksepp and his colleagues (1985) have shown amelioration of social distress through opioid injection in a number of species including rats, chicks, and dogs. We have replicated these findings using albino rats to show a marked increase in pain threshold following morphine injection and a marked decrease in ultrasonic vocalization during maternal and sibling separation. That both these effects were naltrexone reversible (as they were in Panksepp's studies) constitutes behavioral evidence for opioid mediation. Moreover, from the perspective of the present series of studies, it provides a behavioral bioassay against which to evaluate the effects of an experimental manipulation and its mechanism of action.

II. BEHAVIORAL STRESS REDUCTION

Operationally, behavioral stress reduction is defined as the lessening of crying in isolated infants (rat and human) through the experimental manipulations described below.

Figure 1 presents number of distress vocalizations measured on a minute-by-minute basis in control rats that had received a jaw cannula (per Hall, 1979) and in rats that had received infusions of milk (half and half), sucrose, vegetable oil (fat), or Polycose through the cannula. A number of points are made. First, all infused solutions effectively reduced vocalization levels. Second, the effects were imme-

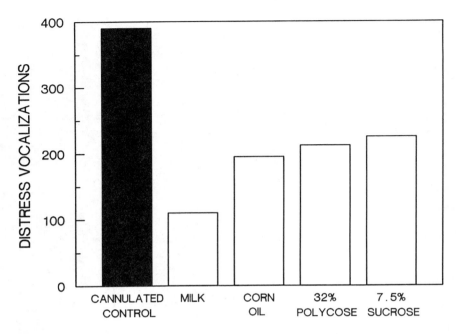

FIGURE 1 Incidence of ultrasonic vocalization in 10-day-old rats receiving intraoral infusions of various substances while isolated from dam and siblings for 8 minutes. (Milk data from Blass and Fitzgerald, 1988; sucrose, Polycose and fat data from Shide and Blass, 1989.)

diate in all cases except for milk. Third, the effects were protracted well past infusion termination. And finally, naltrexone, an opioid antagonist, fully reversed these effects (not shown in Fig. 1).

These data support the following conclusions for the mechanism of at least one member of each class of substances (we will return to milk).

1. The substances are comforting and act through opioid pathways (presumably of a central origin).
2. Their immediate effects must be at the level of the oropharynx because of their rapidity of action and because the effect took place well in advance of any metabolic consequences that might have accrued.
3. The effect could not have been primarily due to changes in gastric volume because neither water nor mineral oil changed rates of distress vocalization.
4. Finally, the effect endured well beyond infusion termination and well beyond the time of taste dilution and habituation, suggesting relatively enduring central changes. (It is possible that prolongation was due to postingestional factors.)

Figure 2 presents data from a representative human infant, 1–3 days old, who,

SUCKLING IN MOTHER-INFANT BONDING

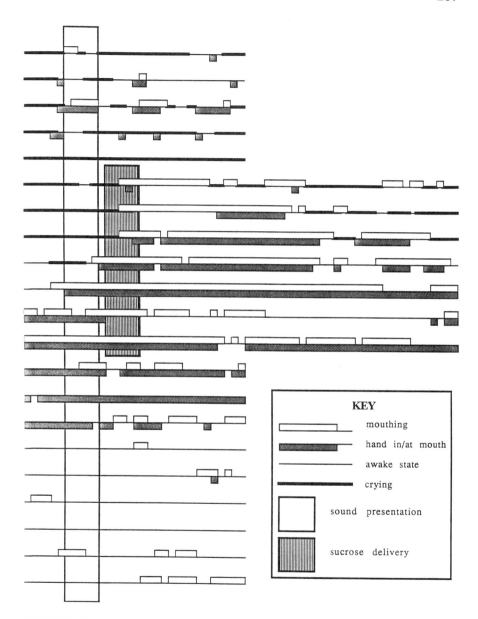

FIGURE 2 Profile of behavioral change of a representative newborn (1–3 days of age) human infant upon receiving 0.2 ml aliquots of 12% sucrose solution. Note the rapidity with which behavior changes occur and their duration even after sucrose is terminated.

after a 5-minute baseline period, received seven 0.2 ml aliquots of a 12% (w/v) sucrose solution; the aliquots were separated by 2 minutes. The transformation in this infant's behavior was quick and remarkable. Crying, which had occupied about 80% of this infant's time prior to sucrose administration, was reduced to naught within minutes and simply did not return after the baby had received a total of only 0.6 ml sucrose. There was an additional, unexpected, and robust phenomenon in almost all crying infants. After calm was achieved, they put their hands in their mouths for as long as the sucrose was forthcoming (indicated by the thick, solid horizontal lines in Fig. 2). A number of control studies indicated that hand-in-mouth behavior occurred only in response to sucrose administration. Group data on the calming effects of sucrose will be presented below in a different context. The behavior of this infant and her age mates have encouraged parallel conclusions concerning the effectiveness of sucrose (glucose too) as a calming agent in human newborns, its short latency, long duration, and possible mechanisms of action.

A. Contact as a Calming Agent

Contact is a well-established source of comfort to infant mammals, and this has been known experimentally at least since the time of Harlow's studies on captive rhesus monkeys. Contact mechanisms of action in rat and human infants aroused our interest because of the paradoxical finding, reported by Kehoe and Blass (1986b), that maternal contact quieted isolated rat pups, as expected, but also markedly reduced the opioid-mediated elevated pain threshold that is associated with isolation. Figure 3, which compares the properties of morphine quieting against that of contact, demonstrates that whereas both morphine and contact markedly reduced distress vocalizations, contact quieting was naltrexone resistant and morphine's action was naltrexone reversible (Blass et al., 1990). This suggests mediation of contact comforting by nonopioid pathways, at least insofar as whole-body contact is concerned, and set the stage for considering the following issues.

Is focal contact, as with a pacifier, also mediated by nonopioid pathways? If it is, do the operating characteristics of this system differ from those of opioid-mediated behavior? Although it is difficult to undertake this comparison across modalities, suckling behavior, with its orotactile and orogustatory components that are normally engaged by the same stimulus layout, allows a conservative assessment of the issue.

Smith et al. (1990), after a 5-minute baseline period, allowed human infants 1–3 days old either to suck a pacifier continuously for 2, 6, 10, or 14 minutes or to receive a 0.1 ml taste of 12% w/v sucrose for 10 seconds each minute for 2, 6, or 10 minutes. The efficacy of the two treatments was evaluated by comparing pre- and posttreatment levels of crying (Fig. 4) and other behaviors.

Even two 0.1 ml tastes of sucrose separated by 60 seconds were sufficient to

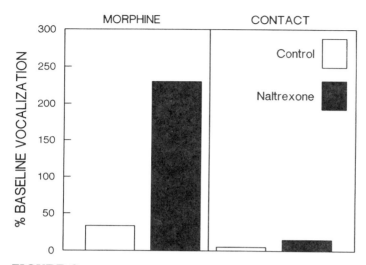

FIGURE 3 Comparison of morphine-induced and contact-induced quieting in rats following saline or naltrexone injections. Both morphine and contact markedly reduce ultrasonic vocalizations. Morphine's quieting effect is reversed by naltrexone; contact's is not.

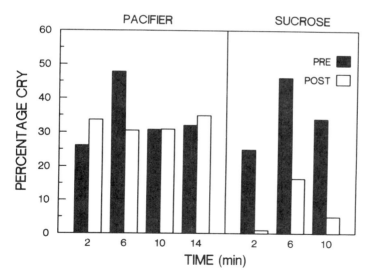

FIGURE 4 Percentage of time spent crying by human infants 1–3 days old prior to and following receipt of a pacifier for either 2, 6, 10, or 14 minutes of continuous suckling on 2, 6, or 10 aliquots (0.1 ml) of 12% sucrose solution delivered once per minute. (From Smith et al., 1990.)

stop crying in the 5-minute posttreatment period. It is of interest that there was no time-related effect, suggesting that the substrate initiating and sustaining quieting in these isolated infants was made available for a protracted period by this modest administration of a 12% sucrose concentration. It is possible that longer posttreatment baseline periods would reveal a time-related differential effect.

Pacifier administration, like sucrose administration, immediately quieted crying infants. When the pacifier was removed, however, crying was reinstated, generally within 15–30 seconds in infants that had sucked a pacifier for 2 to 14 minutes.

B. Summary of Comforting by Suckling Components

We have started to empirically approximate the mechanisms underlying different components and consequences of suckling, the mammals' first behavior for obtaining nutrients. We feel these to be reasonable first steps, since the fluids used approximate different components of mother's milk (we are currently studying this directly), and the pacifier is sufficiently like the nipple to provide the perceptual affordance for suckling. Tactile properties of whole-body and oral contact seem to be mediated by nonopioid mechanisms, whereas fluid components, at least for certain sugars, fats, and polysaccharides, are mediated by opioidlike systems as judged by naltrexone reversal. The original locus of action appears to be oropharyngeal. Milk appears to act both at the level of the mouth and more distally in the stomach and gastrointestinal tract. The functional synergy between opioid and nonopioid systems is currently being explored.

1. Suckling and Its Consequences in Analgesia

As shown in Figure 5, Polycose and corn oil caused an opioid-mediated elevation in paw lift latency from a 48°C hotplate in infant rats, an effect that is naltrexone reversible. Increased pain threshold was obtained whether the intraoral infusions were delivered to isolated infants or to infants housed together as a litter in the mother's absence. This last finding is significant because group isolation from the mother is stressful as measured by increased paw-lift latencies (Kehoe and Blass, 1986b). Thus contact per se does not fully prevent opioid-mediated analgesia induced by infusion of various solutions. It is of considerable interest that returning rats to the dam reversed the isolation-induced elevation within 1 minute upon return. This again speaks to nonopioid mediation of maternal contact. If contact was working through opioid mechanisms, return to the mother should either exaggerate latencies further or, at the least, maintain the isolation level. It should be noted that the current experimental procedure does not fully eliminate the possibility of contact analgesia because the infant, in fact, must be separated from the dam for 10–12 seconds to conduct the paw-lift test. Our methods are currently being modified to allow evaluation of pain threshold while infant animals are in actual contact with their dam.

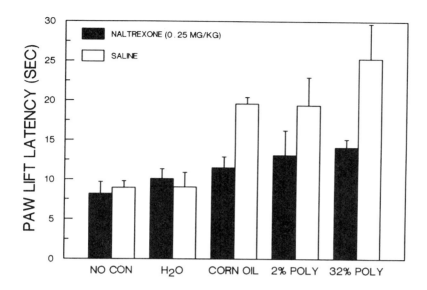

FIGURE 5   The effects of various substances on paw-lift latency in 10-day-old rats.

The importance of this last consideration has been addressed in our studies of pain amelioration in human infants that undergo painful standard hospital procedures such as circumcision or blood collection for diagnostic tests. According to Figure 6, there is an interaction (possibly additive) between the tactile stimulation of a pacifier and gustatory stimulation provided by sucrose. Providing a pacifier alone markedly reduced crying caused by circumcision, a finding previously reported by Gunnar and her colleagues (Stang et al., 1988). Immersing the pacifier in sucrose prior to circumcision further reduced crying significantly (Blass and Hoffmeyer, 1991).

By providing infants with the opportunity to drink 2.0 ml of 12% sucrose solution prior to blood collection via heel lance, Blass and Hoffmeyer (1991) markedly reduced crying during the procedure (Fig. 7). Moreover, infants that had received sucrose stopped crying well before control infants that had received water. Thus sucrose, at least, exerts an antinociceptive effect in rat and human infants that has the same properties as in calming: namely, rapid rise time and slow decay after stimulus termination. Obviously, our conclusion of opioid mediation in humans can be inferred solely on the naltrexone reversal in the rat studies. Studies currently under way in our laboratory are evaluating tactile and sucrose calming of infants of heroin- or morphine-addicted mothers. These studies predict that tolerant infants should not respond normally to sucrose or other substances thought to be opioid mediated but should have normal response profiles to a pacifier.

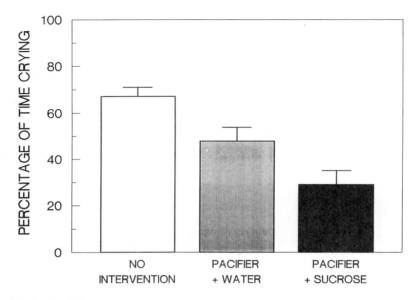

FIGURE 6 Incidence of crying in human infants (2–3 days) undergoing circumcision. Comforting was provided by a pacifier dipped in water or in 12% w/v sucrose. (From Blass and Hoffmeyer, 1991.)

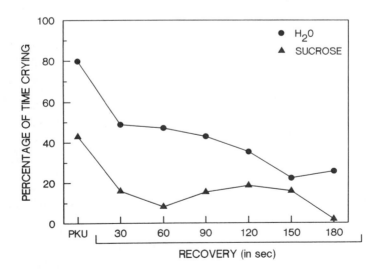

FIGURE 7 Effects of drinking 2.0 ml of 12% sucrose solution upon crying induced by heel-lance blood collection in human infants 2–3 days old. Sucrose markedly reduced crying incidence during blood collection and beyond. (Blass and Hoffmeyer, 1991.)

2. Summary of Analgesic Action of Tactile and Milk Component Substances

As in the case of social stress reduction, opioid and nonopioid mediated treatments reduce responses to physical stress. These findings imply a substantial (direct?) central connection between taste and pain systems and invite exploration of the neural bases of these unexpected connections. At the behavioral level of analysis, because sugars and other tastants reduce infant social and physical stress, their effects on adult stress and coping behavior are of interest. This takes on special significance, to which I will return, because stress can serve as a stimulus for feeding.

3. Suckling-Related Stimulation and Reward

The foregoing factors concerning taste and touch are well-established, unconditioned rewards for infant and adult mammals. Milk is obviously rewarding to human and rat infants whether it is delivered directly into the mouth or via a nipple. Milk has also been established as an unconditioned stimulus for classical conditioning in infant rats (Johanson and Teicher, 1980).

Sucrose and vegetable oil also serve as unconditioned stimuli in classical conditioning. There is good evidence that these effects are opioid mediated. According to Shide and Blass (1990), 8-day-old rats that had associated an astringent orange odor with either sucrose or vegetable oil infusions significantly increase time spent over orange in a subsequent preference test. If the animals were treated with naltrexone prior to conditioning, the preference was never formed, implying that the sweet or fatty substances were rewarding because of opioid release. If orange became rewarding because of its association with an endogenous release, it follows that the orange scent sustained behavior because upon future presentation, the scent itself caused endogenous opioid release. We now have evidence to support this claim.

As shown in Figure 8 (upper portion), naltrexone administered prior to testing completely blocked preference formation for orange associated with either sucrose or vegetable oil. This also held for morphine (Kehoe and Blass, 1986a). The lower panel demonstrates that for orange previously associated with corn oil or sucrose, naltrexone injected at the time of preference testing eliminates the expression of the acquired orange preference. This finding was also obtained when morphine was the unconditioned stimulus (Kehoe and Blass, 1989). This implies, for vegetable oil or morphine, that stimuli associated with sucrose become effective as instruments of motivation, sustain preference, and allow the animal to overcome the offensive sensory qualities of the orange odor, by virtue of the conditioned stimulus causing an endogenous opioid release. Support for this interpretation has been offered by Kehoe and Blass (1989), who demonstrated a 50% increase in paw-lift latencies in rats that had associated orange with morphine and were tested in the presence of orange.

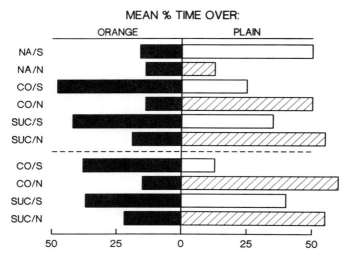

FIGURE 8 Preference for orange odor established through its association with sucrose (suc) or corn oil (co). Naltrexone (0.5 mg/kg) blocks preference formation when either corn oil or sucrose are the unconditioned stimuli. Naltrexone blocks preference expression when corn oil but not sucrose was the unconditioned stimulus. (D. Shide and E. M. Blass, unpublished observations, 1990.)

C. Milk

We now turn to the most interesting of the substances tested so far. Blass and Fitzgerald (1988) demonstrated that milk quieted infant rats and raised their pain thresholds. There had already been demonstrations of milk's efficacy as a reinforcing agent in both operant (Hall, 1979) and classical (see Johanson and Teicher, 1980; Johanson and Terry, 1988, for review) conditioning paradigms. Milk delivered through nipples has incentive properties to 12-day-old rats (Amsel et al., 1976) and dictates preference in rats 15 days of age until weaning (Kenny et al., 1979). Whether the latter behavioral manifestations are opioid mediated remains open to experimental inquiry.

We can speak of the mechanisms though which milk may affect quieting behavior. Two pathways have been identified. From the data of Blass and Fitzgerald we learn that milk's site of action may be postingestional. Milk did not start to exert its quieting effect on isolated infant rats until the infusions had been maintained for 2.5–3.0 minutes. Preliminary data from our laboratory demonstrate that infusions of milk directly into the stomach or the intestine are also quieting. In contrast, the locus of action for sucrose appears to be oral, as gastric infusions do not reduce isolation-induced vocalization.

The putative gastric–intestinal locus of action for milk is of considerable

SUCKLING IN MOTHER–INFANT BONDING

interest because substances of milk's composition may cause the release of cholecystokinin (CCK), which acts on peripheral receptors in the control of feeding. In keeping with our general approach of determining the consequences of suckling behavior, the effect of CCK administration on the behavior of isolated rats was assessed.

Weller and Blass (1988) demonstrated that exogenous CCK in low doses caused a marked and selective reduction in distress vocalization. CCK did not affect either gross locomotor activity or pain threshold, suggesting independence of mechanisms mediating stress of behavioral and somatic origins.

Weller and Blass (1989) then demonstrated an age-linked preference function for an odor classically associated with CCK. According to Figure 9, rats 5 and 11 days old that had been exposed to orange odor prior to CCK injection preferred the side of a maze scented with the conditioned odor. This is a striking reversal from the normal avoidance of the orange scent. Equally striking is the decline in preference over age, so that by the end of weaning, CCK is neutral and, in the experimental paradigm that we utilize, remains so in adult rats.

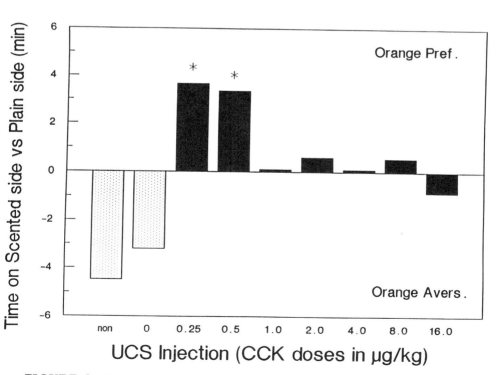

FIGURE 9 Preference for an orange odor linked with exogenous cholecystokinin (CCK) injections in rats 5, 11, 21, and 28 days of age. (Weller and Blass, 1989.)

Based on the indirect evidence presented above, there is reason to suggest calming and positively reinforcing roles for CCK during early development. The effective doses are very low (0.25 μg/kg) and probably within the physiological range. CCK effects during development, like those in adult rats concerning ingestive behavior, are of peripheral origin. They can be blocked by the selective peripheral antagonist L364,718. The possibility of endogenous CCK causing quieting is currently being evaluated in our laboratory by pitting L364,718 against infusions of various substances known or thought to reduce crying, including mother's milk.

Another potential source of quieting and reinforcement provided by mother's milk is in the form of a family of exorphins. Mammalian milk contains the substance β-casomorphine, which is cleaved into various fragments, some of which contain morphiceptin (Teschemacher, 1987). There are now three reports that morphiceptin delivered systemically has analgesic effects (Greckschet al., 1981; Chang et al., 1982; Brantl and Teschemacher, 1983; Teschemacher, 1987).

D. Interactions Between Opioid and Nonopioid Control Systems

The studies presented in this report have started to segregate the motivational properties and mechanisms of the stimulus array presented to infants by their mother during contact and nursing. In particular, the infant receives two classes of stimulation during suckling: thermotactile (from the mother's skin in general, nipple in particular) and gustatory–nutritive (from the milk that she deposits in the infants before short-term deprivation). The evidence presented here suggests that the surface stimulation affects infant state and future behavior through nonopioid systems that have operating characteristics that differ from those of opioid-mediated mechanisms. In contrast, gustatory stimulation by fats and sugars (at least within the substances we have explored) utilize pathways that have opioid components.

Under natural circumstances, of course, the arrays are presented simultaneously. How do they interact? Is nipple attachment, which comes under control of feeding factors after 2 weeks of age, mediated through opioid mechanisms or via tactile controls? What about milk intake from the nipple?

To assess these interactions, Blass et al. (1990) injected 8-hour suckling-deprived and nondeprived rats 9–18 days of age with a large dose of naloxone (5 mg/kg) and evaluated nipple attachment and milk intake. Figures 10 and 11 demonstrate no effect of the opioid antagonist on nipple attachment in either deprived or nondeprived rats (Fig. 10) and a modest, but statistically nonsignificant effect on milk intake. This was not due to a general insensitivity on the part of infant rats to opioid antagonist effects on ingestive behavior because naltrexone at 0.25–1.0 mg/kg reduced milk intake by 25–30% in 10-day-old rats that were eating away from the dam.

This initial study assessing the characteristics of systems with multiple mecha-

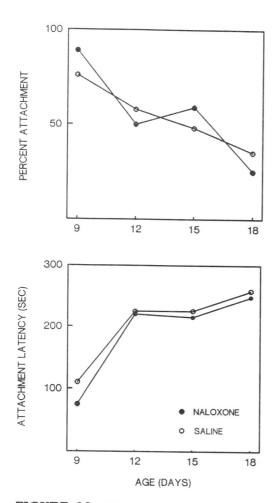

FIGURE 10 Mean nipple attachment latencies in deprived and nondeprived rats following pretreatment with isotonic saline or naloxone (5 mg/kg). (Blass et al., 1990).

nisms suggests that when the behavioral pattern is guided by cutaneous factors, control may be mediated by nonopioid mechanisms even though the consequences of the act cause changes that are opioid-mediated.

III. GENERAL DISCUSSION

The data presented in this chapter should be viewed as an initial analysis of the mechanisms underlying certain facets of mother (caretaker)–infant relationships,

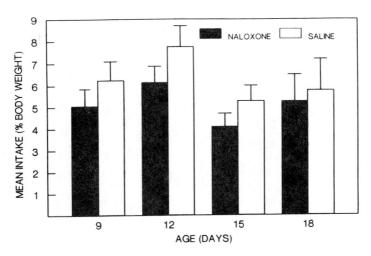

FIGURE 11 Mean milk intake via suckling of rats 9–18 days of age following isotonic saline or naloxone injections (Blass et al., 1990.)

especially those concerning suckling–nursing dynamics. Portions of the data are substantive and can be analyzed and further investigated in their own right. Included in this category are the effects of sugars, fats, and polysaccharides on distress and pain reactions. These data reveal an untold chapter in the dynamics and determinants of food intake. Some of the data are promissory notes—in particular, those concerning the effects of milk infusions on behavior. The data are of interest primarily as they provide direction for the analysis of mother's milk on the behavior of her young.

This line of analysis in conjunction with identifying the properties of thermotactile comfort and its underlying mechanism(s) of action presents a potentially manageable challenge for gaining insights into how mother–infant bonds are formed (i.e., their underlying mechanisms) and how, through these bonds, future acts are influenced through the developmental process. Ingestive behavior can well serve as a model for identifying and evaluating these effects.

Identifying the consequences of milk infused into the mouths of infants that are away from their mothers, although of considerable interest, is preliminary to the task of assessing the effects of milk injected by the dam through her nipples. It is already established that suckling a milk-yielding nipple is more rewarding (Amsel et al., 1976) and preferred (by 2 weeks of age) to a nonlactating nipple (Kenny et al., 1979). Will milk derived from the nipple cast the same calming and analgesic influence as milk infused into the mouth? Will it add to the effects of whole-body contact and nonnutritive suckling per se? This is not a trivial issue because of the rapid reversal of analgesia upon return of isolated pups to their mother. A second source of complexity lies in the resistance of the effects of nipple attachment and milk intake to opioid antagonists. Yet milk intake away from the nipple can be reduced by naltrexone. This is important because it presents the possibility that maternal contact prevents or diminishes the efficacy of opioid mechanisms. These possibilities are currently being evaluated.

We still find it extraordinary that fat, polysaccharide, and sugar calmed distressed rat and human (sugars so far) infants and served as analgesics. These findings were unanticipated both functionally and anatomically. They have potential for advancing the functional neurology linking pleasure and pain through gustatory afferents. This also holds for advancing the behavioral reciprocities as well (see Blass et al., 1989; Shide and Blass, 1989, for further discussion). The domain of stress-related feeding, the particular attractiveness of foods high in fat content, and the preferential intake of fat-rich and sweet foods when feeding under duress, can now be studied from a more rational perspective.

Finally, these data may lead to an additional broadening of perspective in regard to suckling behavior as it concerns the interactions between tactile and gustatory influences. Based on the initial findings of Smith, Fillion, and Blass concerning orotactile and orogustatory mediation, we tentatively conclude that there exist two separate mechanisms of action whose characteristics differ. Both mechanisms reduce stress and pain. They seem to do so differently, however. Sucrose has a rapid rise time and a slow decay time, presumably reflecting opioid involvement. Tactile mechanisms also feature a rapid rise time but have a rapid off time relative to stimulus presentation and withdrawal. The interactions between these two systems and the opioid-mediated ones will be revealing. According to Blass and Hoffmeyer (1991), they appear to be additive insofar as analgesia during circumcision is concerned.

In summary, these studies have provided a framework for assessing the biological bases of the developing mother–infant bonds and in particular for the contribution of the suckling setting to this process. To the extent that infants can recognize different substances in the mother's milk, these and other studies (Galef and Sherry, 1973) provide a means for assessing a developmental basis for long-term food and other preferences.

DISCUSSION

Cabanac: A comment and a question. The comment is anecdotal. When we first started to enter the operating room in pediatric surgery, we had the great surprise to discover that newborn babies were operated without anesthesia even for atresia of the pylorus. The babies—these young boys—started to cry only when the duodenum was eviscerated and pulled out. The head nurse had discovered that a very nice way to calm down the babies was to let them suck sweet water, which confirms anecdotally your experiments.

And now the question. In regards to what you call the pacifiers, a colleague from Tunis has checked the Harlow experiment with the furred mother versus the wire mesh with the bottle. He has shown that the wire mesh was as attractive and was more attractive than the furred mother, provide the wire mesh was warmed. Therefore, there is a temperature regulation component in the pacifier. Would that integrate with your results?

Blass: It certainly could. The pacifier that we offered, of course, was at room temperature, and so it was not supplying heat. It would certainly be easy enough for us to do; to cause a circulation in the pacifier of heated fluid and see if that would make any difference. That's an interesting point that you raise, and I will do that. It's an easy manipulation.

Grill: I come from a long line of Eastern European skeptics, probably deriving from a surgical intervention that occurred at 8 days of age. I'm thinking about alternatives, weighing some of the effects that you see other than those that might be a function of some change in opioid level. Particularly, I'm thinking about the fact that sucrose in water as opposed to water alone gives rise to a set of oral reflexes that are sustained for a period of time, are to some degree correlated with its concentration or its aftertaste, and may be incompatible, motorically, with the infant's capacity to cry.

Blass: The point is well taken. It was one that caused me considerable hesitancy in even undertaking these experiments. I can say the following. When a pacifier is inserted into the infant's mouth, the infant can spit it out and it often does. The fact of the matter is that we've had to reject many infants from these experiments because they spit pacifiers out. So the fact that we put it in, as I'm remembering from the past few months of occasional pacifier use for our baby, does not mean that the baby is going to take it. I think it is a voluntary act on the part of the infant. The point in a more general way still stands—that perhaps what the sugar was doing was engaging a motor pattern that the water on the pacifier did not engage. I don't have a satisfactory answer to that, but to the weaker form of the criticism, I'm not concerned about that because the babies can certainly spit it out and they do.

REFERENCES

Amsel, A., Burdette, D. R., and Letz, R. (1976). Appetitive learning, patterned alternation and extinction in 10 day old rats with non-lactating suckling as a reward. *Nature,* **262**:816–818.

Blass, E. M., and Fitzgerald, E. (1988). Milk-induced analgesia and comforting in 10-day-old rats: Opioid mediation. *Pharmacol., Biochem. Behav.* **29**:9–13.

Blass, E. M., and Hoffmeyer, L. B. (1991). Sucrose as an analgesic in newborn humans. *Pediactrics* (in press).
Blass, E. M., and Teicher, M. H. (1980). Suckling, *Science,* **210**:15–22.
Blass, E. M., Fillion, T. J., and Weller, A. and Brunson, L. (1990). Separation of opioid from nonopioid mediation of affect in neonatal rats: Nonopioid mechanisms mediate maternal contact influences. *Behav. Neurosci.* **104**:625–636.
Blass, E. M., Shide, D. J., and Weller, A. (1989). Stress-reducing effects of ingesting milk, sugars and fats: A developmental perspective. *Ann. N.Y. Acad. Sci.* **575**:292–305.
Brantl, V., and Teschemacher, H. (1983). Opioids in milk. *Trends Pharmacol. Sci.* **4**:193.
Chang, K.-J., Cuatrecasas, P., Wei, E. T., and Chang, J.-K. (1982). Analgesic activity of intracerebroventricular administration of morphiceptin and β-casomorphins: Correlation with the morphine (u) receptor binding affinity. *Life Sci.* **30**:1547–1551.
Galef, B. G., and Sherry, D. G. (1973). Mother's milk: A medium for the transmission of cues reflecting the flavor of mother's diet. *J. Comp. Physiol. Psychol.* **83**:374–378.
Grecksch, G., Schweigert, C., and Matthies, H. (1981). Evidence for analgesic activity of β-casomorphin in rats. *Neurosci. Lett.* **27**:325–328.
Hall, W. G. (1979). The ontogeny of feeding in rats: I. Ingestive and behavioral responses to oral infusions. *J. Comp. Physiol. Psychol.* **93**:977–1000.
Johanson, I. B., and Teicher, M. H. (1980). Classical conditioning of an odor preference in 3-day-old rats. *Behav. Neural Biol.* **29**:132–136.
Johanson, I. B., and Terry, L. M. (1988). Learning in infancy. A mechanism for behavioral change during development. In *Handbook of Behavioral Neurobiology,* Vol. 9, *Developmental Psychobiology and Behavioral Ecology,* E. M. Blass (Ed.). Plenum Press, New York, pp. 245–281.
Kehoe, P., and Blass, E. M. (1986a). Behaviorally functional opioid systems in infant rats: II. Evidence for pharmacological, physiological and psychological mediation of pain and stress. *Behav. Neurosci.* **100**(5):624–630.
Kehoe, P., and Blass, E. M. (1986b). Opioid-mediation of separation distress in 10-day-old rats: Reversal of stress with maternal stimuli. *Dev. Psychobiol.* **19**:385–398.
Kehoe, P., and Blass, E. M. (1989). Conditioned opioid release in ten-day-old rats. *Behav. Neurosci.* **103**:423–428.
Kenny, J. T., Stoloff, M. L., Bruno, J. P., and Blass, E. M. (1979). The ontogeny of preference for nutritive over nonnutritive suckling in albino rats. *J. Comp. Physiol. Psychol.* **93**:752–759.
Panksepp, J., Siviy, S. M., and Normansell, L. A. (1985). Brain opioids and social emotions. In *Biology of Social Attachments,* M. Reite and T. Fields (Eds.). Academic Press, New York.
Shide, D. J., and Blass, E. M. (1989). Opioid-like effects of intraoral infusions of corn oil and Polycose on stress reactions in 10-day-old rats. *Behav. Neurosci.* **103**:1168–1175.
Smith, B. A., Fillion, T. J., and Blass, E. M. (1990). Orally mediated sources of calming in one to three day-old human infants. *Dev. Psychol.* **26**:731–737.
Stang, H. J., Gunnar, M. R., Snellman, L., Condon, L. M., and Kestenbaum R. (1988). Local anesthesia for neonatal circumcision: Effects on distress and cortisol response. *J. Am. Med. Assoc.* **259**:1507–1511.

Teschemacher, H. (1987). Casein-derived opioid peptides: Physiological significance? *Adv. Biosci.* **65**:41–48.

Weller, A., and Blass, E. M. (1988). Behavioral evidence for cholecystokinin–opiate interactions in neonatal rats. *Am. J. Physiol.* **255**:R901–R907.

Weller, A., and Blass, E. M. (1990). Cholecystokinin conditioning in rats: Ontogenetic determinants. *Behav. Neurosci.* **104**:199–206.

15
Children's Experience with Food and Eating Modifies Food Acceptance Patterns

Leann L. Birch

University of Illinois at Urbana–Champaign
Urbana, Illinois

I. INTRODUCTION

A young child contemplates a plate of vegetables. Will he eat them? In a recent satirical paper entitled "The Etiology and Treatment of Childhood," Smoller (1985) describes five defining characteristics of the "childhood" syndrome:
1. Congenital onset.
2. Dwarfism.
3. Emotional lability and immaturity.
4. Knowledge deficits.
5. Legume anorexia.

We know a considerable amount about the first four characteristics of childhood; less about the last one. Over the past dozen years we have been conducting research on why children eat what they do. No one has yet been able to locate the legume anorexia gene, despite the pervasive and enduring nature of the characteristic. The results of the research reviewed below suggest that such a gene does not exist, and that we must look to the child's early experience with food and eating to understand this phenomenon. In addition to the influence of prior experience, the child's acceptance or rejection of a food will depend on several aspects of the child's situation at that moment in time, including the inherent flavor characteristics of the presented food, such as its sweetness, which may predispose the child to accept or reject the food, the child's physiological state, and other foods available and consumed at that meal. However, these factors are not the focus of this review.

This chapter is devoted to reviewing what is known about experiential influences on food acceptance in humans during the first years of life. In particular, the theme of this chapter is that parents and other adult caretakers, as purveyors of culture and cuisine, play a central role in the development of food acceptance patterns. As a result of repeated experience with food, and the contexts and consequences of eating, we learn what to like, when to eat, and even how much to eat. The cultural context in which the individual lives places major constraints on food acceptance patterns, but most of the individual differences in food acceptance patterns that emerge within a cultural group are the result of individual differences in experiential history. To the extent that young children have common experiences with food and eating, we can expect similar patterns of food acceptance to emerge. Individual differences in food acceptance patterns and in intake regulation can result from individual differences in early experience with child feeding practices; the limited evidence on this point is presented below. For the child, food acceptance or rejection is primarily a function of whether the food is liked or disliked. Unlike the adult, young children are not concerned with caloric content, cost, availability, ease of preparation, or other factors that are determinants of our intake.

II. THE CONTROL OF FOOD INTAKE: THE INTERACTION OF CARETAKER INFLUENCE AND INTERNAL CUES

Weingarten (1985) has suggested that the infant may be the only depletion-driven eater. In contrast to the adult and even the older child, whose meal patterns are constrained by cultural patterns dictating when and what to eat, the newborn infant's feeding patterns occur in response to internal cues signaling depletion. The lack of postnatal experiential history dictates that unlearned, depletion-based cues should play a crucial role in the control of food intake in the newborn infant, although there is very little research on this point. Interestingly, the little research that there is suggests that as early as the first few months of life, learning has begun to influence the control of feeding.

Young infants can be responsive to caloric density cues, and can use these cues to control the quantity of food consumed. Fomon (1974) reported data indicating that when infants were fed formulas varying in caloric density, prior to 40 days of age, caloric intake was correlated with caloric density of the formulas. However, after 40 days, evidence of caloric compensation was noted, with infants of the low caloric density formulas consuming larger volumes and those on high caloric density formulas consuming smaller volumes than infants on standard, 67 kcal/100 ml formulas. The fact that this pattern did not emerge until about 6 weeks suggests either the maturation of some innate mechanism, or that responsiveness to caloric density was modified by experience and learning, based on the caloric consequences of consuming the formulas. Fomon does not detail how the infants accomplished this caloric compensation; did they, for example, alter meal size, meal interval, or some combination of the two? He also takes care to point out that the volume of a feeding taken by the infant, especially the bottle-fed infant, can be readily influenced by the attitude of the caretaker. Therefore, even in the young infant, depletion cues do not operate in isolation, and the caretaker's approach to feeding begins to influence feeding patterns.

When Wright (1987) and colleagues studied the feeding patterns of breast-fed and formula-fed infants longitudinally, they found distinctively different patterns for formula-fed and breast-fed infants. These differences were clear over the first 6 months of life. He argued that there is an essential difference between breast and bottle feeding techniques with respect to the baby's establishing control of intake in the feeding situation. In comparing the effects of the two methods of feeding, he noted that mothers of bottle-fed babies tended to feed in response to the amount of formula remaining in the bottle and to be less aware of their infants' cues signaling hunger and satiety. In contrast, breast-feeding mothers were more likely to be aware of variations in their infants' hunger across the course of the day, and more aware that these feeding patterns were changing with development. Wright reported developmental changes in diurnal patterning of meal size in the breast-fed infant, who had a higher degree of control over the feeding interaction than the bottle-fed infant. This changing pattern is represented in Figure 1. At 2 months, breast-fed infants were taking their largest meal in the morning, following the longest interval without a feed. By 6 months, this pattern had reversed, with the largest meal being taken in the evening, in anticipation of the long overnight fast. He argues that this pattern could represent a gradual movement from a reactive pattern, in which the infant responds to a deficit signal and feeds accordingly, to adopting an anticipatory strategy of planning ahead and taking larger meals before the overnight fast. This latter pattern is characteristic of the older infant, at least in Western cultures. Taken together, these results suggest that by the second month of life, infants' intake patterns are beginning to be modified as a result of experience with a particular caretaker feeding style, which contributes to individual differences in diurnal feeding patterns.

The behavior of the caretaker is critical in determining meal size. Especially in the case of formula feeding, infants' depletion cues are often ignored or overridden by the beliefs and actions of the caretaker. Fomon reports that differences in the attitudes of the caretakers (whether babies were fed the least amount that would appear to relieve hunger or the largest volume that the infant would consistently accept) resulted in differences of about 20 ml/kg/day, even though the number of meals did not differ between the two groups. In addition, the frequency of feeding is also determined to some extent by the caretaker. Fomon's comments on maternal feeding style have been directed at the relatively large contribution that can be made by the formula feeding mother to the determination

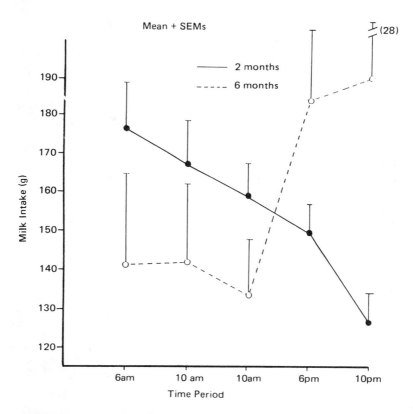

FIGURE 1 Diurnal variation in meal size in breast-fed infants at 2 and 6 months of age. (From Wright, 1987.)

of volume of a particular feeding. While Wright would not take issue with this, he also argues that the caretaker's behavior is crucial in determining what the infant learns about the control of food intake over repeated feeding experiences. Caretaker behavior is important both because of the opportunities it affords for the infant to express hunger and satiety and because of the significance of the caretaker's subsequent responsiveness to these signals. Wright speculates that what may be critical for the infant is the opportunity to signal hunger and satiety, and to be responded to appropriately.

The now classic research of Clara Davis (1928, 1933, 1939) is probably the most widely quoted and frequently misinterpreted research on the topic of young children's eating behavior (see Story and Brown, 1987, for a discussion of this point). One of the unique features of Davis's research is that she was able to investigate what infants would eat without the imposition of adult controls on feeding. Davis's research was designed to provide evidence on several points, including the following:

Could infants, at weaning, in the absence of adult intervention, choose their own foods in sufficient quantities to maintain themselves?

Would infants choose few or many of the various foods offered to them?

Would they show definite preferences?

Would they maintain themselves in a state of health, and would their weight gain and growth be comparable to infants fed in the normal manner?

Davis chose infants who had been breast-fed and had no prior experience with solid foods. The infants participated for a minimum of 6 months and some for several years. The infants self-selected their diets from an array of about three dozen foods, presented at each meal in sets of 10–12 items. All foods were simply prepared, without any additional seasoning. Food was brought to the child by a nurse who was specially instructed not to prompt, encourage, coerce, or otherwise interfere with the child's eating. All foods were weighed before and after meals to determine consumption.

Davis's results provided evidence that in the absence of adult control, infants could choose their own foods to maintain growth and health. On a variety of measures, the children were very healthy and showed few episodes of illness and virtually no feeding problems. Food selection and preferences were seen to develop over time so that the children showed clear patterns of preference by selecting some foods and avoiding others. Davis interpreted the patterns of food selection that resulted in nutritionally adequate diets as evidence for an innate, automatic mechanism: "Such successful juggling and balancing of the more than 30 essential nutrients . . . suggests at once the existence of some innate, automatic mechanism." However, her descriptions of what occurred when the children were first exposed to the self-selection situation suggests the possibility of learning effects:

> When their large trays were placed before them at their first meals, there was not the faintest sign of instinct-directed choice. On the contrary, their choices were apparently wholly random; they tried not only food but chewed hopefully on the clean spoon, dishes, the edge of the tray, or a piece of paper on it. Their faces showed expressions of surprise, followed by pleasure, indifference, or dislike. All the articles on the list . . . were tried by all, but within the first few days, they began to reach eagerly for some and to neglect others. Never again did any child eat so many of the foods as in the first weeks of the experimental period. (1939, pp. 260–261)

The children's food intake was not random; they manifest clear patterns of food acceptance and rejection. These patterns, however, were not present from the beginning of the experiment but developed over time, suggesting that experience was playing an essential role. It now seems more likely that what is innate is the ability to learn to associate food cues with the contexts and physiological consequences of eating. In the absence of adult controls, the children obtained an adequate diet, sufficient to maintain growth and health.

In contrast, a recent study provides an extreme example of the role of parental beliefs about feeding on children's food intake. Pugliese et al. (1987) reported nonorganic failure to thrive in seven children 7–22 months old. This nonorganic failure to thrive was a direct result of parental health beliefs. These children, from middle and upper middle income families, were all below the 5th percentile for weight. Parental health concerns included concern that the child would become obese, concerns that the child might develop cardiovascular disease, and pursuit of a "healthy" diet. Some examples of the parents' feeding behavior in response to these concerns are presented in Table 1. In these cases, access to food in general, or to particular calorically dense foods, was restricted by parents, resulting in devastating consequences for the children. Although these parental responses were extreme, recent evidence indicates that they are not terribly discrepant from those of many concerned parents. For example, in a recent informal survey of 2000 parents of infants calling the Gerber Products' "800" number, 57% said they were concerned that their babies would become overweight, and 70% indicated that the foods they served to their babies were influenced by articles about *adult* nutrition and dieting (Johnson, personal communication). As indicated above, the naive application of information about adult nutrition to the feeding of infants and young children can have potentially devastating results.

The study described above provides evidence that parents often have very inaccurate ideas about food that infants and young children should be consuming. In some cases, as described above, parents think that their children are eating too much or at least too much of the wrong foods. Anecdotal evidence suggests that more frequently, parents underestimate the quality of their children's diets and

TABLE 1 Parental Health Misconceptions

Health concern	Response to health concern
Mother concerned that child will be obese	Formula diluted with water
Mother concerned about obesity	Sweets restricted, high calorie foods restricted
Mother concerned about obesity and cardiovascular disease	Only low-fat milk and lean meat used
Father was obese child and concerned his child would become obese	High-fat foods restricted, high complex carbohydrates encouraged
Parents believed breast milk was superior and avoided fatty foods and red meat	Prolonged dependence on breast-feeding; foods of animal origin restricted
Parents did not want to create "Junk food addicts"	Snacking not allowed

Source: Adapted from: Pugliese et al. (1987).

overestimate the quantities of food that young children need to consume. Largely as a result of these inaccurate beliefs, many adults assume that young children cannot adequately control their intake. They believe that part of their responsibility as parents is to make sure the children eat in amounts consistent with parental beliefs. Such well-intentioned parents try to exert external control over their children's food intake. However, the research on the relationship between parenting style and child outcomes suggests that these efforts can backfire, leading to unintended, negative consequences in the long term.

In general, the imposition of rigid, authoritarian parenting styles and a high degree of parental involvement typically interferes with the development of adequate self-control in the child. Most of this research on parenting style and child outcome has focused on the domains of cognitive and social development. However, there is some evidence that this general relationship between parenting style and the development of self-control also applies to the control of food intake. Recently, Costanzo and colleagues (Costanzo and Woody, 1985) have developed a model of the development of obesity proneness in children that can help to explain how the psychosocial environment of feeding may interact with genetic predisposition to produce obesity in susceptible individuals.

Parenting style is pivotal in this model. Parents are not consistent in the use of a parenting style across all the domains of children's development. Rather, parents tend to monitor and control their children particularly forcefully in areas of development if (a) they see the child as potentially deviant (the obese parent fears that the child will be obese in a world that values thinness), (b) the parents themselves are particularly invested in an aspect of development (health, slimness, and beauty consciousness), or (c) the parent doubts that the child's naturally occurring opportunities for learning will produce the desired outcome. However, as stated above, this imposition of the heavy external control typical of authoritarian parenting styles actually *impedes* the development of adequate self-control in the very areas the parent is most concerned about.

In support of the foregoing hypothesis, Morgan and Costanzo (1985) report a study of college students who were classified on the restraint scale (Herman and Polivy, 1980), which discriminates among individuals in terms of how diet-conscious they are, how much their weight fluctuates, and how much they chronically diet and are concerned about their weight. Herman and colleagues have argued that restraint rather than degree of overweight should be most useful in determining the individuals who eat inappropriately and who are most "out of touch" with their physiological state. Morgan and Costanzo obtained retrospective data from their subjects, who varied in degree of restraint. They included questions regarding the subjects' parents' child feeding practices and the extent to which the parents were dieters and were concerned about their weight. They found that one factor that separated the restrained eaters from the rest of the sample was high parental control and influence on eating. They also noted that parental restraint and control over food intake was significantly and positively related to degree of overweight. These retrospective data are provocative with respect to the relationships among early feeding and later individual differences in intake control, and the pattern of results emphasizes the need for longitudinal data on the etiology of individual differences in the control of food intake. We have obtained some experimental evidence relevant to this issue, which is described below. This research suggests that, left to their own devices, and in the absence of adult coercion and control, children have some capacity to adequately control intake.

In our research with children 2-4 years old, we have reported that in the absence of adult control on eating, young children can show responsiveness to caloric density cues that result from differences in carbohydrate content, eating more following low-density than high-density preloads (Birch and Deysher, 1985, 1986). This responsiveness has both unlearned and learned components. Subsequent research has revealed that caretakers' behavior can also influence whether children show responsiveness to such caloric density cues (Birch et al., 1987b).

To obtain evidence on children's responsiveness to caloric density cues, 21 children, ages 30 months to 5 years old, and 26 adults ate two lunches, approximately one week apart. The two lunches consisted of a fixed volume of a pudding preload, followed by the opportunity to eat an ad libitum lunch consisting of sandwich quarters, fresh fruits, and vegetables. The two meals differed in the caloric density of the preload, which had either 40 or 150 kcal/100 ml. The pattern of ad libitum consumption following the two preloads is presented in Figure 2. The stacked bars show the total consumption in the different conditions, including the preload and the ad libitum consumption. The children's total consumption across the two meals was nearly identical, with the children consuming significantly fewer kilocalories ad libitum following the high- than the low-density preload. In contrast, the adults consumed considerably more total kilocalories on the occasion of the high-density preload, showing no evidence of caloric compensation.

In the absence of the imposition of adult control strategies, children can also learn to associate food cues with the physiological consequences of eating those foods, and modify intake accordingly. Forming such associations could allow the organism to learn to anticipate the caloric consequences of eating familiar foods, and make the appropriate adjustments to intake. The existence of such "conditioned satiety" was suggested by the fact that eating usually stops well before postingestive cues signaling satiety have the opportunity to develop (Le Magnen, 1955; Stunkard, 1975). Experimental evidence for such conditioning of satiety has been obtained by Booth (1972) for the rat and for the human adult (Booth et al., 1976, 1982). In the experiments we have conducted with young children (Birch and Deysher, 1986; Birch et al., 1987b), 3- to 5-year-olds consumed a series of pairs of two-part snacks, again consisting of a fixed-volume preload, followed by the opportunity to eat a variety of foods ad libitum. In this case, the preloads again differed in caloric density, but also were distinctively flavored.

CHILDREN'S FOOD ACCEPTANCE PATTERNS 311

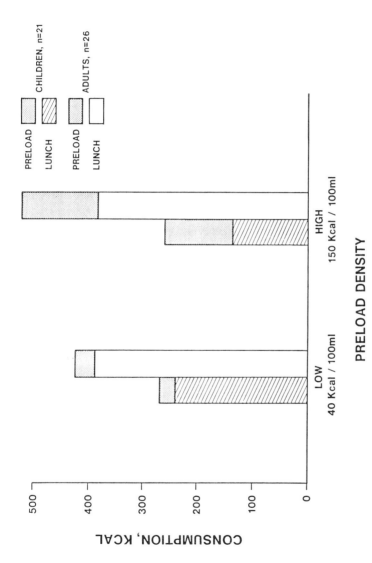

FIGURE 2 Ad libitum lunch consumption and high and low caloric density preload consumption by children and adults.

Following conditioning, the children were given pairs of extinction test trials in which the two preloads continued to differ in flavor but were isocaloric. Results from one of these experiments (Fig. 3) reveal that the children continued to eat more ad libitum following the flavor that had previously been associated with low than with high caloric density.

In cases in which parents believe that their children are consuming too little, they may try to increase the child's food intake using a variety of strategies. Some of these strategies, such as the use of contingencies, can effect immediate and short-term increases in consumption. However, research from our laboratory indicates that they can also have other, more negative long-term effects on intake. The evidence that these strategies, usually chronically imposed, can alter preferences for specific foods is reviewed below. The imposition of such coercive child feeding strategies may also influence which cues children learn to use in controlling intake and could contribute to individual differences in the control of food intake.

To investigate the effects of differences in degree of external control imposed by adults on children's food intake, we replicated the conditioned satiety experiments described above, assigning the children to one of two child feeding contexts: one designed to focus the children on internal cues of hunger and satiety, and the other designed to focus the children on external controls, such as time of day and amount of food remaining on the plate. Children had repeated experience with consuming fixed volumes of two distinctively flavored preloads that differed in caloric density, followed by the opportunity to eat ad libitum. Following pairs of conditioning trials that occurred in one social context or the other, the children in both contexts were given identical extinction test trials. In these trials, the preloads were of the same flavors as during conditioning, but were isocaloric. All the test trials were conducted in a nondirective, relatively neutral social context.

The results (Fig. 4) show that the patterns of intake in the two conditions are quite different. For the children in the internal condition, the results are very similar to those reported previously, with children showing evidence of responsiveness to caloric density cues, eating more following the low than the high caloric density paired flavor. In contrast, the children in the external condition showed no evidence that their eating was being influenced by the caloric density of the preloads during conditioning, or that they had learned any association between the flavor of the preload and its caloric consequences. The evidence is consistent with the view that, in addition to learning associations between food cues and the contexts and consequences of ingestion, children are learning through experience which of the many cues are relevant in the control of food intake.

Evidence presented above suggests that at least under some circumstances, learning influences meal size; satiety can be conditioned in children. This conditioning involves the association of food cues with the caloric consequences of eating those foods. Recently, we have obtained evidence that learning can also influence children's initiation of meals. This research was inspired by Wein-

CHILDREN'S FOOD ACCEPTANCE PATTERNS

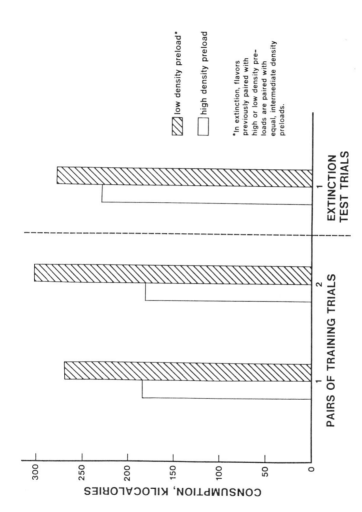

FIGURE 3 Ad libitum consumption following high and low caloric density preloads during conditioning and extinction.

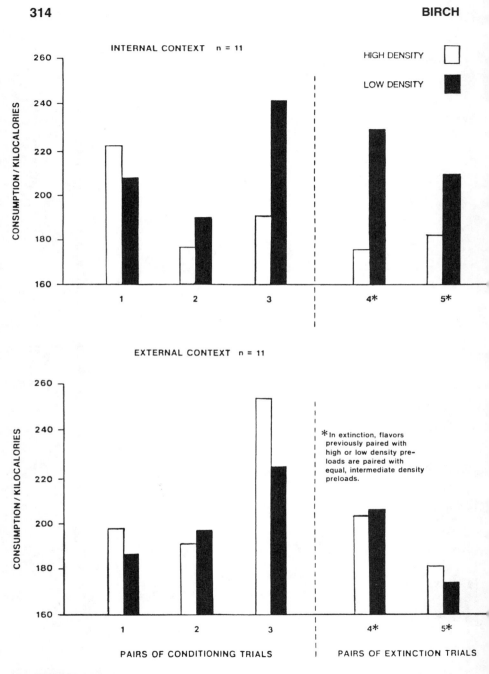

FIGURE 4 Ad libitum consumption following high and low caloric density preloads during conditioning and extinction; internal and external social contexts.

garten's (1982, 1984, 1985) findings indicating that meals could be initiated in sated rats in the presence of cues that had been associated with food and eating. In our research, children were also seen in a series of conditioning and test sessions. A series of four experiments, which differ somewhat in the details, has now been completed. In general, the children were given repeated opportunities to eat in the presence of distinctive cues (CS^+) when moderately hungry. On other occasions, they had free play sessions that were repeatedly associated with another set of cues (CS^-), and in which food was never presented. Following a series of pairs of such trials, we assessed the extent to which the CS^+ cues could elicit eating in the children when relatively sated. To do this, in a series of test sessions over several days, we fed them ice cream preloads that were roughly equivalent to their ad libitum caloric intake during the CS^+ training sessions ($\approx$ 260 kcal). Immediately following consumption of the preloads, they were presented with the CS^+ cues on some days, CS^- cues on other days. In both cases, food was present. Results provide evidence that in the presence of the CS^+ cues, the children ate more food and had shorter latencies to eating than in the CS^- conditions (see Fig. 5). Thus the environmental cues previously associated with food and eating came to facilitate consumption when the children were relatively sated, providing additional evidence for how learning can modify the control of food intake.

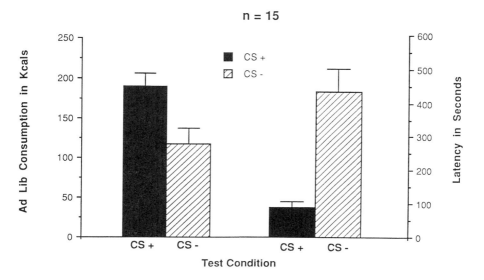

FIGURE 5 Conditioned meal initiation: ad libitum consumption and latency to eating following CS^+ and CS^- cues, as discussed in text.

III. THE ACQUISITION OF SPECIFIC FOOD LIKES AND DISLIKES: CARETAKER INFLUENCE AND PHYSIOLOGICAL FEEDBACK

Omnivores are predisposed to learn to associate food cues with the physiological consequences of consuming those foods. Negative consequences of ingestion such as nausea are readily associated with food cues, resulting in aversions after only one pairing of food and consequence. In contrast, documentation of learned preferences for foods is rarer in the animal literature (see Rozin, 1976, for a discussion of this issue). Thus, it appears that animals learn more readily to dislike and avoid foods than to like and accept them. However, especially for humans, the social context in which foods are presented is particularly central in establishing food acceptance patterns. Depending on the affective tone of the social context, experience can lead to either acceptance or rejection.

Conditioned aversions have been the subject of extensive study. For obvious ethical reasons, relatively few studies have been conducted with humans and particularly with children. Bernstein and her colleagues have conducted an excellent series of experiments with both child and adult cancer patients receiving chemotherapy (see Bernstein and Borson, 1986, for a review of this literature). When ingestion of food was followed by chemotherapy that produced illness, typically including nausea and vomiting, children develop aversions, in a manner similar to other organisms. While survey data reveal that most individuals have such an aversion to at least one food (Garb and Stunkard, 1974; Pelchat and Rozin, 1982), and that many of these aversions are formed during childhood, I have argued elsewhere (Birch, 1987) that for normal individuals, such aversions have a negligible effect on the quality of the diet.

In contrast to this view, Bernstein and Borson (1986) argue persuasively that at least in the animal literature on food intake, the effects of conditioned aversions have been underestimated; many of the treatments used with animal models to produce general anorexia exert their effects not by producing a general loss of appetite but by inducing a specific aversion to the available diet. To the extent that such aversions generalize to other foods perceived as similar to the target food, and can be produced by general malaise and nausea associated with chronic conditions in humans, such as depression and anorexia nervosa, we may have underestimated the effects of conditioned aversions in determining food acceptance patterns in humans. In addition, given the frequency of illness during early childhood, and childhood amnesia for early events, the influence on food acceptance patterns of aversions formed during this early period may be especially underestimated.

Children are neophobic in a manner similar to other omnivores. In general, they show a preference for familiar foods. Rozin (1976) has pointed out that, as omnivores, we need variety in the diet and must seek it out. On the other hand, new, novel foods are potentially dangerous and are approached and sampled with

CHILDREN'S FOOD ACCEPTANCE PATTERNS 317

caution. Other omnivores also show a degree of neophobia and, in general, prefer the familiar. For the young child, who is just being introduced to many of the foods of the adult diet of their culture, many foods are novel and many occasions arise when the child is asked to sample a new food. One of the major concerns voiced by parents of young children is that children do not consume an adequate variety of foods. Parents' misunderstanding of children's transitory dislike of novel foods contributes to the restricted range of foods accepted by the child. There are certainly individual differences in children's temperaments that can influence the strength of this neophobia. However, rather than viewing the child's rejection of a novel food as normal and adaptive, parents frequently interpret this initial rejection as an indication of a fixed and immutable dislike of the food in question. As a result of this view, parents may adopt a strategy that effectively ensures that the food will not become an accepted part of the child's diet: the food is not presented again in the near future. This provides another example of ways in which caretaker behavior profoundly influences children's food acceptance patterns.

In contrast to the commonly held parental belief that the rejection of novel foods reflects a relatively fixed dislike for these foods, there are data indicating that children's initial neophobia is reduced through repeated exposure (Birch and Marlin, 1982; Birch, McPhee et al., 1987a). These findings are consistent with those of other investigators, who have reported exposure effects on liking in nonhuman organisms (cf. Burghardt, 1967; Weiskrantz and Cowey, 1963) and in human adults (Pliner, 1982). In our research into the effects of exposure on the reduction of neophobia, 2-year-old children were given repeated experience with initially novel fruits or cheeses (Birch and Marlin, 1982). Each child had 20 exposures to one item, 15 to a second, 10 to a third, and 5 exposures to a fourth food, while the fifth remained novel. These exposures occurred over a period of about 6 weeks. Following these exposures, paired comparisons were obtained to determine whether preference for the foods were related to frequency of exposure.

The results reveal a clear relationship between frequency of exposure and preference; correlations of .90 or higher were obtained between preference and frequency of exposure. Unfortunately for parents, these data indicate that about 10 exposures are necessary to begin to obtain a clear exposure effect, and this is typically a greater number of exposures than parents are willing to endure. In subsequent work, we have investigated the question of what kind of exposure is necessary to get such exposure effects on preference. One question investigated was whether there were cross-modal effects on liking: Would exposure to the sight and smell of a food result in increased liking of the taste of a food? Such a finding would certainly have practical implications for reducing neophobia in young children. In a second experiment (Birch et al., 1987a), children received 15, 10, or 5 exposures to six initially novel foods, and one food remained novel. For three of the foods, an exposure consisted of tasting the food, while the children looked at and smelled the other three foods. Following the exposures, children made two types of judgment about all of the foods: taste judgments, based on actually tasting the foods, and visually based judgments. The results of this research indicated that taste experience was necessary to obtain significant exposure effects on taste preference. No evidence of cross-modal effects was noted; visual exposure resulted in enhanced liking as indicated by the visually based judgments, but no change was noted for the taste judgments.

Therefore, with repeated exposure, foods that are initially novel and rejected come to be accepted. In such cases, if ingestion of the food is followed by negative consequences, an aversion readily results. In most cases, ingestion of a food is followed by physiological consequences of minimal salience. For children, eating typically occurs in a social context, populated by other eaters who can serve as models and by adults attempting to control children's food intake. I speculated earlier on the possible effects of such practices on the development of individual differences in the control of food intake.

What are the effects on food acceptance patterns of repeatedly eating foods in a particular social context? With respect to the effects of selected child feeding practices on the acceptance and rejection of specific foods, the data are clear. We have conducted research that speaks to limited aspects of this question by investigating the effects of peers as models, and we have also examined the effects of one of the most commonly used child feeding techniques. One way that parents attempt to increase children's consumption of nutritionally desirable foods is to have children eat such foods as a prerequisite to obtaining extrinsic rewards. In other cases, parents use highly palatable foods as rewards when children perform desired behaviors. In some of our earlier research we investigated the effects of such practices (Birch et al., 1982, 1984) on preference for the foods so consumed.

In these experiments, the general approach was to assess the children's preferences for a set of foods, and to select one that was neither strongly preferred nor disliked for repeated presentation in the social context in question. Following a number of exposures to the target food in the social context, preferences were reassessed. Results indicate that repeated experience with a food in a particular social context can shift preferences systematically, either in a positive or negative direction, depending on the social context employed. In two experiments (Birch et al., 1982, 1984), we noted that having children consume a food in order to obtain rewards (in a manner similar to that used to induce children to "eat their vegetables") resulted in relatively large and significant negative shifts in liking. In contrast, when foods are repeatedly presented as rewards, or in conjunction with positive adult attention, preference for the food increased. These changes in preference were maintained for at least several weeks following the cessation of the presentations. The results indicate that the use of such parental control strategies can alter children's preferences for specific foods. In general, these practices work in opposition to establishing food acceptance patterns that result in a healthy diet. The use of these external controls tends to potentiate liking for highly palatable foods used as rewards. Such foods are often high in sugar and fat.

In contrast, foods eaten to obtain rewards are typically foods that parents believe are nutritious and part of a healthy diet, such as vegetables. The evidence, which is consistent with the literature on extrinsic motivation (Lepper and Greene, 1978), indicates that having children consume foods as a prerequisite to obtaining extrinsic rewards reduces their liking for the foods so eaten, and certainly these feeding practices are contributing factors to the pervasiveness of legume anorexia.

In an early study to investigate social learning effects on the formation of food preferences, Duncker (1938) investigated why certain likes and dislikes rather than others come to be dominant in a group. He was interested in the mechanisms by which likes and dislikes were instilled, and he believed that children's acquisition of food preferences provided a setting in which to study these questions. In a series of experiments, Duncker had young children observe other children who were making food choices that were distinctly different from the observers' preferences. He noted that the exposure to peers with differing preferences resulted in shifts in the observers' preferences in the direction of the models' preferences. Surprisingly, he noted that children were more effective models than adults. The effects were greatest when the observing children were younger than the models, or when the children were friends. Fictional heroes were also effective agents of change.

We subsequently investigated the effects of peer models' food choices and consumption patterns on children's preferences for vegetables (Birch, 1980). To do this, we obtained preference data from the children and used them as the basis for seating children at lunch so that one "target" child who, for example, loved carrots and hated peas, was seated with three or four peers who showed the opposite preference pattern. The target child, who was in the minority, then watched the other children selecting and eating one of the two vegetables for lunch. Results revealed that after four successive luncheon exposures in this situation, the target children showed clear shifts in preference, selecting and eating their initially nonpreferred food.

Cultural rules of cuisine are rules about appropriateness. Cuisine rules provide information about which edible substances are and are not food within a culture, and provide guidelines regarding food combinations and food–flavor combinations. They specify the times of day when a food should be eaten, who can eat a food, and the appropriate social contexts for eating. Such rules also tell us which foods are appropriate food contexts for particular flavorants (Should a food be salty, sweet, sour, or piquant? Flavored with cumin or ginger?); see E. Rozin (1973) for discussions of these issues. Although we know relatively little about how such rules are acquired, we have begun to obtain some information that can provide a framework for understanding the acquisition of such rules. The results of this work (and that of Beauchamp, see Chapter 6, this volume) indicate that although explicit transfer of information about these rules undoubtedly occurs, much of what is learned seems to be inferred from repeated experience with what is appropriate. The limited evidence suggests that through repeated exposure and the reduction of neophobia, and the associative conditioning of food cues to the social contexts of eating, rules of cuisine are gradually being constructed by the child. It is my view that these rules of cuisine are supported and maintained by the affective reactions we acquire through repeated experience. This relationship between affect and cognition certainly needs investigation. Although much remains to be done, several pieces of evidence in support of this view are presented below.

At least in affluent Western cultures, we have rules about which foods are consumed at particular mealtimes. We have conducted some research to investigate the acquisition of such cuisine rules. This work was designed to determine whether a food's acceptability changes with time of day and degree to which a food was seen as appropriate for a given mealtime. The preferences of young children (3- and 4-year-olds) and adults were obtained both in the morning and the late afternoon for a set of foods that included items appropriate "for breakfast" or "for dinner" for middle-class Americans (e.g., for breakfast: Cheerios, scrambled eggs; for dinner: pizza, macaroni and cheese) (Birch et al., 1984). All participants were seen twice, once in the morning and once in the afternoon. Even the young children had no difficulty in categorizing the items in a manner consistent with conventional rules of cuisine. The preferences of both the adults and the children showed significant shifts with time of day, with breakfast items more preferred in the morning than in the afternoon, and dinner items more preferred in the afternoon than in the morning. This research indicates that by the preschool period, children are already acquiring these uniquely human cuisine rules regarding food appropriateness. The children's affective reactions to foods also indicate that affect consistent with the cuisine rules also is in place by 3–4 years of age. One central question that remains concerns the relationship between the acquisition of affect and knowledge.

A second piece of evidence on children's learning about contextual appropriateness comes from a recently completed thesis project in our laboratory (Sullivan and Birch, 1990). This research was designed to examine the role of experience in the development of food–flavorant combinations. Cowart and Beauchamp's (1986) previous research revealed that by about 2 years of age, children were showing a preference for salted over unsalted soup, while preferring unsalted to salted water, suggesting that by this time preference for salt was becoming context specific as a result of the child's greater experience with salted soup and unsalted water. It was our intent to see whether repeated exposure to one version of a novel food was sufficient to produce this contextual appropriateness. Preschool children were assigned to receive a series of exposures to only one of three flavored versions of tofu, a food that was unfamiliar to all the participants. The tofu was presented either salty, sweetened, or plain. Preferences for the three versions of the tofu, and for three analogous versions of ricotta cheese, were assessed before any exposures, after 8 exposures, after 8 exposures and a one-week delay, and after 15 exposures. Results indicated a clear exposure effect for the presented food, whether flavored or unflavored. However, we saw no evidence of a generalized increase in preference for tofu across the flavor versions; in fact, quite the opposite was true. The acquired preference was very specific to a food–flavor combination. Children seemed to be extracting information about an appropriate food context–flavor combination, and this combination, through exposure, also comes to be the preferred one. When foods were categorized as flavored or unflavored, exposure produced increases in preference for the exposed flavor–food version and concomitant decreases in preference for the unexposed version. For example, children who had repeated experience with salted (or sweet) tofu increased their preference for that version, while simultaneously significantly decreasing their preference for the unexposed, plain version. Conversely, children who had experience with the plain version came to prefer that version more with experience, while showing a decline in preference for the flavored versions. In general, when the presentation context is not aversive, children come to like what we give them, and this research suggests that they are also learning to dislike the "inappropriate" version of the food.

IV. SUMMARY AND CONCLUSIONS

The transmission of food acceptance patterns from one generation of humans to the next is obviously a complex process, and it involves a number of forms of learning. Although our knowledge is far from complete, several features of the acquisition of food acceptance patterns can be drawn from the literature.

 1. Much of the transmission of food acceptance patterns occurs without, or even in spite of, explicit instruction by adults. Rather, children are forming associations between food cues and the social contexts and consequences of eating through the repeated, redundant nature of their meals and snacks.

 2. Explicit instruction is involved in delimiting food from nonfood ("Please don't eat the daisies"), although social referencing is also probably important. The explicit transfer of information is also involved in the acquisition of information regarding where certain substances come from (hence whether they are candidates for "disgusts," and the potential consequences of eating foods that are "good for you," "bad for you," or that "make you sick"; see Fallon et al., 1984, for a discussion of children's acquisition of categories of food rejection).

 3. Likes and dislikes for specific foods are acquired through associative conditioning to other powerful stimuli in the feeding context, through repeated exposure to novel foods, the influence of other eaters who serve as models, and as a result of caretakers' direct attempts to control children's food intake.

 4. Children learn rules of cuisine regarding what is appropriate: about the appropriateness of foods for particular mealtimes, and about foods as appropriate or inappropriate contexts for particular flavorants. Again, much of this information is inferred by children from the repetitive nature of eating occasions, not transferred via explicit instruction.

5. Children learn to like what is familiar (at least in the absence of negative contexts and consequences), and this liking of the familiar seems to provide a powerful way to perpetuate the rules of cuisine. Conversely, parental beliefs about the immutability of initial food rejections, and food taboos that tag foods as inappropriate for young children, set limits on the array of edibles children can learn to like through exposure and associative conditioning.

6. Children can learn to associate food cues with environmental antecedents, social contexts, and physiological consequences of ingestion. Meal size and meal interval, and the variety of foods consumed, can be influenced by learning and experience.

7. Many of the implicit messages children infer and the food likes and dislikes children acquire from parental imposition of common child feeding practices are not those intended by parents.

8. Although the results of research with animal models reveal that it may be easier to acquire aversions than food preferences, there is evidence that children can learn either to accept or to reject foods, depending on the affective quality of the social context.

9. The feeding strategies imposed by parents provide children with feedback regarding which cues are relevant to the control of food intake.

REFERENCES

Bernstein, I., and Borson, S. (1986). Learned food aversion: A component of anorexia syndromes. *Psych. Rev.* **93**:462–472.

Birch, L. L. (1980). Effects of peer models' food choices and eating behaviors on preschoolers' food preferences. *Child. Dev.* **51**:489–496.

Birch, L. L. (1987). Children's food preferences: Developmental patterns and environmental influences. In *Annals of Child Development*, Vol. 4, R. Vasta (Ed.). JAI Press, Greenwich, CT, pp. 131–170.

Birch, L. L., and Deysher, M. (1985). Conditioned and unconditioned caloric compensation: Evidence for self-regulation of food intake by young children. *Learn. Motiv.* **16**:341–355.

Birch, L. L., and Deysher, M. (1986). Caloric compensation and sensory specific satiety: Evidence for self-regulation of food intake by young children. *Appetite*, **7**:323–331.

Birch, L. L., and Marlin, D. W. (1982). I don't like it; I never tried it: Effects of exposure on two-year-old children's food preferences. *Appetite*, **3**:353–360.

Birch, L. L., Birch, D., Marlin, D., and Kramer, L. (1982). Effects of instrumental eating on children's food preferences. *Appetite*, **3**:125–134.

Birch, L. L., Billman, J., and Richards, S. (1984). Time of day influences food acceptability. *Appetite*, **5**:109–112.

Birch, L. L., Marlin, D., and Rotter, J. (1984). Eating as the "means" activity in a contingency: Effects on young children's food preference. *Child. Dev.* **55**:432–439.

Birch, L. L., McPhee, L., Shoba, B. C., Pirok, E., and Steinberg, L. (1987a). What kind of exposure reduces children's food neophobia? Looking vs. tasting. *Appetite*, **9**:171–178.

Birch, L. L., McPhee, L., Shoba, B. C., Steinberg, L., and Krehbiel, R. (1987b). "Clean up your plate": Effects of child feeding practices on the development of intake regulation. *Learn. Motiv.* **18**:301–317.

Booth, D. A. (1972). Conditioned satiety in the rat. *J. Comp. Physiol. Psychol.* **81**:457–471.

Booth, D. A., Lee, M., and McAleavey, C. (1976). Acquired sensory control of satiation in man. *Br. J. Psychol.* **67**:137–147.

Booth, D. A., Mather, P., and Fuller, J. (1982). Starch content of ordinary foods associatively conditions human appetite and satiation, indexed by intake and eating pleasantness of starch-paired flavors. *Appetite,* **3**:163–184.

Burghardt, G. M. (1967). The primary effect of the first feeding experience in the snapping turtle. *Psychonomic Sci.* **7**:383–384.

Costanzo, P. R., and Woody, E. Z. (1985). Domain-specific parenting styles and their impact on the child's development of particular deviance: The example of obesity proneness. *J. Soc. Clin. Psychol.* **3**:425–430.

Cowart, B. J., and Beauchamp, G. K. (1986). Factors affecting acceptance of salt by human infants and children. In *Interaction of the Chemical Senses with Nutrition,* M. R. Kare and J. G. Brand (Eds.). Academic Press, New York, pp. 25–44.

Davis, C. (1928). Self-selection of diet by newly weaned infants. *Am. J. Dis. Child.* **36**:651–679.

Davis, C. (1939). Results of the self-selection of diets by young children. *Can. Med. Assoc. J.* **41**:257–261.

Davis, C. (1933). A practical application of some lessons of the self-selection of diet study to the feeding of children in hospitals. *Am. J. Dis. Child.* **46**:743–750.

Duncker, K. (1938). Experimental modification of children's food preferences through social suggestion. *J. Abnorm. Soc. Psychol.* **33**:489–507.

Fallon, A. E., Rozin, P., and Pliner, P. (1984). The child's conception of food: The development of food rejection, with special reference to disgust and contamination sensitivity. *Child. Dev.* **55**:566–575.

Fomon, S. J. (1974). *Infant nutrition,* 2nd ed. Saunders, Philadelphia.

Garb, J. L., and Stunkard, A. J. (1974). Taste aversions in man. *Am. J. Psychol.* **131**:1204–1207.

Herman, P., and Polivy, J. (1980). Restrained eating. In *Obesity,* A. J. Stunkard (Ed.). Saunders, Philadelphia, pp. 208–225.

Le Magnen, J. (1955). Sur le mécanisme d'établissement des appétits caloriques. *C.R. Acad. Sci.* **240**:2436–2438.

Lepper, M., and Greene, D. (Eds.). (1978). *The Hidden Costs of Reward: New Perspectives on the Psychology of Human Motivation.* Erlbaum, Hillsboro, NJ.

Morgan, J., and Costanzo, P. R. (1985). Factors discriminating restrained and unrestrained eaters: A retrospective self-report study. Unpublished manuscript.

Pelchat, M. L., and Rozin, P. (1982). The special role of nausea in the acquisition of food dislikes by humans. *Appetite,* **3**:341–351.

Pliner, P. (1982). The effect of mere exposure on liking for edible substances. *Appetite,* **3**:283–290.

Pugliese, M. T., Weyman-Daum, M., Moses, N., and Lefshitz, M. (1987). Parental health beliefs as a cause of nonorganic failure to thrive. *Pediatrics,* **80**:185–182.

Rozin, E. (1973). *The Flavor Principle Cookbook.* Hawthorn Books, New York.

Rozin, P. (1976). The selection of foods by rats, humans, and other animals. *Adv. Stud. of Behav.* **6**:21–76.
Smoller, J. W. (1985). The etiology and treatment of childhood. *J. Polymorphous Perversity.*
Story, M., and Brown, J. E. (1987). Do you know children instinctively know what to eat? *New Engl. J. Med.* **316**(2):103–106.
Stunkard, A. J. (1975). Satiety is a conditioned reflex. *Psychosom. Med.* **37**:383–389.
Sullivan, S. A. and Birch, L. L. (1990). Pass the sugar, pass the salt: Experience dictates preference. *Dev. Psych.* **26**:546–551.
Weingarten, H. (1983). Conditioned cues elicit eating in sated rats: A role for learning in meal initiation. *Science,* **220**:431–433.
Weingarten, H. (1984). Meal initiation controlled by learned cues: Basic behavioral properties. *Appetite,* **5**:147–158.
Weingarten, H. (1985). Stimulus control of eating: Implications for a two-factor theory of hunger. *Appetite,* **6**:387–401.
Weiskrantz, L., and Cowey, A. (1963). The aetiology of food reward in monkeys. *Animal Behav.* XI, **2–3**:225–234.
Wright, P. (1987). Hunger, satiety and feeding behavior in early infancy. In *Eating Habits: Food, Physiology and Learned Behavior,* R. A. Boakes, D. A. Popplewell, and M. J. Burton (Eds.). Wiley, Chichester.

16
Changes in Appetitive Variables as a Function of Pregnancy

Judith Rodin

Yale University
New Haven, Connecticut

Norean Radke-Sharpe

Bowdoin College
Brunswick, Maine

I. INTRODUCTION

It is thought that the prevalence of dietary cravings and aversions during pregnancy is quite high. Taggert (1961) reported that cravings and aversions occur in approximately two-thirds of pregnant women, while Dickens and Trethowan (1971) and Walker et al. (1985) found the figure to be closer to three-quarters. Additional studies have described the incidence of cravings and aversions with respect to specific foods (Hook, 1978; Baylis et al., 1983; Fidanza and Fidanza, 1986).

A. Possible Functions of Cravings and Aversions

Cravings for food and for nonfood items (pica) during pregnancy have been attributed to the increased metabolic energy requirement associated with the developing fetus and resulting increased maternal needs for calories and calcium, although such conclusions, perhaps surprisingly, are not well supported (Hook, 1980). For example, cravings for sweets and dairy products per se are not ubiquitous, and it may be that cultural and geographic factors influence the expression of cravings in pregnancy. As another example, cravings among Southern black women for clay were not reported by black women in the Northeast (Hook, 1978).

It has been suggested (Hook, 1978) that aversions to certain foods and/or beverages during pregnancy may be due to homeostatic mechanisms, such as nausea and vomiting during pregnancy (NVP), which evolved for fetal protection. This is supported, in part, by the observation that the absence of NVP is identified as a risk factor for early fetal death, and that, therefore, the presence of NVP appears to serve some protective function. The NVP may cause the pregnant woman to avoid specific foods that are toxic to the fetus (Hook, 1980).

Although the prevalence of NVP is well documented (Fairweather, 1968; Soules et al., 1980; Jarnfelt-Samsioe et al., 1983; Fitzgerald, 1984), few studies have actually investigated the relationship between NVP and dietary aversions during pregnancy. It has been hypothesized that the occurrence of vomiting during pregnancy is a response to a fetal reaction to dietary antigens (Baylis et al., 1983). Evidence for intrauterine sensitization has been given in a study showing high agglutination antibody titers to some dietary antigens in amniotic fluid (Matsumura et al., 1975). It is known that maternal immunoglobulin G (IgG) antibodies can be transferred across the placenta to provide the infant with a passive immunity in the neonatal period. Measurements of IgG antibodies to β-lactoglobulin in cow's milk in atopic maternal and cord sera indicate transport across the placenta, and 50% of the infants studied developed atopy by the age of 2 years (Dannaeus and Johansson, 1979). It has been hypothesized that impaired placental permeability exists in atopic pregnant women, allowing large molecules such as milk protein to reach the fetus (Baylis et al., 1983), since it has been shown that placental transmission of dietary antigens occurs in rats and rabbits (Hemmings and Kulangara, 1978).

It is also the case that a hormone produced by the placenta, human chorionic gonadotrophin (hCG), has been associated with severe vomiting during pregnancy (Kauppila et al., 1979; Soules et al., 1980; Masson et al., 1985) and is thought to play a role in fetal protection by reducing lymphocyte sensitivity to antigens (Stern, 1976). A recent study questioned the association between the immune response of the fetus and the intake of food allergens by the mother during pregnancy, but this analysis was based on data obtained in the later portion of pregnancy only (Lilja et al., 1988).

The study of the relationship between NVP and aversions can also be considered in terms of learned associations. The nausea and/or vomiting experienced during pregnancy can be thought of as an unconditioned stimulus (US). Through conditioning, foods associated with nausea and/or vomiting may also be experienced as noxious (Garcia et al., 1985), leading to the strong aversions to specific foods often reported in pregnancy. Although the rat does not vomit, behavior typically associated with emesis is observable in the poisoned rat (Grill and Norgren, 1978; Garcia et al., 1985); and Lett (1984) has recently shown that it is specifically this "emetic" reaction in the rat that induces taste aversions. Captive coyotes vomiting after being given toxic food have been known to develop an aversion for that particular food (Gustavson et al., 1974). This same pattern of vomiting and food aversion has been observed among birds (Brower and Brower, 1964, 1974) and insects (Berenbaum and Miliczy, 1983). As a result of these studies, one may hypothesize that the internal homeostatic mechanism for taste aversion learning predominantly manifests itself in the act of vomiting as the US, at least in animals that are able physiologically to vomit.

B. Problems with Past Research

While providing informative material, prior studies of cravings and aversions have suffered from three major interpretive problems. First, they have largely been retrospective, thus relying on memory from 2 months to 5 years after delivery. We have no way of knowing how accurate these retrospective accounts were. Second, since they were retrospective, they asked about pregnancy in general, rather than focusing on specific events potentially unique to each trimester of pregnancy. Third, they lacked a control group of nonpregnant women. It is essential to demonstrate that these experiences are unique to pregnancy per se and not due to women's relationship to, and experience with, food in general.

To address these problems, we have been conducting a large-scale study using a prospective design to identify cravings and aversions prior to and during each of the trimesters of pregnancy. We also attempted to examine the association of reported dietary aversions during pregnancy and the expressions of the symptoms of nausea and/or vomiting. Over a similar temporal period, we have also been examining the likes and dislikes of foods in a group of nonpregnant women. Since there is some evidence that sweet preference may be affected by menstrual cycle phase (Wright and Crow, 1973; Aaron, 1975; Pliner and Fleming, 1983; Cohen et al., 1987), we analyzed the incidence of cravings and aversions for subjects in the control sample by phase of the menstrual cycle (luteal and follicular).

II. THE YALE PREGNANCY STUDY

A. Procedure

The data set we report on here includes complete measures on 80 pregnant women who have been followed from baseline (prior to becoming pregnant), over the course of pregnancy, until delivery. The data set also includes complete measures on a sample of 80 nonpregnant women who have been followed over a similar time period to control for sensitization and random changes over time, and to determine the generality of the experience of cravings and aversions among nonpregnant women.

Subjects were recruited from two demographically similar health maintenance organizations in New Haven, that serve a broad cross section of the community. The women for the pregnant group were selected on the basis of their intention to become pregnant within 2 years. The women for the control group were selected on the basis of their intention to avoid becoming pregnant for at least 2 years, but without the use of birth control pills.

The presence of cravings and aversions was determined by each subject's responses to a series of standardized questions. The subjects were asked whether they had craved, or avoided, any particular foods over the past 2 weeks, and then asked why. If the reason for the increase or decrease in a particular food was clearly intentional—for example, due to medical advice, religious practice, convenience, seasonal availability, or weight control—that food was excluded from the data analysis. Foods that were deliberately avoided because of concern for fetal toxicity (e.g., alcohol) were also excluded. After the dietary portion of the interview was completed, a list of a wide range of foods, covering all the major nutritional and sensory food groups, was shown to each subject. The subject was then asked if these foods reminded her of any additional cravings or aversions previously not mentioned. Her responses were added to the data set if they did not satisfy any of the exclusion criteria.

The dietary cravings and aversions reported by the subjects were then divided into groups using two different methods: nutritional value of the food and sensory perception of the food. The nutritional value of the food was based on the major food groups established by the U.S. Department of Agriculture: milk products, eggs, grain products, fruits, and vegetables, and on the following subgroups: red meats, poultry, fish, sugars and sweets, and nonalcoholic beverages. The sensory components of the food were evaluated by an expert in sensory taste psychophysics for the following qualities: fatty, salty, sweet, sour, bitter, and spicy. Several foods qualified (by containing at least a moderate amount of the taste quality) to be placed in two sensory food groups. Thus the sensory groups were not independent. However, no foods appeared in more than two groups; and based on the distribution of the foods, it is not likely that any single food could account for significance in more than one sensory group.

The cravings and aversion data were analyzed using a categorical variable representing the presence and absence of cravings and aversions during pregnancy. The incidence of cravings and aversions was measured by the number of subjects reporting at least one craving or aversion within a particular food group at each interview.

B. Dietary Cravings

A large number of subjects in the Yale study reported cravings and aversions during pregnancy. We found the proportion of pregnant women reporting cravings and aversions to range between 71 and 87% and 62 and 95%, respectively, over the three trimesters. These findings paralleled those of other studies (Taggert, 1961; Dickens and Trethowan, 1971; Walker et al., 1985). The value of a prospective study, however, is that the incidence of cravings and aversions can also be studied in the same subjects before they become pregnant. Importantly, the incidence of cravings in this sample was almost as high at baseline as during pregnancy. Moreover, the high prevalence of cravings reported by the control group further indicates that the experience of dietary cravings in nonpregnant women is not uncommon.

This result might lead one to speculate on a possible relationship between menstrual cycle phase and the preference and/or avoidance of certain foods or food "types." Our analysis of the control sample, however, revealed no such relationship between either the luteal or follicular phase and a desire or disdain for particular foods. Neither the measure based on the nutritional value of the food, nor the one on the sensory perception of the food, was associated with menstrual cycle phase. Although it has been suggested that food intake (Pliner and Fleming, 1983; Cohen et al., 1987), and specifically carbohydrate consumption (Dalvit-McPhillips, 1983), increases in the luteal, or postovulatory, period, the results of prior studies examining the association between preference for a specific food "type" and menstrual phase have been inconclusive. Both Wright and Crow (1973) and Aaron (1975) found that sweet solutions tasted significantly less pleasant to women in the postovulatory phase of their cycle, while Tomelleri and Grunewald (1987) reported no significant difference in women's cravings for high-sugar or high-starch foods over the stages of the menstrual cycle. The absence of a relationship between food "type" and menstrual cycle phase in this sample of nonpregnant women supports the latter study by demonstrating the lack of evident change in cravings over the menstrual cycle.

Although our data suggest that the existence of general dietary cravings is common in nonpregnant women, it is instructive to consider the data for the specific foods that have been reported as unique to the condition of pregnancy in previous studies. It has been suggested that among pregnant women, the most commonly reported preferred food is fruits (Edwards et al., 1954; Taggert, 1961; Dickens and Trethowan, 1971; Hooks, 1978; Fidanza and Fidanza, 1986). As shown in Table 1, craving for fruit did not increase significantly in our study among pregnant women, relative to baseline, but the number of pregnant women craving fruit in the second trimester was significantly greater than the number of nonpregnant women at the same time point. Thus, cross-sectional comparisons and within-subjects contrasts appear to provide somewhat different pictures.

Milk and dairy products (such as ice cream) have also been reported as preferred foods (Taggert, 1961; Hook, 1978), yet our study revealed no conclusive increase in milk cravings in pregnant women when compared to their own rates prior to pregnancy, or to the rate in the controls. A craving for fish among pregnant women has also been reported (Edwards et al., 1954; Hook, 1978), yet the number of women experiencing a fish craving in this research was not sufficiently different, either within the pregnant group from baseline to delivery or between the pregnant and nonpregnant subject samples, to indicate the presence of a significant change.

TABLE 1 Incidence of Cravings by Nutritional Food Group

Food group	Study group	Baseline	First trimester	Second trimester	Third trimester
Milk products	P	21	30	18	26
	C	20	18	21	14
Eggs	P	1	5	6	4
	C	0	0	1	0
Grain products	P	17	26	18	24
	C	26	17	15	17
Fruits	P	14	21	24	19
	C	20	12	11*	11
Vegetables	P	18	26	13	3
	C	15	20	16	9
Red meats	P	8	16	12	8
	C	10	4*	11	8
Poultry	P	3	5	3	0
	C	3	0	1	1
Fish	P	11	10	14	6
	C	8	6	7	6
Sugars and sweets†	P	30[a]	9[b]	17	18
	C	26	24*	17	17
Nonalcoholic beverages	P	6	9	4	9
	C	13	5	5	6

*Chi-square between pregnants (P) and controls (C), $p \leq .01$.
†Interaction of main effects, $p < .01$.
[a,b]Linear contrast, $p < .001$.

The effect of pregnancy on the desire for sweets has been the source of much disagreement. Although a prevalence of cravings among pregnant women for sweet foods has been reported (Dickens and Trethowan, 1971; Hook, 1978; Walker et al., 1985), other studies have found a decreased desire for sweets during pregnancy (Dippel and Elias, 1980; Fidanza and Fidanza, 1986). Our study, which reveals a significant decline in the number of pregnant women craving sugars and sweets in their first trimester, supports the latter group of experimental results (see Table 2). The number of women craving sweets in the second and third trimesters was consistent with the rate prior to pregnancy (baseline).

Although our study did show a decline in the desire for salty foods in the last trimester of pregnancy, this trend was also observed in the control sample. Thus, this decline may have been due to some cohort shift over time, since the incidence of salty cravings in pregnant women was not significantly different from that in nonpregnant women at both the baseline and third trimester interviews.

TABLE 2 Incidence of Cravings by Sensory Food Group

Food group	Study group	Baseline	First trimester	Second trimester	Third trimester
Fatty	P	19	19	17	24
	C	20	16	23	15
Sweet†	P	51	36	45	51
	C	51	45	42	38
Salty‡	P	21^a	30^a	12	6^b
	C	19	15*	19	11
Sour	P	7	19	12	9
	C	9	7*	7	5
Bitter	P	2	0	0	4
	C	5	3	2	3
Spicy	P	18	26	13	3
	C	15	20	16	9

*Chi-square between pregnants (P) and controls (C), $p \leq .01$.
†Interaction of main effects, $p < .05$.
‡Interaction of main effects, $p < .01$.
^{a,b}Linear contrast, $p < .001$.

C. Dietary Aversions

Unlike the data for cravings, the prospective within-subjects analysis for aversions showed clear differences, especially in the first trimester of pregnancy. The most commonly reported food aversion for pregnant women has been red meats (Dickens and Trethowan, 1971; Hook, 1978; Walker et al., 1985; Fidanza and Fidanza, 1986), and the results of the Yale study strongly support these findings. Significantly more pregnant women were found to express an aversion to red meats during the first term of pregnancy (see Table 3). Also during the first trimester, more pregnant women described an aversion to vegetables and nonalcoholic beverages. The increase in avoidance of beverages included in the nonalcoholic group was primarily accounted for by coffee and tea. Coffee, in particular, appears to be among the most commonly avoided foodstuffs during pregnancy (Taggert, 1961; Dickens and Trethowan, 1971; Walker et al, 1985; Fidanza and Fidanza, 1986).

Differences in the incidence of dietary aversions among pregnant women appeared to be even more apparent when the foods were divided into groups based on the sensory perception of the food (see Table 4). Earlier studies have shown that pregnant women tend to avoid fatty or fried foods (Taggert, 1961; Dickens and Trethowan, 1971; Walker et al., 1985; Fidanza and Fidanza, 1986). Our results reinforce this theory by showing an increase in the number of women avoiding fat-tasting foods in the first term of pregnancy. It has also been suggested

TABLE 3 Incidence of Aversions by Nutritional Food Groups

Food group	Study group	Baseline	First trimester	Second trimester	Third trimester
Milk products	P	12	16	4	4
	C	10	6	2	2
Eggs	P	2	3	6	3
	C	4	6	3	1
Grain products	P	7	10	5	6
	C	4	5	1	2
Fruits	P	5	15	4	7
	C	6	3*	1	4
Vegetables†	P	19^a	38^b	25	14^a
	C	15	7*	5*	4*
Red meats†	P	17^c	30^d	17	13^c
	C	24	10*	4*	6
Poultry	P	1	10	5	2
	C	4	2	0	0
Fish	P	13	20	13	8
	C	7	6*	8	3
Sugars and sweets	P	6	16	3	3
	C	7	4*	8	1
Nonalcoholic beverages†	P	6^a	27^b	10	5^a
	C	7	4*	3	1

*Chi-square between pregnants (P) and controls (C), $p \leq .01$.
†Interaction of main effects, $p < 0.1$.
^{a,b}Linear contrast, $p < .001$.
^{c,d}Linear contrast, $p < .01$.

(Hook, 1978) that pregnant women tend to avoid spicy foods. Our data support this finding and reveal that this aversion occurs primarily during the first trimester. The presence of a high incidence of a dislike for sweets in the first trimester supports the corresponding decline in sweet cravings at the same time point.

D. Symptoms of Nausea and Vomiting

The symptoms of nausea and vomiting have a long history of being reported during pregnancy. As early as the nineteenth century, Giles (1893) postulated that a relationship existed between the occurrence of cravings and/or aversions and NVP, but found no correlation. A more recent study confirmed the work of Giles by finding no relationship between the presence of cravings or aversions and NVP (Dickens and Trethowan, 1971). However, other workers have continued to suggest that aversions develop in response to a homeostatic mechanism, such as nausea and vomiting, for fetal protection, or as a result of mechanisms of associative conditioning.

TABLE 4 Incidence of Aversions by Sensory Food Group

Food group	Study group	Baseline	First trimester	Second trimester	Third trimester
Fatty†	P	15^a	40^b	20	16^a
	C	25	10*	4*	2*
Sweet†	P	15^a	35^b	11^a	12^a
	C	16	7*	10	6
Salty	P	16	24	16	10
	C	23	12	6	3
Sour	P	8	18	9	12
	C	7	2*	1*	4
Bitter	P	3	19	10	4
	C	6	2*	1*	0
Spicy†	P	13^a	39^b	24^c	24^c
	C	15	6*	6*	1*

*Chi-square between pregnants (P) and controls (C), $p \leq .01$.
†Interaction of main effects, $p < .001$.
^{a,b}Linear contrast, $p < .001$.
^{b,c}Linear contrast, $p < .01$.

In our subject sample, a total of 63 women, or 79% of the 80 pregnant subjects, reported experiencing nausea of some degree during the first trimester. The incidence of the symptoms of nausea and vomiting is shown in Table 5. As expected, the number of women reporting NVP decreased markedly by the third trimester; and not surprisingly, women who did not experience NVP during the first trimester did not develop it later.

Even using this prospective design, we lack sufficient data to test the hypoth-

TABLE 5 Incidence of Nausea and Vomiting Reported During Pregnancy

Incidence of symptom	First trimester	Second trimester	Third trimester
Nausea			
None	17	60	64
Infrequent	15	13	14
Moderate–frequent	48	7	1
Vomiting			
None	54	64	72
Infrequent	17	13	6
Moderate–frequent	9	3	2

esis that the specific foods eaten prior to the occurrence of nausea or vomiting are those to which aversions developed. This of course would be the strongest test of the conditioned taste aversion phenomenon. To begin to address the question with the available data, however, we considered whether increased incidence of nausea or vomiting was associated with higher rates of aversions to particular types of foods.

The results of this analysis seem to support the presence of a link between vomiting and the expression of aversions to certain foods or food "types." The results of the Wilcoxon rank tests are shown in Tables 6 and 7. Aversions reported during the first trimester that appeared to be strongly associated with vomiting include poultry, fruits, and sugars and sweets. Vomiting experienced in the second trimester was significantly associated with aversions to red meats, poultry, and salty, sour, and spicy foods in the second term of pregnancy. For the most part, the food groups that were more commonly avoided among pregnant than nonpregnant women were strongly related to the symptoms of NVP.

Nausea in the first trimester of pregnancy was significantly related to the incidence of aversions reported to vegetables and sour foods only in the first trimester. Second trimester nausea was strongly associated with the number of women reporting aversions to poultry, vegetables, and sour foods.

It is interesting that aversions to poultry and sour and spicy foods were

TABLE 6 Relationship Between Aversions and Frequency of Vomiting

Food group	Aversions (yes or no)*	
	First trimester	Second trimester
Vomiting in First Trimester		
Red meats	ns	1.95
Poultry	2.04	1.97
Fruits	2.13	ns
Sugars and sweets	2.97**	ns
Sweet	2.68**	ns
Salty	ns	2.85**
Spicy	ns	2.28
Vomiting in Second Trimester		
Red meats		2.72**
Poultry		2.48**
Salty		2.23
Sour		3.11**
Spicy		2.21

*All Z scores given are significant at $p \leq .05$.
**$p \leq .01$.

TABLE 7 Relationship Between Aversions and Frequency of Nausea

	Aversions (yes or no)*	
Food group	First trimester	Second trimester
Nausea in First Trimester		
Vegetables	2.46**	ns
Sour	1.95	ns
Spicy	ns	2.21
Nausea in Second Trimester		
Poultry		3.00**
Vegetables		2.30
Sour		2.33

*All Z scores given are significant at $p \leq .05$.
**$p \leq .01$.

associated with both nausea and vomiting, while a dislike for red meats, fruits, and sweet and salty foods seemed to be strongly related only to the symptom of vomiting. Since it is true that women who experienced vomiting also experienced nausea, it is difficult to discern independent effects. Both symptoms of vomiting and nausea were significantly associated with the absolute number of aversions reported by the pregnant women during the first and second trimesters. However, since a number of different foods may be contained in one food group, this is not indicative of the number of avoided food "types" associated with the symptoms.

Since it is hypothesized that nausea or vomiting precedes the development of specific aversions, one might also expect a relationship between first trimester symptoms and second trimester aversions. Tables 6 and 7 also show these analyses. First trimester vomiting was associated with a high incidence of aversions reported in the second trimester to red meat, poultry, and salty and spicy foods. First trimester nausea related only to aversions to spicy food in the second trimester. Thus, the strongest associations were between symptoms and aversions expressed during the same trimester.

III. RESEARCH NEEDS

Far more research needs to be done on the changes in appetitive behavior during pregnancy and the mechanisms underlying these changes. The influence of maternal nutrition on infant birthweight and health has been widely studied (see, e.g., Higgins et al., 1979; Klein et al., 1976; Ravelli et al., 1976; Stein et al., 1978). Much more attention has been focused on maternal undernutrition, since it is related to low infant birthweight and poor health; but it also appears reciprocally that maternal weight and perhaps overnutrition, both prior to and during pregnancy, may be related to fatter infants at birth (Udall et al., 1978). Much more work needs to be done to relate maternal appetitive behavior during pregnancy to neonatal variables other than weight such as taste responsiveness or rate of weight gain after birth. The Yale study was designed, in part, to address these issues.

A most important research need will be to measure and relate actual food intake and chemosensory changes to cravings and aversions and hormonal status. Experiments with a variety of animals have shown a clear relationship between levels of steroids and food intake (Czaja, 1975; Czaja and Goy, 1975; Blaustein and Wade, 1976). There is a reduction in food intake at ovulation when estrogen levels are at their peak. In menstruating human females, a cyclic pattern of food intake has also been observed (Dalvit, 1981). Food consumption is higher during a 10-day interval after ovulation than during a 10-day interval before. When the macronutrients are analyzed individually over the menstrual cycle, cycling females show fluctuations in carbohydrate consumption but not in protein or fat (Dalvit-McPhillips, 1983). Carbohydrate consumption is lower before ovulation and greater after ovulation.

Another body of literature has considered the relationship between hormone levels and chemosensory changes. Animal studies have suggested depressed preference for sweet taste that is related to diminished availability or effectiveness of estrogen or estradiol in the activation of taste mechanisms (Richter and Barelare, 1938; Wade and Zucker, 1969; Wade, 1972). In pregnancy and at certain times in the menstrual cycle, this may be due to increased progesterone levels (Dippel and Elias, 1980), since at least in humans progesterone reduces the number of estradiol receptors and increases the activity of estradiol dehydrogenase (Gurpide and Tseng, 1976).

Different investigators have been interested in assessing the relationship between chemosensory changes and food intake. More molar measures, looking at specific taste changes associated with self-reported cravings and aversions, support the hypothesis that these relate to decreased food intake in cancer patients (Brewin, 1980, 1982). Surprisingly, however, such analyses have not been done prospectively for pregnancy.

To link the related areas of inquiry just reviewed, studies must consider together changes in placental hormones, taste perception, food preferences and aversions, and food intake. Such data will enrich our understanding of the determinants of food intake and other aspects of appetitive behavior, as well as provide specific information regarding how pregnancy-induced changes occur. This is an area that studies an important health event and one that has been subjected to minimal biobehavioral research.

IV. CONCLUSIONS

It is clear that the presumption of extensive cravings during pregnancy is overstated. There is a strikingly high incidence of cravings in women prior to their becoming pregnant, and in nonpregnant controls. These data can be viewed in the context of the current preoccupation of many women with their eating and weight. The National Center for Health Statistics (1985) survey reported that 46% of the women were dieting at the time of the survey. With this high incidence of dieting, women may experience numerous cravings and a frank sense of deprivation. Quite likely, these women are more likely to act on their cravings and eat the desired food during pregnancy. It is an occasion for sanctioned weight gain for many, and a chance to divert from restrained eating patterns. One could imagine that when these women are asked retrospectively about their cravings while pregnant, those that are most memorable are those that they acted on (i.e., ate). Therefore, retrospective accounts may give an inflated incidence rate.

Women in general may have a high incidence of cravings, not only because they are deprived, but because dieting behavior is often associated with cycles of weight loss and regain. Repeated gain–loss cycles are associated with an elevated preference for fat and sugar in both animals and humans (Drewnowski et al., 1985; Reed et al., 1988; Kayman et al., in press). In the Yale study, by far the highest number of cravings reported by the nonpregnant women was in the sweet category. All these speculations merit further and more direct empirical analysis.

By contrast, the case for aversions appears to be real and substantial. The aversions were found to occur primarily in the first trimester, when the development of the fetus is most vulnerable. The occurrence of the highest rates of the symptoms of nausea and vomiting and of dietary aversions all during the first trimester could reflect the same underlying biological protective mechanisms and/or could be related through associative conditioning processes. Both are tenable hypotheses.

Finally, it is worth repeating that research designs using prospective analyses yield different results from studies either using cross-sectional comparisons or retrospectively testing only women who have been pregnant, with no control population. Various methodologies exist for experimental inquiry, and it is not always possible or appropriate to use prospective designs. The latter approach, however, does afford a high degree of control by keeping extraneous subject variables constant, and in this instance, has provided us with a far clearer understanding of cravings and aversions in pregnancy.

ACKNOWLEDGMENTS

This research has been supported by the National Institutes of Health grant NS 16993-08.

REFERENCES

Aaron, M. (1975). Effect of the menstrual cycle on subjective ratings of sweetness. *Percept. Mot. Skills,* **40**:974.

Baylis, J. M., Leeds, A. R., and Challacombe, D. N. (1983). Persistent nausea and food aversions in pregnancy. *Clin. Allergy,* **13**:263–269.

Berenbaum, M. R., and Miliczy, E. (1983). Mantids and milkweed bugs: Efficacy of aposematic coloration against invertebrate predators. *Am. Mid. Natl.* **111**:64–68.

Blaustein, J. D., and Wade, G. N. (1976). Ovarian influences on the meal patterns of female rats. *Physiol. Behav.* **17**:201–208.

Brewin, T. B. (1980). Can a tumour cause the same appetite perversion or taste change as a pregnancy? *Lancet,* **25**:907–908.

Brewin, T. B. (1982). Appetite perversions and taste changes triggered or abolished by radiotherapy. *Clin. Radiol.* **33**:471–475.

Brower, L. P., and Brower, J. V. Z. (1964). Birds, butterflies and plant poisons: A study in ecological chemistry. *Zoologica,* **49**:137–159.

Brower, L. P., and Brower, J. V. Z. (1974). Palatability dynamics of cardenolides in the monarch butterfly. *Nature,* **249**:280–283.

Cohen, I. T., Sherwin, B. B., and Fleming, A. S. (1987). Food cravings, mood, and the menstrual cycle. *Horm. Behav.* **21**:457–470.

Czaja, J. A. (1975). Food rejection by female rhesus monkeys during the menstrual cycle and early pregnancy. *Physiol. Behav.* **14**:570–587.

Czaja, J. A., and Goy, R. W. (1975). Ovarian hormones and food intake in female guinea pigs and rhesus monkeys. *Horm. Behav.* **6**:329–349.

Dalvit, S. P. (1981). The effect of the menstrual cycle on patterns of food intake. *Am. J. Clin. Nutr.* **34**:1811–1815.

Dalvit-McPhillips, S. (1983). The effect of the human menstrual cycle on nutrient intake. *Physiol. Behav.* **31**:209–212.

Dannaeus, A., and Johansson, S. G. (1979). A follow-up study of infants with adverse reactions to cow's milk: I. Serum IgE, skin test reactions and RAST in relation to clinical course. *Acta Paediatr. Scand.* **68**:377–382.

Dickens G., and Trethowan, W. H. (1971). Cravings and aversions during pregnancy. *J. Psychosom. Res.* **15**:259–268.

Dippel, R. L., and Elias, J. W. (1980). Preferences for sweet in relationship to use of oral contraceptives and pregnancy. *Horm. Behav.* **14**:1–6.

Drewnowski, A., Brunzell, J. D., Sande, K., Iverius, P. H., and Greenwood, M. R. C. (1985). Sweet tooth reconsidered: Taste responsiveness in human obesity. *Physiol. Behav.* **35**:617–622.

Edwards, C. H., McSwain, H., and Haire, S. (1954). Odd dietary practices of women. *J. Am. Diet. Assoc.* **30**:976–980.

Fairweather, D. V. I. (1968). Nausea and vomiting in pregnancy. *Am. J. Obstet. Gynecol.* **102**:135–175.

Fidanza, A. A., and Fidanza, R. (1986). A nutrition study involving a group of pregnant women in Assisi, Italy. Part 1: Anthropometry, dietary intake and nutrition knowledge, practices, and attitudes. *Int. J. Vitam. Nutr. Res.* **56**:373–380.

Fitzgerald, C. M. (1984). Nausea and vomiting in pregnancy. *Br. J. Med. Psychol.* **57**:159–165.

Garcia, J., Lasiter, P. S., Bermudez-Rattoni, F., and Deems, D. A. (1985). A general theory of aversion learning. In *Experimental Assessments and Clinical Applications of Conditioned Food Aversions,* N. S. Braveman and P. Bronstein (Eds.). New York Academy of Sciences, New York, pp. 8–21.

Giles, A. (1893). The longings of pregnant women. *Trans. Obstet. Soc.* **35**:242.

Grill, H., and Norgren, R. (1978). The taste reactivity test: II. Mimetic responses to gustatory stimuli in chronic thalamic and chronic decerebrate rats. *Brain Res.* **143**:281–297.

Gurpide, E., and Tseng, L. (1976). Estrogen in normal human endometrium. In *Receptors and Mechanisms of Action of Steroid Hormones*, J. R. Pasgualini (Ed.). Dekker, New York, pp. 109–155.

Gustavson, C. R., Garcia, J., Hankins, W. G., and Rusiniak, K. W. (1974). Coyote predation control by aversive conditioning. *Science,* **184**:581–583.

Hemmings, W. A., and Kulangara, A. O. (1978). Dietary antigens in breast milk. (Letter.) *Lancet,* **9**:575.

Higgins, A. C., Moxley, J. E., Pencharz, P. B., and Maughan, G. B. (1979). Prenatal nutrition and birthweight: Experiments and quasi-experiments in the past decade. (Letter.) *J. Reprod. Med.* **22**:67.

Hook, E. B. (1978). Dietary cravings and aversions during pregnancy. *Am. J. Clin. Nutr.* **31**:1355–1362.

Hook, E. B. (1980). Influence of pregnancy on dietary selection. *Int. J. Obesity,* **4**:338–340.

Hytten, F. E., and Leitch, I. (1964). *The Physiology of Human Pregnancy*. Blackwell, Oxford.

Jarnfelt-Samsioe, A., Samsioe, G., and Velinder, G. M. (1983). Nausea and vomiting in pregnancy—A contribution to its epidemiology. *Gynecol. Obstet. Invest.* **16**:221–229.

Kauppila, A., Huhtaniemi, I., and Ylikerkala, O. (1979). Raised serum human chorionic gonadotrophin concentrations in hyperemesis gravidarum. *Br. Med. J.* **1**:1670–1671.

Kayman, S. G., Bruvold, W., and Stern, J. S. (in press). Maintenance and relapse after weight loss in women. *Am. J. Clin. Nutr.*

Klein, R. E., Arenales, P., Delgado, H., et al. (1976). Effects of maternal nutrition on fetal growth and infant development. *Bull. Pan. Am. Health Org.* **10**:301–306.

Lett, B. T. (1985). The painlike effect of gallamine and naloxone differs from sickness induced by lithium. *Behav. Neurosci.* **99**:145–150.

Lilja, G., Dannaeus, A., Falth-Magnusson, K., Graff-Lonnevig, V., Johansson, S. G., Kjellman, N. I., and Oman, H. (1988). Immune response of the atopic woman and foetus: Effects of high- and low-dose food allergen intake during late pregnancy. *Clin. Allergy,* **18**:131–142.

Masson, G. M., Anthony, F., and Chau, E. (1985). Serum chorionic gonadotrophin (hCG), schwangerschaftsprotein 1 (SP1), progesterone and oestradiol levels in patients with nausea and vomiting in early pregnancy. *Br. J. Obstet. Gynaecol.* **92**:211–215.

Matsumura, T., Kuroume, T., Oguri, M., Iwasaki, I., Kanbe, Y., Yamada, T., Kawabe, S., and Negishi, K. (1975). Egg sensitivity and eczematous manifestations in breast-fed newborns with particular reference to intrauterine sensitization. *Ann. Allergy,* **35**:221–229.

National Center for Health Statistics. (1985). Provisional data from the Health Promotion and Disease Prevention supplement to the National Health Interview Survey. *Advancedata,* pp. 2–5.

Pliner, P., and Fleming, A. S. (1983). Food intake, body weight, and sweetness preferences over the menstrual cycle in humans. *Physiol. Behav.* **30**:663–666.

Ravelli, G. P., Stein, Z., and Susser, M. (1976). Obesity in young men after famine exposure in utero and early infancy. *New Engl. J. Med.* **295**:349–353.

Reed, D. R., Contreras, R. J., Maggio, C., Greenwood, M. R. C., and Rodin, J. (1988).

Weight cycling in female rats increases dietary fat selection and adiposity. *Physiol. Behav.* **42**:389–395.

Richter, C., and Barelare, B. (1938). Nutritional requirements of pregnant and lactating rats studied by the self-selection method. *Endocrinology,* **25**:15–24.

Rodin, J., Silberstein, L., and Striegel-Moore, R. (1985). Women and weight: A normative discontent. In *Psychology and Gender: Nebraska Symposium on Motivation,* T. B. Sonderegger (Ed.). University of Nebraska Press, Lincoln, pp. 267–307.

Soules, M. R., Hughes, C. L., Garcia, J. A., Livengood, C. H., Prystowsky, M. R., and Alexander, E. (1980). Nausea and vomiting of pregnancy: Role of human chorionic gonadotrophin and 17-hydroxyprogesterone. *Obstet. Gynecol.* **55**:696–700.

Stein, Z., Susser, M., and Rush, D. (1978). Prenatal nutrition and birth weight: Experiments and quasi-experiments in the past decade. *J. Reprod. Med.* **21**:287–297.

Stern, C. M. (1976). Feto-maternal relationships. In *The Immune System—A Course on the Molecular and Cellular Basis of Immunity,* M. J. Hobart and I. McConnell (Eds.). Blackwell, Oxford, pp. 306–316.

Taggert, N. (1961). Food habits in pregnancy. *Proc. Nutr. Soc.* **20**:35–40.

Tomelleri, R., and Grunewald, K. K. (1987). Menstrual cycle and food cravings in young college women. *J. Am. Diet. Assoc.* **87**:311–315.

Udall, J. N., Harrison, G. G., Vaucher, Y., Walson, M. D., and Morrow, G., III. (1978). Interaction of maternal and neonatal obesity. *Pediatrics,* **62**:17–21.

Wade, G. (1972). Gonadal hormones and behavioral regulation of body weight. *Physiol. Behav.* **8**:523–534.

Wade, G., and Zucker, I. (1969). Hormonal and developmental differences on rat saccharin preferences. *J. Comp. Physiol. Psychol.* **69**:291–300.

Walker, A. R. P., Walker, B. F., Jones, J., Verardi, M., and Walker, C. (1985). Nausea and vomiting and dietary cravings and aversions during pregnancy in South African women. *Br. J. Obstet. Gynaecol.* **92**:484–489.

Wright, P., and Crow, R. (1973). Menstrual cycle: Effect on sweetness preferences in women. *Horm. Behav.* **4**:387–391.

17
Changes in Taste and Smell Over the Life Span: Effects on Appetite and Nutrition in the Elderly

Susan S. Schiffman and Zoe S. Warwick

Duke University
Durham, North Carolina

I. INTRODUCTION

The senses of smell and taste decline with advancing age. These decrements result from a combination of conditions including normal aging, disease states, and drug therapy (see Schiffman, 1983, 1987a,b,c for reviews). This chapter briefly summarizes the effects of aging, disease, and drugs on the chemical senses. In addition, five new studies are presented that further confirm chemosensory losses in the elderly. Potential methods for treating olfactory and gustatory dysfunction are discussed.

A. Summary of Perceptual Losses of Smell

There is a general decline in olfactory functioning with age (see Table 1). Elevated detection and recognition thresholds have been reported for a wide range of odors. Magnitude estimation experiments in which numbers are assigned to odors in proportion to their perceived intensities suggest that olfactory and nasal trigeminal stimuli are perceived to be less intense by the elderly. Older persons are also less proficient at odor identification and have diminished capacity to discriminate among suprathreshold odorants.

TABLE 1 Compounds for Which Olfactory Losses Have Been Reported in the Elderly Using a Variety of Measurement Techniques

Measurement technique	Compound	Ref.
Threshold	n-butanol	Kimbrell and Furchtgott, 1963
	Coal gas	Chalke and Dewhurst, 1957; Chalke et al., 1958
	Coffee and citral	Megighian, 1958
	Food odors	Schiffman, 1979; Schiffman et al., 1976
	Menthol	Murphy, 1983
	Pyridine and thiophene	Perry et al., 1980
	18 purified odorants	Venstrom and Amoore, 1968
Magnitude estimation	Benzaldehyde, D-limonene (pleasant), pyridine (foul), and ethyl alcohol and isoamyl alcohol (neutral)	J. C. Stevens and Cain, 1985
	CO_2 (stimulated trigeminal nerve)	J. C. Stevens et al., 1982
	Isoamyl butyrate	J. C. Stevens et al., 1982; J. C. Stevens et al., 1984; J. C. Stevens and Cain, 1985
	Menthol	Murphy, 1983
Identification	Coffee, peppermint, coal tar, and oil of almonds	Anand, 1964
	Foods—a wide range	Murphy, 1985; Schiffman, 1977
	40 common substances	Schemper et al., 1981
	Microencapsulated battery of 40–50 odors	Doty et al., 1984a, b
Discrimination using multidimensional scaling techniques	Food odors	Schiffman and Pasternak, 1979
	Common odors	D. A. Stevens and Lawless, 1981
	Pyrazines	Schiffman and Leffingwell, 1981

The diminished odor perception can result from a variety of anatomic and physiological losses (Table 2), medical conditions (Table 3), and drugs (Table 4).

B. Summary of Perceptual Losses of Taste

Small losses in taste acuity have consistently been reported in the elderly (see Table 5), but the decrements are not nearly as great as those found in olfaction. On the average, taste thresholds are 2–2.5 times higher in persons over 65 years of

TABLE 2 Age-Related Modification of the Olfactory System

Age-related change	Ref.
Reduced protein synthesis and structural alterations in olfactory epithelium	Dodson and Bannister, 1980; Naessen, 1971
Atrophy in olfactory bulb and nerve	Brizzee et al., 1975; Hinds and McNelly, 1981; Liss and Gomez, 1958; Smith, 1942
Presence of senile plaque and neurofibrillary tangles in hippocampus and amygdaloid complex	Scheibel and Scheibel, 1975; Tomlinson and Henderson, 1976
Hypothalamic degeneration	Machado-Salas et al., 1977
Altered calcium homeostasis	Landfield and Pitler, 1984
Neural–endocrine alterations	Landfield et al., 1978

TABLE 3 Disorders That Affect Smell

Endocrine	*Nervous*
Adrenal–cortical insufficiency	Alzheimer's disease
Cushing's syndrome	Head trauma
Diabetes mellitus	Korsaskoff's syndrome
Gonadal dysgenesis (Turner's syndrome)	Multiple sclerosis
Hypogonadotropic hypogonadism (Kallman's syndrome)	Parkinson's disease
	Tumors and lesions
Hypothyroidism	*Nutritional*
Primary amenorrhea	Chronic renal failure
Pseudohypoparathyroidism	Liver disease, including cirrhosis
Local	Vitamin B_{12} deficiency
Adenoid hypertrophy	*Viral and infectious*
Allergic rhinitis, atopy, and bronchial asthma	Acute viral hepatitis
Leprosy	Influenzalike infections
Ozena	*Other*
Sinusitis and polyposis	Familial (genetic)
Sjögren's syndrome	Laryngectomy
	Olfactory sarcoidosis

Source: Adapted from Schiffman (1983, 1987b, c).

age when compared with the young. Magnitude estimation experiments suggest that modest losses in perceived intensity also occur for suprathreshold tastants. Suprathreshold losses for the amino acids glutamic acid and aspartic acid were relatively larger than for other amino acids in magnitude estimation experiments (Schiffman and Clark, 1980). This is noteworthy because alterations in glutamate binding have recently been implicated in Alzheimer's disease (Greenamyre et al.,

TABLE 4 Drugs That Affect Smell

Classification	Drug
Anesthetics, local	Cocaine hydrochloride, tetracaine hydrochloride
Antimicrobial agents	Streptomycin, tyrothricin
Antithyroid agents	Carbimazole, methimazole, methylthiouracil, propylthiouracil
Opiates	Codeine, hydromorphone hydrochloride, morphine
Sympathomimetic drugs	Amphetamines, phenmetrazine theoclate with fenbutrazate hydrochloride
Vasodilator	Diltiazem
Other	Acetylcholinelike substances; industrial chemicals, including insecticides, menthol, strychnine

Source: Adapted from Schiffman (1983, 1987b, c).

TABLE 5 Tastes for Which Losses Have Been Reported in the Elderly Using a Variety of Measurement Techniques

Measurement technique	Taste	Ref.
Threshold	Salty	Grzegorczyk et al., 1979
		Murphy, 1979
		Weiffenbach et al., 1982
	Sweet	Moore et al., 1982
		Murphy, 1979
		Schiffman et al., 1981
	Sour	Glanville et al., 1964
		Murphy, 1979
	Bitter	Cooper et al., 1959
		Glanville et al., 1964
		Kalmus and Trotter, 1962
		Murphy, 1979
		Weiffenbach et al., 1982
	Amino acids (including those with unusual tastes such as glutamic acid and aspartic acid)	Schiffman et al., 1979
	Weak galvanic current	Hughes, 1969
Magnitude estimation	Sweet, sour, salty, bitter	Cowart, 1983
	Tomato juice	Little and Brinner, 1984
	Sweeteners	Schiffman et al., 1981
	Amino acids (especially glutamic and aspartic acids)	Schiffman and Clark, 1980
Identification	Sweet, sour, salty, bitter	Byrd and Gertman, 1959
		Hermel et al., 1970
		Schiffman, 1977, 1979

1985). Identification tasks also reveal loss in the ability to recognize common tastes, as well as foods that require cooperative functioning of taste and smell.

The anatomic and physiological changes in the gustatory system that occur during normal aging are not well understood. Early studies reported losses in the number of papillae and number of taste buds per papilla with age (Arey et al., 1935; Mochizuki, 1937, 1939; Moses et al., 1967; Conger and Wells, 1969). More recent studies (Arvidson, 1979; Mistretta and Baum, 1984; Bradley et al., 1985) have found no age-related losses in taste buds or papillae. Age-related decrements in taste perception can result from medical disorders (Table 6) or drugs (Table 7) in addition to normal aging.

II. FIVE NEW STUDIES THAT SHOW CHEMOSENSORY LOSS IN THE ELDERLY

Five recent studies provide further evidence that losses in taste and smell occur in old age. The first two studies report data on elevated taste thresholds for sodium salts and acids with different anions. The third study indicates that the elderly have reduced ability to identify blended foods on the basis of odor alone. The fourth study investigates the role of an intracellular mechanism in age-related olfactory losses. The final study compares four different psychophysical procedures for assessing olfactory changes over the life span.

TABLE 6 Disorders That Affect Taste

Endocrine	*Nervous*
Adrenal cortical insufficiency	Alzheimer's disease
Congenital adrenal hyperplasia	Bell's palsy
Cretinism	Damage to chorda tympani
Cushing's syndrome	Familial dysautonomia
Diabetes mellitus	Head trauma
Gonadal dysgenesis (Turner's syndrome)	Multiple sclerosis
Hypothyroidism	Raeder's paratrigeminal syndrome
Panhypopituitarism	*Nutritional*
Pseudohypoparathyroidism	Cancer
Local	Chronic renal failure
Facial hypoplasia	Liver disease, including cirrhosis
Glossitis and other oral disorders	Niacin (vitamin B_3) deficiency
Leprosy	Thermal burn
Oral Crohn's disease	Zinc deficiency
Radiation therapy	*Other*
Sjögren's syndrome	Hypertension
	Influenzalike infections
	Laryngectomy

Source: Adapted from Schiffman (1983, 1987b, d).

TABLE 7 Drugs That Affect Taste

Classification	Drug
Agents for dental hygiene	Sodium lauryl sulfate
Amoebicides and anthelmintics	Metronidazole, niridazole
Anesthetics (local)	Benzocaine, procaine hydrochloride (Novocain), and others
Anticholesteremics	Clofibrate
Anticoagulants	Phenindione
Antihistamines	Chlorpheniramine maleate
Antimicrobial agents	Amphotericin B, ampicillin, bleomycin, cefamandole, griseofulvin, ethambutol hydrochloride, lincomycin, sulfasalazine, tetracyclines
Antiproliferatives, including immunosuppressive agents	Doxorubicin and methotrexate, azathioprine, carmustine, vincristine sulfate
Antirheumatic, analgesic–antipyretic, anti-inflammatory drugs	Allopurinol, colchicine, gold, levamisole, D-penicillamine, phenylbutazone, 5-thiopyridoxine
Antiseptics	Hextidine
Antithyroid agents	Carbimazole, methimazole, methylthiouracil, propylthiouracil, thiouracil
Diuretics and antihypertensive agents	Acetazolamide, amiloride, captopril, diazoxide, enalapril, ethacrynic acid, nifedipine
Hypoglycemic drugs	Glipizide, phenformin and derivatives
Muscle relaxants and drugs for treatment of Parkinson's disease	Baclofen, chlormezanone, levodopa
Psychopharmacologic agents, including antiepileptic drugs	Carbamazepine, lithium carbonate, phenytoin, psilocybin, trifluoperazine
Sympathomimetic drugs	Amphetamines
Vasodilators	Bamifylline hydrochloride, diltiazem, dipyridamole, oxyfedrine
Others	Germine monoacetate, idoxuridine, insecticides, iron sorbitex, metal ions, vitamin D

Source: Adapted from Schiffman (1983, 1987b, d).

A. Study 1: Elevated Thresholds in the Elderly for 10 Sodium Salts with Different Anions

Previous studies have shown that sensitivity to the taste of NaCl decreases with age (see Table 5). The purpose of this study was to determine whether loss in sensitivity also occurs for other sodium salts with different anions.

Three groups of subjects were tested: young inexperienced tasters ($n = 13$;

mean age = 19.9 years), young experienced tasters ($n = 14$; mean age = 23.1 years), and elderly inexperienced tasters ($n = 11$; mean age = 75.1 years). Detection and recognition thresholds were found for 10 salts with different anions at pH 7 using the forced-choice method described by Schiffman et al. (1979a).

An analysis of variance was performed on the detection thresholds for each salt, and F tests revealed significant differences by group for every salt at the .01 level of significance. The mean detection thresholds for each group are given in Table 8, along with the results of Duncan's multiple-range test. Mean threshold values for a given salt with identical alphabetic superscripts are statistically equivalent.

The detection thresholds for young subjects (both experienced and inexperienced) were significantly different from elderly subjects in all cases. In addition, the thresholds for young subjects with different levels of experience with taste tests were equivalent. The losses in threshold sensitivity varied widely by salt. Table 8 gives the ratio of $\text{threshold}_{\text{elderly}}/\text{threshold}_{\text{young}}$ for inexperienced tasters: it ranged from 2.7 for monosodium glutamate to 26.7 for sodium sulfate.

A recognition threshold for saltiness was found for some but not all subjects. This was because some subjects never found some of these stimuli to taste salty at any of the concentrations tested. The recognition thresholds for saltiness were also elevated in the elderly subjects who did detect a salty component. These data suggest that thresholds for sodium salts are elevated in older persons and that the degree of loss with age varies with the anion. There was considerable individual variation among elderly subjects in thresholds for sodium salts.

TABLE 8 Mean Detection Thresholds at pH 7.0

Substance	Mean detection threshold (M)			Ratio of elderly, inexperienced, to young, inexperienced
	Young, experienced	Young, inexperienced	Elderly, inexperienced	
Monosodium glutamate	0.000163^a	0.00236^a	0.00638^b	2.7
Na acetate	0.00250^a	0.00235^a	0.0190^b	8.1
Na ascorbate	0.00618^a	0.00190^a	0.0250^b	13.2
Na carbonate	0.00260^a	0.00177^a	0.00829^b	4.7
Na chloride	0.00200^a	0.00277^a	0.01850^b	6.7
Na citrate	0.000479^a	0.000583^a	0.0130^b	22.3
Na phosphate	0.00325^a	0.00289^a	0.0160^b	5.5
Na succinate	0.000932^a	0.000775^a	0.0138^b	17.8
Na sulfate	0.000903^a	0.00106^a	0.0283^b	26.7
Na tartrate	0.00230^a	0.000727^a	0.0159^b	21.8

Mean threshold values for a given salt with identical alphabetic superscripts are statistically equivalent.

TABLE 9 Mean Detection Thresholds for Acids

	Mean detection thresholds (M)			
Acid	Young	Elderly	Significance level	Ratio
Acetic	0.000106	0.000273	*	2.58
Ascorbic	0.000281	0.000725	ns	2.58
Citric	0.0000498	0.000375	**	7.53
Glutamic	0.0000920	0.000463	ns	5.03
Hydrochloric	0.0000179	0.0002	***	11.17
Succinic	0.000132	0.000188	ns	1.42
Sulfuric	0.0000468	0.000100	*	2.14
Tartaric	0.0000864	0.000163	ns	1.89

ns = not significant.
*$p < .05$.
**$p < .01$.
***$p < .001$.

B. Study 2: Elevated Thresholds in the Elderly for Acids

Previous studies have shown that the thresholds for some acids are elevated in older persons (see Table 5). The purpose of this experiment was to extend these studies using a wider range of acids.

Two groups of subjects participated: young ($n = 20$, mean age = 20.5 years) and elderly ($n = 12$, mean age = 83.9 years). Detection thresholds were found for the eight acids shown in Table 9. A concentration series for each acid was made such that each dilution differed by a factor of 2. The highest concentration used was 3.2×10^{-2} M; the lowest was 4.9×10^{-7} M. Thresholds were determined by triangle tests using the method described by Schiffman et al. (1979a).

A significant difference was found using t tests for four acids: acetic, citric, hydrochloric, and sulfuric. The average ratio of detection threshold$_{elderly}$/detection threshold$_{young}$ was 4.29. Because some subjects never identified these acids as sour at any concentration tested, recognition thresholds for sourness were not established for all subjects. A significant difference in recognition threshold was found for three acids (citric, glutamic, and hydrochloric) for the subjects who did detect a sour component. The average ratio of recognition threshold$_{elderly}$/recognition threshold$_{young}$ was 6.81.

These data indicate that the thresholds for acids, like those for salts, are found to be elevated in many elderly persons. In addition, the degree of reduced sensitivity appears to vary with the anion.

C. Study 3: Reduced Food Recognition by Odor Cues in Elderly Subjects

Schiffman (1977) found that elderly subjects were significantly less able than young subjects to identify blended foods based on a combination of taste and odor cues. The purpose of this study was to compare youthful and elderly subjects' ability to identify the same blended foods based on odor cues alone.

Two groups of subjects were tested: young ($n = 24$; mean age = 19.4 years) and elderly ($n = 19$; mean age = 79.2 years). The stimuli consisted of a group of 21 blended common foods previously used by Schiffman (1977). They included fruits, vegetables, meats, fish, nuts, dairy products, and coffee. The fruits and vegetables were steamed just enough to give a soft texture when blended. The meats and fish were baked in aluminum foil after the fat and bones had been removed. The eggs were fried without butter. The foods were tested at 160°F.

Subjects were blindfolded and asked to smell a container of blended food presented at the level of the nostrils. (In the earlier experiment they not only smelled the food but tasted one spoonful as well.) At no time did subjects see the colors of the stimuli. The subjects were then asked to identify each food on the basis of its odor.

The percentages of young and elderly subjects who correctly identified each of the blended foods on the basis of odor alone are given in columns b and d of Table 10. The percentages reported earlier by Schiffman (1977) for correct identification of these foods on the basis of both odor and taste are given in columns a and c for comparison. Twenty of the 21 foods were correctly identified more frequently (or as often, in the case of potato) by youthful subjects on the basis of odor. This difference is highly significant ($\chi^2 = 17.1, p < .001$).

The use of the senses of both taste and smell provided some advantage in the majority of the cases in the correct identification of blended foods when compared to odor cues alone. The data in Table 10 also suggest that odor plays a relatively greater role than taste in correct identification of these foods.

The guesses for both elderly and youthful groups were examined to determine how many of the incorrect identifications were reasonable, that is, that they fell into the appropriate food group; 64.7% of the young subjects and 44.0% of the elderly subjects made reasonable identifications based on odor cues alone. This compares with 82.0% of young subjects and 66.5% of elderly subjects when both taste and smell cues were used.

D. Study 4: Investigation of Mechanism of Olfactory Losses in the Elderly

Olfactory changes with age have been presumed to result from degeneration in the olfactory epithelium, the olfactory bulb and nerves, the hippocampus, and the amygdaloid complex (see Table 2). However, it is possible that age-related

TABLE 10 Percentage of Subjects Correctly Identifying Each of the Blended Foods

Food	Young (%)		Elderly (%)	
	Taste and smell (a)	Smell only (b)	Taste and smell (c)	Smell only (d)
Fruits				
Apple	81	46	55	42
Banana	41	25	24	0
Pear	41	4	33	0
Pineapple	70	63	37	47
Strawberry	78	67	33	26
Walnut	33	33	21	0
Vegetables				
Acorn squash	19	8	11[a]	5
Broccoli	30	37	0	16
Cabbage	4	8	7	0
Carrot	63	13	7	5
Celery	59	33	24	21
Corn	67	42	38	32
Green bean	30	17	14	0
Green pepper	19	25	11	0
Potato	19	42	38	42
Tomato	52	33	69	53
Other				
Beef	41	42	28	26
Fish	78	83	59	53
Pork	15	8	7	0
Egg	41	21	16[a]	0
Coffee	89	75	70	74
Mean	46.2%	34.5%	28.7%	21.0%

[a]Determined in pretesting. The remaining values in columns a and c from Schiffman (1977).

modifications in the transduction mechanism may play a role. Two signal transduction pathways have been described in the olfactory system: the adenylate cyclase system and the phosphatidylinositol system. The first transduction system consists of alterations in intracellular levels of cyclic adenosine 5′-monophosphate (cAMP) through activation or inhibition of adenylate cyclase via a regulatory protein that binds guanosine 5′-triphosphate (GTP). This results in the activation of cAMP-dependent protein kinase. Recent studies in frog and rat suggest that odorants vary in their ability to stimulate GTP-dependent adenylate cyclase (Sklar et al., 1986). The purpose of this study was to determine whether there are differential losses with age for odors that differ in their capacity to activate adenylate cyclase. Both threshold measurements and dose–response curves for odorants that vary in their capacity to stimulate adenylate cyclase were obtained in young and elderly subjects.

Two groups of subjects were tested: young ($n = 20$; mean age = 22.9 years) and elderly ($n = 20$; mean age = 82.5 years). The odorants selected for this experiment span a range of effectiveness in stimulating GTP-dependent adenylate cyclase in animals (see column e of Table 11A). The pyrazines and acetic acid were diluted in deionized water. The other odorants were diluted in an odorless grade of diethyl phthalate. Thresholds were determined by a forced-choice technique described earlier by Schiffman et al. (1979a) for tastants. The direct scaling procedure called magnitude estimation was used to determine the relationship between perceived intensity and concentration (see Schiffman and Clark, 1980).

The detection thresholds for young and elderly subjects are given in columns a and b in Table 11A, along with their ratios, which are shown in column c. For seven of the eight odorants, the numerical value of the threshold for the elderly exceeded that for the young. The mean ratio of $\text{threshold}_{\text{elderly}}/\text{threshold}_{\text{young}}$ over all odorants was 6.06. Student's t tests were performed to determine whether young and elderly subjects had significantly different detection thresholds for any of the odorants. Column d indicates that only the detection threshold for citralva was significantly higher in the elderly.

The magnitude estimates of intensity for each odorant for a given subject were plotted against concentration in log–log coordinates, and a regression line was fit to the points using a simple power function $s = kc^n$ (or $\log(s) = \log(k) + n \log(c)$)

TABLE 11A Detection Thresholds for Young and Elderly Subjects

Compound	Mean detection thresholds (M)		Mean threshold ratio, elderly/ young (c)	Significance level (d)	Stimulation of adenylate cyclase (%)[a] (e)
	Young (a)	Elderly (b)			
Citralva	3.86×10^{-4}	4.34×10^{-3}	11.20	**	100
Geraniol	2.35×10^{-3}	4.20×10^{-3}	1.79	ns	58 ± 3
Citronellal	2.83×10^{-4}	2.75×10^{-3}	9.72	ns	56 ± 5
2-Methoxy-3-isobutylpyrazine	1.41×10^{-9}	8.47×10^{-10}	0.60	ns	53 ± 4
Benzaldehyde	2.17×10^{-4}	4.26×10^{-4}	1.96	ns	27 ± 4
Limonene	1.82×10^{-4}	3.39×10^{-3}	18.60	ns	5 ± 4
2-Methoxypyrazine	7.90×10^{-7}	1.80×10^{-6}	2.28	ns	-5 ± 5
Acetic acid	1.71×10^{-4}	3.95×10^{-4}	2.31	ns	-14 ± 33

**$p < .01$.
[a] As reported by Sklar et al. (1986).

TABLE 11B Slopes for Young and Elderly Subjects

	Mean slopes		Mean slope ratio, young/elderly	Significance level	Stimulation of adenylate cyclase (%)[a]
	Young	Elderly			
Compound	(a)	(b)	(c)	(d)	(e)
Citralva	0.39	0.21	1.86	**	100
Geraniol	0.35	0.17	2.06	*	58 ± 3
Citronellal	0.50	0.17	2.94	***	56 ± 5
2-Methoxy-3-isobutylpyrazine	0.35	0.14	2.50	***	53 ± 4
Benzaldehyde	0.48	0.25	1.92	**	27 ± 4
Limonene	0.29	0.11	2.64	****	5 ± 4
2-Methoxypyrazine	0.32	0.18	1.78	**	−5 ± 5
Acetic acid	0.39	0.21	1.86	**	−14 ± 33

*$p < .05$.
**$p < .01$.
***$p < .001$.
[a] As reported by Sklar et al. (1986).

as a model; t tests between the slopes for young subjects and for elderly subjects were performed for each odorant. The slopes based on the magnitude estimates for young and elderly subjects are given in columns a and b in Table 11B, along with their ratios, which are shown in column c. For all eight odorants, the numerical value of the slope for the young subjects exceeded that for the elderly. The mean ratio of threshold$_{young}$/threshold$_{elderly}$ over all odorants was 2.20. Student's t tests were performed to determine whether young and elderly subjects had significantly different slopes for any of the odorants. Column d indicates that the slopes for young and elderly were significantly different for all odors.

These data are similar to those reported by other experimenters. Thresholds tend to be higher for elderly subjects although, in this study, the threshold for only one odorant was significantly greater. Suprathreshold studies, however, show highly significant losses for every compound. It is probable that the elderly still have enough receptors to detect odorants but not enough to discriminate among various concentrations (Schiffman et al., 1979b). For these odorants, there were no significant correlations between the degree of loss (i.e., the ratios for thresholds and slopes between young and elderly) and stimulation of adenylate cyclase. In addition, losses in threshold and losses in slopes were uncorrelated.

E. Study 5: Comprehensive Screening of Smell Over the Life Span

The purpose of this study was to compare four methods for assessing olfactory and trigeminal functioning to determine the most effective means of detecting deficits.

Three of the tasks (a discrimination experiment, adjective ratings, and identification scores) evaluated the ability to discern differences among suprathreshold odors. The fourth task measured detection thresholds. Although an overview of the results suggests that a systematic decrement in the sense of smell begins in the sixties, the data from each test type are not highly correlated over individuals. The finding that subjects often performed well on one test and poorly on another suggests that comprehensive screening of the sense of smell may require a variety of experimental tasks.

The subjects were 143 individuals distributed into seven decades, beginning at 10, 20, 30, 40, 50, 60, and 70 years of age. The seven groups were approximately equivalent in intelligence, socioeconomic status, race, gender, and cultural eating patterns. At least 20 persons in each decade were tested. All subjects were healthy and relatively drug-free. Every subject participated in the four separate experiments that comprised the study: the discrimination experiment (confusabilities with similarity judgments), the ratings on adjective scales, the odor identification task, and the detection threshold experiment.

The stimuli for the suprathreshold experiments were nine odorants that ranged widely in hedonicity, chemical structure, and olfactory quality. The odorants (benzaldehyde, n-butanol, caproic acid, citral, citronellal, geraniol, guaiacol, menthol, and methyl salicylate) were diluted in a special odorless grade of diethyl phthalate until they were approximately equal to each other in moderate odor intensity.

Using diethyl phthalate as a diluent, detection thresholds were determined for geraniol, guaiacol, and benzaldehyde. These three odorants were chosen because they have been reported to vary in their ability to stimulate the trigeminal nerve (Doty, 1985). Experiments with patients lacking olfactory nerve function but with intact trigeminal nerve function suggested that benzaldehyde has a pronounced trigeminal component, whereas guaiacol and geraniol do not.

1. Discrimination Experiment: Quantification of Confusability and Degree of Similarity

The two-step discrimination task used an approach similar to that described previously by Schiffman and Dackis (1976). The first was a confusability task in which the subjects' ability to distinguish between any two odors was determined. Subjects sniffed three bottles presented one at a time at the level of the nostrils; two of the bottles contained the same odorant. One of the six possible ways of presenting the stimuli (AAB, ABA, ABB, BAA, BAB, and BBA) was selected randomly for each subject. This was done to balance cross-adaptation effects. The subjects then decided which two bottles contained the same odorant and which contained a different odorant. In the second step, subjects made judgments of the degree of similarity in quality for the two odorants that were different by circling a number along a 9-point scale labeled "identical" at *1* and "completely different" at *9*.

a. **Confusability task** Figure 1 shows the percentage of the 36 triads correctly identified by decade. The mean is indicated with an asterisk; the standard deviations are given by lines. The top and the bottom of the rectangles span the range from the 25th to the 75th percentiles. The performance of the subjects in the seventh decade was significantly worse than those in the other six decades.

The patterns of decline over the life span were similar for both sexes. Both groups showed a slight but now significant decline in the percentage of correct discriminations in the fourth decade. These slight decrements represent the data for two males and three females aged 47–49 years whose scores ranged from 33 to 52% correct. The standard deviations for the females were larger than for males for all decades except for the first and seventh decades.

Only 13 of the 143 subjects in this experiment were smokers. They were approximately evenly distributed in each of the decades. The percent correct scores for 10 of the 13 smokers fell below the mean for their age group, suggesting that there may be a smell decrement for smokers relative to nonsmokers. The scores for 11 of the 23 persons who reported allergies fell above the mean; 12 fell below the mean.

Table 12 gives the percentiles in which a person in each decade would fall, given a certain percent correct score on the confusability task. It can be seen that a score of 54.2% correct would classify a 75-year-old in the 75th percentile; however, it would relegate an 18-year-old to a position below the 1st percentile. The highest score achieved here by the elderly, 77.8% correct, would be rated as average (50th percentile) for the group as a whole.

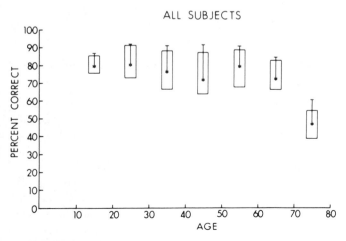

FIGURE 1 Percentage of the 36 triads correctly identified by decade. The mean is designated by an asterisk; the standard deviations are given as well. The top and the bottom of the rectangles span the range from the 25th to the 75th percentile.

CHANGES IN TASTE AND SMELL OVER THE LIFE SPAN

TABLE 12 Percentile in Which a Person Would Fall by Decade and Overall, Based on Percent Correct Score

Percentile	10s	20s	30s	40s	50s	60s	70s	Overall
99th	91.7	100.0	97.2	100.0	97.2	91.7	77.8	100.0
95th	91.6	99.7	96.9	99.4	96.7	91.7	77.8	94.4
90th	88.9	94.4	91.7	93.9	91.1	91.7	72.8	91.7
75th	85.4	91.0	88.4	87.4	88.9	83.3	54.2	86.1
50th	80.6	79.2	83.3	77.8	86.1	69.4	47.2	77.8
25th	75.7	72.9	66.7	63.9	68.1	66.7	38.9	61.1
10th	66.9	61.8	53.1	42.2	61.6	58.3	30.3	45.5
5th	64.0	61.1	50.1	29.2	58.6	50.0	23.0	39.5
1st	63.9	61.1	50.0	27.8	58.3	50.0	22.2	24.7

b. Judgments of the degree of similarity Whereas the confusability experiment suggests that subjects in their sixties retain the ability to choose the stimulus of a triad that differs from the other two, the analysis of similarity judgments indicates that persons in their sixties demonstrate diminished ability to discriminate the *degree* of difference among odorants in ratings along similarity scales. The individual differences option of ALSCAL, a multidimensional scaling procedure, was applied to the mean similarity matrices for each decade computed by averaging the ratings of each pair along the 9-point scale ranging from "identical" to "completely different." The two-dimensional space representing

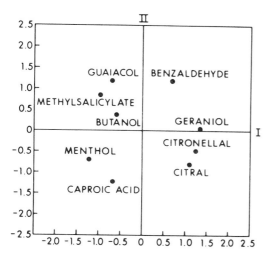

FIGURE 2 Two-dimensional space representing the similarities among the odorants determined from direct ratings of similarity along the 9-point scale ranging from "identical" to "completely different."

the similarities common to all decades is shown in Figure 2. Stimuli located close to each other in the space were rated more similar to one another than odorants located distant from one another.

The individual differences option of ALSCAL provides weights for each decade on each of the two dimensions of the multidimensional space common to all subjects (see Fig. 3). Weights give insight into the perceptual space for a specific decade; they also indicate the salience (or importance) of the two dimensions that is idiosyncratic for each decade. The configuration of Figure 2 better represents decades with weights distant from the origin than weights close to the origin. The weight space in Figure 3 clearly suggests that decades 6 and 7 are not well represented by the space in Figure 2. Individual multidimensional spaces for decades 6 and 7 indicate that subjects in their sixties and seventies have difficulty rating the degree of similarity between two different odors. However, the loss is considerably greater in the seventh decade than in the sixth.

2. Adjective Ratings

After all discrimination comparisons had been made, subjects were asked to rate each of the nine stimuli on the following adjective scales: floral, burnt, sweet, animal, citrus, minty, fuel, almond, and pleasant. The adjective data that describe the odor experience in words were collected to compare with the "nonverbal" confusability and similarity judgments.

The ratings of each odorant on the adjective scales were analyzed by the general linear models procedure in SAS to determine whether there were differ-

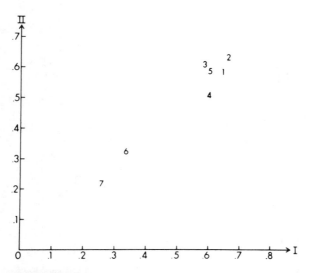

FIGURE 3 Weights provided by ALSCAL, a multidimensional scaling program, indicating that direct scaling of differences declines in the sixties and seventies.

ences among the decades for each of the nine odorants. The multiple analysis of variance (MANOVA) test, Hotelling–Lawley trace, indicated that there was a statistical difference at the $p < .01$ significance level or better for all odorants. Individual F tests revealed statistical differences among the decades at $p < .05$ for at least one adjective for every odorant.

Qualitative aspects of each odorant were rated less intense by the oldest subjects. In general, with increasing age, subjects tended to perceive benzaldehyde as less sweet; n-butanol, less burnt; caproic acid, less animal; citral, less sweet, less citrusy, and less pleasant; citronellal and geraniol, less citrusy; menthol, less minty; and methyl salicylate, less sweet and less minty. No statistically significant differences were found between males and females.

3. Odor Identification

Subjects were asked to identify each of the nine odorants. Subjects had made the ratings on adjective scales prior to the odor identification experiment, so they were familiar with these descriptors. If subjects were uncertain of the identity of the odor, they were required to make their best guess to identify the odor. Subjects were not permitted to answer "don't know."

An odor identification score was calculated by assigning the value 1 to a highly reasonable guess, ½ to a moderately reasonable guess, and 0 to an unreasonable guess. An analysis of variance indicated that there were no significant differences in identification scores for persons from 10 to 59 years of age. Persons in their sixties were significantly worse than those in the first five decade groups. Persons in their seventies had the worst performance and had significantly lower scores than subjects in their sixties and subjects in the first five decades. The odor identification scores were moderately correlated with the percent correct scores on triadic comparisons ($r = .47, p < .05$). The greatest losses in odor identification were for geraniol, guaiacol, citronellal, and caproic acid. The smallest losses were for the lemon (citral) and mint (methyl salicylate).

4. Detection Thresholds

The detection thresholds were determined using the forced-choice method also employed in study 4. An analysis of variance revealed no significant differences over the decades, although the oldest decade had mean thresholds that were higher in all cases than the youngest decade. The thresholds from persons in their seventies were 1.87, 2.42, and 2.01 times higher for geraniol, guaiacol, and benzaldehyde, respectively, when compared with teenagers. Detection thresholds were not significantly correlated with the percent correct score for triads.

5. Overview of the Four Tasks

An overview of the four tasks for comprehensive screening of olfaction suggests that a systematic loss in the ability to discriminate odors from one another begins in the sixties. Although the subjects in their sixties retained the ability to choose that stimulus which differed from the other two in triadic testing, they demonstrated a diminished ability to discriminate the *degree of difference* among odorants in ratings along similarity scales. There was also a tendency in the sixties to begin rating the qualitative aspects of odors as less intense on adjective scales and to have more difficulty identifying odors.

The finding that the ability to discriminate among suprathreshold odors is not well correlated with detection thresholds is not surprising. The receptors (olfactory, I, and trigeminal, V) stimulated at threshold concentrations may not be identical to those stimulated at suprathreshold intensities. In addition, discrimination tasks may require more cortical processing than threshold tasks, and central losses may play a greater role.

A. Pharmacological Modes of Enhancement

Four chemicals have been found to enhance taste sensations: caffeine and other methyl xanthines, 5′-ribonucleotides including inosine-5′-monophosphate, inosine, and bretylium tosylate. Each of these compounds was found to potentiate certain tastes in signal detection experiments in which half of the tongue was adapted to the chemical for approximately 4 minutes and the other side was treated with a water control (see Schiffman et al., 1983, for testing method). Caffeine at 10 μM has been found to enhance sweeteners with bitter components, quinine HCl, NaCl, KCl, and certain amino acids (Schiffman et al., 1985, 1986a). Addition of 10 μM adenosine to the caffeine reverses the potentiation. These findings suggest that the enhancement properties of methyl xanthines may be due to their inhibition of the adenosine receptor.

5′-Ribonucleotides, including inosine-5′-monophosphate (IMP), are known to enhance the intensity of the taste of monosodium glutamate (Yamaguchi and Kimizuka, 1979). IMP also enhances sucrose and aspartame, neither of which is enhanced by caffeine. Inosine, a breakdown product of IMP and adenosine, also enhances sucrose and aspartame as well as NaCl. An IMP or inosine receptor may operate in a manner that is opposite to that described earlier for the gustatory adenosine receptor.

III. METHODS OF ENHANCING CHEMOSENSORY SIGNALS

Clinical studies suggest that reduced taste and smell acuity can play a significant role in the nutritional status of the elderly (Schiffman and Warwick, 1988). For this reason, it would be helpful to develop means for enhancing both taste and odor sensations to prevent or reverse the effects of malnutrition. Three approaches to chemosensory enhancement have been described: (a) pharmacological alterations of the transduction and modulation systems of chemosensory cells (Schiffman, 1987d), (b) flavor amplification by addition of more odorous molecules to food (see Schiffman and Warwick, 1988), and (c) behavioral changes in eating patterns (Schiffman, 1983).

CHANGES IN TASTE AND SMELL OVER THE LIFE SPAN

Bretylium tosylate, a quaternary ammonium compound, has been found to enhance taste responses to NaCl in humans and rats. It is possible that the mechanism of this enhancement is the opening of amiloride-sensitive sodium channels in the tongue (Schiffman et al., 1986b). Amiloride is a potassium-sparing diuretic that blocks the taste of NaCl by inhibiting sodium from entering the cell (Schiffman et al., 1983).

Although each of these drugs has enhancing properties when a small area of the tongue is tested using a signal detection experiment, it has been difficult to demonstrate enhancement for most of these substances (except IMP for glutamic acid) in whole-mouth stimulation. This is because there is a great deal of redundancy of receptors on the tongue. For example, a stimulus is often perceived to be just as strong when a small area of the tongue is stimulated. Persons with only half of the tongue functioning due to neural damage often make threshold and suprathreshold judgments that are statistically equivalent to those of persons with totally functional tongue. Thus, taste enhancers must be quite potent to achieve potentiation in whole-mouth experiments.

B. Flavor Amplification

Addition of flavors to foods for the elderly has been found to greatly increase the preference in older persons for the food to which they are added (Schiffman, 1979; Schiffman and Warwick, 1988). Simulated potato flavor can be added to mashed potatoes, and chicken can be enhanced with chicken aroma. These flavors

TABLE 13 Percentage Preference for Flavor-Enhanced Foods

Foods	Unenhanced	Enhancer 1	Enhancer 2[a]
Vegetables			
Carrots	25.0%	12.5% (carrot-1)	62.5% (carrot-2)
Green beans	16.7%	38.9% (bacon-1)	44.4% (bacon-2)
Green peas	6.2%	81.3% (pea)	12.5% (bacon)
Potatoes (mashed)	38.9%	16.7% (potato)	44.4% (bacon)
Meats			
Beef (ground)	5.6%	72.2% (bacon)	22.2% (beef)
Beef casserole (low sodium)	37.5%	18.7% (tomato)	43.8% (bacon)
Chicken	40.0%	20.0% (chicken)	40.0% (bacon)
Soups (low sodium)			
Chicken noodle	22.2%	44.4% (chicken)	33.3% (bacon)
Tomato	15.8%	36.8% (bacon)	47.4% (tomato)
Vegetable	16.7%	55.5% (bacon)	27.2% (tomato/bacon/pea)
Apple juice	44.4%	33.3% (apple-1)	22.2% (apple-2)

[a]The concentration of flavor is greater than that for foods in the Enhancer 1 column.
Source: Schiffman and Warwick (1988).

are mixtures of odorous molecules selected by chromatographic analysis of the natural products. Flavors may also contain compounds that stimulate the taste system as well.

Flavor amplification has been found to increase not only the hedonic value of foods to which it is added (Table 13), but also the intake of nutrient-dense food in older persons (Schiffman and Warwick, 1988). Flavor amplification is not always effective, however. For elderly persons who are totally anosmic, additional flavors will not be detected independent of the concentration.

C. Behavioral Changes in Eating

Persons with diminished taste and smell acuity often benefit from increased chewing of food and alternating among foods as they eat. Mastication breaks food apart and often releases more molecules that stimulate the chemical senses. Alternating among foods can reduce the effects of sensory adaptation. Addition of some textural properties to food can also improve overall acceptability of food for older persons.

IV. CONCLUSIONS

The senses of smell and taste decline with advancing age. These losses can result from anatomical and physiological changes associated with normal aging as well as from disease states and drug therapy. Chemosensory losses are a significant factor in the reduced food intake and subsequent malnutrition often seen in the elderly. Use of flavor amplification by addition of more chemosensory molecules to foods can increase food intake in the elderly. Pharmacological methods of taste and olfactory enhancement may play a future role in treating chemosensory losses.

ACKNOWLEDGMENTS

This research was supported in part by grant AG 00443 from the National Institute on Aging. The authors thank Mrs. Edna Bissette for the typing of the manuscript.

DISCUSSION

Plata-Salaman: In your experiments in the elderly where the thresholds increased, did you divide the population into heavy smokers and nonsmokers?

Schiffman: They're all nonsmokers. The issue of smoking is an interesting one because most of the data would not say that smoking decreases your sense of smell and taste very much. The data are quite equivocal. In that one longitudinal study we did, we found that smokers were slightly worst than nonsmokers, but it's not near as bad as you think.

Plata-Salaman: I have seen patients who say they can't taste, and when they give up smoking it's fine.

Schiffman: I'm sure there are many people who smoke who do have a diminished sense of smell, but statistically it's hard to document it in a large study. There are some studies that say "yes" and some that say "no."

Rolls: Can you tell us what's known about taste and smell in anorexia, bulimia, and obesity?

Schiffman: I love the question. I had honor students each term for about four terms in a row test anorectics at Duke. We found their sense of smell was slightly better than average, which surprised me because they were compromised nutritionally. But their taste thresholds for salt were quite a bit higher. It was the same with bulimics. I didn't see any difference in terms of smell—in fact, they were slightly better—but they couldn't stand putting something in their mouth. That was one of the problems. I don't know if the taste thresholds were higher because they didn't like putting taste in their mouth or whether it's actually higher. I think the salt thresholds are higher. I have never seen a difference in taste thresholds, dose–response curves, or any difference between a thin person and an obese person. What I do see is this tremendous preference for high-fat foods, high flavor, and a lot of texture. There's no difference in terms of perceived intensity. In terms of being able to identify smells in blended food experiments or when they're blindfolded, they [obese persons] are much better. Actually, we did one study and we're just doing an analysis of it. In terms of smell and taste together, obese people do better identifying blended foods while blindfolded, but in terms of smell alone, there wasn't a statistical difference.

REFERENCES

Anand, M. P. (1964). Accidents in the home. In *Current Achievements in Geriatrics*, W. F. Anderson and B. Isaacs (Eds.). Cassell, London, pp. 239–245.

Arey, L. E., Tremaine, M. J., and Monzingo, F. L. (1935). The numerical and topographical relations of taste buds to human circumvallate papillae throughout the life span. *Anat. Rec.* **64**:9–25.

Arvidson, K. (1979). Location and variation in number of taste buds in human fungiform papillae. *Scand. J. Dent. Res.* **87**:435–442.

Bradley, R. M., Stedman, H. M., and Mistretta, C. M. (1985). Age does not affect numbers of taste buds and papillae in adult rhesus monkeys. *Anat. Rec.* **212**:246–249.

Brizzee, K. R., Klara, P., and Johnson, J. E. (1975). Changes in microanatomy, neurocytology, and fine structure with aging. In *Advances in Behavioral Biology*, Vol. 16, J. M. Ordy and K. R. Brizzee (Eds.). Plenum Press, New York.

Byrd, E., and Gertman, S. (1959). Taste sensitivity in aging persons. *Geriatrics*, **14**:381–384.

Chalke, H. D., and Dewhurst, J. R. (1957). Accidental coal-gas poisoning: Loss of sense of smell as a possible contributory factor with old people. *Br. Med. J.* **2**:915–917.
Chalke, H. D., Dewhurst, J. R., and Ward, C. W. (1958). Loss of sense of smell in old people. *Public Health, London,* **72**:223–230.
Conger, A. D., and Wells, M. A. (1969). Radiation and aging effect on taste structure and function. *Radiat. Res.* **37**:31–49.
Cooper, R. M., Bilash, I., and Zubek, J. P. (1959). The effect of age on taste sensitivity. *J. Gerontol.* **14**:56–58.
Cowart, B. J. (1983). Direct scaling of the intensity of basic tastes: A life span study. Paper presented at the annual meeting of the Association of Chemoreception Sciences, Sarasota, FL.
Dodson, H. C., and Bannister, L. H. (1980). Structural aspects of ageing in the olfactory and vomeronasal epithelia in mice. In *Olfaction and Taste,* Vol. VII, H. van der Starre (Ed.). IRL Press, Oxford.
Doty, R. L. (1975). Intranasal trigeminal detection of chemical vapors by humans. *Physiol. Behav.* **14**:855–859.
Doty, R. L., Shaman, P., Applebaum, S. L., Giberson, R., Siksorski, L., and Rosenberg, L. (1984a). Smell identification ability: Changes with age. *Science,* **226**:1441–1443.
Doty, R. L., Shaman, P., and Dann, M. (1984b). Development of the University of Pennsylvania Smell Identification Test: A standardized microencapsulated test of olfactory function. *Physiol. Behav.* **32**:489–502.
Glanville, E. V., Kaplan, A. R., and Fischer, R. (1964). Age, sex, and taste sensitivity. *J. Gerontol.* **19**:474–478.
Greenamyre, J. T., Penney, J. B., Young, A. B., D'Amato, C. J., Hicks, S. P., and Shoulson, I. (1985). Alterations in L-glutamate binding in Alzheimer's and Huntington's diseases. *Science,* **227**:1496–1498.
Grzegorczyk, P. B., Jones, S. W., and Mistretta, C. M. (1979). Age-related differences in salt taste acuity. *J. Gerontol.* **34**:834–840.
Hermel, J., Schonwetter, S., and Samueloff, S. (1970). Taste sensation and age in man. *J. Oral Med.* **25**:39–42.
Hinds, J. W., and McNelly, N. A. (1981). Aging in the rat olfactory system: Correlation of changes in the olfactory epithelium and olfactory bulb. *J. Comp. Neurol.* **203**:441–453.
Hughes, G. (1969). Changes in taste sensitivity with advancing age. *Gerontol. Clin.* **11**:224–230.
Kalmus, H., and Trotter, W. R. (1962). Direct assessment of the effect of age on PTC sensitivity. *Ann. Hum. Genet.* **26**:145–149.
Kimbrell, G. M., and Furchtgott, E. (1963). The effect of aging on olfactory threshold. *J. Gerontol.* **18**:364–365.
Landfield, P. W., and Pitler, T. A. (1984). Prolonged Ca^{2+}-dependence after hyperpolarizations in hippocampal neurons in aged rats. *Science,* **226**:1089–1092.
Landfield, P. W., Waymire, J. C., and Lynch, G. (1978). Hippocampal aging and adrenocorticoids: Quantitative correlations. *Science,* **202**:1098–1102.
Liss, L., and Gomez, F. (1958). The nature of senile changes of the human olfactory bulb and tract. *Arch. Otolaryngol.* **67**:167–171.
Little, A. C., and Brinner, L. (1984). Taste responses to saltiness of experimentally prepared tomato juice samples. *J. Am. Diet. Assoc.* **21**:1022–1027.

Machado-Salas, J., Scheibel, M. E., and Scheibel, A. B. (1977). Morphologic changes in the hypothalamus of old mouse. *Exp. Neurol.* **57**:102–111.

Megighian, D. (1958). Variazioni della soglia olfattiva nell'ets senile. *Minerva Otorinolaringol.* **9**:331–337.

Mistretta, C. M., and Baum, B. J. (1984). Quantitative study of taste buds in fungiform and circumvallate papillae of young and aged rats. *J. Anat.* **138**:323–332.

Mochizuki, Y. (1937). An observation on the numerical and topographical relations of the taste buds to circumvallate papillae of Japanese. *Okajimas Folia Anat. Jpn.* **15**:595–608.

Mochizuki, Y. (1939). Studies on the papillae foliata of Japanese: II. The number of taste buds. *Okajimas Folia Anat. Jpn.* **18**:355–369.

Moore, L. M., Nielsen, C. R., and Mistretta, C. M. (1982). Sucrose taste thresholds: Age-related differences. *J. Gerontol.* **37**:64–69.

Moses, S. W., Rotem, Y., Jagoda, N., Talmor, N., Eichhorn, F., and Levin, S. (1967). A clinical, genetic and biochemical study of familial dysautonomia in Israel. *Isr. J. Med. Sci.* **3**:358–371.

Murphy, C. (1979). The effect of age on taste sensitivity. In *Special Senses in Aging: A Current Biological Assessment,* S. S. Han and D. H. Coons (Eds.). University of Michigan Institute of Gerontology, Ann Arbor, pp. 21–33.

Murphy, C. (1983). Age-related effects on the threshold, psychophysical function, and pleasantness of menthol. *J. Gerontol.* **38**:217–222.

Murphy, C. (1985). Cognitive and chemosensory influences on age-related changes in the ability to identify blended foods. *J. Gerontol.* **40**:47–52.

Naessen, R. (1971). An inquiry on the morphological characteristics and possible changes with age in the olfactory region of man. *Acta Otolaryngol.* **71**:49–62.

Perry, J. D., Frisch, S., Jafek, B., and Jafek, M. (1980). Olfactory detection thresholds using pyridine, thiophene, and phenethyl alcohol. *Otolaryngol. Head Neck Surg.* **88**:778–781.

Scheibel, M. E., and Scheibel, A. B. (1975). Structural changes in the aging brain. In *Aging,* Vol. 1, *Clinical, Morphological, and Neurochemical Aspects in the Aging Central Nervous System,* H. Brody, D. Harman, and J. M. Ordy (Eds.). Raven Press, New York, pp. 11–37.

Schemper, T., Voss, S., and Cain, W. S. (1981). Odor identification in young and elderly persons: Sensory and cognitive limitations. *J. Gerontol.* **36**:446–452.

Schiffman, S. (1977). Food recognition by the elderly. *J. Gerontol.* **32**:586–592.

Schiffman, S. (1979). Changes in taste and smell with age: Psychophysical aspects. In *Sensory Systems and Communications in the Elderly,* Vol. 10, *Aging,* J. M. Ordy and K. Brizzee (Eds.). Raven Press, New York.

Schiffman, S. S. (1983). Taste and smell in disease. *New Engl. J. Med.* **308**:1275–1279, 1337–1343.

Schiffman, S. S. (1987a). Smell. In *Encyclopedia of Aging,* G. L. Maddox (Ed.). Springer, New York, pp. 618–619.

Schiffman, S. S. (1987b). Taste. In *Encyclopedia of Aging,* G. L. Maddox (Ed.). Springer, New York, pp. 655–658.

Schiffman, S. S. (1987c). The role of taste and smell in nutrition: Effects of aging, disease state, and drugs. In *Food and Health: Issues and Directions,* M. L. Wahlqvist, R. W. F. King, J. J. McNeil, and R. Sewell (Eds.). Libbey, London, pp. 85–91.

Schiffman, S. S. (1987d). Recent developments in taste enhancement. *Food Technol.* **41**:72–73, 124.

Schiffman, S. S., and Clark, T. B. (1980). Magnitude estimates of amino acids for young and elderly subjects. *Neurobiol. Aging,* **1**:81–91.

Schiffman, S. S., and Leffingwell, J. C. (1981). Perception of odors of simple pyrazines by young and elderly subjects: A multidimensional analysis. *Pharmacol. Biochem. Behav.* **14**:787–798.

Schiffman, S. S., Moss, J., and Erickson, R. P. (1976). Thresholds of food odors in the elderly. *Exp. Aging Res.* **2**:389–398.

Schiffman, S., and Pasternak, M. (1979). Decreased discrimination of food odors in the elderly. *J. Gerontol.* **34**:73–79.

Schiffman, S. S., and Warwick, Z. S. (1988). Flavor enhancement of foods for the elderly can reverse anorexia. *Neurobiol. Aging,* **9**:24–26.

Schiffman, S. S., Hornack, K., and Reilly, D. (1979a). Increased taste thresholds of amino acids with age. *Am. J. Clin. Nutr.* **32**:1622–1627.

Schiffman, S., Orlandi, M., and Erickson, R. P. (1979b). Changes in taste and smell with age: Biological aspects. In *Sensory Systems and Communication in the Elderly,* Vol. 10, *Aging,* J. M. Ordy and K. Brizzee (Eds.). Raven Press, New York, pp. 247–268.

Schiffman, S. S., Lindley, M. G., Clark, T. B., and Makino, C. (1981). Molecular mechanism of sweet taste: Relationship of hydrogen bonding to taste sensitivity for both young and elderly. *Neurobiol. Aging,* **2**:173–185.

Schiffman, S. S., Lockhead, E., and Maes, F. W. (1983). Amiloride reduces the taste intensity of Na^+ and Li^+ salts and sweeteners. *Proc. Natl. Acad. Sci. USA,* **80**:6136–6140.

Schiffman, S. S., Gill, J. M., and Diaz, C. (1985). Methyl xanthines enhance taste: Evidence for modulation of taste by adenosine receptor.

Schiffman, S. S., Diaz, C., and Beeker, T. B. (1986a). Caffeine intensifies taste of certain sweeteners: Role of adenosine receptor. *Pharmacol. Biochem. Behav.* **24**:429–432.

Schiffman, S. S., Simon, S. A., Gill, J. M., and Beeker, T. G. (1986b). Bretylium tosylate enhances salt taste. *Physiol. Behav.* **36**:1129–1137.

Sklar, P. B., Anholt, R. R. H., and Snyder, S. H. (1986). The odorant-sensitive adenylate cyclase of olfactory receptor cells. *J. Biol. Chem.* **261**:15538–15543.

Smith, C. G. (1942). Age incidence of atrophy of olfactory nerves in man. *J. Comp. Neurol.* **77**:589–595.

Stevens, D. A., and Lawless, H. T. (1981). Age-related changes in flavor perception. *Appetite,* **2**:127–136.

Stevens, J. C., and Cain, W. S. (1985). Age-related deficiency in the perceived strength of six odorants. *Chem. Senses,* **10**:517–529.

Stevens, J. C., Plantinga, A., and Cain, W. S. (1982). Reduction of odor and nasal pungency associated with aging. *Neurobiol. Aging,* **3**:125–132.

Stevens, J. C., Bartoshuk, L. M., and Cain, W. S. (1984). Chemical senses and aging: Taste versus smell. *Chem. Senses,* **9**:167–179.

Tomlinson, B. E., and Henderson, G. (1976). Some quantitative cerebral findings in normal and demented old people. In *Neurobiology of Aging,* Vol. 3, R. D. Terry and S. Gershon (Eds.). Raven Press, New York, pp. 183–204.

Venstrom, D., and Amoore, J. E. (1968). Olfactory threshold in relation to age, sex, or smoking. *J. Food Sci.* **38**:264–265.

Weiffenbach, J. M. (1984). Taste and smell perception in aging. *Gerontology,* **3**:137–146.

Yamaguchi, S., and Kimizuka, A. (1979). Psychometric studies on the taste of monosodium glutamate. In *Glutamic Acid: Advances in Biochemistry and Physiology,* L. J. Filer, S. Garattini, M. R. Kare, A. R. Reynolds, and R. J. Wurtman (Eds.). Raven Press, New York, pp. 35–54.

18
Disturbances of Thirst and Fluid Balance in the Elderly

Barbara J. Rolls

*The Johns Hopkins University
Baltimore, Maryland*

Paddy A. Phillips

*University of Melbourne and Austin Hospital
Heidelberg, Victoria, Australia*

I. INTRODUCTION

Abnormalities of salt and water homeostasis are seen in all branches of clinical medicine and cause significant morbidity and mortality, especially in the elderly. However, although fluid and electrolyte disorders are common in the elderly, little is known about the changes in control mechanisms of salt and water balance with age in humans (Judge, 1978; R. J. Anderson et al., 1985; Snyder et al., 1987).

Maintenance of sodium and water homeostasis depends on a balance between intake (controlled by thirst and sodium appetite) and output (controlled by the kidney). Renal water excretion is largely dependent on the action of the antidiuretic hormone, vasopressin (AVP). It is secreted by the posterior pituitary in response to (a) increased extracellular fluid tonicity acting on hypothalamic "osmoreceptor" neurons, (b) reduced blood volume and blood pressure acting via pressure/volume receptors in the cardiac atria, aortic arch, and carotid sinus, and (c) possibly by angiotensin II, either bloodborne or from the brain renin–angiotensin system. Thirst is thought to be controlled by similar mechanisms, although there are few studies of human thirst (see Rolls et al., 1980; Rolls and

Rolls, 1982; Phillips et al., 1984a,b, 1985a,b). Sodium excretion is controlled by the renin–angiotensin–aldosterone system, and possibly by atrial natriuretic peptide (ANP). ANP is secreted in response to volume expansion and has prompt and potent diuretic and natriuretic actions (Laragh, 1985; Hodsman et al., 1986; J. V. Anderson et al., 1987).

Changes in several of these control mechanisms might predispose the elderly to disturbances of sodium and water balance.

II. THIRST IN THE ELDERLY

Thirst and water intake are particularly important in the maintenance of fluid and electrolyte balance, since it is through these alone that water deficits can be replenished. Renal water conservation can only minimize further losses.

Since dehydration is a common cause of fluid and electrolyte disturbance in the elderly (Judge, 1978), and it is a common clinical observation that the elderly do not seem to get thirsty despite obvious physiological need, we investigated the effects of dehydration on thirst and plasma and urine variables in both healthy young and healthy old men (Phillips et al., 1984a). To produce dehydration, we deprived them of water for 24 hours and asked them to consume a dry diet. Such dehydration produces both osmotic and hypovolemic stimuli to thirst, as well as vasopressin secretion, and is a noninvasive and natural experimental manipulation.

The subjects were seven healthy, active, community-dwelling elderly men (67–75 years old) and seven healthy young men (20–31 years old) who were matched for the degree of weight loss during the 24 hours of fluid deprivation (1.8–1.9% of predeprivation body weight). The weight loss provided an immediate indication of whether the subjects had complied with the fluid deprivation.

Subjects came to the experimental room the morning the deprivation was to begin. They had a blood sample drawn, and rated their thirst, how pleasant it would be to drink water, mouth dryness, and the taste in the mouth on 100 mm visual analogue scales. Subjects were instructed to go without all fluids and to eat only dry foods (i.e., < 75% water by weight) until the same time the next day. When the subjects returned to the hospital the following morning, blood was again sampled and thirst rated. At this time the elderly group showed large and significant increases in plasma sodium concentration (from 140.2 ± 0.4 to 143.2 ± 0.5 mmol/L) and osmolality (from 288.4 ± 1.3 to 296.3 ± 1.2 mOsm/kg water), whereas the young group showed smaller, nonsignificant increases (from 141.5 ± 0.4 to 142.5 ± 0.4 mmol/L and from 287.7 ± 1.8 to 290.4 ± 1.1 mOsm/kg water, respectively). Thus despite identical weight loss and similar changes in indexes of plasma volume, the elderly subjects showed larger increases in plasma osmolality and sodium concentration (Fig. 1) after fluid deprivation.

During rehydration with room temperature tap water, the young subjects drank enough to dilute their body fluids to at least predeprivation levels; the elderly

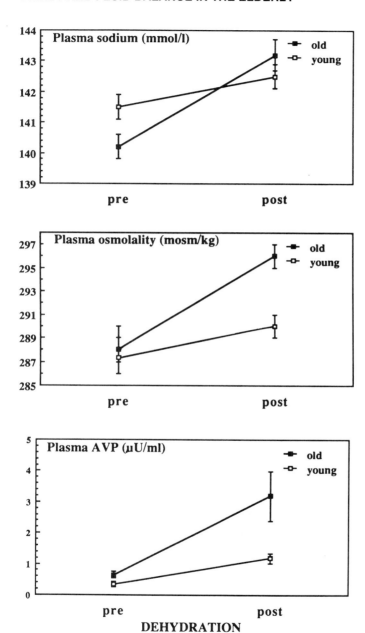

FIGURE 1 Mean (± SEM) changes from before to after the fluid deprivation in plasma sodium, osmolality, and AVP in healthy old and young subjects. (From Phillips et al., 1984a.)

subjects did not, even though water was freely available. The volumes drunk were 3.4 ml per kilogram of body weight in the elderly group and 8.5 ml/kg in the young group (Fig. 2). Whether the elderly group would have consumed enough to dilute body fluids back to predeprivation levels given more time or a more palatable fluid is unknown.

The important finding was that despite their obvious physiological need, the elderly subjects were not markedly thirsty. One elderly subject denied any thirst and drank nothing. This is extraordinary, since 24 hours without fluids evoked a very strong thirst response in young subjects. This was evident in the significantly elevated ratings of thirst, pleasantness of drinking water, mouth dryness, and unpleasantness of the taste in the mouth in the young group following deprivation. In marked contrast to this, the elderly group showed no significant increase in thirst, mouth dryness, or unpleasantness of the taste in the mouth after the deprivation (Fig. 3). Only the rating of the pleasantness of drinking water was elevated postdeprivation (Fig. 3). This relative lack of change in thirst ratings and the low water intake indicate a deficit in thirst and water intake in the elderly group.

This initial finding of reduced thirst in the elderly has been confirmed by us and by others. For example, in another recent study in which we were investigating excretion of a water load in healthy elderly men ($n = 6$, aged 63–80) compared to two groups of young men ($n = 6$ per group, aged 20–32), we confirmed that the elderly do not experience normal thirst (Crowe et al., 1987). Following ingestion

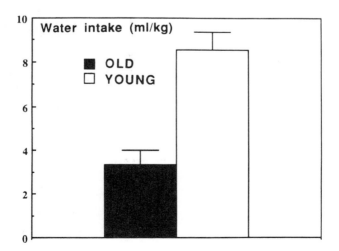

FIGURE 2 Mean ($\pm$ SEM) total water intake (ml/kg) over 2 hours by healthy old ($N = 7$) and young ($N = 7$) subjects following fluid deprivation. (From Phillips et al., 1984a.)

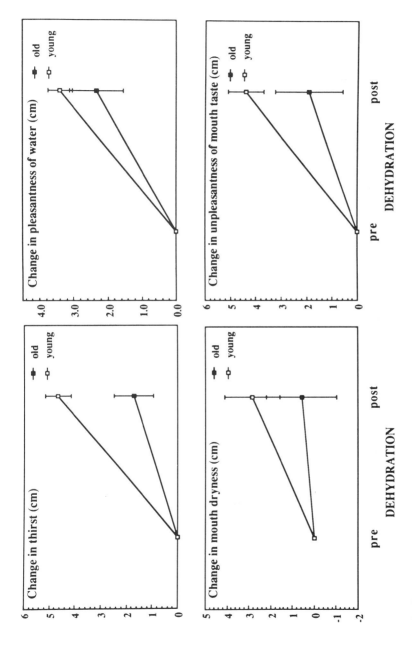

FIGURE 3 Mean (± SEM) changes in four thirst-related variables in healthy old and young subjects. (From Phillips et al., 1984a.)

of the water load, the subjects were not allowed further access to fluids for 7 hours. During this time there was no alteration in thirst ratings in the elderly group (Fig. 4), despite increasing plasma sodium and osmolality indicating that an osmotic thirst stimulus was present. This was in contrast to the younger groups, in which thirst increased significantly after 5 hours without further water.

Thirst in response to heat stress and thermal dehydration also is reduced in the elderly. Miescher and Fortney (Fortney et al., 1989; Miescher and Fortney, 1989) have determined the effects of age on thermal and plasma responses, and subjective ratings of thirst and hotness to dehydration and rehydration in a hot dry environment. Five healthy, active older (62–67 years old) and six younger men (24–29 years old) rested in a hot and dry chamber (45°C, 25% relative humidity) for 240 minutes. For the first 180 minutes no water was allowed (period of dehydration), while for the final 60 minutes cool water (11°C) was offered. Upon leaving the chamber, the men rested an additional 30 minutes in a 25°C room before the final measurements were made.

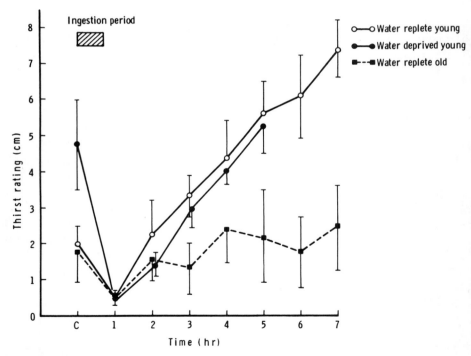

FIGURE 4 Mean (± SEM) thirst ratings in three groups at baseline (B) and 1, 2, 3, 4, and 5 hours following ingestion of a water load (20 ml/kg). Ratings in water-replete young and old groups were also taken at 6 and 7 hours. (From Crowe et al., 1987.)

Although sweating, which is the only effective method of heat loss under these conditions, was similar in both groups, the older men thermoregulated less well than the younger men. Their rectal temperature increased more rapidly and to a higher level than in the younger subjects when water was not allowed. In both groups, drinking water prevented further increases in rectal temperature.

After 180 minutes in the heat chamber, both groups were similarly dehydrated in terms of body weight loss, but the elderly group showed larger changes in plasma variables. Changes in plasma osmolality (+5.0 mOsm/kg) and plasma volume (−11.3%) were greater in the elderly than in the younger group (+1.1 mOsm/kg and −4.9%, respectively). Thus one would expect greater thirst in the elderly. However, thirst ratings of the older men indicated that they were *less* thirsty than their younger counterparts. In view of the greater increase in body temperature and greater magnitude of hemo- and osmoconcentration, it is surprising that the older men did not experience a greater thirst than the younger men. These results support the concept of an age-associated deficit in thirst that may predispose to disturbances of fluid balance.

It is important to realize that given time, water homeostasis is eventually achieved in the elderly, probably through a combination of intake associated with palatable liquids and foods, as well as through renal factors. The important finding from these studies is that even in *healthy* old individuals, thirst is diminished. Should this be accompanied by illness or physical incapacity, which increase water loss or prevent access to water, dangerous dehydration could follow.

A. Mechanisms of the Thirst Deficit

The mechanism of the thirst deficit is unclear. Occult central nervous system disease is a possible explanation. P. D. Miller et al. (1982) described six elderly patients (68–91 years old) with prior strokes who had repeated hospitalizations with dehydration and hypernatremia and who had deficient thirst. There was no evidence for hypothalamic–pituitary dysfunction in these patients, and all were physically able to obtain water if they wanted it. Miller et al. suggested that the thirst deficit might have been due to cerebral cortical dysfunction. In contrast, there was no clinical evidence for a neurological disorder or any other disease in the subjects in our studies.

Thirst and vasopressin secretion depend on the normal functioning of receptors that detect dehydration of the cells (osmoreceptors) and decreases in plasma volume (baroreceptors and volume receptors). Osmoreceptor sensitivity in the elderly has been tested in relation to vasopressin release and has been found to be increased (Helderman et al., 1978). This does not, however, preclude an impairment in thirst osmoreceptor sensitivity, since thirst and vasopressin osmoreceptors may be anatomically and functionally separate in man (Baylis et al., 1981; Hammond et al., 1986). Evidence to support functional and/or anatomical differences in the neuronal pathways involving the thirst and vasopressin osmoreceptors and/or their efferent pathways comes from patients with selective loss of vasopressin responses to osmotic (but not hypovolemic) stimulation, who may become thirsty during hypertonic saline infusions (Baylis et al., 1981), and from a patient with the "reverse" syndrome—that is, selective loss of thirst with normal vasopressin responses following osmotic stimulation (Hammond et al., 1986). The elderly have been found to drink less following intravenous infusions of hypertonic saline (Fish et al., 1985), suggesting that the sensitivity of the thirst osmoreceptors declines with age.

Baroreceptor sensitivity declines with age (Gribbin et al., 1971; Rowe et al., 1982), and this affects the release of vasopressin; but it is not clear whether hypovolemic stimuli such as hemorrhage, salt depletion, vomiting, and diarrhea evoke normal thirst in the elderly.

Oropharyngeal factors such as a dry mouth are also associated with thirst (Rolls and Rolls, 1982), and it may be that the elderly are insensitive to these changes, although this has not been tested. In one of our studies (Phillips et al., 1984a), the elderly group had nonsignificant and much less well-defined changes in mouth dryness in comparison to the young group. It also may be that age-related changes in other oropharyngeal receptors such as those involved in taste (Cowart, 1981) might decrease the pleasure associated with fluid intake.

There is some preliminary evidence that alterations in the endogenous opioid system could account for the hypodipsia of the elderly (Silver and Morley, 1988). After overnight deprivation, older subjects drank less (433 ± 291 ml) than younger subjects (578 ± 310 ml) over 2 hours of access to water and were significantly less thirsty than the younger subjects, again confirming the thirst defect in the elderly. When the opioid antagonist naloxone (100 µg/kg) was administered postdeprivation, the younger group decreased fluid intake by 51%. However, in the older group the same dose of naloxone decreased intake by only 10%. This suggests that the hypodipsia of the elderly may be due to a deficit in the opioid system. More information about the importance of the opioid system in the control of drinking is needed before the significance of this finding can be appreciated.

III. RENAL FUNCTION IN THE ELDERLY

Changes in renal function with age are well documented and have been reviewed elsewhere (Epstein, 1979; Ledingham et al., 1987). Glomerular filtration rate declines (Epstein, 1979), and investigators have found reduced urinary concentrating ability (Phillips et al., 1984a), less efficient sodium conserving capacity (Epstein and Hollenberg, 1976), and reduced ability to excrete a water load (Lindeman et al., 1966; Crowe et al., 1987) in the elderly compared to the young subjects. These changes reduce the ability of the elderly to conserve salt and water in the face of increased losses, predisposing them to hypovolemia and dehydration. The inability to excrete excess water promptly also means that the elderly are at risk of water overload and hyponatremia when excess water is given exogenously such as in intravenous fluid therapy (R. J. Anderson et al., 1985).

IV. NEUROENDOCRINE FUNCTION AND AGE

In addition to alterations in thirst and renal function in the elderly, changes in hormone systems involved in sodium and water homeostasis occur with age (Phillips et al., in press). It is well established that the activity of the renin–angiotensin–aldosterone system declines with age. Angiotensin II is thought to play a role in thirst following depletions of the extracellular fluid compartment. Angiotensin II is a potent dipsogen in a wide variety of species when infused intravenously or when injected into the brain (see Rolls and Rolls, 1982). However, it is not clear that angiotensin II is important in the control of normal thirst and fluid intake in humans. We (Phillips et al., 1985b) found that very high doses of intravenous angiotensin were required to stimulate drinking in healthy young men. Furthermore, 24-hour fluid deprivation did not significantly increase plasma angiotensin II levels in either young or old men (Phillips et al., 1984a). These healthy young and elderly men had similar levels of angiotensin, so it seemed unlikely that angiotensin was a factor in the thirst deficit seen in the older men. However, angiotensin may play a role in the thirst associated with some pathological conditions (see Rolls and Rolls, 1982; Yamamoto et al., 1986; Oldenburg et al., 1988). A recent double-blind placebo-controlled study of the angiotensin-converting enzyme enalapril (which blocks production of angiotensin II) in patients with dialysis-dependent chronic renal failure showed that blockade of the renin–angiotensin system was associated with lower thirst and less fluid intake (Oldenburg et al., 1988). Yamamoto et al. (1988) observed a reduction in endogenous angiotensin II production in nine elderly patients with adipsia, progressive dehydration, impaired consciousness, and hypernatremia. They suggest the following sequence of events in these patients: impaired angiotensin II production caused impairment of thirst perception, renal-concentrating capacity, and AVP secretion, and contributed to the development of hypernatremic dehydration. More studies are needed on the role of angiotensin II in thirst in humans before it can be concluded that it is important for the hypodipsia of the elderly.

Plasma atrial natriuretic peptide (ANP) levels are increased in the elderly (Ohashi et al., 1987a) at least in part because of reduced ANP clearance (Ohashi et al., 1987b). In addition, the elderly demonstrate much greater absolute increases in plasma ANP with volume loading compared with young controls (Ohashi et al., 1987a). Interestingly, there is evidence that ANP receptors and/or ANP–receptor coupling may be diminished in the elderly (Ohashi et al., 1987a). This as well as the decline in glomerular filtration rate and activity of the renin–angiotensin–aldosterone system may be factors in limiting sodium excretion in the elderly. The higher levels of ANP in the elderly may also account for the reduced thirst and lower plasma renin and aldosterone found with advancing age, as ANP is a potent inhibitor of renin and aldosterone (Richards et al., 1988) and thirst (Antunes-Rodrigues, 1985).

In the elderly, diminished renal concentrating ability occurs despite increased AVP levels (see Fig. 1) and AVP osmoreceptor sensitivity (Helderman et al., 1978; Phillips et al., 1984a). Nonosmotic hypovolemic stimulation of AVP secretion, however, is reduced (Rowe et al., 1982). The reduced capacity to excrete a water load is not due to the inability to suppress AVP secretion but rather to the reduction, with age, of the glomerular filtration rate (Lindeman et al., 1966; Crowe et al., 1987).

V. DEHYDRATION IN THE ELDERLY

The elderly are very susceptible to dehydration. Dehydration is the most common cause of renal impairment and failure in elderly patients (Kafetz, 1983), and may be the most common cause of fluid and electrolyte disturbances (Judge, 1978). Frequently, the elderly become dehydrated when exposed to even mild stresses such as water restriction, fever, infection, or diarrhea (Leaf, 1984). Since fluid intake is the only way to replenish water deficits, and the kidneys are paramount in reducing further water losses, diminished thirst and reduced renal concentrating ability are important in predisposing the elderly to dehydration.

What advice should be given to the elderly and to their caretakers to avoid dehydration? Often in hospitals and nursing homes the elderly are encouraged to take large volumes of oral fluids or they are given intravenous fluids. In addition to abundant fluids, elderly individuals are also often given dietary supplements low in sodium (M. Miller, 1987). Excess fluids and a low sodium diet can be potentially dangerous, since not only do the elderly have a tendency to become dehydrated, but they are also susceptible to overhydration, since they have a decreased ability to excrete water and to retain sodium (Kleinfield et al., 1979; Sunderam and Mankikar, 1983; M. Miller et al., 1985).

Thus it is important to be aware of the fluid intake and output of the elderly. Dehydration and hyponatremia can both lead to confusional states, which interfere with fluid intake. Several regimens for maintaining balance in such circumstances have been described (M. Miller, 1987; Reedy, 1988). In dehydration, if a lack of thirst is the problem as it was in patients following strokes (P. D. Miller et al., 1982), it may be necessary to prescribe an obligatory fluid intake to maintain fluid balance. In a study of nonambulatory geriatric patients (Spangler et al., 1984), it was found that a routine of regular prompts and assistance with fluids reduced dehydration as assessed by measures of the specific gravity of the urine. Prior to this intervention, 25% of the residents were mildly dehydrated, and this was reduced to zero.

The extent of dehydration in elderly individuals living in the community has not been established. It is likely that when the elderly have access to a variety of palatable fluids and are feeling well, they have no trouble with fluid balance. But if they are feeling unwell, or if access to fluids is in any way restricted, they may develop undetected dehydration as a result of the diminished perception of thirst.

VI. CONCLUSIONS

Perturbations of fluid and electrolyte balance are a major clinical problem in the elderly. Not only do the kidneys function less efficiently in the elderly, but thirst is also blunted. In dehydration, the lack of thirst will hinder restoration of fluid balance, which must be achieved by taking in more water. Although the lack of thirst has been demonstrated clearly, additional studies are needed to define the mechanisms underlying the disturbance, as well as its etiology and prevalence. Such studies are vital for the prevention of disturbances of fluid balance in the elderly, as well as for the development of new therapeutic interventions.

ACKNOWLEDGMENTS

We gratefully acknowledge the advice and/or collaboration of M. J. Crowe, M. L. Forsling, S. M. Fortney, J. G. G. Ledingham, E. Miescher, J. J. Morton, and L. Wollner.

DISCUSSION

Stricker: Coming just after Sue's presentation, in which there were sensory deficits of a different sort, it makes me think that the problem of detecting thirst by the elderly may not be a specific problem at all but a general sensory problem. There may not be a collection of individual problems, but a more global activational problem. Certainly, as you know, with animal studies you can produce brain lesions that will decrease behavioral arousal, and therefore the animals show up with sensory deficits in all modalities. I wonder what you think of that possibility here—that what you're looking at is a general activational issue rather than something that is specific to thirst.

Rolls: I think that's quite possible; but I also do think it's very possible that some of the specific thirst osmoreceptors may be altered in their sensitivity. The evidence for vasopressin is quite clear; that is, that these receptors are modified. To look at the question of how general it is: there haven't been any equivalent studies done, for example, asking if the elderly get hungry after 24 hours of food deprivation. Quite a simple question, Do they experience hunger? We don't know. There could be global sensory impairments. They don't enjoy drinking as much, that kind of thing. I think all of this is open. It's an area that needs a lot of work.

Schiffman: How do you know if people are dehydrated in a nursing home? How do you determine that 25% of the people are dehydrated?

Rolls: Well, you could look at plasma, which would be the best thing to do. You could also look at urine output and concentration, which would give you some indication, but it's not as precise.

Schiffman: Precise as what?

Rolls: Plasma variables. If you look at plasma you'll have normal levels. In an individual in a nursing home, it would be good to have a baseline, because there is some variance between individuals. Know where they normally fall and look for variations. Blood is being taken in hospitals and nursing homes all the time, so this would be something that could be done quite easily.

Schiffman: Electrolyte balances are frequently affected by various medications, and that's why I would ask the question. If you look at blood, would you know if this is thirst, or would you know if this was a problem related to medication?

Rolls: You would know if they were dehydrated. You wouldn't know whether the dehydration was due to the fact that they might be on a diuretic or something else that could affect their fluid output. But certainly, if you see any signs of dehydration, then you know that you have to encourage them to drink. And one problem with both dehydration and overhydration is that people get very confused. That's one of the first signs. So to say to people they should be aware of this, that, and the other—they're not going to be, once it happens. There are some case reports in the literature about people who had undetected dehydration—elderly people—and the sorts of things that they got put through, including major surgery, looking for the root of this, when all they needed to do was to drink more fluid. Hospitals can be one of the worst places for them. They're put on all sorts of restricted regimes for the various tests, they are given high salt diets. It's a disastrous scenario. But I think doctors are very aware of this. This finding now is everywhere. Everybody knows this.

Morck: I just want to go back to one of those points about the plasma measurements. I noticed that one difference in your baseline measurements between the young and the elderly was a lower protein level in the elderly. Do you think that actually colloid osmotic pressure may be a separate factor from just simply osmotic pressure?

Rolls: The protein difference wasn't significant, so I don't know about that. In terms of baseline measures—whether the elderly are starting off at a different level on any of these variables—in one of our studies, the elderly looked a bit more dehydrated than the young, and in the other one they didn't. They started off at the same level across the board. And that's why I'm saying we need more normative data—people out in the community—to see where they are falling. I suspect in the community when the elderly are healthy, they've got plenty of tea to drink, lots of other nice soft drinks. When they've got lots of good palatable fluids, I don't think they're so likely to get into trouble. It's when they get sick, they get a bit feverish, they're on various drugs, that these problems can arise very rapidly and then be misdiagnosed.

Blundell: This has to do with the demand you created in your two groups by withholding water for 24 hours, which is a good way to match. On the other hand, if the old and young are naturally drinking different amounts, they're imbalanced because you're depriving the young, say, of their three glasses a day and the elderly their two glasses a day. So that since the old don't show so much drinking, they don't actually have the same degree of dehydration.

Rolls: You can't assume that normally they're not showing so much drinking. What I'm saying is that in a natural community-dwelling situation, I suspect you wouldn't see any difference. In fact, the elderly have got more time to sit around drinking tea. They may drink more. You can only try to match in terms of weight changes and plasma variables.

Blundell: I think there is another way to do it, but if the elderly are drinking as much as the young anyway, what is the problem? I thought it was the case they weren't drinking so much.

Rolls: This is in a rehydration situation. When they are deprived of water and you give them an opportunity to rehydrate with just water—no foods available—they are not responding appropriately because they are not feeling thirsty. But thirst in our everyday lives is probably not why we're drinking. We've done some studies where we followed young people around all day as they go about their business to see why they actually drink. Do they drink in response to fluid variable changes or what? We find they normally anticipate fluid deficits. The people who are wandering around drinking freely get a little bit of a dry mouth sensation, which precedes drinking. That's why they subjectively say that they're drinking, but they certainly are not drinking in response to changes in plasma variables.

Blundell: I thought it was the case, but I guess I misunderstood. Are there or are there not data in the natural circumstances showing that the old drink less than the young?

Rolls: I don't know of such data. That's why I'm saying we need more data on community-dwelling individuals.

Blundell: The point is still valid. If they do drink less, then depriving the old and the young for the same length of time is not going to balance the two motivational demands.

Rolls: The point is, we were dehydrating them to an equal weight loss; they were weight-loss matched. That's the standard technique that's used in thirst and body fluid physiology. You deprive to equal weight loss. These groups lost 1.8–1.9 kilograms overnight. It's an objective measure. That's what we were aiming for. Now there are some problems with that in terms of body composition and what not, but we couldn't get into all that. You could go into what's their normal fluid intake and do such-and-such a percentage of that, but we didn't do it that way. We just deprived both groups of subjects of water for 24 hours, and we got very big differences both in intake and subjective reports, and in their ability to get back into fluid balance.

Bieber: This isn't really a question, it's just a comment. There's so much water in food that if they're eating healthy diets with a lot of food, their actual need for water is probably diminished. We're eating about 80% water in our foods.

Rolls: The point I want to leave you with is that thirst is probably just a problem when the elderly get into trouble, when they get sick. When they're normally living in the community, they're probably doing fine.

REFERENCES

Anderson, J. V., Donckier, J., Payne, N. N., Beecham, J., Slater, J. D. H., and Bloom, S. R. (1987). Atrial natriuretic peptide: Evidence of action as a natriuretic hormone at physiological concentrations. *Clin. Sci.* **72**:305–312.

Anderson, R. J., Chung, H. M., Kluge, R., and Schrier, R. (1985). Hyponatremia: A prospective study of its epidemiology and the pathogenetic role of vasopressin. *Ann. Intern. Med.* **102**:164–168.

Antunes-Rodrigues, J., McCann, S. M., Rogers, L. C., and Samson, W. K. (1985). Atrial natriuretic factor inhibits dehydration- and angiotensin-II-induced water intake in the conscious, unrestrained rat. *Proc. Natl. Acad. Sci. USA,* **82**:8720–8723.

Baylis, P. H., Gaskill, M. B., and Robertson, G. L. (1981). Vasopressin secretion in primary polydipsia and cranial diabetes insipidus. *Q. J. Med.* **50**:345–358.

Cowart, B. J. (1981). Development of taste perception in humans: Sensitivity and preference throughout the life span. *Psychol. Bull.* **90**:43–73.

Crowe, M. J., Forsling, M. L., Rolls, B. J., Phillips, P. A., Ledingham, J. G. G., and Smith, R. F. (1987). Altered water excretion in healthy elderly men. *Age Ageing,* **16**:285–293.

Epstein, M. (1979). Effects of aging on the kidney. *Fed. Proc.* **38**:168–172.

Epstein, M., and Hollenberg, N. K. (1976). Age as a determinant of renal sodium conservation. *J. Lab. Clin. Med.* **87**:411–417.

Fish, L. C., Minaker, K. L., and Rowe, J. W. (1985). Altered thirst threshold during hypertonic stress in aging man. *Gerontologist,* **25**:118a.

Fortney, S., Miescher, E., and Rolls, B. (1989). Body hydration and aging. *Progess in Biometeorol.* **7**:105–115.

Gribbin, B., Pickering, T. G., Sleight, P., and Peto, R. (1971). Effect of age and high blood pressure on baroreflex sensitivity in man. *Circ. Res.* **29**:424–431.

Hammond, D. N., Moll, G. W., Robertson, G. L., and Chelmicka-Schorr, E. (1986). Hypodipsic hypernatremia with normal osmoregulation of vasopressin. *New Engl. J. Med.* **315**:433–436.

Helderman, J. H., Vestal, R. E., Rowe, J. W., Tobin, J. D., Andres, R., and Robertson, G. L. (1978). The response of arginine vasopressin to intravenous ethanol and hypertonic saline in man: The impact of aging. *J. Gerontol.* **33**:39–47.

Hodsman, G. P., Phillips, P. A., Ogawa, K., and Johnston, C. I. (1986). Atrial natriuretic factor in normal man: Effects of posture, exercise and haemorrhage. *J. Hypertension,* **4**:S503–S505.

Judge, T. G. (1978). The *milieu interieur* and aging. In *Textbook of Geriatric Medicine and Gerontology,* J. C. Brocklehurst (Ed.). Churchill & Livingstone, Edinburgh.

Kafetz, K. (1983). Renal impairment in the elderly: A review. *J. R. Soc. Med.* **76**:398–401.

Kleinfield, M., Casimir, M., and Borra, S. (1979). Hyponatremia as observed in a chronic disease facility. *J. Am. Geriatr. Soc.* **27**:156–161.

Laragh, J. H. (1985). Atrial natriuretic hormone, the renin–aldosterone axis, and blood pressure–electrolyte homeostasis. *New Engl. J. Med.* **313**:1339–1340.

Leaf, A. (1984). Dehydration in the elderly. *New Engl. J. Med.* **311**:791–792.

Ledingham, J. G. G., Crowe, M. J., Forsling, M. L., Phillips, P. A., and Rolls, B. J. (1987). Effects of aging on vasopressin secretion, water excretion and thirst in man. *Kidney Int.* **32**:S90–S92.

Lindeman, R. D., Lee, D. T., Yiengst, M. J., and Shock, N. W. (1966). Influence of age, renal disease, hypertension, diuretics and calcium on antidiuretic responses to suboptimal infusions of vasopressin. *J. Lab. Clin. Med.* **68**:202–223.

Miescher, E., and Fortney, S. (1989). Responses to dehydration and rehydration during heat exposure in young and older men. *Am. J. Physiol.* **257**:R1050–R1056.

Miller, M. (1987). Fluid and electrolyte balance in the elderly. *Geriatrics,* **42**:65–76.

Miller, M., Morley, J. E., Rubenstein, L. Z., Ouslander, J., and Strome, S. (1985). Hyponatremia in a nursing home population. *Gerontologist,* **25**:118.

Miller, P. D., Krebs, R. A., Neal, B. J., and McIntyre, D. O. (1982). Hypodipsia in geriatric patients. *Am. J. Med.* **73**:354–356.

Ohashi, M., Fujio, N., Nawata, H., Kato, K.-I., Ibayashi, H., Kangawa, K., and Matsuo, H. (1987a). High plasma concentrations of human atrial natriuretic polypeptide in aged men. *J. Clin. Endocrinol. Metab.* **64**:81–85.

Ohashi, M., Fujio, N., Nawata, H., Kato, K.-I., Matsuo, H., and Ibayashi, H. (1987b). Pharmacokinetics of synthetic -human atrial natriuretic polypeptide in normal men; effect of aging. *Regul. Peptides,* **19**:265–272.

Oldenburg, B., MacDonald, G. J., and Shelley, S. (1988). Controlled trial of enalapril in patients with chronic fluid overload undergoing dialysis. *Br. Med. J.* **296**:1089–1091.

Phillips, P. A., Rolls, B. J., Ledingham, J. G. G., Forsling, M. L., Morton, J. J., Crowe, M. J., and Wollner, L. (1984a). Reduced thirst following water deprivation in healthy elderly men. *New Engl. J. Med.* **311**:753–759.

Phillips, P. A., Rolls, B. J., Ledingham, J. G. G., and Morton, J. J. (1984b). Body fluid changes, thirst and drinking in man during free access to water. *Physiol. Behav.* **33**:357–363.

Phillips, P. A., Rolls, B. J., Ledingham, J. G. G., Forsling, M. L., and Morton, J. J. (1985a). Osmotic thirst and vasopressin release in man: A double-blind cross-over study. *Am. J. Physiol.* **248**:R645–R650.

Phillips, P. A., Rolls, B. J., Ledingham, J. G. G., Morton, J. J., and Forsling, M. L. (1985b). Angiotensin-II-induced thirst and vasopressin release in man. *Clin. Sci.* **68**:669–674.

Phillips, P. A., Hodsman, G. P., and Johnston, C. I. (in press). Neuroendocrine mechanisms and cardiovascular homeostasis in the elderly. *Cardiovasc. Drugs Ther.*

Reedy, D. F. (1988). How can you prevent dehydration? *Geriatr. Nurs.* **9**:221–224.

Richards, A. M., Tonolo, G., Montorsi, P., Finlayson, J., Fraser, R., Inglis, G., Towrie, A., and Morton, J. J. (1987). Low dose infusions of 26- and 28-amino acid human atrial natriuretic peptides in normal man. *J. Clin. Endocrinol. Metab.* **66**:465–472.

Rolls, B. J., and Rolls, E. T. (1982). *Thirst.* Cambridge University Press, Cambridge.

Rolls, B. J., Wood, R. J., Rolls, E. T., Lind, H., Lind, W., and Ledingham, J. G. G. (1980). Thirst following water deprivation in humans. *Am. J. Physiol.* **239**:R476–R482.

Rowe, J. W., Minaker, K. L., Sparrow, D., and Robertson, G. L. (1982). Age-related failure of volume-pressure-mediated vasopressin release. *J. Clin. Endocrinol. Metab.* **54**:661–664.

Silver, A. J., and Morley, J. E. (1988). Role of the opioid system in hypodipsia of aging. *Clin. Res.* (abstract).

Snyder, N. A., Feigal, D. W., and Arieff, E. G. (1987). Hypernatremia in elderly patients. *Ann. Intern. Med.* **107**:309–319.

Spangler, P. F., Risley, T. R., and Bilyew, D. D. (1984). The management of dehydration and incontinence in nonambulatory geriatric patients. *J. Appl. Behav. Anal.* **17**:397–401.

Sunderam, S. G., and Mankikar, G. D. (1983). Hyponatremia in the elderly. *Age Ageing,* **12**:77–80.

Yamamoto, T., Shimizu, M., and Morioka, M. (1986). Role of angiotensin II in the pathogenesis of hyperdipsia in chronic renal failure. *JAMA,* **256**:604–608.

Yamamoto, T., Harada, H., Fukuyama, J., Hayashi, T., and Mori, I. (1988). Impaired arginine–vasopressin secretion associated with hypoangiotensinemia in hypernatremic dehydrated elderly patients. *JAMA,* **259**:1039–1042.

Part IV Discussion

Hirsch: For Dr. Rodin: In looking at pregnant women, have you noticed anything in regards to colors? We'd been taught in OB/GYN residency that, for instance pregnant women had a very aversive response to green.

Rodin: No.

Morck: I have one question for you, Judy, regarding the pica measurements. It's been reported that pica has been associated with iron-deficiency states. In your correlations, have you seen any relationships between iron parameters, such as hemoglobin, with your incidence of pica?

Rodin: We had no incidence of pica in our sample. We intended to look at that. We indeed have measures of iron and iron deficiency, but in our Northeastern population, although we had both blacks and whites, we had not a single case.

Morck: I guess maybe the description of pica as just eating clay or laundry starch. . . .

Rodin: No, not even ice chewing. We looked for all forms of pica, and we had no case.

Schneider: For Dr. Birch: I'm interested if parental control, in terms of how food was handled by infants, was taken into account; whether it was imposed that utensils be used when foods were being acquired.

Birch: Certainly, in the work we do in the lab, kids are often eating foods that they can eat as finger foods. We always try to give them things in our intake tests that are easily eaten by them, even with 2-, 3-, and 4-year-olds. We try not to have that impede their intake. With respect to parents, I understand what you're saying, but I don't think we have any information on it.

Schneider: But in your infant studies, you just let the children eat?

Birch: In the preloading studies, we didn't have much trouble getting them to eat. That was always our manipulation, not our dependent measurement of interest.

Lorenz: I have a question for Dr. Rodin. Would you comment on the parity of the mothers and on the gender of the fetus in terms of preferences and aversions?

Rodin: The latter is a wonderful question. I don't have the data divided that way so I can't answer, but I obviously have the data and could look at that. With respect to your first question, all the women in this particular subset are primiparous, but I have as well women who have more than one pregnancy in another data set.

Lorenz: Did that change the response?

Rodin: No, it didn't, which is really interesting. I might say one of the reasons we were so interested in this is that when I interviewed a lot of clinicians, and also women who were friends, they often reported that the same woman had different aversions with different pregnancies. That's what got me interested in the question of whether there really was some interaction between maternal state and that specific fetal need. Now the craving seems to be something that the mother brings with her. The reports were always about aversions. I hope that our biological data will allow us to uncover what it is that is motivating those aversions in that particular pregnancy. I don't think it's experience with pregnancy per se.

Birch: I have a question for Elliott. It has to do with what those babies were doing following sucrose. I ask because of your early slide where you show the baby stopping crying, and then you had the last part that showed hand-to-mouth behavior. I bring that up because that's one of the things that gets scored on the Brazelton as a self-quieting behavior. Do you want to comment on that?

Blass: The hand-to-mouth behavior is something that for us was completely unexpected, and is very robust. We believe that it doesn't have anything to do with the quieting because of two reasons. First of all, it never appeared until the babies had quieted. And secondly, the babies remained quiet after it disappeared. So I think that the hand-to-mouth behavior is irrelevant from the point of view of the general issues I was discussing earlier. I think it's very relevant from the point of view of what was seen here. I think it's a possible precursor of the feeding system—certainly a possible precursor of the babies bringing things to their mouths—because this is the only case that I know of in infants at this age in which there is sustained hand-to-mouth behavior in a very orderly fashion. They do get their hands into their mouths sometimes in sleep. They certainly do as fetuses. But this is the only circumstances that I am aware of in which they get them into the mouth in a sustained fashion.

Beauchamp: You emphasized very clearly that experience is a major component in altering children's food preferences. Yet, it's remarkable how little correlation there is between parent and child in food preferences. I think many of us who have had two children often remark on the completely opposite preferences of the two kids. Given that, how do you explain those types of differences within and between the generations?

Birch: I think Bob Plomin has done some recent work that suggests the family environment is not really very similar for two kids in the same family; that it could be quite different. One thing I didn't mention at all today is the temperamental differences that kids seem to bring into the world with them. You might get kids who are much more willing to try new foods. That sets up one kind of dynamic with the parents, whereas the child with a different initial approach to things can probably set up a different dynamic. That's really all I can say because there isn't any research that speaks to it.

Rodin: Wouldn't you also say, Leann, that birth order obviously will make a difference? There's another reason why your second child is different, which is that you think you learn something from the first. You may really change a variety of your behaviors. You think you do things differently and, in fact, you may, that change the nature of that child's experience, maybe with a lot of things, but I would bet especially with food intake.

Birch: So that in dealing with the first one, the second one gets ignored.

Rodin: It would be interesting to look at the birth order in terms of the impact of experiential variables.

Stricker: The question for Elliott I would raise is this: Of course, tasting something very sweet is going to set up a variety of potential physiological changes. We've heard some of them described during the past two days here. I'm thinking specifically about potential changes in insulin that might affect amino acid transport and, ultimately, serotonin metabolism in the brain. I wonder what you think of that possibility?

Blass: Insofar as other transmitters are concerned, I don't have an opinion because we haven't tested anything other than naloxone. My position is a very conservative one, which is to say that as part of the enormous cascade of neurotransmitters that are figuring in the transduction of sugar or fat into the appropriate taste or flavor sensation and into the alleviation of pain or the reduction of stress, there are somewhere in that chain of events some opioid synapses. In the case of the tactile calming and possibly analgesia, based on the evidence we presented, the conclusion that we're coming to is that such synapses are probably absent. I have nothing else to say about what may or may not be in there on empirical grounds, and I'm not sufficiently familiar with the classes of literature to offer an intelligent opinion, so I pass.

Plata-Salaman: I have two questions. One is for Dr. Rolls. Is there any evidence that any of the hormones for hydromineral balance are involved in the response of the elderly?

Rolls: There's a lot of evidence for alterations in various hormonal responses—atrial natriuretic peptide, vasopressin, etc.—but I don't know that there's been any work on what you're asking about.

Plata-Salaman: I have a question for Dr. Rodin. Dr. Schiffman described some of the deficiencies in trace elements or vitamins which may induce some increase in the threshold for taste sensitivity. Is there any evidence of craving in pregnancy for a specific food because of some trace element or vitamin deficiency, which occurs when these deficiencies are not compensated?

Rodin: I can answer with two facts. The first is that we do chemosensory testing in these women and there's no disturbance, at least using whole-mouth, sip-and-spit procedures for salty, sweet, sour, and bitter. So we expected, in fact, a change during pregnancy. There is in hedonics, but not in the perceived intensity. The second is that we have the data on minerals, vitamins, and trace elements. I have not looked at them yet, but my guess is that they are certainly not undernourished. They're all on vitamin supplementation, they come from two very well-staffed HMOs and they're under continuous care by the HMO physician. Now, I'm sure that there could be an isolated case, or two or ten, but as a rule I would guess that they're in pretty good shape.

Plata-Salaman: I know from my region in Mexico that what we see sometimes depends on the culture.

PART IV DISCUSSION

Rodin: Oh, undoubtedly. Yes, that's why I wanted you to know that the socioeconomic status of this group was quite heterogeneous. But they were all being treated at the same HMOs and receiving the same kind of care. We have all the food intake and know that they were under vitamin supplementation. They could have differed significantly in what they were eating by socioeconomic status, and we'll be able to know that. But unless they weren't taking the vitamin supplements—and we have no reason to think that they wouldn't because they're free—then I wouldn't think in this particular sample that would be an issue.

Torii: People who have hypertension at an elderly age sometimes use a calcium blocker to decrease blood pressure. Taste perception originates from opening of membrane calcium channels. Did you use those kinds of drugs?

Schiffman: Calcium blockers, right? I've tried them all. I can tell you every drug I've tried. I've tried verapamil, nifedipine; they do not block calcium. One drug does block calcium, though; it's trifluoperazine, which blocks calmodulin inside of cells. Calmodulin is four-pronged protein that binds calcium on three-prongs and magnesium on another. It does in fact then come around and bind the calcium ATPase in the bottom side of the cell. That does in fact diminish calcium taste. The question is: Is trifluoperazine binding to calmodulin or to a calmodulin-type protein in the apical membrane? I don't know the answer to that, but yes, that is the only one I know that blocks.

Plata-Salaman: Did you apply this to the surface of the tongue?

Schiffman: Yes.

Plata-Salaman: There are some records of patients receiving oral or parenteral administration of calcium blockers and those also decrease the. . . .

Schiffman: That's right, they're all in the literature. Verapamil is one that seems to have a lot of taste problems associated with it. But the question is, Is that at the peripheral level or is it at the central level? That I don't know.

Bieber: How do we choose a market-approach for foods for the elderly with a different type of taste, etc. without saying it's a special food for the elderly?

Schiffman: I think that's a good question. I can tell you one that would appeal to me, as I'm just starting to have trouble reading a menu in a dark restaurant. You can have large lettering on packages. You can give nutritional information; the elderly do in fact read nutritional information more than younger people. I think enhancing the flavor of the product would certainly help as well. I think the way in which they sell fur coats in Hawaii, with these magnificent-looking women and silver-haired men—this is exactly what you need. I think there are many ways of doing that. I don't think you have to say "Are you old?" and "Do you want a food for the elderly?"

Friedman: I had a question for Elliott Blass. You mention that milk had special properties in contrast to, say, corn oil, with respect to the effects on ultrasonic vocalization. I was wondering, since milk and corn oil differ in fatty acid composition, if you have had a chance to look at that variable? Specifically, I'm wondering whether the response to a medium-chain oil would look more like the response to milk.

Blass: No, we haven't done that; it's clearly in the cards. The special property that I was alluding to was really the presence of β-casomorphine. All mothers, all mammals, secrete β-casomorphine and excrete it in their milk. There have now been three studies that show in animals that β-casomorphine does have analgesic properties. It's potentially a very nice system: feeding the baby and calming it down.

Part V
Neural Integration

19
Neuroanatomical Bases of Cephalic Phase Reflexes

Terry L. Powley and Hans-Rudolf Berthoud

*Purdue University
West Lafayette, Indiana*

I. INTRODUCTION

At a global level, the neuroanatomical substrates of the cephalic phase reflexes are well understood. For example, a considerable literature has established that the cephalic phase insulin response (CPIR), the reflex we focus on in this chapter, is mediated by the vagus nerve. A number of other cephalic responses are also expressed via the vagus (for review see Powley, 1977; Brand et al., 1982; etc.). For some purposes, this level of analysis—that is, the identification of the nerve mediating the response—is a sufficiently precise analysis. For such applications, the presently available information is generally comprehensive.

At a more cellular level, however, the detailed circuitry is not known for any of the cephalic responses. Which particular motor neurons within the larger pool of preganglionics forming the nerve participate in the response is unknown. The size of the motor neuron pool controlling the response is unestablished. The particular peripheral branches and paths of the axons innervating the effector organ are not fully known. For many neurobiological analyses and/or experimental manipulations of the cephalic responses, this more complete characterization of the circuitry involved is required.

Although such a detailed profile is not yet available for any cephalic response, some progress has been made in delineating the neural mechanisms of the CPIR. In fact, more is probably known about the architecture of the CPIR than is known about the circuitry of any other cephalic response. This emerging outline of the neural substrate of the CPIR is summarized in the present chapter. We attempt to sketch in as much of this profile as possible, indicating in passing which of the conclusions are still particularly tentative and which of the remaining questions need most immediate experimental attention. We also summarize a readily testable hypothesis that can serve to focus further analyses of the circuitry of cephalic responses.

This chapter deals only with the neuroanatomy of cephalic responses such as the CPIR. Illustrations of the CPIR as well as discussions of its physiology and its behavioral significance are covered in other chapters in this volume and in earlier reviews (Louis-Sylvestre and Le Magnen, 1980; Powley and Berthoud, 1985). Our earlier review also identifies a number of experimental issues pertaining to physiological and behavioral analyses of the cephalic responses that are not repeated here.

II. BACKGROUND

Parasympathetic preganglionic neurons innervating the abdominal viscera, including those that influence the pancreatic B cells and their release of insulin, have their perikarya in the dorsal motor nucleus of the vagus (dmnX) of the medulla oblongata and issue their axons through the different branches of the peripheral vagus. As our laboratory group has recently established (see Fox and Powley, 1985; Powley et al., 1987a), the elongated, fusiform dmnX (extended sagittally in the medulla) is comprised of several separate, spatially discrete longitudinal columns of cells (see Fig. 1). Each of these columns of somata gives rise to a different primary branch of the subdiaphragmatic vagus. Each columnar subnucleus spans the height (dorsal–ventral dimension) of the dmnX, and the different columns are positioned medial or lateral to each other.

Furthermore, the columns of preganglionic perikarya are organized basically into symmetrical (or homotypical) pairs of columns, with one of the pair on either side of the midline. Thus both the left and the right dmnX contain a medial column of cells that project through one of the two corresponding "gastric" branches of the vagus. In addition, both the left and the right dmnX contain a lateral column of perikarya that project through one of the corresponding "celiac" branches of the vagus. In addition to this symmetrical ground plan, a fifth, more diffuse column of cells responsible for the efferents in the unpaired hepatic branch is found in the left dmnX, where it overlaps the gastric column of cells.

This vagal system has fundamentally an uncrossed and unilateral plan of projections (in the rat, the system is virtually entirely uncrossed). Thus the neurons of the left dmnX project through the left cervical vagus, the anterior (or ventral) abdominal vagal trunk, and then, as the axons move more distally, they separate into the hepatic, the anterior (also called accessory) celiac, and the anterior gastric branches of the vagus. The right dmnX projects through the right cervical vagus, posterior (or dorsal) abdominal vagus, and then going progressively more distally, the celiac and posterior gastric branches (see Fig. 1).

NEUROANATOMY AND CEPHALIC PHASE REFLEXES

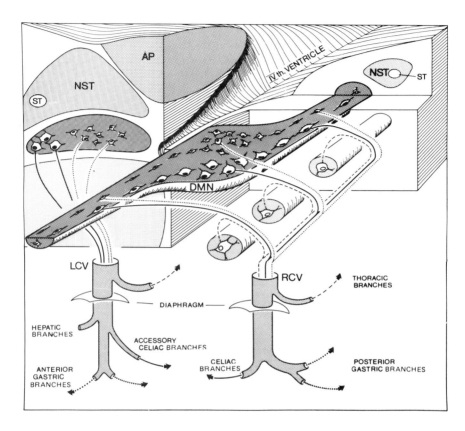

FIGURE 1 Schematic view of both the dorsal vagal complex of the medulla oblongata viewed from a caudal and lateral (right) perspective and the abdominal vagal trunks with their respective branches. The dorsal motor nucleus of the vagus (labeled DMN) can be seen in a horizontal orientation on the right-hand side of the brain stem and in a frontal orientation on the left-hand side of the brain stem. The medial cell column on either side is indicated with smaller perikarya, and the lateral cell column on each side is indicated with larger somata. This view and the different line symbols used for axons illustrate that each column of perikarya in the dorsal motor nucleus gives rise to a separate branch of the subdiaphragmatic vagus (see text; hepatic branch axons are not shown). The nucleus ambiguus, which innervates supradiaphragmatic and striated tissues, is also illustrated on the right, where it is positioned ventrolateral and parallel to the dorsal motor nucleus of the vagus. Other abbreviations: AP, area postrema; LCV, left cervical vagus; NST, nucleus of the solitary tract; RCV, right cervical vagus; ST, solitary tract.

Knowing this organization, it becomes feasible to determine experimentally which (or whether any at all) of the abdominal vagal branches contain the axons mediating any particular cephalic response. By extension, the same information then establishes which column(s) of the preganglionic perikarya in the dmnX contain the somata specifically controlling that cephalic response. We have recently determined such branch and column information for the CPIR.

A. CPIR Is Mediated by Gastric and Hepatic Branches of the Vagus

In a recent experiment, we (Berthoud and Powley, 1990) measured the CPIR in a standard test situation [immunoreactive insulin (IRI) increase above basal during first 2 minutes of chow ingestion] of different groups of rats that had previously been subjected to a selective subdiaphragmatic vagotomy that spared different branches or combinations of branches. As a means of verifying the selective vagotomies, we used a protocol that provides a thorough assessment of the condition of each of the branches (Powley et al., 1987b). The results were clear, although they were somewhat contrary to traditional expectation. Animals with the gastric branches spared and/or the hepatic branch spared exhibited reliable CPIRs. In contrast, animals with only the celiac branches spared did not evidence a CPIR. Put another way, a CPIR could be expressed by the gastric branches or the hepatic branch, but not by either celiac branch.

Exactly the same pattern of branch dependencies was observed when we elicited insulin secretion in the selectively vagotomized animals by direct electrical stimulation of the cervical vagus (Berthoud et al., 1990).

B. CPIRs Are Controlled by Neurons in Medial Cell Columns of the dmnX

As indicated above in the rationale for the selective branch cut experiment just described, the same studies also allowed us to identify the specific pool (or column) of motor neurons in which the preganglionics controlling the CPIR can be found. The selective vagotomies (done a number of days before testing) left viable only those dmnX motor neurons projecting through the spared branches. Thus the dmnX columns of cell bodies associated with the effective branches must be the source of the CPIR.

By this logic, the preganglionics organizing CPIR must be found in the medial cell columns situated on each side of the dmnX (see Fig. 1). The medial column of the left dmnX is the source of the effective axons in the anterior gastric branch; the medial column of the right dmnX is the source of the axons effecting the CPIR through the posterior gastric branch. The cell bodies projecting through the hepatic branch occupy the sparse and somewhat diffuse column that is basically coincident with the medial cell column in the left dmnX. In contrast, the lateral cell columns on either side of the dmnX that are the source of efferent axons in the celiac branches do not contain the motor neurons controlling the CPIR. [Direct electrical stimulation of the dmnX in the lateral part of the nucleus is effective in releasing insulin (Laughton and Powley, 1987), but this phenomenon may be explained by the fact that all dmnX efferent axonal outflow leaves the nucleus from the lateral pole. Alternatively it could also suggest that the preganglionics influencing insulin release may be laterally situated within the medial column.]

Although this process of elimination is useful in delimiting the preganglionics that are potentially responsible for the CPIR, it still leaves unresolved which particular neurons within the larger pools of cells forming the effective columns they are. Assuming that not all cell bodies in the medial cell columns of the dmnX are responsible for endocrine pancreas secretion (and indeed they are not, since this pool of neurons also contains the large population of preganglionics that control gastric secretion and motility), a more complete determination of the neurons is still needed. Although we cannot yet specify which particular neurons mediate the response, we can address the question of the number of neurons involved in the response.

C. CPIRs Can Be Expressed by Even a Few Preganglionic Neurons

In one of our experiments addressed to the question of which branches of the vagus mediate the CPIR (Berthoud and Powley, 1990), we have obtained some information on the number of preganglionics needed to support a cephalic response. The case in point concerns the hepatic branch of the vagus. Specifically, we have found that the hepatic branch, when it is the single spared branch of the vagus, is capable of supporting a considerable CPIR. In our experiment, the group of animals with only the hepatic branch spared evidenced a CPIR that was, in terms of absolute values, 54% of the response magnitude of the controls.

From an independent set of experimental analyses done in our lab, we know that in total the hepatic branch of the vagus contains the axons of only about 200 preganglionic motor neurons (Fox and Powley, 1985, 1989; see also Norgren and Smith, 1988). Furthermore, the majority of these 200 preganglionics are presumed to project to the liver and/or adjacent portal tissues (although this remains to be firmly established). The corollary would be that a minority of the 200 motor fibers project to the pancreas. Nonetheless, this limited complement of preganglionics is able to support a significant CPIR.

A corresponding analysis for the gastric branches is less telling. Each of the gastric branches carries a large number of motor axons ($\approx$ 2500); therefore, although animals with only the gastric branches spared exhibit a very well developed CPIR, this response cannot be attributed to any particular small pool of preganglionics within the branch. If the argument for the handful of hepatic branch efferents mediating the CPIR is sound, then it may well be appropriate to extrapolate to the preganglionics within the gastric columns that innervate the pancreas B cells and conclude that they are a very small complement of cells as well. (One further possible delimitation of these neurons is discussed below in terms of the sensory–motor lattice hypothesis we summarize.)

D. CPIR Vagal Efferents Must Diverge Widely

The observation that a few vagal preganglionic fibers can support a well-developed CPIR (at least in the case of those in the hepatic branch) suggests another tentative conclusion about the neural architecture of the CPIR as well. The rat pancreas has been estimated to contain several thousand islets of Langerhans. Each islet in turn contains a large number of B cells and cells of other types.

Since in the case of the hepatic branch, a few axons are available to mediate the significant CPIR, one must conclude that the efferent projections are widely divergent. Either the distal terminals of the preganglionics, or the postganglionic neurons, or some final paracrine, neurocrine, or endocrine signal must be distributed widely enough to reach the necessary number of B cells.

E. CPIR Is Effected by a Preganglionic Subpopulation with Larger C-Fiber Axons Within the Vagal Branches

Another inference concerning the efferents carrying the parasympathetic control of the pancreas is also possible. This inference, based on two sets of observations we have made recently, suggests that the CPIR may be expressed by a subpopulation of larger caliber axons within the vagus.

In the first set of observations, we have characterized the spectra of axon diameters found in the different branches of the subdiaphragmatic vagus (Prechtl and Powley, 1990). The subpopulations of unmyelinated fibers making the branches vary somewhat from branch pair to branch pair in terms of the fiber caliber histograms, but they have mean diameters of essentially 0.76 μm with ranges of 0.1–1.8 μm. These diameters would be consistent with conduction velocities in the C-fiber and smaller B-fiber range.

In the second set of observations (Berthoud and Powley, 1987), we used an electrophysiological strategy of parametrically varying stimulation frequency to determine half-maximal response values, maximal following frequencies, and other frequency–response characteristics. In this experiment, the axons innervating the pancreas could be differentiated from those controlling gastric acid secretion (which had lower half-maximal frequencies and also quit following at lower frequencies) and those controlling bradycardia (which had even higher half-maximal values and followed higher frequencies). Based on these frequency–response curves, it appeared that the pancreatic islets might be innervated by axons in the large C-fiber or small B-fiber range, whereas the gastric parietal cells might be controlled by smaller C-fiber axons. While alternative explanations of these findings must be considered (cf. the discussion in Berthoud and Powley, 1987), the larger fibers implicated in control of the endocrine pancreas by this technique might correspond to the right side of the fiber caliber spectra described above.

F. CPIR May Be Expressed by Rostral Pole of Each Medial dmnX Column, While Intestinal and Postabsorptive Insulin Reflexes May Be Expressed by the Caudal Pole of the Same Columns of Vagal Efferents

Consideration of the observations we have made on the role of different elements of the vagal system in the expression of the CPIR suggests another issue and offers at least a tentative conclusion regarding this question. The issue is the following. Although the question has not been explicitly examined before (as far as we have determined), it is conceivable that all vagal preganglionics that influence the B cell release of insulin are involved in organizing the CPIR or, alternatively, it is also possible that only some of these preganglionics are involved in expressing the CPIR, whereas another independent subset of the preganglionics are involved in gastric–, hepatic–, or intestinal–pancreatic reflexes. Although direct data are not yet available to resolve this issue, indirect observations favor the latter alternative.

While it is highly inferential at the present time, we have developed a working hypothesis that incorporates currently available information and specifies where within the dmnX the preganglionics mediating the CPIR can be found. When the significance of the longitudinally organized subnuclei or motor cell columns is considered in conjunction with other information on the structure of the dorsal vagal complex, we think a potential structural substrate for all vagal reflexes emerges. We have referred to this substrate as a *sensory–motor lattice* and described it more fully elsewhere (Powley et al., 1987a). The basic point emerges from the observation that the topographic organizations of the nucleus of the solitary tract (NST) inputs and the dmnX outputs are basically orthogonal to one another. If one assumes that the dmnX subnuclei are approximately longitudinal columns, by a similar token the different afferent subfields in the NST are organized as a series of roughly frontal bands that correspond to the different inputs. Gustatory and other oral inputs project to the rostral NST, while more caudal abdominal inputs project to the more caudal NST (see Fig. 2; also see, e.g., Hamilton and Norgren, 1984; Norgren and Smith, 1988). Since (a) the NST has the same basic horizontal shape as the dmnX, (b) the two nuclei are effectively fused, and (c) the two are fused in a manner that has them in register, the vagal sensory and motor nuclei effectively form a lattice. In this lattice, the gustatory inputs that support cephalic responses such as the CPIR would overlie the rostral end of each of the medial cell columns associated with one of the branches effective in releasing insulin. Conversely, the afferents controlling vagal reflexes triggered by gastric or intestinal stimuli would terminate over the caudal end of each of the cell columns. Given the details of the dendroarchitecture of the dmnX preganglionics (Shapiro and Miselis, 1985; Fox and Powley, 1988), one would

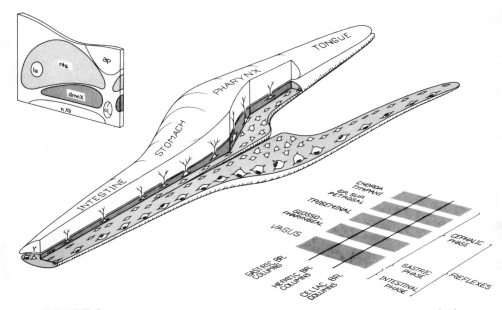

FIGURE 2 Schematic illustration of the structural basis of the sensory–motor lattice hypothesis. The dorsal vagal complex (in the middle, as in Fig. 1) is viewed from a dorsal caudal and lateral (right) perspective. This horizontal view shows the dorsal motor nucleus [the more ventral nucleus (shaded) with different sized cells illustrating the separate columns] on both left and right of the brain stem, while the nucleus of the solitary tract (more dorsal unshaded nucleus) is represented only on the left-hand side of the brain. A corresponding frontal view of a section through the left-hand side of the vagal trigone region is included in the upper left corner of the figure. The lower right corner of the figure illustrates the formal idea of a lattice in which the longitudinally arrayed columns of the dorsal motor nucleus of the vagus are orthogonal to the incoming afferent terminals of the several nerves projecting to the nucleus of the solitary tract. This formal lattice sketch illustrates nerves supplying the afferents on the left, with the corresponding reflexes labeled on the right. The afferent inputs are also translated into a schematic viscerotopic map and expressed in terms of organs on the dorsal surface of the nucleus of the solitary tract in the middle view of the figure.

expect that preganglionics in the rostral end of the columns would be primarily influenced by the overlying gustatory inputs and, conversely, the preganglionics in the caudal end would be largely affected by the overlying inputs of the more distal viscera.

Although we have only recently formulated the lattice hypothesis and direct tests of it are not yet available, our earlier results do contain another observation that could be construed as evidence that indeed only a fraction of the preganglionics projecting to the pancreas through each of the three positive vagal branches actually express the CPIR. In our experiments, electrical stimulation of the vagus has consistently produced much larger insulin responses than those we are able to elicit with a cephalic phase test. Although this observation has several possible alternative explanations, it is consistent with the hypothesis that only a subset of vagal/pancreatic preganglionics that influence the B cells are involved in the CPIR.

A second observation we have reviewed above is also consistent with the lattice hypothesis and might be construed as support for the idea, insofar as it is inconsistent with one alternative plan of organization. Assuming that only some of the dmnX motor neurons projecting to the endocrine pancreas actually participate in the CPIR, the clearest alternative to the lattice idea (which postulates the CPIR-positive preganglionics distributed rostrally within the longitudinal columns) would be a plan in which the CPIR was compartmentalized within one or two of the three dmnX columns projecting to the pancreas. Our results, however, do not fulfill this alternative expectation. We found identical branch dependencies for electrical stimulation and for the natural cephalic phase test stimulation. This suggests that, if the CPIR is indeed expressed by only some of the preganglionics influencing the B cell, this subpopulation is distributed across all three insulin-positive columns and their corresponding branches as the lattice hypothesis would predict.

G. CPIR Preganglionics Projecting Through Different Vagal Branches May Innervate Different Populations of Islet Cells

The pattern of branch dependencies we have observed also brings up another question concerning the neural substrates of cephalic responses into focus and provides a tentative inference concerning that question.

Conceivably the three separable pools of dmnX motor neurons projecting directly or indirectly to the pancreatic B cells (i.e., those in the medial left dmnX column projecting through the hepatic branch, those in the medial left dmnX projecting through the anterior gastric, and those in the medial right dmnX projecting through the posterior gastric) might distribute in an overlapping or diffuse pattern to control the collective islet mass of the pancreas or, alternatively, the three pools might innervate three separate regions or pools of endocrine tissue in the pancreas. While this question has not been asked before, our branch cut experiments provide one means of looking at the issue.

The results of our experiment employing electrical stimulation of the cervical vagus suggest the second alternative—namely, that the different branches may innervate different regions—is correct. The relevant observation comes from the series we performed examining stimulation of the left cervical vagus done in conjunction with selective vagotomies of the axons issuing through the different branches of this trunk. We obtained from each branch partial responses (hepatic:

103% increase from basal; anterior gastric: +177%) that summed to the equivalent of the response seen in sham animals that had both branches spared (shams: +272%). While once again other explanations of these results suggesting additivity cannot be ruled out at the present time, the observations are consistent with a model in which each of these branches influences a different zone or region of pancreas. Such a model would be consistent with the pattern of vagal innervation of a different target tissue, the gastric parietal cell mass, which is organized into a series of separate domains (Pritchard et al., 1968). Such a model would also be compatible with the fact that embryologically, the pancreas develops from more than one bud (Spooner et al., 1970).

III. ADDITIONAL QUESTIONS CONCERNING THE NEURAL ARCHITECTURE OF THE CPIR

As suggested in Section I, our understanding of the circuitry of the CPIR (and other cephalic responses) is incomplete. Even the preliminary outline of the circuit that we do have is tentative in most details. Besides the issues just reviewed, where some conclusions can be drawn, there are several other questions for which either there is simply no information available yet or else the available information is so contradictory that it is hard to draw even a tentative conclusion. Some of these more problematic issues are particularly relevant to identifying the neural circuitry of the CPIR and should be at least identified.

One such question on which information is critically needed concerns the neurochemistry of the neural pathways mediating the CPIR. Not knowing the pathways has made any attempt to characterize the relevant neurochemistry impractical. As the pathways become progressively better delineated, however, such attempts should become practical—and they certainly are needed. What transmitters are specifically involved in the CPIR has not been established. Which neuromodulators can influence the CPIR is unknown. Whether there is a form of neurochemical coding for the CPIR has similarly not been determined.

A second issue concerns the identification of the higher CNS connections or circuits that project to the reflex loops in the caudal brain stem and influence and/or alter the bias on these basic circuits. Even assuming for argument's sake that the lattice hypothesis accurately summarizes the structural basis of the cephalic reflexes, it describes only the lowest level of the circuitry involving the primary and secondary afferents and the preganglionic outflow. This lattice organization might account for the basic reflexive architecture of the CPIR, but it does not suggest the full CNS circuitry involved in all the subtle modulation of the cephalic reflexes as they operate in their characteristically labile physiological and behavioral roles (cf. Powley, 1977; Louis-Sylvestre and Le Magnen, 1980; Grill and Berridge, 1985; Powley and Berthoud, 1985). Though we already have a lot of information, in broad strokes, about CNS projections to the vagal complex, we do not have any information about the details of the descending projections that articulate specifically with the lower brain stem circuits of the CPIR.

Another issue that requires considerably more elucidation is the exact locus and the nature of the linkage between the vagal preganglionic terminals and the pancreatic B cells. The classic observations concerning the neural innervation of the pancreas (e.g., dissections or degeneration) cannot differentiate between the innervation of the exocrine and endocrine elements, nor can they differentiate between the connections mediating cephalic as opposed to gastric or intestinal reflexes. More recent experiments with retrograde tracers have the same fundamental problems. Although some retrograde tracer studies employing injections of the compound into the pancreas have labeled primarily the medial columns of the dmnX (see, e.g., Powley and Laughton, 1981; Rinaman and Miselis, 1987), not all have. Furthermore, retrograde tracer studies involving the abdominal viscera have also been notoriously plagued by diffusion and contamination, which lead to false-positive labeling (see Fox and Powley, 1989). If, as outlined above, only a very few preganglionic neurons are needed to express the CPIR, then any false-positive artifact will significantly distort the critical pattern.

Finally, even where and how the preganglionic axons synapse to express the CPIR is unknown. A possibility raised by physiological and tracing studies is that the vagal preganglionics actually synapse on a postganglionic element before the pathway enters pancreatic tissue.

IV. CONCLUSION

The profile of the neural circuitry controlling the cephalic phase insulin response is beginning to emerge. The perikarya of the preganglionics controlling the response are located in the medial columns or subnuclei of the dmnX. These neurons send their axons through the two gastric and the hepatic branches of the abdominal vagus. Each of these branches can apparently mediate a CPIR independently of the other two branches. A CPIR can also be effected by even a very small complement of preganglionics within the pool of neurons that innervate the B cells of the pancreas. Although similar analyses have yet to be done for other cephalic responses as well (with the partial exception the gastric acid cephalic response), the profile for the CPIR may provide an analogue for other vagally mediated cephalic responses.

Additional—although more tentative—extrapolations that require more experimental evaluation can also be drawn concerning the CPIR. The response may be expressed by some of the larger unmyelinated fibers within the vagus, different preganglionics organizing the response may innervate different regions of the pancreas or different subpopulations of the islet cell mass, and the structural substrate of the CPIR, as well as of other cephalic responses, may be the sensory–motor lattice formed in the vagal trigone by the nucleus of the solitary tract and the dorsal motor nucleus of the vagus.

ACKNOWLEDGMENTS

We thank the several other members of the laboratory who have performed one or more of the experiments summarized here: Elizabeth Baronowsky, Edward A. Fox, James C. Prechtl, and Watson Laughton.

This work was supported by National Institutes of Health grants DK27627 and NS26632.

DISCUSSION

Weingarten: In the olden days when people used to do ventromedial hypothalamic (VM) lesions and vagotomy, it was the selective celiac vagotomy apparently that blocked the effects. The classic interpretation was that you eliminate cephalic insulin, and that is why you reverse the VM syndrome. I assume this anatomical analysis forces the reinterpretation, and I guess my question is, How should we now interpret those data?

Powley: The data are troublesome, although the data are not as troublesome—and you put your finger on it—as the interpretation. In fact, in those early analyses insulin wasn't measured. The interpretation was that because it was a celiac vagotomy and because on the basis of morphological observations—actually dissection studies done on the vagus nerve—it was assumed that the celiac branch went to the pancreas, a celiac branch vagotomy would eliminate cephalic phase insulin responses. Other candidate responses were not considered or measured, and insulin responses were generally not measured. There is one study in the literature which did, but for the most part, in the studies you're talking about, in particular, there were no measurements made; it was an interpretation.

Our data certainly force a reinterpretation. We worried quite a bit about it; in fact, worried about it to the point that we've done this study a number of ways. One of the reasons we looked not only at the behavioral responses, but then went on and stimulated those same nerves, is to see if we can drive any insulin secretion out of the celiac branches, for example. No matter how we do it, standing on our head or upright, we can't get any insulin responses out of the celiac branches. It may very well be the case that the celiac branches do go to the pancreas. There's the whole business of the exocrine pancreas to consider. But we can't get insulin or for that matter glucagon responses—endocrine pancreatic responses—out of the celiac branches of the vagus.

Friedman: I just have one question. You mentioned that there were 200 out of 2700 nerves that were motor. What proportion are motor versus sensory in the other vagal branches?

Powley: In Jim Prechtl's inventories of the different nerves, in fact what you find is that the sensory–motor ratios do differ in each of the nerves. Also the axon calibers differ in each of the branches of the vagus. The vagus branches are not random samples of the overall population of axons that are coming down the line. The numbers get up to something on the order of about 30% motor in some of the branches; they're obviously less than 10% motor in the case of the hepatic. So we do see a large number of differences. One other caveat: not all axons in the vagus are vagal if you insist that "vagal" means it has a central connection. So when I say that 200 are motor neurons, the other 2700 axons are largely sensory, but some of those actually are probably purely peripheral extrinsic axons. So the total complement of vagal sensory neurons in the hepatic branch is actually less than 2500.

Naim: In relation to gastric acid secretion, how did you eliminate the vagovagal reflex in this pathway? Have you eliminated the sensory input from the stomach?

Powley: In which particular measurement?

Naim: In the cephalic phase of insulin. In the saccharin experiment, for example, since we know from vagovagal reflexive gastric secretion that it could be quite weak.

Powley: We haven't definitively eliminated it in the case of the cephalic response. One of the reasons we check all of that with the electrical stimulation is because in the case of electrical stimulation we can cut the vagus, eliminating any sort of vagovagal feedback or any sort of long sensory-to-motor loops. In the case of the saccharin consumption, we suspect that the gastric response is not important, but we do that for indirect and incomplete reasons such as the volumes of the saccharin we're administering. But the arguments there are weak arguments. We can't eliminate entirely the possibility that the cephalic response involves an afferent limb from the stomach.

REFERENCES

Berthoud, H.-R., and Powley, T. L. (1987). Characteristics of gastric and pancreatic responses to vagal stimulation with varied frequencies: Evidence for different fiber calibers? *J. Auton. Nerv. Syst.* **19**:77–84.

Berthoud, H.-R., and Powley, T. L. (1990). Identification of vagal preganglionics that mediate cephalic phase insulin response. *Am. J. Physiol.* **258**:R523–R530.

Berthoud, H.-R., Fox, E. A., and Powley, T. L. (1990). Localization of vagal preganglionics that stimulate insulin and glucagon secretion. *Am. J. Physiol.* **258**:R160–R168.

Brand, J. G., Cagan, R. H., and Naim, M. (1982). Chemical senses in release of gastric and pancreatic secretions. **2**:249–270.

Fox, E. A., and Powley, T. L. (1985). Longitudinal columnar organization within the dorsal motor nucleus represents separate branches of the abdominal vagus. *Brain Res.* **341**:269–282.

Fox, E. A., and Powley, T. L. (1988). Dendritic fields and morphology of identified neurons in the dorsal motor nucleus of the vagus. *Soc. Neurosci. Abstr.* **14**:315.

Fox, E. A., and Powley, T. L. (1989). False-positive artifacts of tracer strategies distort autonomic connectivity maps. *Brain Res. Rev.* **14**:53–77.

Grill, H. J., and Berridge, K. C. (1985). Taste reactivity as a measure of the neural control of palatability. In *Progress in Psychobiology and Physiological Psychology*, Vol. 11, J. M. Sprague and A. N. Epstein (Eds.). Academic Press, New York, pp. 1–61.

Hamilton, R. B., and Norgren, R. (1984). Central projections of gustatory nerves in the rat. *J. Comp. Neurol.* **222**:560–577.

Laughton, W. B., and Powley, T. L. (1987). Localization of efferent function in the dorsal motor nucleus of the vagus. *Am. J. Physiol.* **252**:R13–R25.

Louis-Sylvestre, J., and Le Magnen, J. (1980). Palatability and preabsorptive insulin release. *Neurosci. Biobehav. Rev.* **4**(Suppl. 1):43–46.

Norgren, R., and Smith, G. P. (1988). Central distribution of subdiaphragmatic vagal branches in the rat. *J. Comp. Neurol.* **273**:207–223.

Powley, T. L. (1977). The ventromedial hypothalamic syndrome, satiety, and a cephalic phase hypothesis. *Psychol. Rev.* **84**:89–126.

Powley, T. L., and Berthoud, H.-R. (1985). Diet and cephalic phase insulin responses. *Am. J. Clin. Nutr.* **42**:991–1002.

Powley, T. L., and Laughton, W. B. (1981). Neural pathways involved in the hypothalamic integration of autonomic responses. *Diabetologia*, **20**:378–381.

Powley, T. L., Fox, E. A., Baronowsky, E., Keller, D. L., and Berthoud, H.-R. (1987a). Longitudinal column of gastric branch neurons in the dorsal motor nucleus of the vagus is composed of subcolumns corresponding to distal divisions of gastric branch. *Soc. Neurosci. Abstr.* **13**:386.

Powley, T. L., Fox, E. A., and Berthoud, H.-R. (1987b). Retrograde tracer technique for assessment of selective and total subdiaphragmatic vagotomies. *Am. J. Physiol.* **253**:R361–R370.

Powley, T. L., Berthoud, H.-R., Fox, E. A., and Laughton, W. The dorsal vagal complex forms a sensory-motor lattice: The circuitry of gastrointestinal reflexes. In: *Vagal Afferents*, Ritter, R. C., Ritter, S., and Barnes, C. D. (in press).

Prechtl, J. C., and Powley, T. L. (1990). The fiber composition of the abdominal vagus of the rat. *Anat. Embryol.* **181**:101–115.

Pritchard, G. R., Griffith, C. A., and Harkins, H. N. (1968). A physiologic demonstration of the anatomic distribution of the vagal system to the stomach. *Surg. Gynecol. Obstet.* **126**:791–798.

Rinaman, L., and Miselis, R. R. (1987). The organization of vagal innervation of rat pancreas using cholera toxin–horseradish peroxidase conjugate. *J. Auton. Nerv. Syst.* **21**:109–125.

Shapiro, R. E., and Miselis, R. R. (1985). The central organization of the vagus nerve innervating the stomach of the rat. *J. Comp. Neurol.* **238**:473–488.

Spooner, B. S., Walther, B. T., and Rutter, W. J. (1970). The development of the dorsal and ventral mammalian pancreas in vivo and in vitro. *J. Cell Biol.* 47:235–246.

20
Neuroendocrine Activity During Food Intake Modulates Secretion of the Endocrine Pancreas and Contributes to the Regulation of Body Weight

Anton B. Steffens, Jan H. Strubbe,
Anton J. W. Scheurink, and Börk Balkan

University of Groningen
Haren, The Netherlands

I. INTRODUCTION

It is widely recognized that hormone release from the cells of the islets of Langerhans is a rather complex process. When the blood glucose concentration increases above a basal level of 100 mg/dl, insulin is released. Insulin release from the B cells shows a sigmoidal course during a linear increase of blood glucose from 100 to 160 mg/dl (Malaisse et al., 1967), whereas glucagon release from the A cells is suppressed in this situation. When blood glucose declines below 100 mg/dl, glucagon release rises, with a concomitant suppression of insulin release, which is, however, never completely suppressed. Increases of blood levels of amino acids, especially arginine and ornithine, elicit glucagon release, whereas insulin is released too, provided blood glucose levels are basal—that is, 100 mg/dl (Müller et al., 1971). Besides the A and B cells, the islets of

Langerhans contain D and F cells, which are capable of releasing somatostatin and pancreatic polypeptide (PP), respectively. The D cells are strategically situated between the B cells in the center and the A cells at the rim of the islets of Langerhans (Unger et al., 1978). Somatostatin might suppress insulin and glucagon release from A and B cells. Hence, modulation of D-cell function might be partly involved in the control of insulin and glucagon release. In addition, somatostatin might diminish the motility of the gastrointestinal tract and the resorption of nutrients (Schusdziarra et al., 1980). The function of pancreatic polypeptide regarding metabolism is less clear and is not discussed in this chapter.

Several hormones released from the alimentary tract during the process of digestion [e.g., cholecystokinin (CCK) and gastrin] stimulate insulin release from the B cells. This might explain the larger increment in plasma insulin levels during an oral glucose tolerance test compared to an intravenous glucose tolerance test, even if the rise in blood glucose is the same (Strubbe and Bouman, 1978).

In the past two decades it has become clear that besides the classical stimuli just mentioned, the central nervous system (CNS) also plays an important role in the regulation of islet of Langerhans function. We present evidence below that the CNS control of islet function is most important in the modulation of insulin and glucose profiles over the day–night cycle. Because of that, nutrient flows are led into the appropriate body tissues so that body weight and body composition remain fairly constant over prolonged time periods in adulthood. In addition, the CNS continuously receives information about basal plasma insulin levels via a feedback mechanism by which food intake is regulated.

II. CONNECTIONS BETWEEN THE CENTRAL NERVOUS SYSTEM AND THE ISLETS OF LANGERHANS

The islets of Langerhans receive a rich sympathetic and parasympathetic innervation (Luiten et al., 1987). The membranes of the A, B, and D cells contain many α- and β-adrenergic and muscarinic receptors so that these cells can respond to immediate sympathetic and parasympathetic activity. Also norepinephrine (NE) and epinephrine (E) released into the blood circulation can affect the cellular response of the islet cells. Furthermore, peptidergic mechanisms seem to be involved in the release of insulin, glucagon, and pancreatic polypeptide (Miller, 1981; Edwards, 1984). Thus, the islet cells are provided with such sets of receptors that a change in activity of the autonomic nervous system may profoundly influence the islet of Langerhans. Now we inquire which autonomic pathways are involved in the CNS control of islet function.

In several studies (see, for a review, Luiten et al., 1987), it has been shown that distinctive areas in the hypothalamus play an important role in CNS control of peripheral metabolism. The hypothalamic paraventricular nucleus (PVN) has direct connections with the motor centers of both the parasympathetic and sympathetic nervous system [dorsal motor nucleus of the vagus (DMV) and intermediolateral column in the spinal cord (IML), respectively]. These connections of the PVN with the motor areas of the autonomic nervous system make direct contact with preganglionic autonomic neurons innervating the pancreatic islets as well as with the adrenal medulla. The lateral hypothalamic area (LHA) has direct connections with the DMV and also indirect ones via the reticular formation (RET). The latter structure has connections with the parts of the IML affecting pancreatic islet and adrenal medullary functions. The ventromedial hypothalamic area (VMH) mainly projects on the IML via the periaqueductal gray (PAG) and RET. Finally, the dorsomedial hypothalamic (DMH) area projects on both these sympathetic and parasympathetic pathways. Recent research (Steffens et al., 1988) showed the existence of intricate intrahypothalamic connections between the VMH, LHA, DMH, and PVN. There are major unidirectional pathways from the VMH and LHA to the DMH and from there to the PVN. Minor unidirectional connections exist from the LHA to the PVN and to the VMH. Evidently, the hypothalamus receives major inputs from many other parts of the CNS, for example, the limbic system and the brain stem via the ventral and dorsal noradrenergic bundle (see, for a review, Luiten et al., 1987). A review of these extrahypothalamic afferent connections of the hypothalamus is beyond the scope of this chapter.

Of special interest are the effects of electrical and chemical stimulation of the areas just mentioned on pancreatic islet function and homeostasis of blood glucose and plasma free fatty acid (FFA) levels. Special attention needs to be given to the NE innervation of the hypothalamus because noradrenergic stimulation greatly affects food intake (Lichtenstein et al., 1985; Shimazu et al., 1986) and glucose and FFA homeostasis (Steffens et al., 1984; Scheurink et al., 1988), which are discussed further on in this chapter.

III. REGULATION OF PANCREATIC ISLET FUNCTION

As already mentioned, neuroendocrine factors might affect the function of the pancreatic islets in addition to such classical stimulatory substances as glucose, amino acids, and gut factors. More than 20 years ago, Kaneto showed for the first time (Kaneto et al., 1967) that electrical stimulation of the vagus to the pancreas elicits insulin release. This insulin release is caused by a muscarinic mechanism. Soon afterward it was demonstrated that either sympathetic activation or the administration of catecholamines affects islet activity (Porte and Robertson, 1973). Stimulation of α_2-adrenoceptors suppresses insulin release from the B cells and induces glucagon release from the A cells. Stimulation of β_2-adrenoceptors enhances insulin release and can also affect A-cell function (Samols and Weir, 1979). It has to be mentioned that intravenous infusion of physiological quantities of NE (20 ng/min) does not suppress insulin release, whereas the same quantity of E does, even though E and NE are both α_2-adrenoceptor agonists and only E exerts β_2-adrenoceptor agonistic activity (Scheurink et al., 1989). There is no satisfactory explanation for this. It might be that peripheral E can reach the α_2-adrenoceptors more easily than NE. The suppression by splanchnic stimulation of insulin release can be explained by the high NE concentration in the synaptic clefts between postsynaptic sympathetic nerve endings and the B-cell membrane. In some species (e.g., calf and dog), vagal activation can elicit glucagon release (Kaneto et al., 1974; Edwards, 1984). It is doubtful, however, whether this occurs in man and rat.

Somatostatin, released from the D cells, which counteracts insulin and glucagon secretion, is also affected by sympathetic and parasympathetic activity. Vagal and α_2-adrenergic stimulation inhibit somatostatin release, whereas β_2-adrenergic activity stimulates somatostatin release (Samols et al., 1981).

In addition to muscarinic and adrenergic influences on islet cell activity, peptidergic mechanisms act on the islets. Vasoactive intestinal peptide (VIP), gastrin, and CCK present in nerve endings in the islets are capable of stimulating insulin and glucagon release (Miller, 1981).

In summary, the autonomic nervous system can profoundly affect the pancreatic islets. Next to this, intrapancreatic hormonal effects are described, including the suppressive activity of somatostatin on the A and B cells, the stimulatory effect of glucagon on the B and D cells, and the suppressive activity of insulin on the A cells (Unger et al., 1978). Consequently, the islet is a complex neuroendocrine unit that can be controlled both by such classical stimuli as glucose, amino acids, and gut hormones, and the autonomic nervous system, and by paracrine interactions of the different islet cell types.

Since glucose is the major insulinogenic substance, it is useful to pay attention to the mechanisms involved in its absorption and production. Glucose can be absorbed from the alimentary tract if carbohydrates are ingested. The second source of blood glucose is the liver, from which glucose can be released by glycogenolysis and gluconeogenesis. These processes in the liver are under hormonal and neural control. Insulin transforms inactive liver glycogen synthetase into the active component, whereas inactive liver glycogen phosphorylase conversion into the active form is prevented, with the result that glycogen synthesis is promoted and glycogenolysis with concomitant glucose release is inhibited. Glucagon has an opposite effect and also promotes gluconeogenesis. Therefore the insulin–glucagon ratio determines whether glycogenesis or glycogenolysis in the liver occurs. Sympathetic activity and circulating catecholamines cause glycogenolysis and concomitant glucose release by an α_2-adrenergic mechanism (Blair et al., 1979). Simultaneously, insulin output from the B cells is suppressed by an α_2-adrenergic mechanism. Peptidergic mechanisms seem to be involved too (Shimazu, 1986). Finally, prolonged rises in plasma glucocorticoids leads to increased blood sugar levels by increased protein degradation and gluconeogenesis and diminished glucose utilization.

Since glucose utilization by tissue cells means a continuous drain on the glycogen stores and blood glucose levels as such, it is remarkable that the daily fluctuations in blood sugar are extremely small. Hence, the processes just described must exert a fine-tuned control on blood glucose concentrations to maintain these at about 100 mg/dl. Before discussing the effects of the CNS on the autonomic outflow to the endocrine pancreas and the liver, the release of free fatty acids merits some attention because the FFAs are metabolic substrates for most tissues except neurons, hence they can have a glucose-sparing effect. Increased blood glucose and insulin levels (e.g., after food ingestion, especially of carbohydrates) result in fat synthesis. Decreased plasma insulin causes diminished glucose transport over the fat cell membrane and results in FFA release. Lipolysis is further enhanced by circulating catecholamines via a β-adrenergic mechanism, whereas α_2-adrenoceptor stimulation inhibits lipolysis (Smith, 1983). However, α_2-adrenoceptors are absent in rat adipocytes (Lafontan et al., 1985). It is remarkable that the white adipocytes do not receive autonomic innervation, even though release of FFA is controlled by catecholamines. NE released from the sympathetic nerve endings is probably responsible for FFA release because intravenous infusion of physiological quantities of NE (20 and 50 ng/min) leads to increases of FFA without effects on insulin and glucose levels (Steffens et al., 1984a; Scheurink et al., 1989). In this situation NE presumably acts as a hormone.

The above-mentioned actions of glucose, insulin, the autonomic nervous system, and catecholamines are depicted in Figure 1. For the sake of simplicity, only an islet of Langerhans B cell is presented.

In the next section, we discuss the extent to which central nervous system

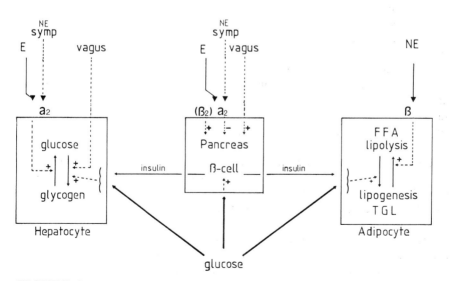

FIGURE 1 Model showing influences of autonomic nervous system and circulating catecholamines on the hepatocytes, adipocytes, and B cells of the islet of Langerhans. The effects of glucose on B cells and the combined effects of glucose and insulin on glycogenesis and lipogenesis are also indicated.

control is active on the pancreatic islets, hepatocytes and adipocytes, via the autonomic nervous system in the freely moving animal, showing its normal ongoing behavior.

IV. HYPOTHALAMIC INFLUENCES ON ISLETS OF LANGERHANS, HEPATOCYTES, AND ADIPOCYTES

This section presents only essential points of hypothalamic influences on the islets of Langerhans, hepatocytes, and adipocytes, because detailed data can be found in recent reviews (Rohner-Jeanrenaud et al., 1983; Steffens and Strubbe, 1983 and Shimazu, 1986). The neural pathways between several hypothalamic areas and motor neuron pools of the autonomic nervous system were described above. These networks represent the anatomical hardware of the central nervous control of metabolism. Lesioning of the VMH results in an immediate increase in activity in pancreatic vagal nerves and a decrease in splanchnic nerve activity at the onset of the lesion, whereas lesioning of the LHA leads to exactly the reverse pattern (Yoshimatsu et al., 1984). Electrical stimulation of the VMH leads to an increased conversion of inactive phosphorylase b to active phosphorylase a in the liver, resulting in glycogenolysis (Shimazu, 1986). In addition, the conversion of inactive liver glycogen synthetase in the active form is inhibited. The activity of phosphoenol pyruvate carboxylkinase, a key enzyme in gluconeogenesis is also augmented. Finally, an increase in glucagon release from the islets of Langerhans is observed, along with a concomitant rise in blood glucose (Shimazu, 1986). Electrical stimulation of the LHA results in the opposite changes (i.e., glycogenesis, inhibition of gluconeogenesis, and an increase in plasma insulin) (Shimazu, 1986).

Electrical stimulation of the VMH elicits lipolysis in the anesthetized rat (Takahishi and Shimazu, 1981). Lipolysis does not occur if stimulation is applied in conscious rats. This discrepancy might be explained, as the authors do, by suggesting that VMH stimulation causes arousal in conscious rats, which, in turn, leads to increased catecholamine output, hence to increased lactate formation. Lactate inhibits FFA release, probably because of increased reesterification.

Local application of neurotransmitters and/or neuropeptides for which receptors are present in the area under investigation may provide more specific information with respect to the function of the hypothalamus in the regulation of metabolism. The existence of a number of neurotransmitters and neuropeptides involved in the regulation of metabolism as well as food intake has been demonstrated (Morley et al., 1985; Nieuwenhuijs, 1985; Leibowitz, 1986). The most noteworthy one is norepinephrine. Injection or infusion of NE into the hypothalamus affects food ingestion and blood component levels related to metabolism (Steffens et al., 1984a,b; Shimazu, 1986). NE administration into the VMH and PVN causes an increase in blood glucose levels (Steffens et al., 1984a; Chafetz et al., 1986), which is the result of direct activation of glycogenolysis in the liver

(Shimazu, 1986). Furthermore, noradrenergic stimulation of the VMH causes glucagon release in rats (De Jong et al., 1977) and rabbits, with concomitant increases in blood glucose (Shimazu and Ishikawa, 1981). Infusion of NE into the VMH also results in an increase in plasma FFA and insulin (Fig. 2). The rise of plasma FFA can be observed in spite of a simultaneous increase of blood glucose and plasma insulin. This suggests that the VMH can directly stimulate lipolysis independently of the levels of glucose and insulin. Infusion of NE into the LHA

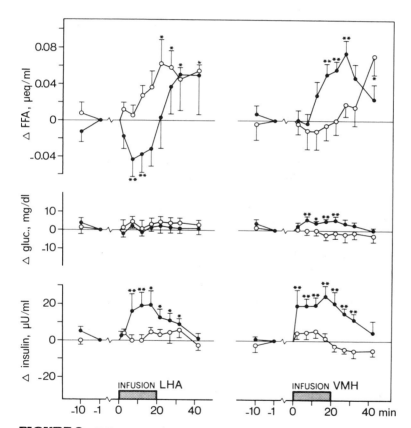

FIGURE 2 Effects of norepinephrine (NE) infusion (25 ng/min per cannula at a rate of 0.25 μl/min during 20 min) (●) and diluted solvent infusion (at a rate of 0.25 μl/min) (○) in lateral hypothalamus (LHA) (left) and in ventromedial hypothalamus (VMH) (right) on plasma FFA, blood glucose, and plasma insulin. Data are changes in means ± SE, μEq/ml plasma, mg/dl blood, and μU/ml plasma, respectively, from preinfusion levels of FFA, glucose, and insulin, which were average of those of minute 1 before start of infusion. One asterisk denotes a significant change from preinfusion level. Two asterisks denote a significant change from diluted solvent infusion. (From Steffens et al., 1984a.)

leads to an increase in plasma insulin, and a decrease in plasma FFA, whereas blood glucose and plasma glucagon are unaltered (Steffens et al., 1984a,b). The increase in insulin is known to be parasympathetically mediated because prior atropinization completely prevents the rise of insulin due to infusion of NE into the LHA (Steffens et al., 1984b).

In conclusion, both neuroanatomical and physiological data provide evidence for a role of the hypothalamus via autonomic pathways in the control of the function of the islets of Langerhans, glycogenolysis in the liver, and lipolysis. In the next section, we discuss how these hypothalamic actions are important in fine-tuning glucose and insulin profiles during the day and night. By this action, the flow of absorbed nutrients is directed in such a way that all body tissues receive an adequate amount. Thus, body weight and body composition (i.e., the ratio of fat to fat-free body mass) are maintained nearly constant over prolonged periods in adult animals.

V. CEPHALIC RESPONSES DURING FOOD INTAKE

Food intake generally causes an increase of blood glucose and insulin and a decline of FFA (Fig. 3) (Strubbe et al., 1977). These results can be satisfactorily explained by the increase of blood glucose caused by absorption of glucose from the alimentary tract. Insulin is then released after stimulation of the B cells of the islets of Langerhans by the rising blood glucose. The increased blood glucose and plasma insulin levels enhance glucose transport across the fat cell membranes, promoting lipogenesis and also causing a suppression of lipolysis, which ultimately results in a decline of FFA.

A different picture emerges if blood samples are withdrawn every minute during food intake and in the period thereafter (Fig. 4) (Strubbe and Steffens, 1975). Plasma insulin rises in the first minute, whereas blood glucose starts to increase only in the third minute after the onset of food ingestion. This preabsorptive insulin release (PIR) must be elicited by a vagally mediated reflex because either atropinization or sectioning of the vagus nerves to the pancreas abolish the PIR (Strubbe and Steffens, 1975; Louis-Sylvestre, 1978). Not only parasympathetic, but also sympathetic, activity increases at the onset of feeding, as can be seen in the rising plasma NE, E, and glucagon concentrations (Fig. 5) (De Jong et al., 1977; Le Blanc et al., 1984; Steffens et al., 1986). Thus, the cephalic phase of food intake elicits a shift of sympathetic and parasympathetic activity in the same direction, which is not in agreement with the generally accepted idea that the two divisions of the autonomic nervous system are acting antagonistically.

One might wonder first why these actions occur, and second whether these cephalic responses are similar to those elicited by either electrical or chemical stimulation of different areas in the hypothalamus. Regarding the first point, the cephalic phase might prepare the alimentary tract to digest the ingested food as described by Powley (1977). Thus, the PIR, and also other phenomena of the autonomic nervous system, might be part of a system designed to induce a smooth

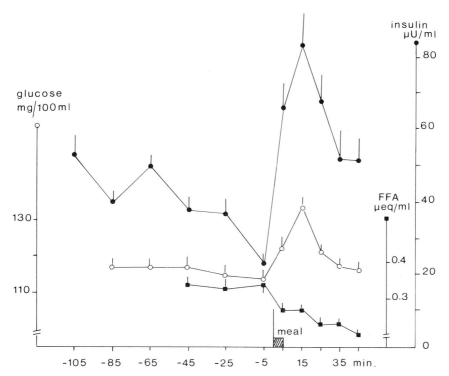

FIGURE 3 Mean blood glucose (○), FFA (■), and insulin (●) responses after ingestion of carbohydrate-rich food in rats. Vertical bars are SEM. (From Strubbe et al., 1977.)

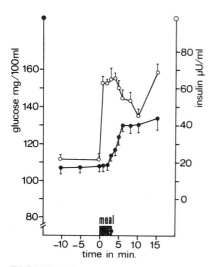

FIGURE 4 Blood glucose and insulin levels during eating carbohydrate-rich food, ad libitum. (From Strubbe and Steffens, 1975.)

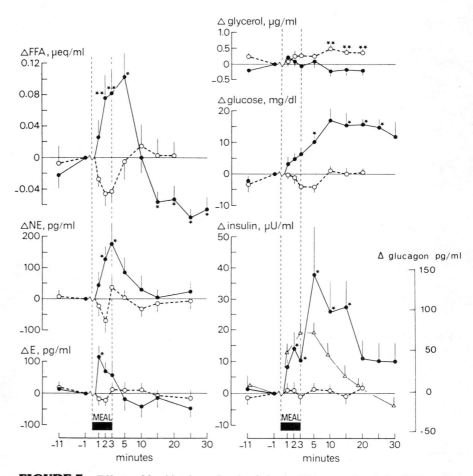

FIGURE 5 Effects of food intake on levels of plasma FFA, norepinephrine (NE), and epinephrine (E) (left) and glycerol, glucose, glucagon, and insulin (right) (●); control test without food (○). Results are means ± SEM. Asterisk denotes significant change in relation to premeal level. Two asterisks denote significant change in relation to levels in controls. (From De Jong et al., 1977; Steffens et al., 1986.)

course of digestion and conversion of absorbed nutrients into energy stores, primarily into liver glycogen.

One might suppose that the PIR is essential to the transformation of inactive liver glycogen synthetase into the active form. This seems to be an attractive explanation at first sight because the absence of a PIR leads to exaggerated glucose and insulin profiles later on during food intake and in the period immediately following food ingestion. This exaggeration is observed in experiments in which PIR has been suppressed by elimination of either the cephalic phase of food intake via direct administration of food into the stomach (Fig. 6) (Steffens, 1976) or denervation of the vagus to the islets of Langerhans (Strubbe and van Wachem, 1981). In the absence of PIR, the absorbed glucose might pass through the liver without conversion into glycogen so that too much glucose reaches the general circulation, resulting in an exaggerated blood glucose level. Then, the higher blood glucose concentration stimulates the B cells of the islets of Langerhans to release a larger amount of insulin. The simultaneous increase in plasma levels of E, NE, and glucagon might counteract the action of insulin on the activation of liver glycogen synthetase and will even result in an activation of liver glycogen

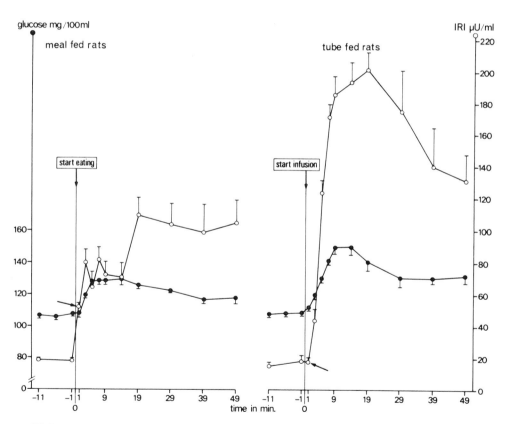

FIGURE 6 Left: Blood glucose (●) and plasma insulin (○) levels during oral ingestion of carbohydrate-rich fluid food. Start of eating at time zero. Right: Glucose (●) and insulin (○) levels during intragastric infusion of carbohydrate-rich fluid food. Entrance of fluid into the stomach at time zero. Arrow indicates insulin level 1 minute after start of ingestion or stomach infusion of food. (From Steffens, 1976.)

phosphorylase. To investigate the role of a PIR as such on a possible diminution of leakage of glucose through the liver, we carried out the following experiment.

Rats were provided with catheters in the stomach and portal vein. Rats received a meal of liquid diet directly into the stomach. Simultaneously, at the beginning of food infusion, insulin (2.5 mU) was infused into the portal vein over a period of one minute, to mimic a PIR. In control experiments saline was infused into the portal vein starting at the beginning of food infusion. An artificial PIR in the first minute suppressed only slightly the increase of blood glucose in the fifth minute. Insulin and glucose levels later on were not significantly different in the experiment with an artificial PIR and in the control experiment (Fig. 7). This result shows that whenever there is an effect of the PIR, the effect is relatively small in the short term.

The cephalic response might serve other purposes, as already suggested in literature (Cox and Powley, 1981), and as we explain below.

Regarding a possible similarity of cephalic responses to those elicited by either electrical or chemical stimulation of different areas of the hypothalamus, the fol-

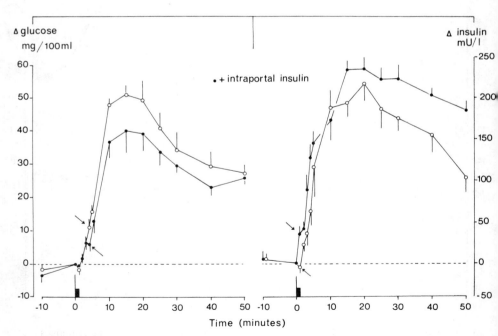

FIGURE 7 Levels of blood glucose (left) and plasma insulin (right) during intragastric infusion of carbohydrate-rich fluid food. Start of infusion at time zero: (●) with intraportal insulin infusion (2.5 mU during 1 minute, starting at time zero); (○) with intraportal saline infusion.

lowing remarks can be made. Neuroanatomical and electrophysiological studies reveal the existence of pathways between the taste receptors in the oral cavity and several areas of the hypothalamus via the solitary tract nucleus (Oomura, 1983; Rohner-Jeanrenaud et al., 1983). Note here that ingestion of foods leads to an immediately increased NE turnover in several hypothalamic and preoptic areas (van der Gugten and Slangen, 1977; McCaleb et al., 1979) and, accordingly, noradrenergic mechanisms in the hypothalamus profoundly affect the islets of Langerhans hepatocytes and adipocytes, as discussed above.

VI. HYPOTHALAMIC INFLUENCES ON GLUCOSE AND INSULIN PROFILES AND BODY WEIGHT

Rats show a clear circadian rhythmicity in food intake. A test meal at night causes a much larger increase in both glucose and insulin levels than it does during the day (Strubbe et al., 1987). Atropinization of rats before the start of a test meal at night reduces the glucose and insulin responses to those seen during the daytime, whereas daytime responses are scarcely affected by atropinization except the inhibited PIR (Fig. 8). Destruction of the suprachiasmatic nucleus (SCN) in the hypothalamus results in a complete loss of circadian rhythmicity in food intake. A test meal in this situation produces a nighttime glucose and insulin profile irrespective of the time. These results show that the SCN controls vagal activity either directly or via other hypothalamic structures, determining the circadian variation in blood glucose and plasma insulin responses after food intake.

Changes in vagal activity might alter glucose and insulin responses via its influence on motility of the gastrointestinal tract and on absorption processes. Even a contribution of the hypothalamus in controlling the sensitivity of the islets of Langerhans to a glucose challenge cannot be excluded in this situation. That this might occur has been shown in an experiment in which a glucose meal was offered during NE infusion into the LHA. An exaggerated plasma insulin response could be observed in spite of a normal blood glucose response (Fig. 9) (Steffens et al., 1984b). A similar exaggerated response was obtained when an intravenous glucose tolerance test was performed during NE infusion into the LHA. These exaggerated insulin responses could be eliminated by atropinization of the animals. Because of the higher plasma insulin levels in these circumstances, glucose transport across fat cell membranes might be increased so that more fat can be synthesized. The result of the experiments clearly shows that the hypothalamus can affect the sensitivity of the B cells to a glucose challenge.

With this result in mind, one might wonder whether a change in the sensitivity to glucose of the B cells might contribute to the obesity syndrome after a VMH lesion. It is generally accepted that the obesity occurring after a VMH lesion is the result of hyperphagia. Hyperphagia as such markedly increases the amount of nutrients being absorbed in the alimentary tract. In addition, higher gut hormone

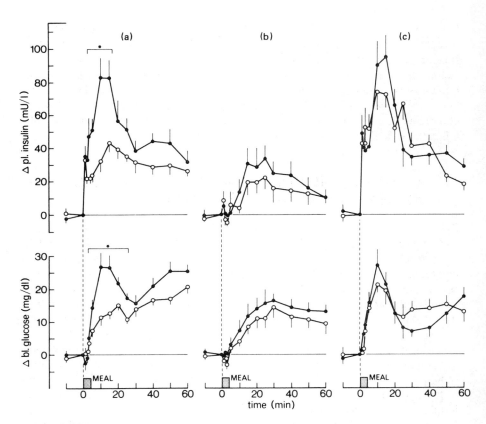

FIGURE 8 Comparison of plasma insulin and blood glucose responses to food intake (mean ± SEM) between the light (○) and the dark phase (●): (a) control condition, (b) after atropine administration, and (c) SCN lesion. Asterisk denotes a significant change ($p < 0.05$). (From Strubbe et al., 1987.)

output stimulates the release of insulin. These factors would enhance insulin secretion by the B cells of the islets of Langerhans, which could explain the high plasma insulin levels in VMH obesity. However, there are data in the literature that are not in agreement with this idea.

In 1970 Han and Frohman (1970) reported that strict pair feeding of VMH-lesioned and control rats by tube feeding led to larger fat deposition in the VMH-lesioned animals than in the nonlesioned ones. Cox and Powley (1981) described essentially the same results. In their experiments also, VMH-lesioned and control rats were strictly pair-fed. However, the VMH-lesioned rats either could eat the diet or received it directly in the stomach. Even though the VMH-lesioned animals received the food in exactly the same amount and according to the same time

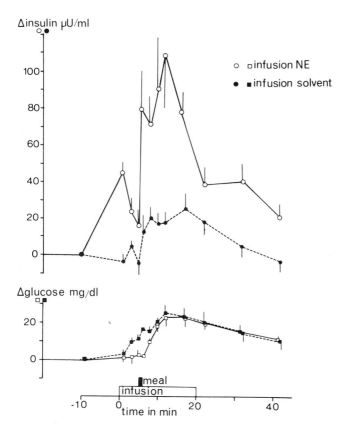

FIGURE 9 Mean changes ± SE of blood glucose (mg/dl) and plasma insulin (μU/ml) during ingestion of 135 mg glucose, which was ingested during either buffer infusion at a rate of 0.25 μl/min during 20 minutes (●) or NE infusion in the LHA (12.5 ng/min at a rate of 0.25 μl/min) during 20 minutes (○). (From Steffens et al., 1984b.)

schedule as sham-lesioned controls, the quantity of fat deposition was much larger in VMH-lesioned animals than in controls. The authors also reported a much higher basal insulin level in the pair-fed VMH-lesioned animals irrespective whether they received their food intragastrically or orally. The origin of the increased insulin level is unknown.

To obtain more information, we designed the following experiment. Two groups of rats were offered a liquid diet in seven meals a day, one at the end of the light period and six separated by equal intervals during the night period, to mimic their normal meal pattern. Each meal was eaten completely. After one week, one group received a VMH lesion, whereas the other group was sham-operated. Then the food regime was continued for 4 weeks. During the period before and after lesioning of the VMH, an intravenous glucose tolerance test (IVGTT) was performed each week in both groups of animals.

The IVGTT led to glucose profiles that were exactly identical in both groups of animals and during each test irrespective of when the IVGTT was performed. However, the insulin profiles during the IVGTT started to deviate in the first week after VMH lesion. In the course of 4 weeks, the insulin profiles were worsening in VMH-lesioned animals during the IVGTT (i.e., in the fourth week after a VMH lesion, the insulin profile was highly exaggerated in spite of a normal glucose profile). In sham-operated rats, the insulin profiles during the IVGTT were identical over time. In the VMH-lesioned animals, basal insulin levels also slowly increased over time. After termination of the experiment, the increases in weight and fat mass in VMH-lesioned animals were comparable to those reported by Cox and Powley (1981).

Our hypothesis to explain the phenomenon observed is as follows. The islets of Langerhans are continuously controlled by a balance between sympathetic and parasympathetic activity. This balance determines the sensitivity of the islets of Langerhans to a glucose load. In favor of this idea is the exaggerated insulin release that occurs after oral glucose ingestion during an intravenous fentolamine infusion (Fig. 10). Apparently fentolamine eliminates a continuous noradrenergic antagonistic effect on the B cells (Fig. 1). This result is even more convincing because blood glucose levels are lower than in the control situation, indicating that glucose is not involved at all. On the other hand, atropinization of the animal, which eliminates muscarinic activity on the B cells during an oral glucose load, leads to diminished insulin release (Strubbe and Steffens, 1975). In addition, a glucose challenge leads to an exaggerated insulin release if this balance is disrupted by either NE infusion into the LHA or destruction of the VMH. Because of that response, glucose transport across the fat cell membranes is enhanced, resulting in increased lipogenesis and consequently leading to obesity. Thus, we suppose that a VMH lesion alters the sensitivity to glucose of the B cells of the islets of Langerhans. It is not clear at present how the VMH exerts this influence. It might be that the VMH controls the LHA because increased noradrenergic activity in LHA neurons affects insulin release. Because direct pathways between the VMH and LHA are absent, indirect pathways (e.g., via the DMH) might be responsible. Alternatively, extrahypothalamic pathways might be involved. It is evident that not only the B cells but also the A cells are affected by a VMH lesion. A VMH lesion leads to an immediately enhanced glucagon release after an arginine load as compared to a control situation (Rohner-Jeanrenaud and Jeanrenaud, 1984). The role of a VMH lesion on somatostatin release from the D cells is less clear. It could be argued that increased vagal activity on the D cells leads to decreased somatostatin release, which in its turn might contribute to increased insulin and glucagon release.

Evidently our hypothesis can explain the absence of obesity in VMH-lesioned

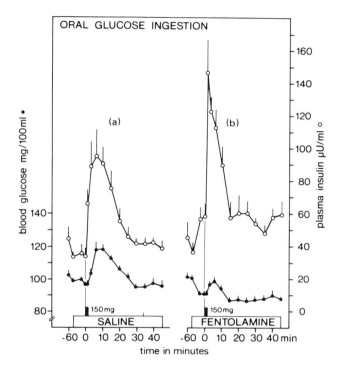

FIGURE 10 Mean changes ± SE of blood glucose (mg/dl) (●) and plasma insulin (μU/ml) (○) during ingestion of 150 mg glucose, which was ingested during intravenous infusion of either saline at a rate of 0.1 ml/min or fentolamine (10 μg/min) at a rate of 0.1 ml/min.

vagotomized rats (Powley and Opsahl, 1974) because in that situation a change of sensitivity of the islets of Langerhans to glucose and amino acids cannot occur.

VII. CONCLUDING REMARKS

Neuroanatomical, electrophysiological, and histochemical techniques have revealed that the islets of Langerhans receive dense innervation by both divisions of the autonomic nervous system. Several areas in the hypothalamus and in other brain structures can affect islet activity via these autonomic pathways, which can be considered to be the relay station and efferent parts of a reflex arc. The afferent part could consist of receptors in the oral cavity and neural pathways to the hypothalamus via the nucleus solitary tract. The original idea (Powley, 1977) that this reflex arc acts only during food ingestion is incomplete and needs revision. At present there are many indications that the islets of Langerhans are continuously controlled by a balance between the sympathetic and parasympathetic system. In addition, the hypothalamus might be involved in the regulation of the sensitivity of islet cells to glucose and amino acid loads. Because of this, the hypothalamus might contribute to the regulation of insulin and glucose profiles during the day–night cycle so that adequate amounts of absorbed nutrients are directed to different tissues in the body, ensuring that body mass and body composition are maintained reasonably constant over long periods in adulthood. Disturbances in the hypothalamus (e.g., caused by lesions) may affect glucose and insulin profiles. This chapter presents experimental evidence that these disturbances lead to the increase of body weight and obesity independently of hyperphagia and that this can be prevented by denervation of the islet of Langerhans. Apparently obesity is a syndrome that can be separated from hyperphagia.

DISCUSSION

Grill: Just a comment on your anatomy of cephalic phase. I just wanted to point out there, in addition to the data you're presenting on hypothalamic noradrenaline infusion, that the action of an infusion like that on hypothalamus is clearly mediated by caudal brain stem in order to be producing these effects in the periphery. In the absence of the forebrain in the decerebrated animal, it is possible to produce the cephalic response to oral stimulation.

REFERENCES

Blair, B., James, M. E., and Foster, J. L. (1979). Adrenergic control of glucose output and adenosine $3':5'$-monophosphate levels in hepatocytes from juvenile and adult rats. *J. Biol. Chem.* **254**:7579–7584.

Chafetz, M. D., Parko, K., Diaz, S., and Leibowitz, S. F. (1986). Relationships between medial hypothalamic α_2-receptor binding, norepinephrine, and circulating glucose. *Brain Res.* **384**:404–408.

Cox, J. E., and Powley, T. L. (1981). Intragastric pair feeding fails to prevent VMH obesity or hyperinsulinemia. *Am. J. Physiol.* **240**:E566–E572.

De Jong, A., Strubbe, J. H., and Steffens, A. B. (1977). Hypothalamic influence on insulin and glucagon release in the rat. *Am. J. Physiol.* **233**:E380–E388.

Edwards, A. V. (1984). Neural control of the endocrine pancreas. In *Recent Advances in Physiology*, Vol. 10, P. F. Baker (Ed.). Churchill & Livingstone, London, pp. 277–315.

Han, P. W., and Frohman, L. A. (1970). Hyperinsulinemia in tube-fed hypophysectomized rats bearing hypothalamic lesions. *Am. J. Physiol.* **219**:1632–1636.

Kaneto, A., Kosaka, K., and Nakao, K. (1967). Effects of stimulation of the vagus nerve on insulin secretion. *Endocrinology*, **80**:530–536.

Kaneto, A., Miki, E., and Kosaka, K. (1974). Effects of vagal stimulation on glucagon and insulin secretion. *Endocrinology*, **95**:1005–1010.

Lafontan, M., Berlan, M., and Carpene, C. (1985). Fat cell receptors: Inter- and intraspecific differences and hormone regulation. *Int. J. Obesity,* **9**:117–128.

Le Blanc, J., Cabanac, M., and Samson, P. (1984). Reduced postprandial heat production with gavage as compared with meal feeding in human subjects. *Am. J. Physiol.* **246**:E95–E101.

Leibowitz, S. F. (1986). Brain monoamines and peptides: Role in the control of eating behavior. *Fed. Proc.* **45**:1396–1403.

Lichtenstein, S., Marinescu, C., and Leibowitz, S. F. (1985). Chronic infusion of norepinephrine and clonidine into the hypothalamic paraventricular nucleus. *Brain Res. Bull.* **13**:591–595.

Louis-Sylvestre, J. (1978). Relationship between two stages of prandial insulin release in rats. *Am. J. Physiol.* **235**:E119–E125.

Luiten, P. G. M., ter Horst, G. J., and Steffens, A. B. (1987). The hypothalamus, intrinsic connections and outflow of pathways to the endocrine system in relation to the control of feeding and metabolism. *Prog. Neurobiol.* **28**:1–54.

Malaisse, W. J., Malaisse-Lagae, F., Wright, P. H., and Ashmore, J. (1967). Effects of adrenergic and cholinergic agents upon insulin secretion in vitro. *Endocrinology,* **80**:975–978.

McCaleb, M. L., Meyers, R. D., Singer, G., and Willis, G. (1979). Hypothalamic norepinephrine in the rat during feeding and push–pull perfusion with glucose, 2DG, or insulin. *Am. J. Physiol.* **236**:R312–R321.

Miller, R. E. (1981). Pancreatic neuroendocrinology: Peripheral neural mechanisms in the regulation of the islets of Langerhans. *Endocr. Rev.* **2**:471–494.

Morley, J. E., Levine, A. S., Gosnell, B. A., and Krahn, D. D. (1985). Peptides as central regulators of feeding. *Brain Res. Bull.* **14**:511–519.

Müller, W. A., Faloona, G. R., and Unger, R. H. (1971). The influence of the antecedent diet upon glucagon and insulin secretion. *New Engl. J. Med.* **285**:1450–1454.

Nieuwenhuijs, R. (1985). *Chemoarchitecture of the Brain.* Springer-Verlag, Berlin.

Oomura, Y. (1983). Glucose as a regulator of neuronal activity. In *Advances in Metabolic Disorders,* Vol. 10, A. J. Szabo (Ed.). Academic Press, New York, pp. 31–65.

Porte, D., Jr., and Robertson, R. P. (1973). Regulation of insulin secretion by catecholamines, stress and the sympathetic nervous system. *Fed. Proc.* **32**:1792–1796.

Powley, T. L. (1977). The ventromedial hypothalamic syndrome, satiety, and a cephalic phase hypothesis. *Psychol. Rev.* **84**:89–126.

Powley, T. L., and Opsahl, C. A. (1974). Ventromedial hypothalamic obesity abolished by subdiaphragmatic vagotomy. *Am. J. Physiol.* **226**:25–33.

Rohner-Jeanrenaud, F., Bobbioni, E., Ionescu, E., Sauter, J. F., and Jeanrenaud, B. (1983). Central nervous system regulation of insulin secretion. In *Advances in Metabolic Disorders,* Vol. 10, A. J. Szabo (Ed.). Academic Press, New York, pp. 193–220.

Rohner-Jeanrenaud, F., and Jeanrenaud, B. (1984). Oversecretion of glucagon by pancreases of ventromedial hypothalamic-lesioned rats: A reevaluation of a controversial topic. *Diabetologia,* **27**:535–539.

Samols, E., and Weir, G. C. (1979). Adrenergic modulation of pancreatic A, B, and D cells. α-adrenergic suppression of β-adrenergic stimulation of somatostatin, α-adrenergic stimulation of glucagon secretion in the perfused dog pancreas. *J. Clin. Invest.* **63**:230–238.

Samols, E., Stagner, J. I., and Weir, G. C. (1981). Autonomic function and control of pancreatic somatostatin. *Diabetologia,* **20**:388–392.

Scheurink, A. J. W., Steffens, A. B., and Benthem, L. (1988). Central and peripheral adrenoceptors affect glucose, free fatty acids, and insulin in exercising rats. *Am. J. Physiol.* **255**:R547–R556.

Scheurink, A. J. W., Steffens, A. B., Bouritius, H., Dreteler, G. H., Bruntink, R., Remie, R., and Zaagsma, J. (1989). Sympathoadrenal influence on glucose, FFA, and insulin levels in exercising rats. *Am. J. Physiol.* **256**:R161–R168.

Schusdziarra, V., Zyznar, E., Rouiller, D., Boden, D., Brown, J. C., Arimura, A., and Unger, R. H. (1980). Splanchnic somatostatin: A hormonal regulator of nutrient homeostasis. *Science,* **207**:530–532.

Shimazu, T. (1986). Neuronal control of intermediate metabolism. In *Neuroendocrinology,* S. Lightman and B. Everitt (Eds.). Blackwell, Oxford, pp. 304–330.

Shimazu, T., and Ishikawa, K. (1981). Modulation by the hypothalamus of glucagon and insulin secretion in rabbits: Studies with electrical and chemical stimulations. *Endocrinology,* **108**:605–611.

Shimazu, T., Noma, M., and Saito, N. (1986). Chronic infusion of norepinephrine into the ventromedial hypothalamus induces obesity in rats. *Brain Res.* **369**:215–223.

Smith, U. (1983). Adrenergic control of lipid metabolism. *Acta Med. Scand. Suppl.* **672**:41–47.

Steffens, A. B. (1976). The influence of the oral cavity on the release of insulin in the rat. *Am. J. Physiol.* **230**:1411–1415.

Steffens, A. B., Scheurink, A. J. W., Luiten, P. G. M., and Bohus, B. (1988). Hypothalamic food intake regulating areas are involved in the homeostasis of blood glucose and plasma FFA levels. *Physiol. Behav.* **44**:581–589.

Steffens, A. B., and Strubbe, J. H. (1983). CNS regulation of glucagon secretion. In *Advances in Metabolic Disorders,* Vol. 10, A. J. Szabo (Ed.). Academic Press, New York, pp. 221–257.

Steffens, A. B., Damsma, G., van der Gugten, J., and Luiten, P. G. M. (1984a). Circulating free fatty acids, insulin, and glucose during chemical stimulation of hypothalamus in rats. *Am. J. Physiol.* **247**:E765–E771.

Steffens, A. B., Flik, G., Kuipers, F., Lotter, E. C., and Luiten, P. G. M. (1984b). Hypothalamically induced insulin release and its potentiation during oral and intravenous glucose loads. *Brain Res.* **301**:351–361.

Steffens, A. B., van der Gugten, J., Godeke, J., Luiten, P. G. M., and Strubbe, J. H. (1986). Meal-induced increases in parasympathetic and sympathetic activity elicit simultaneous rises in plasma insulin and free fatty acids. *Physiol. Behav.* **37**(1):119–122.

Strubbe, J. H., Alingh Prins, A. J., Bruggink, J., and Steffens, A. B. (1987). Daily variation of food-induced changes in blood glucose and insulin in the rat and the control by the suprachiasmatic nucleus and the vagus nerve. *J. Auton. Nerv. Syst.* **20**:113–119.

Strubbe, J. H., and Bouman, P. R. (1978). Plasma insulin patterns in the unanesthetized rat during intracardial infusion and spontaneous ingestion of graded loads of glucose. *Metabolism,* **27**:341–351.

Strubbe, J. H., and Steffens, A. B. (1975). Rapid insulin release after ingestion of a meal in the unanesthetized rat. *Am. J. Physiol.* **229**(4):1019–1022.

Strubbe, J. H., and van Wachem, P. (1981). Insulin secretion by the transplanted neonatal pancreas during food intake in fasted and fed rats. *Diabetologia,* **20**:228–236.

Strubbe, J. H., Steffens, A. B., and de Ruiter, L. (1977). Plasma insulin and the time pattern of feeding. *Physiol. Behav.* **18**:81–86.

Takahishi, A., and Shimazu, T. (1981). Hypothalamic regulation of lipid metabolism in the rat: Effect of hypothalamic stimulation on lipolysis. *J. Auton. Nerv. Syst.* **4**:195–205.

Unger, R. H., Dobbs, R. E., and Orci, L. (1978). Insulin, glucagon, and somatostatin secretion in the regulation of metabolism. *Annu. Rev. Physiol.* **40**:307–343.

Van der Gugten, J., and Slangen, J. L. (1977). Release of endogenous catecholamines from the rat hypothalamus related to feeding and other behaviors. *Pharmacol. Biochem. Behav.* **7**:211–219.

Yoshimatsu, H., Niijima, A., Oomura, Y., Yamabe, K., and Katafuchi, T. (1984). Effects of hypothalamic lesion on pancreatic autonomic nerve activity in the rat. *Brain Res.* **303**:147–152.

21
Central Nervous System Pathways and Mechanisms Integrating Taste and the Autonomic Nervous System

David F. Cechetto

The John P. Robarts Research Institute
London, Ontario, Canada

I. INTRODUCTION

Historically, investigators involved in experimental work aimed at elucidating mechanisms integrating gustatory or general autonomic responses have focused almost exclusively on one or the other. However, many recent electrophysiological, immunohistochemical, and neuroanatomical investigations demonstrate that gustatory and general autonomic functions are very closely related in the central nervous system. For example, the sensory representation for special visceral (gustatory) and general visceral receptors has been shown to be in close apposition throughout the neuraxis, beginning in the nucleus of the solitary tract, the primary site of termination of many visceral afferents, to the cortex, the highest site of central nervous system integration.

There are a number of ways in which gustatory and/or feeding responses may interact with central control of the autonomic nervous system. It is not the intent here to review all these mechanisms, but to briefly examine some of the more important interactions. In the latter, and major, portion of this chapter, I discuss recent results that may provide the neuroanatomical and functional substrate for the integration of general visceral sensation and autonomic responses with feeding and gustatory mechanisms.

II. GUSTATORY, FEEDING, AND GENERAL VISCERAL INTERACTIONS

There are at least two possible ways in which gustatory and feeding responses interact with general visceral afferent and efferent responses: general visceral inputs may alter gustatory responses and feeding behaviors, and gustatory information may change the central neuronal responses to general visceral receptor input or induce changes in autonomic output. Let us begin with the first mode of response.

General visceral input can be shown to influence gustatory responses. It is possible for visceral afferents to affect the perceived quality of a gustatory input. Fasting human subjects presented with a sweet stimulus rated a sample as unpleasant following the ingestion of a certain quantity of sucrose or glucose (Cabanac, 1971). This rating contrasted with the pleasant rating prior to the sucrose and glucose intake. This suggests that the change in internal state resulted in an alteration of the gustatory response. A later investigation indicated that this effect is likely mediated by afferent impulses in duodenal chemoreceptors (Cabanac and Fantino, 1977). The neural basis of the changes in the quality of taste following increases in nutrient intake has been provided by electrophysiological investigations. These studies have demonstrated that central gustatory neuronal responsiveness is altered by increases in blood glucose levels mediated by central glucoreceptors or vagal afferents from the gut (Giza and Scott, 1983). Evidence for vagal afferents, and not central glucoreceptors, modulating gustatory responses has been provided by the demonstration that gustatory neurons in the brain stem receive converging taste and hepatic inputs (Hermann et al., 1983).

In addition to changes in blood glucose levels, gustatory neuronal responses are altered by changes in extracellular sodium concentration (Contreras, 1977; Contreras and Frank, 1979). These investigations demonstrated that afferent fibers in gustatory sensory nerves responding best to sodium chloride taste stimulation were most affected by sodium deprivation. This suggests a possible mechanism for elucidating the importance of gustatory stimulation in the initiation of sodium ingestion and the satiation of sodium appetite (Contreras, 1977).

Activation of general visceral receptors may not only change gustatory responses but may also induce behavioral changes related to feeding. Early experimental evidence demonstrated that the oral intake of food or water is inhibited by the introduction of food or water into the stomach (Janowitz and Grossman, 1949; Towbin, 1949). The introduction of an inert substance or balloons directly into the stomach was also shown to inhibit feeding (Janowitz and Grossman, 1949; Share et al., 1952). These studies and others have indicated that there is good correlation of events in the stomach with the control of food intake and that this is likely to be a mechanical mechanism (Deutsch et al., 1978; Deutsch, 1982; Novin, 1983). A number of investigations have examined the properties of gastric mechanoreceptors in the splanchnic and vagal nerves (Paintal 1954a,b; Iggo, 1955; Ranieri et al., 1973). The gastric mechanoreceptors carried in the vagus enter the brain stem and terminate in the nucleus of the solitary tract, while the afferents in the splanchnic nerves enter the spinal cord and ascend via spinal–thalamic tracts or, as discussed in more detail later, a spinal–parabrachial pathway.

Both changes in the caloric content of food administered intragastrically and removal of portions of an oral meal were shown also to alter the amount of food ingested (Share et al., 1952; Le Magnen, 1983). This suggests that there is another component of satiation by the gut that may be mediated by chemoreceptors (Le Magnen, 1983). There is evidence that vagal afferents from chemoreceptors exist in the gut (Jeanningros, 1982). It has been shown that hypertonic solutions injected into the duodenum are effective in reducing the meal size in rats and pigs (Snowdon, 1975; Houpt et al., 1983; Mei, 1985). On the other hand, nutritive rather than hypertonic infusions into the rat stomach have been shown to be effective in reducing meal size (Snowdon, 1975).

In addition to the stomach and duodenum, the liver has been implicated in the control of food intake (Novin, 1976, 1983). In 1971 Russek demonstrated that intraperitoneal infusions of glucose and adrenaline were significantly more effective in inhibiting food intake than direct intravenous injection. The intraperitoneal injections would presumably pass through the hepatic vasculature before being distributed systemically, suggesting that the liver has an important chemoreceptor function in the control of food intake. However, in this regard, the infusion of nutrients directly into the hepatic–portal vein has yielded equivocal results. Some of the investigations have demonstrated an inhibition of feeding (Novin et al., 1974; Novin, 1976, 1983; Anil and Forbes, 1980), while others have not been able to demonstrate an effect of the liver directly on feeding (Stephens and Baldwin, 1974; VanderWeele et al., 1974; Bellinger et al., 1976; Strubbe and Steffens, 1977). On the other hand, it has been shown that stimulation of hepatic osmoreceptors with low concentrations of glucose increases afferent discharge in the hepatic vagus and that changes in both sodium chloride and glucose concentration alter drinking behavior (Niijima, 1969; Blake and Lin, 1978).

The evidence discussed above indicates that the vagus appears to be the afferent projection for visceral information from the gut and liver involved in control of food or fluid intake mechanisms (Mei, 1985). It is reasonable to expect that vagotomy will alter the ingestive responses to activation of visceral receptors. Subdiaphragmatic vagotomy effectively blocks the inhibition of feeding in response to stimulation of both hepatic and duodenal chemoreceptors (Novin, 1983). In addition, subdiaphragmatic vagotomy alters dietary selection in rats (Fox et al., 1976; Sclafani and Kramer, 1983). Finally, abdominal vagotomy has been demonstrated to decrease water intake (Kraly et al., 1975; Kraly, 1978; Smith and Jerome, 1983).

The preabsorptive reflex responses to the sensory properties of food are referred to as the cephalic phase (Berridge and Grill, 1981; Norgren, 1984). These are responses to taste, odor, and tactile stimulation of the food and result in a number of autonomic responses including changes in salivation, arterial blood pressure, regional vascular flow, heart rate, cardiac output, respiration, gastric motility, gastric acid secretion, and exocrine and endocrine secretions.

The cardiovascular changes associated with food intake include increases in cardiac output and arterial pressure (Vatner et al., 1970a,b). The transient increases in cardiac output appear to be due to an increase in heart rate accompanied by a fall in stroke volume (Fronek and Stahlgren, 1968; Vatner et al., 1970a,b; Kelbaek et al., 1987). The initial arterial pressure increases were the result of transient increases in renal and mesenteric resistance and transient decreases in iliac and coronary resistance (Vatner et al., 1970a,b). A more prolonged increase in mesenteric blood flow began 5–15 minutes after the presentation of food. Recently, a careful analysis of the time course of cardiovascular changes in response to food presentation and intake has demonstrated that the transient changes are coincident with the beginning of eating but not with the presentation of the food (Matsukawa and Ninomiya, 1985, 1987). These changes then may be mediated by the gustatory stimulus, or they may be of a central origin associated with the eating behavior.

An interesting phenomenon that has frequently been investigated is the larger increase in insulin release following oral administration of glucose compared to an intravenous glucose route (Scow and Cornfield, 1954; Elrick et al., 1964; Perley and Kipnis, 1967; Fischer et al., 1972). The insulin release occurs prior to changes in blood glucose levels. This has often been explained as a neurally mediated release of insulin in response to meal-related stimuli (Louis-Sylvestre, 1976). In support of this contention are the findings indicating that an early phase of insulin release is eliminated following an intragastric administration of food compared to an oral route (Hommel et al., 1972). The neurogenic release of insulin is supported by the demonstration of extensive innervation of the pancreatic islets (de Castro, 1923) and the release of insulin in response to electrical stimulation of the vagus nerve (Daniel and Henderson, 1967). In addition to meal-related stimuli for insulin release presumably mediated by taste, odor, or tactile sensation, there is recent evidence that general visceral afferents such as arterial chemoreceptors are capable of eliciting an increase in the output of hepatic glucose (Alvarez-Buylla and de Alvarez-Buylla, 1988).

In summary, it is clear that activation of visceral receptors is capable of eliciting a number of alterations in central gustatory responses and feeding behaviors, including satiation. The evidence indicates that these effects are mediated by vagal nerves with a primary site of termination in the nucleus of the solitary tract. However, these results do not preclude other visceral inputs of a spinal origin. In addition, meal-related stimuli, such as gustatory activation, activate a number of autonomic changes, including cardiovascular and neurogenic or endocrine metabolic responses. These interactions presumably must be integrated by inputs to the appropriate sites in the brain for gustatory, feeding, and autonomic responses.

III. CENTRAL NERVOUS SYSTEM MECHANISMS AND PATHWAYS

A. Insular Cortex

The possibility that gustatory and feeding responses can be integrated with general visceral afferent and efferent responses at the level of the insular cortex is provided by the evidence that all these are represented in this cortical region. The evidence to be reviewed indicates that sensory representation of taste and general visceral input occurs in the insular cortex and that the autonomic responses and ingestive behaviors are integrated in this site.

A role of the insular cortex in gustatory behaviors has been known since 1901 (Norgren, 1984; Augustine, 1985). Numerous investigations have since demonstrated a gustatory sensory area in the rostral insular cortex (Benjamin and Pfaffmann, 1955; Benjamin and Akert, 1959; Emmers, 1966; Ganchrow and Erickson, 1972; Yamamoto and Kawamura, 1972; Norgren and Wolf, 1975; Yamamoto et al., 1980b, 1981a,b). Early investigations, using evoked potentials, suggested the possibility that general visceral information also might be projecting to the region of the insular cortex. Stimulation of the cut central end of the vagus in the encéphale isolé cat, resulted in an increase in the rate and amplitude of EEG activity in the lateral frontal cortex in the vicinity of the insular cortex (Bailey and Bremer, 1938). In 1951 Dell and Olson, also stimulating the central end of the severed cervical vagus, elicited short latency evoked potentials in the insular cortex of the cat. More recently, Yamamoto et al. (1980b) stimulated the cranial nerves involved in gustatory responses and mapped the insular cortex for the distribution of evoked responses from these nerves. The results of this investigation indicated an anterior–posterior topographical gradient to the representations of the facial, glossopharyngeal, and vagus nerves and that the glossopharyngeal representation was considerably larger than the area specifically devoted to gustatory representation (Yamamoto et al., 1980b). These electrophysiological investigations highly suggest that there is general autonomic representation in the insular cortex.

Recent studies have supported this conclusion by recording single neuron activity in the insular cortex in response to activation of specific visceral receptors (Cechetto and Saper, 1987a). The visceral inputs included taste receptors, arterial baroreceptors, arterial chemoreceptors, and gastric mechanoreceptors. The results of the extracellular recording demonstrated that there is a viscerotopic organization of the inputs. Gustatory responses were obtained primarily from a rostral dysgranular region of the insular cortex. Gastric mechanoreceptors projected to neurons in granular insular cortex immediately dorsal to the taste inputs.

Finally, the cardiorespiratory information terminated on neurons in the posterior granular insular cortex dorsal and posterior to the taste region. The neurons in the insular cortex responded primarily to only one visceral input. These results demonstrate that there is a region of the cortex with a sensory representation for both taste and general visceral inputs, including gastric afferents. This suggests that these inputs could be integrated at this level to produce the integration of behaviors outlined previously.

The loss of taste discrimination was one of the earliest demonstrations that the insular cortex is involved in gustatory functions (Campbell, 1905). Since that time, other investigators have provided evidence for a role of the insular cortex in taste discrimination (Benjamin and Akert, 1959; Yamamoto et al., 1980b, 1981a,b; Lasiter et al., 1985a,b,c). Gastric mechanoreceptors have been implicated in the short-term satiation of hunger and thirst (as discussed previously: Janowitz and Grossman, 1949; Towbin, 1949; Share et al., 1952). In the experiments outlined above, involving recording of neurons in the insular cortex in response to visceral inputs, the gastric mechanoreceptors were shown to be located immediately adjacent to the taste-responsive units. In view of the role of the insular cortex in gustatory behaviors such as conditioned taste aversions and discriminating taste quality (Campbell, 1905; Benjamin and Akert, 1959; Grill and Norgren, 1978; Yamamoto et al., 1980b, 1981a; Lasiter et al., 1985a,c), it is likely that the close proximity of the gustatory and gastric regions in the insular cortex may have functional significance. Thus, the insular cortex could regulate ingestion by integrating signals for both the perceived quality and the quantity of food or water consumption. Although the response of neurons to gastric chemoreceptors was not tested, the possibility exists that the vagal afferents for this sensation also project to the level of the insular cortex. In this case, the nutritive content of food could also be involved in modulating ingestive behaviors at the level of the cortex. However, this possibility remains to be tested.

The demonstration of neuroanatomical connections of the insular cortex with subcortical areas that have been implicated in gustatory, feeding, and autonomic control also supports the possibility that this cortical region has an important role in the integration of these behaviors. Injections of horseradish peroxidase at sites of physiologically characterized neurons in the insular cortex demonstrated a topographical organization from the ventral basal thalamus (Cechetto and Saper, 1987a). The gustatory dysgranular insular cortex was shown to be reciprocally connected with the ventroposteriomedial parvocellular nucleus of the thalamus. This nucleus has been demonstrated to receive taste inputs in a variety of species (as outlined in more detail below; Norgren, 1984). The general visceral, posterior granular region of the insular cortex is reciprocally connected with a parvocellular part of the ventral basal thalamus immediately lateral to the gustatory thalamus. This general visceral parvocellular thalamic region is termed the ventroposteriolateral parvocellular thalamic nucleus (Cechetto and Saper, 1987a).

The insular cortex also has extensive connections with the nucleus of the solitary tract, the parabrachial nucleus, the lateral hypothalamic nucleus, and the amygdala (Ross et al., 1981; Saper, 1982a,b; Shipley, 1982; van der Kooy et al., 1982, 1984; Ruggiero et al., 1987). The involvement of all these sites in both gustatory responses and central autonomic control has been demonstrated, as described later.

The methods of combined immunohistochemical staining and retrograde tract tracing have demonstrated that the afferent pathways to the insular cortex are composed of neurons containing a variety of neuropeptides (Mantyh and Hunt, 1984; Shimada et al., 1985a,b). One of these investigations indicated that the insular cortex receives a calcitonin gene-related peptide (CGRP) projection from the ventral basal thalamus (Shimada et al., 1985b). However, the injection site of the retrograde tracer in this study is situated caudal to the insular cortex in the perirhinal and temporal cortices. Because there appears to be a pronounced projection of CGRP neurons in many levels of the ascending gustatory and general visceral afferent system, a study was done to reinvestigate the precise sources of CGRP-like immunoreactive fibers to the insular cortex (Yasui et al., 1988). The CGRP terminals in the lateral frontal cortex are most dense in the perirhinal cortex, although a lighter pattern of staining is also observed in the insular cortex. The positively stained cell bodies in the thalamus are seen to be medial to the parvocellular gustatory and general visceral region in the thalamus and contained primarily in the subparafascicular nucleus of the thalamus. This group of CGRP-like neurons continues more posteriorly in the thalamus, migrating laterally immediately dorsal to the medial lemniscus. At the most posterior level of these CGRP neurons, there is a cluster of positively stained cell bodies medial and ventral to the medial geniculate nucleus. This lateral group of CGRP neurons is contained in the lateral subparafasciculus nucleus. Injections of fluoral gold into the insular cortex combined with immunostaining with CGRP demonstrated that only a few of the CGRP neurons in the lateral parafasciculus nucleus project to this cortical area (Yasui et al., 1988). There were no retrogradely labeled neurons in the gustatory, or general visceral parvocellular, relay region of the thalamus positively stained for CGRP. However, the injections of fluoral gold into the insular cortex demonstrated a CGRP-like projection from the ipsilateral ventral lateral parabrachial subnucleus (Yasui et al., 1988). Injections of fluoral gold into the perirhinal cortex resulted in numerous neurons in the subparafascicular and lateral subparafascicular nuclei that were both retrogradely labeled and CGRP-like. Thus, it appears that CGRP neurons are not involved in the relay of visceral information from the parvocellular region of the ventral basal thalamus to the insular cortex.

Stimulation and ablation investigations have indicated that the insular cortex is an important cortical region in central autonomic regulation. Earlier studies demonstrated that electrical stimulation of the insular cortex evokes a variety of autonomic responses such as changes in blood pressure, heart rate, piloerection, pupillary dilatation, gastric motility, peristaltic activity, and adrenaline secretion in cats, dogs, monkeys, and humans (Kaada, 1951; Hoffman and Rasmussen, 1953; Von Euler and Folkow, 1958; Delgado, 1960; Hall et al., 1977). In a more recent investigation, microstimulation of the insular cortex elicited increases in arterial pressure (Ruggiero et al., 1987). Thus, the insular cortex not only receives multiple visceral sensations, including taste, but in turn is able to influence a number of autonomic responses.

The pathways mediating the autonomic responses from the insular cortex were examined using microstimulation of the insular cortex to elicit changes in sympathetic nerve discharge combined with synaptic with cobalt in subcortical sites (Cechetto et al., 1988). These results demonstrated that cobalt injections into the lateral hypothalamic area effectively blocked the sympathetic response evoked by stimulation of the insular cortex. Interestingly, the lateral hypothalamic area has long been implicated in the central control of feeding.

B. Ventral Basal Thalamus

In 1951 Dell and Olson demonstrated that electrical stimulation of the central end of the transected cervical vagus in the cat elicited short-latency evoked potentials in the ventral basal thalamus. Subsequent investigations have indicated that gustatory information terminates on neurons in the ventroposteriomedial parvocellular nucleus of the thalamus (Ganchrow and Erickson, 1972; Norgren and Wolf, 1975). However, until recently there has been little evidence that visceral inputs other than taste terminate in this thalamic region. In rats, changes in activity of single-unit recordings of neurons in the ventral basal thalamus were made in response to activation of a variety of visceral receptors, including arterial baroreceptors and chemoreceptors, gustatory receptors, and gastric mechanoreceptors (Cechetto and Saper, 1987b). The results of these experiments confirmed the taste input to the ventroposteriomedial parvocellular nucleus of the thalamus. In addition, the projection of general visceral information to an adjoining parvocellular nucleus, the ventroposteriolateral nucleus, was demonstrated. Similar to the cortex, gastric afferents terminated in close proximity to the taste region, while the arterial baroreceptor and chemoreceptor responsive neurons were more lateral to the taste neurons (Cechetto and Saper, 1987b).

An earlier investigation using rats also demonstrated a general visceral input to the ventral basal thalamus (Rogers et al., 1979a,b). Responses of extracellularly recorded neurons in the rat thalamus were observed following infusions of small amounts of sodium chloride into the hepatic–portal vein. A band of cells was found that responded to the hepatic osmotic stimulation immediately adjacent to the gustatory ventroposteriomedial parvocellular nucleus of the thalamus. Thus, these electrophysiological recording studies have provided the neural substrate for integration of gustatory and feeding responses with general visceral information at the level of the thalamus. However, the precise interaction of taste input with other visceral information and the mechanisms mediated by this thalamic area remain to be established.

Neuroanatomical and electrophysiological investigations have demonstrated that the primary site of termination of gustatory afferents is in the nucleus of the solitary tract (Norgren and Leonard, 1973; Norgren, 1976, 1984). The rostral part of the nucleus, in turn, has an extensive projection to the ventroposteriomedial parvocellular nucleus of the thalamus (Norgren, 1984). Earlier investigations have provided evidence for a projection from the parabrachial nucleus to the ventroposteriomedial parvocellular nucleus of the thalamus and to other thalamic nuclei (Norgren, 1984; Ogawa et al., 1984a,b, 1987; Hayama and Ogawa, 1987). The exact source of gustatory inputs to the thalamus has not been investigated until recently. General visceral primary afferents are organized in a manner parallel to the gustatory system. The primary site of termination of general visceral afferents is in the caudal part of the nucleus of the solitary tract (Calaresu et al., 1984). There are extensive projections from this region of the solitary tract to the parabrachial nucleus (Loewy and Burton, 1978; Ricardo and Koh, 1978). Recently, the precise relationship of the projection from the parabrachial nucleus to the taste and general visceral relay nuclei of the thalamus has been investigated (Cechetto and Saper, 1987a,b). To identify specific parabrachial subnuclei projecting to the parvocellular ventral basal thalamus, injections of wheat germ agglutinin conjugated to horseradish peroxidase (WGA-HRP) were injected into the visceral relay nuclei of the thalamus to retrogradely label neurons in the parabrachial nucleus. These results indicate that there is a diffuse projection to the gustatory nucleus of the thalamus from many ipsilateral parabrachial subnuclei and a pronounced contralateral projection from the rostral part of the external medial subnucleus (Cechetto and Saper, 1987a,b). However, the injections into the gustatory thalamus also included surrounding intralaminar nuclei. There is a less extensive projection from the ipsilateral parabrachial nucleus to the parvocellular general visceral nucleus of the thalamus. However, there is an intense projection from the caudal part of the contralateral external medial parabrachial subnucleus to the general visceral ventroposteriolateral parvocellular nucleus of the thalamus. The pattern of terminal fields in the thalamus of the projection from the external medial parabrachial subnucleus in the thalamus was demonstrated by injecting three different anterograde tracers, WGA-HRP, phytohemagluttinin-L, and [^{3}H]-amino acids into the parabrachial nucleus (Cechetto and Saper, 1987b). These results confirmed that there is a specific, topographical projection from the contralateral external medial parabrachial subnucleus to the visceral relay nuclei of the thalamus. The ipsilateral projection from the parabrachial nucleus essentially avoids the visceral relay nuclei and terminates in the surrounding intralaminar nuclei (Cechetto and Saper, 1987b).

Immunohistochemical staining with antiserum to CGRP indicates that the parvocellular visceral portion of the thalamus is heavily innervated with terminals containing this peptide (Kawai et al., 1985; Yasui et al., 1988). Destruction of the parabrachial nucleus in the rat resulted in a moderate decrease in the fibers containing CGRP in the ventral basal thalamus (Shimada et al., 1985b). A recent investigation using combined immunohistochemical staining and retrograde labeling has indicated the origin of some of the CGRP terminals in the visceral thalamus (Yasui et al., 1988). Fluoral gold injections into the ventral basal thalamus, which included the visceral relay thalamic nuclei as well as some intralaminar nuclei, result in a number of retrogradely labeled neurons in the parabrachial nucleus that were also positively stained for CGRP. The most prominent CGRP projection is from the contralateral external medial parabrachial subnucleus to the ventral basal thalamus (Yasui et al., 1988). This suggests that the projection of visceral sensation from the parabrachial nucleus to the thalamus is CGRPergic.

C. Brain Stem

As indicated above, a large number of investigations have demonstrated a topographically organized and complex projection from the nucleus of the solitary tract to the parabrachial nucleus. Recording of neuronal activity in the parabrachial nucleus in response to activation of gustatory receptors demonstrates that taste afferents terminate in the medial and lateral parabrachial subnuclei (Bernard and Nord, 1971; Norgren and Pfaffmann, 1975; Perotto and Scott, 1976; Yamamoto et al., 1980a; Schwartzbaum and Block, 1981; Di Lorenzo and Schwartzbaum, 1982a,b; Norgren, 1984; Ogawa and Hayama, 1984, 1987; Travers and Smith, 1984; Hermann and Rogers, 1985; Travers et al., 1987). Several of these investigations have demonstrated neuronal responses laterally in the parabrachial nucleus in a region corresponding to the external medial subnucleus (Norgren and Pfaffmann, 1975; Schwartzbaum and Block, 1981; Di Lorenzo and Schwartzbaum, 1982a,b). The termination of other visceral sensory modalities in the parabrachial nucleus has not been as extensively investigated. Cardiovascular and respiratory afferents have been shown to project to the parabrachial nucleus (Sieck and Harper, 1980; Ward et al., 1980; Hamilton et al., 1981; Cechetto and Calaresu, 1983), but the individual subnuclear organization has not been documented.

Convergence of gustatory and general visceral afferents on neurons within the parabrachial nucleus has been shown (Hermann and Rogers, 1985). In this investigation, units in the parabrachial nucleus of the rat were tested for responsiveness to electrical stimulation of both the cervical vagus and the tongue. Neurons responsive to the tongue were also tested for a specific taste input. These results indicate a separation of vagal and gustatory responsive neurons in the parabrachial nucleus. However, there is a discretely localized group of neurons receiving both types of input. Similarly, within the nucleus of the solitary tract there is evidence for the convergence of general visceral afferents such as gastric distension and blood glucose distension onto gustatory neurons (Bereiter et al., 1981; Travers et al., 1987). These investigations represent some of the clearest evidence for a neural substrate for the integration of gustatory and general visceral responses.

Recent evidence has indicated that there is a possible spinal source of visceral input to the parabrachial nucleus in addition to the gustatory and cardiorespiratory afferents carried in cranial nerves (Cechetto et al., 1985). Injections of WGA-HRP into the parabrachial nucleus resulted in extensive labeling of lamina I neurons throughout the entire length of the spinal cord. This projection was confirmed by anterograde labeling of terminals in the parabrachial nucleus following WGA-HRP injections into the spinal cord (Cechetto et al., 1985). Thus, it is possible that this projection is mediating pain, temperature, or sympathetic afferent inputs to the parabrachial nucleus.

IV. FUTURE PROSPECTS AND CONCLUSIONS

There is a wide variety of interactions of the general autonomic system with gustatory and feeding mechanisms. Many research efforts in the past have provided large amounts of evidence for the pathways and the precise input from each of these systems individually. However, there is little information on the specific central nervous system mechanisms that are involved in integrating central autonomic control with gustation and feeding behaviors. Among the particular questions that remain, we must ask: What is the extent of convergence of sensory modalities in the ascending afferent pathway from the spinal cord to the cortex? Recent evidence has demonstrated a parallel pathway for general visceral sensory representation and taste afferents from the medulla to the cortex. In addition, spinal sensations such as pain, temperature, and sympathetic visceral input may ascend in close proximity to the taste input. There is some indication of convergence of gustatory afferents with general visceral input at brain stem levels. However, there is little experimental evidence for convergence of general visceral and gustatory afferents at thalamic and cortical levels. If, in fact, there is convergence of visceral information at these levels, what function does it have that is different from the converging sites lower in the neuraxis? This question needs to be investigated more thoroughly to permit a determination of the precise behavioral roles these structures play.

REFERENCES

Alvarez-Buylla, R., and de Alvarez-Buylla, E. R. (1988). Carotid sinus receptors participate in glucose homeostasis. *Respir. Physiol.* **72**:347–360.
Anil, M. H., and Forbes, J. M. (1980). Feeding in sheep during intraportal infusions of short-chain fatty acids and the effect of liver denervation. *J. Physiol.* **298**:407–414.
Augustine, J. R. (1985). The insular lobe in primates including humans. *Neurol. Res.* **7**:2–10.
Bailey, P., and Bremer, F. (1938). A sensory cortical representation of the vagus nerve with a note on the effects of low blood pressure on the cortical electrogram. *J. Neurophysiol.* **1**:405–412.
Bellinger, L. L., Trietley, G. J., and Bernardis, L. L. (1976). Failure of portal glucose and adrenaline infusions of liver denervation to affect food intake in dogs. *Physiol. Behav.* **16**:299–304.

Benjamin, R. M., and Akert, K. (1959). Cortical and thalamic areas involved in taste discrimination in the albino rat. *J. Comp. Neurol.* **111**:231–259.

Benjamin, R. M., and Pfaffmann, C. (1955). Cortical localization of taste in the albino rat. *J. Neurophysiol.* **18**:56–64.

Bereiter, D. A., Berthoud, H.-R., and Jeanrenaud, B. (1981). Chorda tympani and vagus nerve convergence onto caudal brain stem neurons in the rat. *Brain Res. Bull.* **7**:261–266.

Bernard, R. A., and Nord, S. G. (1971). A first-order synaptic relay for taste fibers in the pontine brain stem of the cat. *Brain Res.* **30**:349–356.

Berridge, K., and Grill, H. J. (1981). Relation of consummatory responses and preabsorptive insulin release to palatability and learned taste aversions. *J. Comp. Physiol.* **95**:363–382.

Blake, W. D., and Lin, K. K. (1978). Hepatic portal vein infusion of glucose and sodium solutions on the control of saline drinking in the rat. *J. Physiol.* **274**:129–139.

Cabanac, M. (1971). Physiological role of pleasure. *Science,* **173**:1103–1107.

Cabanac, M., and Fantino, M. (1977). Origin of olfacto-gustatory alliesthesia: Intestinal sensitivity to carbohydrate concentration? *Physiol. Behav.* **18**:1039–1045.

Calaresu, F. R., Ciriello, J., Caverson, M. M., Cechetto, D. F., and Krukoff, T. L. (1984). Functional neuroanatomy of central pathways controlling the circulation. In *Hypertension and the Brain,* T. A. Kotchen and C. P. Guthrie (Eds.). Futura, Mount Kisco, NY, pp. 3–21.

Campbell, A. W. (1905). *Histological Studies on the Localization Cerebral Function.* Cambridge University Press, Cambridge, pp. 250–260.

Cechetto, D. F. (1987). Central representation of visceral function. *Fed. Proc.* **46**:17–23.

Cechetto, D. F., and Calaresu, F. R. (1983). Parabrachial units responding to stimulation of buffer nerves and forebrain in the rat. *Am. J. Physiol.* **245** (*Regul. Integrative Comp. Physiol.* **14**):R811–R819.

Cechetto, D. F., and Saper, C. B. (1987a). Evidence for a viscerotopic sensory representation in the cortex and thalamus in the rat. *J. Comp. Neurol.* **262**:27–45.

Cechetto, D. F., and Saper, C. B. (1987b). Organization of visceral sensory thalamus in the rat. *Neurosci. Abstr.* **13**:728.

Cechetto, D. F., Standaert, D. G., and Saper, C. B. (1985). Spinal and trigeminal dorsal horn projections to the parabrachial nucleus in the rat. *J. Comp. Neurol.* **240**:153–160.

Cechetto, D. F., Hachinski, V. C., and Chen, S. J. (1988). Efferent pathways for sympathetic responses from the insular cortex. *Neurosci. Abstr.* **14**:616.

Contreras, R. J. (1977). Changes in gustatory nerve discharges with sodium deficiency: A single unit analysis. *Brain Res.* **121**:373–378.

Contreras, R. J., and Frank, M. (1979). Sodium deprivation alters neural responses to gustatory stimuli. *J. Gen. Physiol.* **73**:569–594.

Daniel, P. M., and Henderson, J. R. (1967). The effect of vagal stimulation on plasma insulin and glucose levels in the baboon. *J. Physiol.* **192**:317–327.

de Castro, F. (1923). Contribution à la connaissance de l'innervation du pancréas. *Trav. Lab. Res. Biol. Madrid,* **21**:423–457.

Dell, P., and Olson, R. (1951). Projections thalamiques corticales et cérébelleuses des afférences viscérales vagales. *C. R. Soc. Biol.* **145**:1084–1088.

Delgado, J. M. R. (1960). Circulatory effects of cortical stimulation. *Physiol. Rev.* **40**:Suppl. 4, 146–171.

Deutsch, J. A. (1982). Controversies in food intake regulation. In *The Neural Basis of Feeding and Reward,* B. Hoebel and D. Novin (Eds.). Haer, Brunswick, pp. 137–148.

Deutsch, J. A., Young, W. G., and Kalogeris, T. J. (1978). The stomach signals satiety. *Science,* **201**:165–167.

Di Lorenzo, P. M., and Schwartzbaum, J. S. (1982a). Coding of gustatory information in the pontine parabrachial nuclei of the rabbit: Magnitude of neural response. *Brain Res.* **251**:229–244.

Di Lorenzo, P. M., and Schwartzbaum, J. S. (1982b). Coding of gustatory information in the pontine parabrachial nuclei of the rabbit: Temporal patterns of neural response. *Brain Res.* **251**:245–257.

Elrick, H., Stimmler, L., Hlad, C. J., Jr., and Arai, Y. (1964). Plasma insulin response to oral and intravenous glucose administration. *J. Clin. Endocrinol.* **24**:1076–1082.

Emmers, R. (1966). Separate cortical receiving areas for gustatory and tongue tactile afferents in the rat. *Anat. Rec.* **154**:460.

Fischer, U., Hommel, H., Ziegler, M., and Michael, R. (1972). The mechanism of insulin secretion after oral glucose administration. *Diabetologia,* **8**:104–110.

Fox, K. A., Kipp, S. C., and VanderWeele, D. A. (1976). Dietary self-selection following subdiaphragmatic vagotomy in the white rat. *Am. J. Physiol.* **231**:1790–1793.

Fronek, K., and Stahlgren, L. H. (1968). Systemic and regional hemodynamic changes during food intake and digestion in nonanesthetized dogs. *Circ. Res.* **23**:687–692.

Ganchrow, D., and Erickson, R. P. (1972). Thalamocortical relations in gustation. *Brain Res.* **36**:289–305.

Giza, B. K., and Scott, T. R. (1983). Blood glucose selectively affects taste-evoked activity in rat nucleus tractus solitarius. *Physiol. Behav.* **31**:643–650.

Grill, H. J., and Norgren, R. (1978). The taste reactivity test: II. Mimetic responses to gustatory stimuli in chronic thalamic and chronic decerebrate rats. *Brain Res.* **143**:281–297.

Hall, R. E., Livingston, R. B., and Bloor, C. M. (1977). Orbital cortical influences on cardiovascular dynamics and myocardial structure in conscious monkeys. *J. Neurosurg.* **46**:638–647.

Hamilton, R. B., Ellenberger, H., Liskowsky, D., and Schneiderman, N. (1981). Parabrachial area as mediator of bradycardia in rabbits. *J. Auton. Nerv. Syst.* **4**:261–281.

Hayama, T., and Ogawa, H. (1987). Electrophysiological evidence of collateral projections of parabrachio-thalamic relay neurons. *Neurosci. Lett.* **83**:95–100.

Hermann, G. E., and Rogers, R. C. (1985). Convergence of vagal and gustatory afferent input within the parabrachial nucleus of the rat. *J. Autonom. Nerv. Syst.* **13**:1–17.

Hermann, G. E., Kohlerman, N. J., and Rogers, R. C. (1983). Hepatic-vagal and gustatory afferent interactions in the brainstem of the rat. *J. Autonom. Nerv. Syst.* **9**:477–495.

Hoffman, B. L., and Rasmussen, T. (1953). Stimulation studies of insular cortex of *Macaca mulatta*. *J. Neurophysiol.* **16**:343–351.

Hommel, H., Fischer, U., Retzlaff, K., and Knofler, H. (1972). The mechanism of insulin secretion after oral glucose administration. *Diabetologia,* **8**:111–116.

Houpt, T. R., Baldwin, B. A., and Houpt, K. A. (1983). Effects of duodenal osmotic loads on spontaneous meals in pigs. *Physiol. Behav.* **30**:787–795.

Iggo, A. (1955). Tension receptors in the stomach and the urinary bladder. *J. Physiol.* **128**:593–607.

Janowitz, H. D., and Grossman, M. I. (1949). Some factors affecting the food intake of normal dogs and dogs with esophagostomy and gastric fistula. *Am. J. Physiol.* **159**:143–148.

Jeanningros, R. (1982). Vagal unitary responses to intestinal amino acid infusions in the anesthetized cat: A putative signal for protein induced satiety. *Physiol. Behav.* **28**:9–21.

Kaada, B. R. (1951). A study of responses from the limbic, subcallosal, orbito-insular piriform and temporal cortex, hippocampus–fornix and amygdala. In *Acta Physiol. Scand.* **24**:Suppl. 83. *Somato-Motor, Autonomic and Electrocorticographic Responses to Electrical Stimulation of "Rhinencephalic" and Other Structures in Primates, Cat and Dog.* A. W. Broggers Boktrykkeri A/S, Oslo, pp. 1–285.

Kawai, Y., Takami, K., Shiosaka, S., Emson, P. C., Hillyard, C. J., Girgis, S., MacIntyre, I., and Tohyama, M. (1985). Topographic localization of calcitonin gene-related peptide in the rat brain: An immunohistochemical analysis. *Neuroscience*, **15**:747–763.

Kelbaek, H., Munck, O., Christensen, N. J., and Godtfredsen, J. (1987). Autonomic nervous control of postprandial hemodynamic changes at rest and upright exercise. *J. Appl. Physiol.* **63**(5):1862–1865.

Kraly, F. S. (1978). Abdominal vagotomy inhibits osmotically induced drinking in the rat. *J. Comp. Physiol. Psychol.* **92**:999–1013.

Kraly, F. S., and Smith, G. P. (1978). Combined pregastric and gastric stimulation by food is sufficient for normal meal size. *Physiol. Behav.* **21**:405–408.

Kraly, F. S., Gibbs, J., and Smith, G. P. (1975). Disordered drinking after abdominal vagotomy in rats. *Nature*, **258**:226–228.

Lasiter, P. S., Deems, D. A., and Garcia, J. (1985a). Involvement of the anterior insular gustatory neocortex in taste-potentiated odor aversion learning. *Physiol. Behav.* **34**:71–77.

Lasiter, P. S., Deems, D. A., and Glanzman, D. L. (1985b). Thalamocortical relations in taste aversion learning: I. Involvement of gustatory thalamocortical projections in taste aversion learning. *Behav. Neurosci.* **99**:454–476.

Lasiter, P. S., Deems, D. A., Oetting, R. L., and Garcia, J. (1985c). Taste discriminations in rats lacking anterior insular gustatory neocortex. *Physiol. Behav.* **35**:277–285.

Le Magnen, J. (1983). Body energy balance and food intake: A neuroendocrine regulatory mechanism. *Physiol. Rev.* **63**:314–373.

Loewy, A. D., and Burton, H. (1978). Nuclei of the solitary tract: Efferent projections to the lower brain stem and spinal cord of the cat. *J. Comp. Neurol.* **181**:421–450.

Louis-Sylvestre, J. (1976). Preabsorptive insulin release and hypoglycemia in rats. *Am. J. Physiol.* **230**:56–60.

Mantyh, P. W., and Hunt, S. P. (1984). Neuropeptides are present in projection at all levels in visceral taste pathways: From periphery to sensory cortex. *Brain Res.* **299**:297–311.

Matsukawa, J., and Ninomiya, I. (1985). Transient responses of heart rate, arterial pressure, and head movement at the beginning of eating in awake cats. *Jpn. J. Physiol.* **35**:599–611.

Matsukawa, K., and Ninomiya, I. (1987). Changes in renal sympathetic nerve activity, heart rate and arterial blood pressure associated with eating in cats. *J. Physiol.* **390**:229–242.

Mei, N. (1985). Intestinal chemosensitivity. *Physiol. Rev.* **65**:211–237.

Nicolaidis, S. (1978). Role des réflexes anticipateurs oro-végétatifs dans la régulation hydrominérale et énergétique. *J. Physiol., Paris,* **74**:1–19.
Niijima, A. (1969). Afferent discharges from osmoreceptors in the liver of the guinea pig. *Science,* **166**:1519–1520.
Norgren, R. (1974). Gustatory afferents to entral forebrain. *Brain Res.* **4**:285–295.
Norgren, R. (1976). Taste pathways to hypothalamus and amygdala. *J. Comp. Neurol.* **166**:17–30.
Norgren, R. (1984). Central neural mechanisma of taste. In *Handbook of Physiology—The Nervous System,* Vol. III, D. Smith (Ed.). Williams & Wilkins, Baltimore, pp. 1087–1128.
Norgren, R., and Leonard, C. M. (1973). Ascending central gustatory pathways. *J. Comp. Neurol.* **150**:217–238.
Norgren, R., and Pfaffmann, C. (1975). The pontine taste area in the rat. *Brain Res.* **91**:99–117.
Norgren, R., and Wolf, G. (1975). Projections of thalamic gustatory and lingual areas in the rat. *Brain Res.* **92**:123–129.
Novin, D. (1976). Visceral mechanisms in the control of food intake. In *Hunger: Basic Mechanisms and Clinical Implication,* D. Novin, W. Wyrwicka, and G. Bray (Eds.). Raven Press, New York, pp. 357–367.
Novin, D. (1983). The integration of visceral information in the control of feeding. *J. Auton. Nerv. Syst.* **9**:233–246.
Novin, D., Sanderson, J. D., and VanderWeele, D. A. (1974). The effect of isotonic glucose on eating as a function of feeding condition and infusion site. *Physiol. Behav.* **13**:3–7.
Ogawa, H., and Hayama, T. (1984). Responsiveness fields of solitario-parabrachial relay neurons responsive to natural stimulation of the oral cavity in rats. *Exp. Brain Res.* **54**:359–366.
Ogawa, H., Hayama, T., and Ito, S. (1984a). Location and taste responses of parabrachio-thalamic relay neurons in rats. *Exp. Neurol.* **83**:507–517.
Ogawa, H., Imoto, T., and Hayama, T. (1984b). Responsiveness of solitario-parabrachial neurons to taste and mechanical stimulation applied to the oral cavity in rats. *Exp. Brain Res.* **54**:349–358.
Ogawa, H., Hayama, T., and Ito, S. (1987). Response properties of the parabrachio-thalamic taste and mechanoreceptive neurons in rats. *Exp. Brain Res.* **68**:449.
Paintal, A. S. (1954a). A study of gastric stretch receptors. Their role in the peripheral mechanism of satiation of hunger and thirst. *J. Physiol.* **126**:255–270.
Paintal, A. S. (1954b). The response of gastric stretch receptors and certain other abdominal and thoracic vagal receptors to some drugs. *J. Physiol.* **126**:271–285.
Perley, M. J., and Kipnis, D. M. (1967). Plasma insulin responses to oral and intravenous glucose: Studies in normal and diabetic subjects. *J. Clin. Invest.* **46**:1954–1962.
Perotto, R., and Scott, T. (1976). Gustatory neural coding in the pons. *Brain Res.* **110**:283–300.
Ranieri, F., Mei, N., and Crousillat, J. (1973). Les afférences splanchniques provenant des mécanorecepteurs gastro-intestinaux et péritoneaux. *Exp. Brain Res.* **16**:291–308.
Ricardo, J. A., and Koh, E. T. (1978). Anatomical evidence of direct projections from the nucleus of the solitary tract to the hypothalamus, amygdala, and other forebrain structures in the rat. *Brain Res.* **153**:1–26.

Rogers, R. C., Novin, D., and Butcher, L. L. (1979a). Electrophysiological and neuroanatomical studies of hepatic portal osmo- and sodium-receptive afferent projections within the brain. *J. Autonom. Nerv. Syst.* **1**:183–202.

Rogers, R. C., Novin, D., and Butcher, L. L. (1979b). Hepatic sodium and osmoreceptors activate neurons in the ventrobasal thalamus. *Brain Res.* **168**:398–403.

Ross, C. A., Ruggiero, D. A., and Reis, D. J. (1981). Afferent projections to cardiovascular portions of the nucleus of the tractus solitarius in the rat. *Brain Res.* **223**:402–408.

Ruggiero, D. A., Marovitch, S., Granata, A. R., Anwar, M., and Reis, D. J. (1987). A role of insular cortex in cardiovascular function. *J. Comp. Neurol.* **257**:189–207.

Russek, M. (1971). Hepatic receptors and the neurophysiological mechanisms controlling feeding behaviors. *Neurosci. Res.* **4**:213–282.

Saper, C. B. (1982a). Convergence of autonomic and limbic connections in the insular cortex of the rat. *J. Comp. Neurol.* **210**:163–173.

Saper, C. B. (1982b). Reciprocal parabrachial-cortical connections in the rat. *Brain Res.* **242**:33–40.

Schwartzbaum, J. S., and Block, C. H. (1981). Interrelations between parabrachial pons and ventral forebrain of rabbits in taste-mediated functions. In *The Amygdaloid Complex, INSERM Symposium No. 20,* Y. Ben-Ari, (Ed.). Elsevier/North Holland Biomedical Press, Amsterdam, pp. 367–382.

Sclafani, A., and Kramer, T. H. (1983). Dietary selection in vagotomized rats. *J. Autonom. Nerv. Syst.* **9**:247–258.

Scow, R. O., and Cornfield, J. (1954). Quantitative relations between the oral and intravenous glucose tolerance curves. *Am. J. Physiol.* **179**:435–438.

Share, I., Martyniuk, E., and Grossman, M. I. (1952). Effect of prolonged intragastric feeding on oral food intake in dogs. *Am. J. Physiol.* **169**:229–235.

Shimada, S., Shiosaka, S., Emson, P. C., Hillyard, C. J., Girgis, S., MacIntyre, I., and Tohyama, M. (1985a). Calcitonin gene-related peptidergic projection from the parabrachial area to the forebrain and diencephalon in the rat: An immunohistochemical analysis. *Neuroscience,* **16**:607–616.

Shimada, S., Shiosaka, S., Hillyard, C. J., Girgis, S. I., MacIntyre, I., Emson, P. C., and Tohyama, M. (1985b). Calcitonin gene-related peptide projection from the ventromedial thalamic nucleus to the insular cortex: A combined retrograde transport and immunocytochemical study. *Brain Res.* **344**:200–203.

Shimada, S., Shiosaka, S., Takami, K., Yamano, M., and Tohyama, M. (1985c). Somatostatinergic neurons in the insular cortex project to the spinal cord: Combined retrograde axonal transport and immunohistochemical study. *Brain Res.* **326**:197–200.

Shipley, M. T. (1982). Insular cortex projection to the nucleus of the solitary tract and brainstem visceromotor regions in the mouse. *Brain Res. Bull.* **8**:139–148.

Sieck, G. C., and Harper, R. M. (1980). Discharge of neurons in the parabrachial pons related to the cardiac cycle: Changes during different sleep–waking states. *Brain Res.* **199**:385–399.

Smith, G. P., and Jerome, C. (1983). Effect of total and selective abdominal vagotomies on water intake in rats. *J. Autonom. Nerv. Syst.* **9**:259–271.

Snowdon, C. T. (1975). Production of satiety with small intraduodenal infusions in the rat. *J. Comp. Physiol. Psychol.* **88**:231–238.

Stephens, D. B., and Baldwin, B. A. (1974). The lack of effect of intrajugular or intraportal injections of glucose or amino acids on food intake in pigs. *Physiol. Behav.* **12**:923–929.

Strubbe, J. H., and Steffens, A. B. (1977). Blood glucose levels in portal and peripheral circulation and their relation to food intake in the rat. *Physiol. Behav.* **19**:303–307.

Towbin, E. J. (1949). Gastric distension as a factor in the satiation of thirst in esophagostomized dogs. *Am. J. Physiol.* **159**:533–541.

Travers, S. P., and Smith, D. V. (1984). Responsiveness of neurons in the hamster parabrachial nuclei to taste mixtures. *J. Gen. Physiol.* **84**:221–250.

Travers, J. B., Travers, S. P., and Norgren, R. (1987). Gustatory neural processing in the hindbrain. *Annu. Rev. Neurosci.* **10**:595–632.

Van Der Kooy, D., McGinty, J. F., Koda, L. Y., Gerfen, C. R., and Bloom, F. E. (1982). Visceral cortex: A direct connection from prefrontal cortex to the solitary nucleus in the rat. *Neurosci. Lett.* **33**:123–127.

Van Der Kooy, D., Koda, L. Y., McGinty, J. F., Gerfen, C. R., and Bloom, F. E. (1984). The organization of projections from the cortex, amygdala, and hypothalamus to the nucleus of the solitary tract in rat. *J. Comp. Neurol.* **224**:1–24.

VanderWeele, D. A., Novin, D., Rezek, M., and Sanderson, J. D. (1974). Duodenal or hepatic–portal glucose perfusion: Evidence for duodenally based satiety. *Physiol. Behav.* **12**:467–473.

Vatner, S. F., Franklin, D., and Van Citters, R. L. (1970a). Mesenteric vasoactivity associated with eating and digestion in the conscious dog. *Am. J. Physiol.* **219**:170–174.

Vatner, S. F., Franklin, D., and Van Citters, R. L. (1970b). Coronary and visceral vasoactivity associated with eating and digestion in the conscious dog. *Am. J. Physiol.* **219**:1380–1385.

Von Euler, U. S., and Folkow, B. (1958). The effect of stimulation of autonomic areas in the cerebral cortex upon the adrenaline and noradrenaline secretion from the adrenal gland in the cat. *Acta Physiol. Scand.* **42**:313–320.

Ward, D. G., Lefcourt, A. M., and Gann, D. S. (1980). Neurons in the dorsal rostral pons process information about changes in venous return and arterial pressure. *Brain Res.* **181**:75–88.

Yamamoto, T., and Kawamura, Y. (1972). Summated cerebral responses to taste stimuli in rat. *Physiol. Behav.* **9**:789–793.

Yamamoto, T., Matsuo, R., and Kawamura, Y. (1980a). The pontine taste area in the rabbit. *Neurosci. Lett.* **16**:5–9.

Yamamoto, T., Matsuo, R., and Kawamura, Y. (1980b). Localization of cortical gustatory area in rats and its role in taste discrimination. *J. Neurophysiol.* **44**:440–455.

Yamamoto, T., Takahashi, T., and Kawamura, Y. (1981a). Access to the cerebral cortex of extra-lingual taste inputs in the rat. *Neurosci. Lett.* **24**:129–132.

Yamamoto, T., Yuyama, N., and Kawamura, Y. (1981b). Cortical neurons responding to tactile, thermal and taste stimulations of the rat's tongue. *Brain Res.* **221**:202–206.

Yasui, Y., Cechetto, D. F., and Saper, C. B. (1988). Calcintonin gene-related peptide immunoreactive innervation of the insular cortex of the rat. *Neurosci. Abstr.* **14**:1319.

22
Metabolic Influences on the Gustatory System

Thomas R. Scott

University of Delaware
Newark, Delaware

I. INTRODUCTION

The sense of taste motivates and directs feeding. Rodents, which have physiological welfare as the primary consideration in feeding, use the chemical senses to select from an inhospitable environment the materials to support a diversity of biochemical processes. But even in human society, where the substance and manner of meals—as well as the social implications of gluttony—assume great cultural significance, widespread obesity and hypertension betray the irresistible attraction of fleeting gustatory pleasure.

As taste guides feeding, the process of feeding influences taste. The satisfaction derived from taste depends not only on the quality and intensity of the stimulus, but on past experience with that taste and on the chronic and acute physiological needs of the organism.

A. Experience

The gustatory experiences of suckling rats establish taste preferences that persist into adulthood (Capretta and Rawls, 1974). Preferences also develop through association of a taste with positive reinforcement (Revusky, 1974), in particular with a visceral reinforcement such as occurs with the administration of a nutrient of which the animal has been deprived (Booth et al., 1974). Gustatory preferences can even be established in humans and other animals by mere familiarity through constant exposure (Capretta et al., 1973). However the most potent effect of experience on subsequent preference obtains from the establishment of a conditioned taste aversion (CTA). This is an especially efficient form of conditioning in which an intense aversion may be developed through a single pairing of conditioned and unconditioned stimuli: a novel taste (CS) and gastrointestinal malaise (US), respectively (Garcia et al., 1955). The aversion to a taste solution paired with a poison is so readily established, so potent, and so resistant to extinction that the CTA protocol has itself become a standard tool for studying physiological processes and taste-related behavior (Bernstein, 1978).

B. Physiological Need

An animal's physiological condition is closely related to its choice of foods. The "body wisdom" demonstrated in cafeteria studies by Richter appears to result from taste-directed changes in food selection. Compensatory feeding behavior has been shown in cases of experimentally induced deficiencies of thiamine (Seward and Greathouse, 1973), threonine (Halstead and Gallagher, 1962), and histidine (Sanahuja and Harper, 1962). It is presumed that the physiological benefits of dietary repletion are paired with the taste that preceded those benefits, creating a conditioned taste preference by which the hedonic value of the taste is enhanced. Since physiological needs are in continuous flux, the hedonic value of a taste experience must be quite labile.

A more abiding preference exists for sodium. Evolving in sodium-deficient environments, most mammals seek out and consume salt wherever it is found and, when plentiful, in excess of need. Both rats and humans select sodium salts in their diets, even when sodium replete (Denton, 1976). This preference becomes exaggerated under conditions of salt deficiency. Humans depleted by pathological states (Wilkins and Richter, 1940) or by experimental manipulation (McCance, 1936) show a pronounced craving for salt. Rodents subjected to adrenalectomy (Epstein and Stellar, 1955), dietary restrictions (Cullen and Harriman, 1973), or injections of formalin (Stricker and Wilson, 1970), cyclophosphamide (Mitchell et al., 1974), aldosterone (Wolf and Handel, 1966), or deoxycorticosterone (Wolf and Quartermain, 1966) show sharp increases in sodium consumption. This compensatory response to the physiological need for salt results from a change in the hedonic value of tasted sodium. Concentrations of salt that had been evaluated negatively and rejected under conditions of sodium repletion evoke a positive hedonic response and acceptance under conditions of sodium depletion.

While changes in amino acid or sodium levels are appreciated over several days, the availability of certain macronutrients, specifically sugars, is of more immediate concern. In just a few hours, the definition of which chemicals are acceptable may be subject to modification as the dangers of malnutrition weigh against those of toxicity. After only minutes of feeding, the definition may reverse again. Thus the digital decision to swallow or reject, as well as the hedonics that guide that decision, change with momentary conditions of need. Our common experience joins with the results of psychophysical studies to confirm this logic:

METABOLIC INFLUENCES ON THE GUSTATORY SYSTEM

with deprivation, foods become more palatable; with satiety, less so (Cabanac, 1971). In the body of this chapter, I review anatomical, physiological, and behavioral evidence to support the assertion that many of these influences on appetite are mediated by specific alterations in gustatory responsiveness.

II. RESULTS AND DISCUSSION

A. Anatomical Considerations

Hindbrain taste nuclei are in close proximity to, and to some degree overlap, regions that receive visceral afferents. They also have reciprocal connections with diencephalic and telencephalic areas that are associated with feeding, hedonics, reinforcement, and emotion.

1. Ascending Fibers

The ascending limb of the taste system in the rat is depicted in Figure 1. Gustatory and visceral afferents to the central nervous system both terminate in the nucleus tractus solitarius (NTS) (Rogers and Novin, 1983). Taste input projects to the rostral one-third of the nucleus (Norgren, 1977), whereas axons of the vagus nerve direct visceral information more caudally. There is no clear demonstration of a monosynaptic connection between these areas (Norgren, 1985); taste cell dendrites do extend into the viscerosensory NTS, however, and, in addition, an

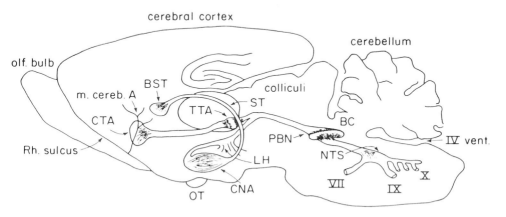

FIGURE 1 Parasagittal section showing the gustatory anatomy of the rat: BC, brachium conjunctivum; BST, bed nucleus of the stria terminalis; CNA, central nucleus of the amygdala; CTA, cortical taste area; LH, lateral hypothalamus; m cereb A, middle cerebral artery; NTS, nucleus of the solitary tract; OT, optic tract; PBN, parabrachial nucleus; ST, stria terminalis; TTA, thalamic taste area; IV vent, fourth ventricle; VII, IX, X, cranial nerves 7, 9, and 10. (Figure supplied by Dr. Carl Pfaffmann and reprinted from Norgren [1977] with permission.)

indirect gustatory–visceral link by way of the subjacent reticular formation has been hypothesized (Hermann and Rogers, 1982). Taste axons from NTS project directly to the dorsal motor nucleus and to the nucleus ambiguus as well as to the salivatory nuclei of the intervening reticular formation (Davis and Jang, 1988). It is hypothesized that only about 20% of NTS taste cells project rostrally to the pons (Ogawa et al., 1982). The remainder either form intranuclear connections or contribute to the pathways serving parasympathetic nuclei (Loewy and Burton, 1978; Norgren, 1978). Therefore these autonomic projections constitute the major output of gustatory NTS.

From both regions of NTS, axons project rostrally to the parabrachial nuclei (PBN), where convergence of gustatory and visceral information onto single neurons has been shown (Hermann and Novin, 1980). Moreover, the gustatory region of the parabrachial nucleus (PBN) sends a substantial projection centrifugally to nucleus ambiguus and the salivatory nuclei. Bifurcating projections from PBN to the thalamocortical axis and to ventral forebrain structures (Norgren and Leonard, 1973) provide the possibility of parallel processing of taste–quality and of taste–hedonic information, respectively.

2. Descending Fibers

The thalamocortical and ventral forebrain taste systems both send centrifugal projections to hindbrain taste relays in PBN and NTS (Saper, 1982; Shipley and Sanders, 1982; Bloom, 1984). In addition, gustatory cortex projects heavily back to the thalamic taste relay; ventral forebrain nuclei reciprocate the connections they receive from hindbrain parasympathetic nuclei (Berk and Finkelstein, 1982; Hosoya et al., 1984). Thus, routes are established by which factors such as motivation or conditioning could at least influence taste-mediated acceptance–rejection reflexes. These connections also imply that the very taste-initiated neural activity that helps guide feeding may be differentially biased at an early synaptic level to accommodate the transient physiological needs of the organism, as suggested in a later section.

B. Physiological Considerations

1. Gustatory Stimulus Characteristics

The relationship between the sense of taste and metabolic state is established at its most basic level by this finding: physiological welfare is the principle underlying the code for taste quality.

No issue is more fundamental to the characterization of a sensory system than a definition of the dimensions on which its perceptions are based. The study of auditory perception may proceed in a logical fashion because it is known that pitch derives largely from the frequency of the incident pressure waves. Similarly with the relationship between color perception and stimulus wavelength, form perception and locus of retinal stimulation, pressure sensation and locus of skin deformation, and temperature perception and the thermal qualities of the impinging stimulus. It is clear from a study of gustatory receptor mechanisms that taste

METABOLIC INFLUENCES ON THE GUSTATORY SYSTEM 449

perceptions are also a product of the physical characteristics of relevant stimuli. However, these are largely independent for each of the basic tastes. Hydrogen ion concentration may partly explain sourness—it relates not at all to saltiness. There is no apparent single physical dimension on which the study of the sense of taste may be organized. This realization has even led to the suggestion that taste is not an integrated sensory system, but a series of independent modalities sensitive to different aspects of the chemical environment.

The introduction of multidimensional scaling techniques into taste research (Erickson et al., 1965; Schiffman and Erickson, 1971) permitted a direct approach to this issue. A wide range of sapid stimuli could be applied and the profile of either neural or behavioral activity elicited by each chemical determined. Similarity measures among the profiles, as provided by correlation coefficients or other statistics, could then be used in a multidimensional scaling program to generate a spatial representation of relative stimulus similarity. The axes of the space in which the chemicals are located must represent the stimulus characteristics that underlie gustatory discriminability, for the amount of each characteristic a stimulus possesses is what determines its taste quality, hence its position in the space. Therefore the dimensions along which taste quality is organized may be determined by finding the optimal match between a stimulus characteristic and each axis of the multidimensional space. The importance of any characteristic in determining taste quality is proportional to the total data variance accounted for by the axis with which it is matched.

This approach has now been used to interpret both psychophysical (Schiffman and Erickson, 1971) and electrophysiological (Scott and Mark, 1987) data. The common result of these studies is that a major dimension along which taste stimuli may be organized relates, not to any one physical feature of the stimulus molecule, but rather to a physiological characteristic: its effect on the welfare of the organism.

Figure 2 presents a two-dimensional space in which 16 chemicals are placed according to the relative similarities of the response profiles they evoked from 42 taste cells in the rat's nucleus tractus solitarius. The two dimensions of Figure 2 (top) account for 95% of the data variance, a preponderance of which—91%—pertains to dimension 1 alone. Thus the stimulus characteristic that corresponds to this dimension is the major factor in permitting these chemicals to be neurally distinguished by the taste system. Stimulus placement on this dimension correlates $+0.83$ ($p < .001$) with stimulus toxicity as indexed here by the rat's oral LD_{50}. Stimulus toxicity, then, provides an excellent basis for predicting relative taste quality across a wide range of chemicals. This dimension is shown in isolation in Figure 2 (bottom).*

Whereas the relationship between LD_{50} and stimulus position on the dominant

---

*Dimension 2, accounting for 4% of the variance, corresponds to differences in the effectiveness of various solutions in driving the system. The mean number of spikes evoked across all 42 neurons during the 5-second response period correlates $+0.76$ ($p < .001$) with placement on this dimension, implying that it is a measure of total response magnitude.

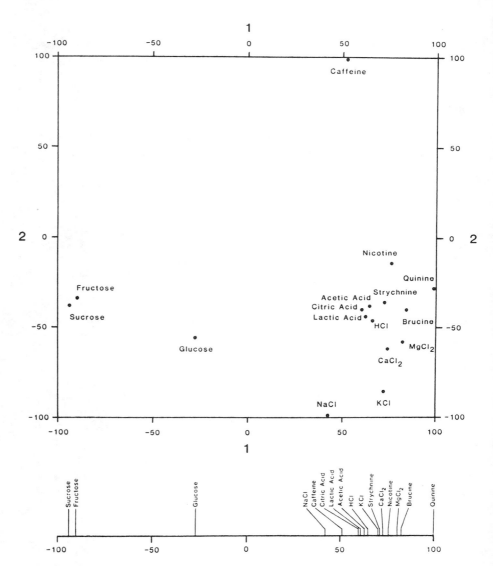

FIGURE 2 Top: Two-dimensional space representing relative similarities among taste qualities as determined by activity profiles across neurons. Dimension 1 accounts for 91% of the data variance, and the position of a stimulus along its length correlates 0.83 ($p < .001$) with stimulus toxicity (rat oral LD_{50}). Dimension 2 accounts for 4% of the variance, and stimulus position along it correlates 0.76 ($p < .001$) with total evoked activity across all neurons. Five percent of the variance is unaccounted for by these two dimensions. Bottom: Dimension 1 shown in isolation. (Reprinted from Scott and Mark [1987] with permission.)

first dimension is highly significant, it does not explain the full discriminative capacity of the rat. The organic acids all generate patterns that correlate about +0.90 with strychnine, which is a thousand times more toxic and which the behaving rat easily discriminates from the acids. Thus the animal has access to more information than is assayed by this analysis. That additional input may derive from the temporal characteristics of the evoked response. In the analysis above, only the total spikes that accumulated from each stimulus application during 5 seconds of evoked activity were considered. Yet the time course of this accumulation not only carries reliable information regarding taste quality (Nagai and Ueda, 1981; DiLorenzo and Schwartzbaum, 1982), but it also may be sufficient to activate appropriate reflexive responses to chemicals in behaving rats (Covey and Erickson, 1980). Therefore, temporal profiles, each composed of fifty 100-ms bins, collapsed across neurons, were generated for every stimulus, and the correlation matrix and multidimensional representation were calculated as before. The result is shown in Figure 3 (top). Dimension 1 is again dominant, accounting for 89% of the variance, and placement on it correlates +0.85 with LD_{50} ($p < .001$). It is represented in isolation in Figure 3 (bottom). Dimension 2, accounting for 6% of the variance, is undefined.

The temporal analysis provides a solution to the confusion between strychnine and the organic acids, all of whose temporal discharge patterns correlate only in the low 0.40s with that of the alkaloid. It also introduces coefficients that would be troublesome (the temporal pattern of salty NaCl correlates nearly +0.90 with those of bitter $MgCl_2$ and $CaCl_2$) if the earlier analysis (in which these coefficients reach only the mid-0.50s) were not available. Certain stimulus pairs, then, are clearly discriminable based on the response distribution either across neurons or across time. Other pairs—quinine and sucrose—are readily discriminable by either means. When neither factor provides separation, however, as with $CaCl_2$ and $MgCl_2$, the rat cannot easily make a behavioral discrimination. The conclusion is that taste quality information in the hindbrain of the rat is carried in a spatiotemporal code, both the spatial and temporal aspects of which are organized predominantly along dimensions that relate to the rat's welfare.

The finding of a dimension of physiological welfare underlying the organization of taste quality coding is not simply a re-creation of a sweet–bitter dichotomy. It is a more fundamental *physiological* dimension upon which the *psychological* perceptions of sweetness and bitterness may be based. Chemicals in the environment promote or disrupt physiological functions in animals, providing nutrition or causing illness or death. Selection among foragers, then, favors the taste system that activates the appropriate hedonic tone (attraction to nutrients, revulsion by toxins) to match the physiological consequences of ingestion.

The evolution of a system designed to distinguish the beneficial from the harmful is hardly unique to taste. The nonchemical senses permit the detection of predator and prey and so promote escape from the former and capture of the latter. But predator and prey are not coded as a survival dimension in the optic, trigemi-

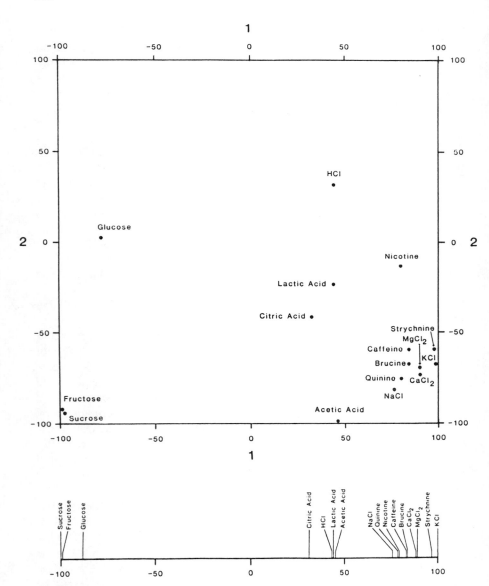

FIGURE 3 Top: Two-dimensional space representing relative similarities among taste qualities as determined by activity profiles across time. Dimension 1 accounts for 89% of the data variance, and the position of a stimulus along its length correlates 0.85 ($p < .001$) with stimulus toxicity. Dimension 2 accounts for 5% of the variance and is undefined. Six percent of the variance is unaccounted for by the two dimensions. Bottom: Dimension 1 shown in isolation. (Reprinted from Scott and Mark [1987] with permission.)

nal, or auditory nerves of mammals as a welfare dimension is in the rat's hindbrain. For this is the primary function of taste: not to provide a continuous record of the surroundings, but to sample the chemical environment discretely, to predict in each case the consequences of ingestion, and to activate both hindbrain reflexes for swallowing or rejection and powerful hedonic components to motivate consumption or withdrawal. The sense of taste looks not just beyond the body, but within it. Its unifying organization is not physical, but physiological.

The mechanisms described above are already evident at lower order levels of the gustatory system. The rich plexus of connections between taste and autonomic nuclei provides the basis for independent control of both reflexive and regulatory processes in the hindbrain. Acceptance–rejection reflexes associated with taste stimuli are stereotypical and unaltered by loss of tissue rostral to the midbrain. While decerebration in rats abolishes most spontaneous behavioral sequences, autonomic reflexes are largely intact (Grill and Norgren, 1978). Moreover, many regulatory mechanisms associated with ingestion survive decerebration. These include the cephalic phase insulin release occasioned by gustatory stimulation (Grill and Berridge, 1981), the sympathoadrenal response to glucoprivation (DiRocco and Grill, 1979), insulin-induced hyperphagia (Flynn and Grill, 1983), and satiety and anorexia induced by the gut hormone cholecystokinin (Grill et al., 1983).

In an extensive series of experiments, Steiner and his colleagues have studied orofacial responses to taste stimuli in normal adult humans (Halbreich and Steiner, 1977), full-term and premature human neonates, anencephalic and hydroanencephalic human neonates (Steiner, 1973), blind adolescents (Steiner, 1976), patients with craniofacial abnormalities (Steiner, 1974), retarded humans (Gonshorowitz, 1977), subhuman primates (Steiner and Glaser, 1985), mammals (Steiner, 1973), and birds (Gentle, 1971). To the extent possible, considering their various limitations, subjects across this phylogenetic, ontogenetic, and pathological range all reacted similarly to the application of basic taste stimuli. Steiner concludes that facial expressions are adaptive both in dealing with the chemical—swallowing if appetitive, clearing from the mouth if aversive—and in communicating a hedonic dimension to other members of the species. Moreover, this "hedonic monitor" is located in the brain stem (Gentle, 1973) and is neurally intact by the seventh gestational month (Pfaffmann, 1978; Steiner, 1979).

2. Plasticity in the Code

The findings of a gustatory code based on physiological welfare, and of the apparent autonomy of that code in regulating ingestive behavior, provides the basis for a system by which nutrients may be selected from a hostile chemical environment. While providing a broad and effective means for maintaining the biochemical welfare of the species, this organization would not recognize the idiosyncratic allergies or needs of the individual, or be sensitive to changes in those needs over time. It was established in the introduction of this chapter that experience and momentary physiological needs are reflected in altered gustatory hedonics, implying that the taste code is plastic. I will now recount the evidence for the postulate that the responsiveness of the taste system reflects physiological condition and may mediate changes in feeding behavior that satisfy biochemical needs.

a. Experience. The neural substrates of conditioning have been studied in scores of experiments, most of which have involved ablating selected structures and testing the capacity of subjects to retain former conditioned responses or to develop new ones. These studies have implicated the cortex, amygdala, hippocampus, thalamus, hypothalamus, olfactory bulb, and area postrema in taste aversion learning. Only occasionally have recordings been made from neurons of conditioned animals to determine the effects of a learned aversion on taste-evoked activity. Aleksanyan et al. (1976) reported that the preponderance of hypothalamic activity evoked by saccharin in rats shifted from the lateral to the ventromedial nucleus with formation of a saccharin CTA. They concluded that the taste signal that formerly would have provided the reinforcement implied by lateral hypothalamus activation had, through conditioning, acquired aversive sensory and motivational properties associated with the ventromedial hypothalamus.

DiLorenzo (1985) recorded the responses evoked by a series of taste stimuli in the pontine parabrachial nucleus of rats, then paired the taste of NaCl with gastrointestinal malaise and repeated the recordings. The response to NaCl increased significantly and selectively in a subset of gustatory neurons. Chang and Scott (1984) recorded gustatory-evoked activity from the NTS of three groups of rats: unconditioned (exposed only to the taste of the saccharin CS with no consequences), pseudoconditioned (experienced only the US, nausea, with no gustatory referent), and conditioned (taste of the saccharin CS paired on three occasions with malaise). They reported that the CS evoked a significantly larger response from conditioned animals than from those of either control group and that the effect was limited to the 30% of neurons that showed a sweet-sensitive profile of responsiveness.

Temporal analyses of the activity evoked from this subgroup of saccharin-sensitive neurons revealed that nearly the entire increase in discharge rate was attributable to a burst of activity that diverged from control group levels 600 ms following stimulus onset, peaked at 900 ms, and returned to control levels by 3000 ms (Fig. 4). Thus the major consequence of the conditioning procedure was to increase responsiveness to the saccharin CS through a well-defined peak of activity. The same enhanced response and temporal pattern were evoked to a lesser extent by other sweet stimuli—fructose, glucose, and sucrose—providing a possible neural counterpart to generalization of the aversion. Since stimuli included not only the CS but also a wider range of chemicals unassociated with malaise, Chang and Scott could evaluate modifications in the entire system resulting from the conditioning procedure. The consequence of the CTA was to disrupt the sweet–nonsweet dichotomy that constitutes the basic distinction among taste

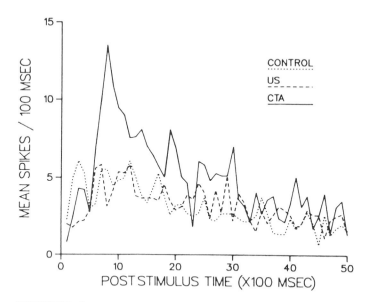

FIGURE 4 Poststimulus time histograms of the mean responses among sweet-sensitive neurons to 0.0025 M Na-saccharin in the three groups of rats. The enhanced activity during the first 3 seconds of the evoked response in conditioned subjects may represent the increased salience of the saccharin taste for these animals. (Reprinted from Chang and Scott [1984] with permission.)

qualities. Moreover, the breakdown occurred differentially, so as to bring the saccharin CS and quinine into rather close apposition. This presumably offers a neural concomitant to the increasingly similar behavioral reaction elicited by quinine and sweet chemicals to which an aversion has been conditioned (Grill and Norgren, 1978).

There are several implications of these findings for the interrelationship of taste and ingestion. First, they reinforce earlier reports that responses of brain stem, and indeed hindbrain taste neurons, are subject to modification by experience. Second, conditioned aversions, which in humans affect primarily the hedonic evaluation of the CS rather than the perceived quality, caused a rearrangement of the neural taste spaces in the NTS of rats. This suggests that the hindbrain neural code in rats combines sensory discrimination with a hedonic component. Finally, activity in these intact animals was modified in ways appropriate for mediation of the behavioral aversion, yet decerebrate rats (with NTS and its vagal afferents intact) are incapable of learning or retaining a CTA (Grill and Norgren, 1978). This implies hypothalamic or amygdaloid involvement in the processes expressed in NTS, given the association of these areas with motivation and hedonics, their required integrity for aversion learning to proceed (McGowan et al., 1972;

Kemble and Nagel, 1973; Roth et al., 1973; Nachman and Ashe, 1974) and their close, reciprocal anatomical relationship with NTS.

A direct test of this implication, however, refuted it. Mark and Scott (1988) treated decerebrate rats with the same conditioning protocol as in the study by Chang and Scott and, based on facial reactions, confirmed the failure of these animals to develop CTAs to the saccharin CS. Subsequent recordings from the NTS of these rats, however, disclosed a peak of evoked activity to the CS with virtually the same relative amplitude and time course as was reported in intact rats. Thus the source of modification must be located caudal to the colliculi (the plane of decerebration). Selective vagotomies and lesions of nuclei such as the area postrema and dorsal motor nuclei may be required to identify that source.

b. Physiological need. The craving for salt and its exaggeration under conditions of sodium deprivation were discussed in the introduction. Preference for high sodium concentrations has been said to follow from the decreased gustatory sensitivity to salt that accompanies depletion. Contreras (1977) analyzed the responses to NaCl of single fibers in the chorda tympani nerve in replete and sodium-deficient rats. Depletion was accompanied by a reduction in salt responsiveness among those 40% of fibers that were most sodium sensitive (Contreras and Frank, 1979). This decreased sensitivity could result in the observed shift of the acceptance curve for sodium to higher concentrations.

This intensity-based interpretation of the effects of sodium depletion has recently been recast by Jacobs et al. (1988). Recording from central taste neurons in the NTS, these researchers confirmed a moderate overall reduction in responsiveness to sodium in salt-deprived rats. Separate analyses of activity among different neuron types, however, revealed that the responsiveness of the salt-oriented group of cells was profoundly depressed and that this effect was partially offset by a dramatic increase in activity among sweet-oriented cells. The net effect was to transfer the burden of coding sodium from salt- to sweet-sensitive neurons. This implies a change not so much in perceived intensity as in perceived quality in sodium-deficient rats: salt should now taste "sweet" or, if sweetness is only a human construct, perhaps "good" to such an animal. This interpretation explains the eagerness with which deprived rats consume sodium, an avidity that is not created simply by reducing salt concentration, as would be implied by the intensity-based interpretation.

In mammals, the term "satiety" encompasses a complex of physical and biochemical mechanisms that may operate through different neural channels. It is clear that alterations in gustatory activity are involved in this process. Glenn and Erickson (1976) recorded multiunit activity from the NTS of freely fed rats as they induced gastric distension and observed a pattern of differential modification in taste responsiveness. Distension depressed activity evoked by sucrose the most, followed in descending order by that to NaCl, HCl, and quinine the responses to which were unmodified. Relief from distension reversed the effect over a 45-minute period. If rats were deprived for 48–72 hours, however, the influence of distension on taste was lost, suggesting that the modulating processes may be sensitive to the overall nutritive state of the animal.

Giza and Scott have investigated the effects of other satiety factors on taste-evoked activity in the NTS. Multiunit responses to the four basic stimuli were recorded before and after intravenous loads of 0.5 g/kg glucose or a vehicle (Giza and Scott, 1983). The glucose infusion had a selectively suppressive effect on taste activity. Elevated blood glucose was associated with a significant reduction in gustatory responsiveness to glucose, with a maximum effect occurring 8 minutes following the intravenous load. Recovery took place over 60 minutes, as blood glucose approached normal levels. Responsiveness to NaCl and HCl was suppressed to a lesser degree and for a briefer period, while quinine-evoked responses were unaffected. Similarly, an intravenous administration of 0.5 u/kg insulin resulted in a transient suppression of taste activity evoked by glucose and fructose (Giza and Scott, 1987b).

Therefore hyperglycemia and modest hyperinsulinemia, both of which result in depression of food intake, are associated with reductions in the afferent activity evoked by hedonically positive tastes. This implies that the pleasure that sustains feeding is reduced, making termination of the meal more likely.

If taste activity in NTS is influenced by the rat's nutritional state, then intensity judgments should change with satiety. Psychophysical studies of human subjects, while not fully consistent among themselves, generally do not support this position. Humans typically report that the hedonic value of appetitive tastes declines with satiety, but that intensity judgments are affected to a lesser extent or not at all (Thompson et al., 1976; Rolls et al., 1981). There are at least three levels of ambiguity that cloud the interpretation of these conflicting results: the neural levels from which the data derive, anesthetic effects, and species differences. The electrophysiological data were taken from the hindbrains of anesthetized rats; the psychophysical responses presumably reflected some involvement of cortical processes in the alert human. A resolution of the implied conflict, then, requires both an analysis of the rat's intensity perceptions and of the human's taste-evoked activity in the hindbrain.

To study intensity perception in the rat, Giza and Scott used behavioral generalization gradients across concentrations in a conditioned taste aversion (CTA) paradigm (Scott and Giza, 1987). Hyperglycemic rats reacted to a range of glucose concentrations as if they were 50% less intense than did conditioned animals with no glucose load (Giza and Scott, 1987a). Thus the neural suppression in the hindbrain that results from an intravenous glucose load appears to be manifested in perceptions of reduced intensity.

While it is reassuring that the rat's behavior conforms to the implications of its neural responses, the original conflict remains unresolved. To complete the puzzle, information is needed on the influence of satiety on taste-evoked activity at various synaptic levels of the human. The closest available approximation to these data may be supplied by subhuman primates. First, the response characteris-

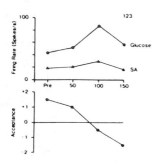
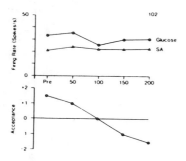
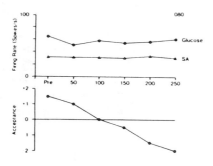

FIGURE 5 The spontaneous activity (SA) and multiunit neural responses (spikes/s) evoked from the NTS by the taste solution on which the monkey was fed to satiety. Each graph represents the results of a separate experiment during which the monkey consumed the satiating solution in 50 ml aliquots, as labeled on the abscissa. Represented below the neural response data for each experiment is a behavioral measure of acceptance of the satiating solution on a scale of +2.0 (avid acceptance) to −2.0 (active rejection). The satiating solution is labeled for each graph (BJ = blackcurrant juice). (Reprinted from Yaxley et al. [1985] with permission.)

METABOLIC INFLUENCES ON THE GUSTATORY SYSTEM 459

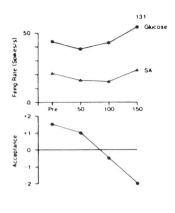
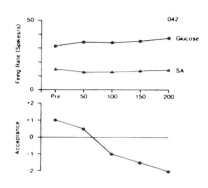
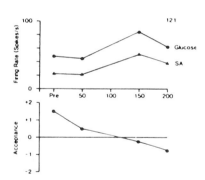
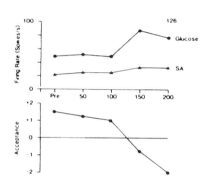
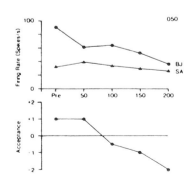
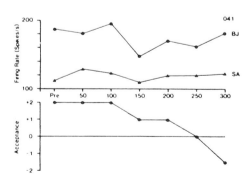

Volume Ingested (ml)

Volume Ingested (ml)

tics of taste neurons in the NTS of cynomolgus monkeys were defined (Scott et al., 1986). Then the activity of small clusters of these cells was monitored as mildly food-deprived monkeys were fed to satiety with glucose (Yaxley et al., 1985). Satiety was measured behaviorally as the monkeys progressed from avid acceptance to active rejection of glucose, typically after consuming 200–300 ml (Fig. 5, bottom of each frame). Despite the effects of gastric distension, elevated blood glucose, and insulin levels this procedure was designed to cause, the responsiveness of NTS neurons to the taste of a range of solutions, including glucose, was unmodified (Fig. 5, top of each frame). These results are in stark contrast to those reported in anesthetized rats.

The same approach has been extended to single neurons in cortical taste areas of the frontal operculum (Rolls et al., 1988) and anterior insula (Yaxley et al., 1988) with similar results. Therefore it appears that the decreased acceptance and reduced hedonic value associated with satiety do not result from a decrement in gustatory responsiveness at any level up to and including primary gustatory cortex. Rather, activity here is related to sensory quality independent of physiological state.

The situation changed when neurons of the monkey's orbitofrontal cortex were studied (Rolls et al., 1985). Taste-responsive cells showed vigorous activity to preferred solutions if the monkey was deprived. As satiety increased—and acceptance turned to rejection—the responsiveness declined to near the spontaneous rate (Fig. 6). Activity elicited by other stimuli, however, was unmodified. Together, these results help clarify the neural mechanisms that underlie sensory-specific satiety (Rolls et al., 1981). Apparently this phenomenon cannot be attributed to receptor adaptation or to reduced activity in peripheral taste nerves or even in central taste cells through primary sensory cortex. Rather, it rests with higher order cortical processes of the lateral orbitofrontal area and beyond. Sensory information is kept separate from hedonic and motivational influences until it reaches the orbitofrontal cortex. There, the association between a stimulus and its reward value could occur, depending on the need state of the animal. Cells in the orbitofrontal cortex have been implicated in behavioral alterations to stimuli that have lost their reinforcing value or have become associated with punishment (Thorpe et al., 1983). Lesions in this area prevent monkeys from learning extinction (Jones and Mishkin, 1972). Perhaps taste-responsive neurons in the orbitofrontal cortex are involved in maintaining the association between a taste stimulus and its reward value. Through alterations in the activity of the orbitofrontal cortex, the animal could modify its behavior according to the availability of environmental resources in relation to its own needs.

Separate populations of neurons in the monkey's lateral hypothalamus that respond to either the taste (Burton et al., 1976) or the sight (Rolls and Rolls, 1982) of preferred foods if the animal is hungry have also been identified. Other cells respond to both taste and vision, implying a convergence of sensory modalities concerned with finding food. As in the orbitofrontal cortex, the induction of satiety suppresses this activity. The electrophysiological evidence, then, supports the position that both visual and gustatory incentives for the initiation and maintenance of feeding may be modulated by momentary physiological needs. In the primate, however, this influence is evident only after several stages of synaptic processing, including cortical relays at which the quality intensity evaluation is held independent of hedonic appreciation. It would not be surprising if macaques, like humans, were found to be able to evaluate the sensory aspects of food separately from its appeal. A resolution of the conflict between rat electrophysiological and human psychophysical data, then, lies not in whether hedonic evaluations are part of the gustatory neural code, but in the neural level at which the interaction occurs.

III. CONCLUSIONS

I have proposed that the sense of taste is intimately and reciprocally involved with metabolic factors associated with ingestion, and that the interaction occurs at three levels.

First, the sense of taste signals the quality and intensity of chemicals through a spatiotemporal code, both the spatial and temporal aspects of which are organized on a dimension of physiological welfare. This analysis offers a first approximation of the appropriateness of consuming a chemical. The capacity to perform the analysis is genetically endowed and, one supposes, derives from evolutionary pressures to avoid chemicals that are toxic and to consume those that provide nutrients. We are, after all, the progeny of those who, among other things, selected wisely from the chemical environment. This analytical process is inherent in the structure and function of the taste system and so applies across all members of a species at all times.

The second level permits tailoring of the inherited code to the physiology of the individual. An enzyme deficiency may render a normally nutritious substance indigestible such that its ingestion causes nausea, vomiting, and diarrhea; hormonal abnormalities may lead to an electrolyte imbalance. The experience resulting from these situations alters gustatory responsiveness in a manner appropriate to accommodate the idiosyncrasy. Thus a chemical that generates a response pattern similar to those of preferred tastes in the naive rat assumes negative characteristics after association with malaise. Conditioning occurs on a time scale that relates to visceral rather than operant processes. In tandem with neophobic behavior, conditioning serves to limit the harm of ingesting a toxin to that caused by a single, mild exposure. While the effects of conditioning apply only to the individual, the mechanism that permits this extraordinarily powerful association between taste and physiological consequences is inherent in the structure of the gustatory–visceral complex.

The third level involves short-term fluctuations in sensitivity that promote or inhibit feeding and encourage consumption of a nutritionally replete diet. Positive

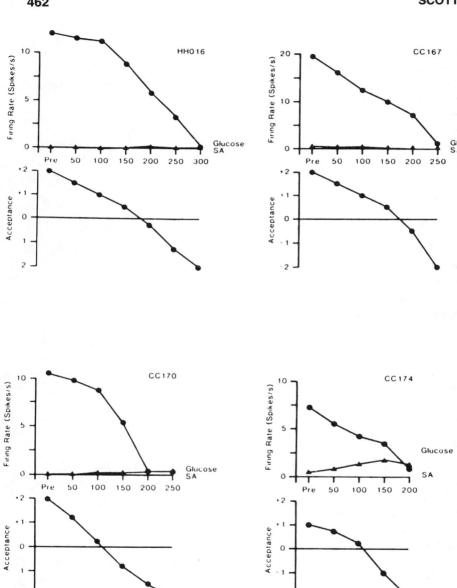

FIGURE 6 The same format as in Figure 5. Responses are derived from single neurons in the orbitofrontal cortex of the monkey. At this level of processing, discharge rate is related to level of satiety rather than the purely sensory aspects of a stimulus.

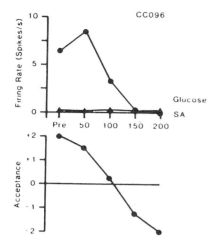
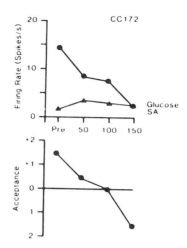
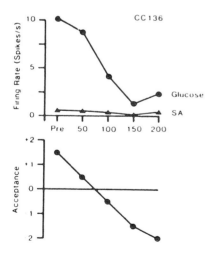
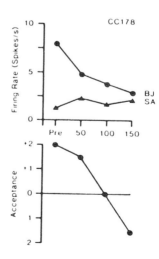

hedonics are suppressed by satiety factors such as gastric distension, hyperglycemia, and moderate hyperinsulinemia. Conversely, a need for sodium arouses enhanced responsiveness to salt among taste neurons associated with positive hedonics.

The taste system is located at the interface between discrimination and digestion. It combines the qualities of rapid stimulus identification and spatiotemporal coding associated with the exteroceptive senses with the slower recognition and still mysterious codes of the visceral senses. The result is a system that reduces the diverse and frequently hostile environment to a chemical subset that effectively satisfies the complex and changing nutritional requirements of the individual it serves.

ACKNOWLEDGMENTS

During preparation of this chapter, the author was supported by research grants from the National Science Foundation, the National Institutes of Health, and the Campbell Institute. Dr. Barbara K. Giza was instrumental in organizing the references, and Judith A. Fingerle and Elsie E. Gibson typed the manuscript.

REFERENCES

Aleksanyan, A. A., Buresova, O., and Bures, J. (1976). Modification of unit responses to gustatory stimuli by conditioned taste aversion in rats. *Physiol. Behav.* **17**:173–179.

Berk, M. L., and Finkelstein, J. A. (1982). Efferent connections of the lateral hypothalamic area of the rat: An autoradiographic investigation. *Br. Res. Bull.* **8**:511–526.

Bernstein, I. L. (1978). Learned taste aversions in children receiving chemotherapy. *Science,* **200**:1302–1303.

Bloom, F. E. (1984). The organization of projections from the cortex, amygdala and hypothalamus to the nucleus of the solitary tract in rat. *J. Comp. Neurol.* **224**:1–24.

Booth, D. A., Stoloff, R., and Nicholls, J. (1974). Dietary flavor acceptance in infant rats established by association with effects of nutrient composition. *Physiol. Psychol.* **2**:313–319.

Burton, M., Rolls, E., and Mora, F. (1976). Effects of hunger on the response of neurons in the lateral hypothalamus to the sight and taste of food. *Exp. Neurol.* **51**:668–677.

Cabanac, M. (1971). Physiological role of pleasure. *Science,* **173**:1103–1107.

Capretta, P. J., and Rawls, L. H. (1974). Establishment of a flavor preference in rats: Importance of nursing and weaning experience. *J. Comp. Physiol. Psychol.* **86**:670–673.

Capretta, P. J., Moore, M. J., and Rossiter, T. R. (1973). Establishment and modification of food and taste preferences: Effects of experience. *J. Gen. Psychol.* **89**:27–46.

Chang, F-C. T., and Scott, T. R. (1984). Conditioned taste aversions modify neural responses in the rat nucleus tractus solitarius. *J. Neurosci.* **4**:1850–1862.

Contreras, R. (1977). Changes in gustatory nerve discharges with sodium deficiency: A single unit analysis. *Brain Res.* **121**:373–378.

Contreras, R., and Frank, M. (1979). Sodium deprivation alters neural responses to gustatory stimuli. *J. Gen. Physiol.* **73**:569–594.

Covey, E., and Erickson, R. P. (1980). Temporal neural coding in gustation. Presentation at the Second Annual Meeting of the Association for Chemoreception. Sciences, Sarasota, FL.

Cullen, J. W., and Harriman, A. E. (1973). Selection of NaCl solutions by sodium-deprived Mongolian gerbils in Richter-type drinking tests. *J. Psychol.* **83**:315–321.

Davis, B. J., and Jang, T. (1988). A Golgi analysis of the gustatory zone of the nucleus of the solitary tract in the adult hamster. *J. Comp. Neurol.* **278**:388–396.

Denton, D. A. (1976). Hypertension: A malady of civilization? In *Systemic Effects of Antihypertensive Agents*, M. P. Sambhi (Ed.). Stratton Intercontinental Medical Books, New York, pp. 577–583.

DiLorenzo, P. M. (1985). Responses to NaCl of parabrachial units that were conditioned with intravenous LiCl. *Chem. Senses,* **10**:438.

DiLorenzo, P. M., and Schwartzbaum, J. S. (1982). Coding of gustatory information in the pontine parabrachial nuclei of the rabbit: Temporal patterns of neural response. *Brain Res.* **251**:245–257.

DiRocco, R., and Grill, H. J. (1979). The forebrain is not essential for sympathoadrenal hyperglycemic response to glucoprivation. *Science,* **204**:1112–1114.

Epstein, A. N., and Stellar, E. (1955). The control of salt preference in adrenalectomized rat. *J. Comp. Physiol. Psychol.* **46**:167–172.

Erickson, R. P., Doetsch, G. S., and Marshall, D. A. (1965). The gustatory neural response function. *J. Gen. Physiol.* **49**:247–263.

Flynn, F. W., and Grill, H. J. (1983). Insulin elicits ingestion in decerebrate rats. *Science,* **221**:188–189.

Garcia, J., Kimmeldorf, D. J., and Koelling, R. A. (1955). Conditional aversion to saccharin resulting from exposure to gamma radiation. *Science,* **122**:157–158.

Gentle, M. J. (1971). Taste and its importance to the domestic chicken. *Br. Poult. Sci.* **12**:77–86.

Gentle, M. J. (1973). Diencephalic stimulation and mouth movements in the chicken. *Br. Poult. Sci.* **14**:167–171.

Giza, B. K., and Scott, T. R. (1983). Blood glucose selectively affects taste-evoked activity in the rat nucleus tractus solitarius. *Physiol. Behav.* **31**:643–650.

Giza, B. K., and Scott, T. R. (1987a). Blood glucose level affects perceived sweetness intensity in rats. *Physiol. Behav.* **41**:459–464.

Giza, B. K., and Scott, T. R. (1987b). Intravenous insulin infusions in rats decrease gustatory-evoked responses to sugars. *Am. J. Physiol.* **252**:R994–R1002.

Glenn, J. F., and Erickson, R. P. (1976). Gastric modulation of gustatory afferent activity. *Physiol. Behav.* **16**:561–568.

Gonshorowitz, J. (1977). Facial expressions in response to taste and food-related odor stimuli in the mentally retarded. Unpublished D.M.D. thesis, Hebrew University, Hadassah School of Dental Medicine.

Grill, H. J., and Berridge, K. C. (1981). Chronic decerebrate rats demonstrate preabsorptive insulin secretion and hyperinsulinemia. *Soc. Neurosci. Abstr.* **7**:29.

Grill, H. J., and Norgren, R. (1978). The taste reactivity test: I. Mimetic responses to gustatory stimuli in neurologically normal rats. *Brain Res.* **143**:263–279.

Grill, H. J., Ganster, D., and Smith, G. P. (1983). CCK-8 decreases sucrose intake in chronic decerebrate rats. *Soc. Neurosci. Abstr.* **9**:903.

Halbreich, U., and Steiner, J. E. (1977). An interaction between hyperbaric pressure and taste in man. *Arch. Oral Biol.* **22**:287–289.

Halstead, W. C., and Gallagher, B. B. (1962). Autoregulation of amino acids intake in the albino rat. *J. Comp. Physiol. Psychol.* **55**:107–111.

Hermann, G., and Novin, D. (1980). Morphine inhibition of parabrachial taste units reversed by naloxone. *Brain Res. Bull.* **5**(Suppl. 4):169–173.

Hermann, G. E., and Rogers, R. C. (1982). Hepatic and gustatory interactions in the brain stem of the rat. *Soc. Neurosci. Abstr.* **8**:200.

Hosoya, Y., Matsushita, M., and Sugihara, Y. (1984). Hypothalamic descending afferents to cells of origin of the greater petrosal nerve in the rat, as revealed by a combination of retrograde HRP and anterograde autoradiographic techniques. *Brain Res.* **290**:141.

Jacobs, K. M., Mark, G. P., and Scott, T. R. (1988). Taste responses in the nucleus tractus solitarius of sodium-deprived rats. *J. Physiol.* **406**:393–410.

Jones, B., and Mishkin, M. (1972). Limbic lesions and the problems of stimulus-reinforcement associations. *Exp. Neurol.* **36**:362.

Kemble, E. D., and Nagel, J. A. (1973). Failure to form a learned taste aversion in rats with amygdaloid lesions. *Bull. Psychonomic Soc.* **2**:155–156.

Loewy, A. D., and Burton, H. (1978). Nuclei of the solitary tract: Efferent projections to the lower brain stem and spinal cord of the cat. *J. Comp. Neurol.* **181**:421–450.

Mark, G. P., and Scott, T. R. (1988). Conditioned taste aversions affect gustatory activity in the NTS of chronic decerebrate rats. *Neurosci. Abstr.* **14**:1185.

McCance, R. A. (1936). Experimental sodium chloride deficiency in man. *Proc. R. Soc. London, Ser. B,* **119**:245–268.

McGowan, B. C., Hankins, W. G., and Garcia, J. (1972). Limbic lesions and control of the internal and external environment. *Behav. Biol.* **7**:841–852.

Mitchell, D., Parker, L. F., and Woods, S. C. (1974). Cyclophosphamide-induced sodium appetite and hyponatremia in the rat. *Pharm. Biochem. Behav.* **2**:627–630.

Nachman, M., and Ashe, J. H. (1974). Effects of basolateral amygdala lesions on neophobia, learned taste aversion, and sodium appetite in rats. *J. Comp. Physiol. Psychol.* **87**:622–643.

Nagai, T., and Ueda, K. (1981). Stochastic properties of gustatory impulse discharges in rat chorda tympani fibers. *J. Neurophysiol.* **45**:574–592.

Norgren, R. (1977). A synopsis of gustatory neuroanatomy. In *Olfaction and Taste,* Vol. VI, J. Le Magnen and P. MacLeod (Eds.). Information Retrieval, London, pp. 225–232.

Norgren, R. (1978). Projections from the nucleus of the solitary tract in the rat. *Neuroscience,* **3**:207–218.

Norgren, R. (1985). Taste and the autonomic nervous system. *Chem. Senses,* **10**:143.

Norgren, R., and Leonard, C. M. (1973). Ascending central gustatory pathways. *J. Comp. Neurol.* **150**:217–238.

Ogawa, H., Imoto, T., and Hayama, T. (1982). Responsiveness of solitario-parabrachial relay neurons to taste and mechanical stimulation applied to the oral cavity in rats. *Exp. Brain Res.* **48**:362.

Pfaffmann, C. (1978). The vertebrate phylogeny, neural code, and integrative processes of taste. In *Handbook of Perception,* Vol. VIA, E. C. Carterette and M. P. Friedman (Eds.). Academic Press, New York, pp. 51–123.

Revusky, S. H. (1974). Retention of a learned increase in the preference for a flavored solution. *Behav. Biol.* **11**:121–125.

Rogers, R. C., and Novin, D. (1983). The neurological aspects of hepatic osmoregulation. In *The Kidney in Liver Disease,* A. N. Epstein (Ed.). Elsevier, New York, pp. 337–350.

Rolls, E. T., and Rolls, B. J. (1982). Brain mechanisms involved in feeding. In *Psychobiology of Human Food Selection,* L. M. Barker (Ed.). AVI, Westport, CT, p. 33.

Rolls, B. J., Rolls, E. T., Rowe, E. A., and Sweeney, K. (1981). Sensory-specific satiety in man. *Physiol. Behav.* **27**:137–142.

Rolls, E. T., Yaxley, S., Sienkiewicz, Z. J., and Scott, T. R. (1985). Gustatory responses of single neurons in the orbitofrontal cortex of the macaque monkey. *Chem. Senses,* **10**:443.

Rolls, E. T., Scott, T. R., Sienkiewicz, Z. L., and Yaxley, S. (1988). The responsiveness of neurons in the frontal opercular gustatory cortex of the monkey is independent of hunger. *J. Physiol.* **56**:876–890.

Roth, S., Schwartz, M., and Teitelbaum, P. (1973). Failure of recovered lateral hypothalamic rats to learn specific food aversion. *J. Comp. Physiol. Psychol.* **83**:184–197.

Sanahuja, J. C., and Harper, A. E. (1962). Effect of amino acid imbalance on food intake and preference. *Am. J. Physiol.* **202**:165–170.

Saper, C. B. (1982). Reciprocal parabrachial-cortical connections in the rat. *Brain Res.* **242**:33.

Schiffman, S. S., and Erickson, R. P. (1971). A psychophysical model for gustatory quality. *Physiol. Behav.* **7**:617–633.

Scott, T. R., and Giza, B. K. (1987). A measure of taste intensity discrimination in the rat through conditioned taste aversions. *Physiol. Behav.* **41**:315–320.

Scott, T. R., and Mark, G. P. (1987). The taste system encodes stimulus toxicity. *Brain Res.* **414**:197–203.

Scott, T. R., Yaxley, S., Sienkiewicz, Z. J., and Rolls, E. T. (1986). Gustatory responses in the nucleus tractus solitarius of the alert cynomolgus monkey. *J. Neurophysiol.* **55**:182–200.

Seward, J. P., and Greathouse, S. R. (1973). Appetitive and aversive conditioning in thiamine-deficient rats. *J. Comp. Physiol.* **83**:157–167.

Shipley, M. T., and Sanders, M. S. (1982). Special senses are really special: Evidence for a reciprocal, bilateral pathway between insular cortex and nucleus parabrachialis. *Brain Res. Bull.* **8**:493.

Steiner, J. E. (1973). The gustofacial response: Observation on normal and anencephalic newborn infants. In *Symposium on Oral Sensation and Perception,* Vol. IV, J. F. Bosma (Ed.). National Institutes of Health–Department of Health, Education, and Welfare, Bethesda, MD, pp. 254–278.

Steiner, J. E. (1974). Testing of the senses of taste and smell in craniofacial abnormalities. Presentation at the Annual Meeting of the American Cleft-Palate Association, Boston.

Steiner, J. E. (1976). Further observations on sensory motor coordinations induced by gustatory and olfactory stimuli. *Isr. J. Med. Sci.* **12**:1231.

Steiner, J. E. (1979). Human facial expressions in response to taste and smell stimulation.

In *Advances in Child Development,* Vol. 13, H. W. Reese and L. Lipsett (Eds.). Academic Press, New York, pp. 257–295.

Steiner, J. E., and Glaser, D. (1985). Orofacial motor behavior-patterns induced by gustatory stimuli in apes. *Chem. Senses,* **10**:452.

Stricker, E. M., and Wilson, N. E. (1970). Salt-seeking behavior in rats following acute sodium deficiency. *J. Comp. Physiol. Psychol.* **72**:416–420.

Thompson, D. A., Moskowitz, H. R., and Campbell, R. G. (1976). Effects of body weight and food intake on pleasantness for a sweet stimulus. *J. Appl. Physiol.* **41**:77–83.

Thorpe, S. J., Rolls, E. T., and Maddison, S. P. (1983). The orbitofrontal cortex: Neuronal activity in the behaving monkey. *Exp. Brain Res.* **49**:93–115.

Wilkins, L., and Richter, C. P. (1940). A great craving for salt by a child with corticoadrenal insufficiency. *JAMA,* **114**:866–868.

Wolf, G., and Handel, P. J. (1966). Aldosterone-induced sodium appetite: Dose response and specificity. *Endocrinology,* **78**:1120–1124.

Wolf, G., and Quartermain, D. (1966). Sodium chloride intake of desoxycorticosterone-treated and of sodium-deficient rats as a function of saline concentration. *J. Comp. Physiol. Psychol.* **61**:288–291.

Yaxley, S., Rolls, E. T., and Sienkiewicz, Z. J. (1988). The responsiveness of neurons in the insular gustatory cortex of the macaque monkey is independent of hunger. *Physiol. Behav.* **42**:223–229.

Yaxley, S., Rolls, E. T. Sienkiewicz, Z. J., and Scott, T. R. (1985). Satiety does not affect gustatory activity in the nucleus of the solitary tract of the alert monkey. *Brain Res.* **347**:85–93.

Part V Discussion

Beauchamp: I've got a question for Tom Scott. I'm wondering whether you or anybody has thought to look at odors to try to relate the pleasantness of odors to their toxicity or some nutritional consequence of ingesting things with that odor.

Scott: We have not. We're not set up to do that type of work and so we're not involved in that at all. I do know that Leslie Wiggins in E. T. Rolls's lab has been recording from cells in the orbito frontal cortex which respond to both olfactory and gustatory stimuli. They report that the hedonic value of the stimuli is closely correlated between the two. When something smells good, it drives the cell; if you put a stimulus that tastes good on the tongue, you will also drive the same cell. There is, of course, the interaction that has been shown in the solitary nucleus of the rat as well between olfactory and gustatory stimuli, but there is no hedonic implication about that interaction.

Cabanac: There was a lot of information in what you presented. I was wondering whether you would cross the Rubicon and pronounce the words "hedonic" and "pleasure." I have the feeling that what you have measured was precisely that: an additional dimension to sensation.

Scott: Yes, exactly. I would say it's related but it's not the same thing. I think we live in a hostile chemical environment where most of the things that we would like to eat are trying to protect themselves chemically. The taste system evolved to allow us to go into that hostile environment and select out the things that we need. Hedonics is the evolutionary outgrowth of that process. We are the offspring of creatures that successfully selected from that environment, and they successfully selected because of the pleasure they derive from it.

Cabanac: I share these ways of expressing things. Therefore I'm glad that you finally did say it because when you say that depriving the rat of sodium moved the salty sensation toward the sweet sensation, [this] can be equally expressed as saying that both share a common dimension, which could be pleasure. Of course, the rat cannot say it.

Scott: I would agree with you and, in fact, would make a more direct statement. My interpretation of that is not necessarily that salt tastes sweet, though I would be very happy if people like Gary [Beauchamp] and others, in asking their human subjects, found that it did taste sweet. I think salt tastes better, and that we're looking at a good–bad dimension hedonically, which corresponds to a nutrition–toxicity dimension biochemically, and an acceptance–avoidance dimension behaviorally, and that they're all essentially the same thing.

Kosar: I have a question for Dr. Cechetto. I want to know something more about the microstimulation of the cortex. Can you evoke oral movements with that microstimulation? What kind of current do you use?

Cechetto: A colleague in Toronto is stimulating in that region and is specifically looking for this. He has found a region and, although I have not seen his histology, it's very close to what I would call the gustatory insular region.

Kosar: In the rat?

Cechetto: He's using cats. For myself, when I'm looking for a specific sympathetic response with the level of stimulation I'm using, I actually have not seen it. But I have another postdoctoral fellow doing some work on just EKGs and cardiac sympathetic nerve activity, and on occasions he's cranked up the stimulation and he does get it. I've seen that happen. But the problem is now that we're not localized to the specific region of the insular cortex, or even if it is insular cortex, is in question. For stimulation I use under 200 μA and usually have a threshold for the sympathetic responses of 30 μA, which is a very small region from which I'm eliciting activity from neural components.

Spector: I have a question for Dr. Scott. The data showing that sweet-sensitive channels seem to decrease responses to sodium in sodium-depleted rats suggest perhaps that the taste quality of sodium does change and may in fact be taking on a sweet character or a sucroselike character. One way of testing that would be to

PART V DISCUSSION

condition a taste aversion to sucrose and then do a conditioned aversion generalization to see if it did or did not generalize to salt when the rats are salt-deprived. Have you tried anything like that or considered it?

Scott: We got as far as the thinking stage on that particular experiment and, yes, that's the first thing that occurred to us. If these things are turning sweet, we should be able to show it behaviorally. Of course, a wonderful paradigm for showing behavioral generalization gradients is using conditioned taste aversion. But we can't because, as you saw before, the conditioned taste aversion procedure itself alters the neural code for taste quality. So we would have been working with a system that was not normal in some way. Therefore, we did try the experiment, but went back to a procedure that had been used 20 years ago, which is to give the two distinctly different stimuli and give the rat a behavioral choice between them. He'd always go left to one, always go right to the other. Then, after getting a greater than 90% correct choice rate from the rat, you start occasionally interposing the third stimulus and see which one the rat confuses it with—Which way does he go? We wanted to do that with deprived and undeprived rats and see if they would confuse salt with sucrose when they were deprived, not when they were undeprived. The experiment did not work.

Elizalde: I have a question for you. I have paired consuming sucrose octaacetate (SOA) with intragastric infusions of Polycose, and rats will prefer that flavor over another flavor not paired with Polycose. Will the bitter SOA become sweet? Does it work the other way around?

Scott: We spent a good deal of last evening in fact working out an experiment in which we will test that. Our colleagues in California (Davis) will generate animals with conditioned preferences and then we will look at them as compared to a control group. The selection of the stimulus will be important. So if SOA works, that's something that we might want to look at as our CS.

Frankmann: With respect to Alan Spector's question to Tom Scott, I have some data that I have always been very unhappy with and I think this may make me happy about them. Sodium depletion can be used to condition a taste aversion in the rat. Unfortunately, it can be used to condition a taste aversion to a sodium chloride solution being used to recover from the depletion. To try and show that I could prevent the aversion to sodium chloride by sodium depletion, I gave the rat a mixture of a saccharin and an isotonic saline solution. When the rat was then given a choice of the saccharin solution and isotonic saline solution, it showed a complete rejection of the isotonic saline solution. It would drink only the saccharin solution, suggesting that in the sodium-depleted state and during the recovery period, the animal was attending to the sodium taste stimulus only, although a sweet stimulus was present. If, in fact, the change in the responsiveness of the sweet cells is not an increase in a responsiveness to sweetness, but a shift in the hedonic value of salt, it may be the animal was not attending to the sweet stimulus. I think that addresses your question.

Scott: That's good. At the same time, the response in the sweet-sensitive fibers to sucrose was still greater in sodium-deprived animals. The response to sucrose at a fairly high concentration was still greater than the response to sodium chloride.

Frankmann: They could be firing not to sweetness but to some hedonic evaluation of the same stimulus.

Blundell: There's a couple of things that I was thinking about during Tom Scott's presentation. One, arising out of the conditioned taste aversion work, that the associative process now has some sort of neurophysiological basis in terms of the shape of the response of the neurons in the solitary tract—a rather interesting thought.

The more troublesome one that I'm trying to sort out—I'm not sure I fully appreciate it yet—is the difference in the response of glucose loads in the rat and monkey—one in the NTS and the other in the orbitofrontal cortex—and your sudden shift from an intensity change only to, in the primate, hedonic response. We talk as if both rats and monkeys have hedonic responses. In the absence of any information to identify the hedonic response, I'm not convinced that either of those changes represents the neural embodiment of pleasure in the sense that Cabanac means it.

Scott: No, we can't say that they represent the neural embodiment of pleasure. We can correlate them with the behavior of the animal and only make inferences as to what is pleasurable to the animal and what is not pleasurable. We certainly are making inferences based on the acceptance of the solution as to what gives the animal pleasure and what doesn't.

Blundell: I don't see any special reason to attribute a hedonic response to changes in the orbitofrontal cortex any more than you would attribute that to the solitary nucleus in the rat.

Scott: I would, in fact, attribute hedonic responses to the solitary nucleus of the rat. Let me expand on what I was saying to Cabanac. I think that in a rat there may very well not be any difference between a quality and intensity judgment and a hedonic evaluation. I think they may be the same thing to a rat. I think they occur at a very low level; in fact, in the caudal hindbrain. For all we can tell, the rat is simply guiding itself by hedonic evaluation. A good–bad dimension.

Blundell: But aren't you saying the same thing about a monkey? Because you didn't identify an intensity response that was different from, say, a hedonic response.

Scott: Of course, yes. I'm saying the same thing about a monkey, except that I do think the processes are distinct. Because in the monkey you can follow the intensity response to a point with no interaction with hedonics and if you ask humans they can say after a glucose load "I don't like it as much," but also "it still tastes just as strong to me." There is some dirtiness in those data—that is, different people have said different things—but the predominant psychophysical data indicate that there is a distinction between intensity evaluation and hedonic evaluation in the human. We would guess, and it's only a guess, that the monkey has the neural apparatus there to make the same distinction between intensity and hedonics. The rat seems not to.

Campfield: I would make a comment on this general issue. In the final part of your presentation relating to infusions of glucose and insulin, when you start to talk about sweet taste, you should be careful about the peripheral consequences. We've seen here quite well-demonstrated cephalic phase responses to saccharin. Saccharin effects have been mentioned throughout the meeting; saccharin has clear effects on glucose production. So as we look at what we think are pure taste stimuli and sweet taste, we perhaps should think that the brain may be very affected by the periphery in this sense: that the response that we measure may be strongly conditioned on the metabolic state. Such things as deprivation, depletion, repletion should be considered. Unfortunately, it seems that because the tongue is closer to the brain, the focus is not to look at the long part of the feedback loop. I would suggest there's a lot of information there. I think your own data show that powerfully. I think the work with saccharin eliciting a cephalic phase insulin release shows another powerful connection between periphery and brain to integrate this very important total body response.

Collier: Your approach is interesting in the sense that you're going back to the evolutionary basis and usefulness of these tastes. But what still puzzles me is that the dimensions the animals really have to make decisions on are the macronutrients and the micronutrients, which aren't represented on a one-to-one basis as sweet, sour, and so on.

Scott: I would agree. As far as macronutrients go, they are largely represented in simple carbohydrates. As far as complex carbohydrates go, that's something that's just being worked out now. Perhaps there's a corresponding system, which may be trigeminal as well as gustatory, that would tie in to the appropriate hedonic areas and would account for the effectiveness of those stimuli. As far as micronutrients go, I agree. It seems perplexing that essential amino acids are pretty much bad tasting to human beings and in these types of plots are bad tasting to rats as well. Perhaps the answer is that rats and humans don't encounter those amino acids in their free forms in their diet. They encounter them in the form of proteins, proteins which are highly appetitive. As those amino acids are freed through the metabolic process, there are receptors for those amino acids. There's no reason for putting a receptor up at the point where you won't encounter that stimulus. The receptors are tuned to the stimuli that will be liberated through the digestive process. The mouth can be considered the beginning of the digestive process, or maybe the nose is in fact.

Collier: That may be a little late in the decision process, though.

Scott: But it can serve as a tag. If something goes wrong later on, you now know what to attribute it to because of the taste experience that you had.

Index

Addison's disease, 104
  depletion-specific salt appetite, 87–89
Adenoid hypertrophy and smell, 343
Adipose tissue, obesity, 164
Adrenal–cortical insufficiency
  smell, 343
  taste, 343
Adrenal insufficiency, nonspecific lowered taste threshold, 91
Adrenal mineralocorticoids, induced sodium appetite, 3
Alcohol
  animal models of, 40
  characteristics of murine alcohol consumption, 42–45
    alcohol intake, 42
    circadian pattern drinking behavior, 42
    conclusions to motivation, 45
    effect of diet on alcohol consumption, 44
    plasma concentrations of alcohol after bouts of drinking, 43
    plasma concentrations of ethanol at midnight, 42, 43
    response of alcohol intake to sucrose challenges, 44, 45
    temporal relationship with feeding behavior, 43, 44
  Korsakoff's syndrome and sense of smell, 343
  metabolic theory of, 48–52
    alcohol-specific appetite, 50
    role of liver, 50
    unique metabolism of ethanol, 51
  motivations disease, 39, 40
  murine motivation for drinking, 41
  operational definitions of alcohol-specific intake, 40, 41
  relevance of alcohol intake to
    alcohol intemperance in animals, 46–48
    modeling the preaddictive state, 45, 46
    peaks in blood alcohol as reinforcing, 46

475

Aldosterone
  induced sodium depletion, 94, 95, 101
  mediation of a specific behavioral response to sodium, 7
  relation between thirst, sodium appetite, and plasma, 7
  stimulus for sodium appetite, 3, 6
  regulation of extracellular fluid volume, 228
  taste-directed changes in food selection, physiological need, 446
  and thirst, 368, 375, 376
Alliesthesia, 134, 137, 138, 142
  depletion-specific sodium appetite, 212
  ponderostat, 153–156
α_2-adrenergic activity
  glucagon, 407, 408
  insulin, 407, 408
Alzheimer's disease
  smell, 343
  taste, 345
Angiotensin
  mediation of drinking response, 5
  relation between thirst, sodium, appetite, plasma, 7
  stimulation of sodium appetite, 6, 7
    regulation of extracellular fluid volume, 228
  thirst, 368, 375
Anorexia, 157, 158, 163, 164
  CCK-induced, 453
    postdecerebration, 453
  due to sodium depletion, 91
  hedonic response profiles, 75, 76
  salivary response, 157
  smell, 361
  taste, 361
Anosmia
  diet, selection and avoidance, 190
  in the elderly, 360

Appetite
  acquired preferences, 177–279
  diet, selection and avoidance, 177–191
  factors in salt hunger, 211–231
  learned food preferences, 239–256, 271, 272
  primate gastronomy, 195–207
  protein- and carbohydrate-specific cravings, 261–279
  during life stages, 283–387
  changes of appetitive variables in pregnancy, 325–337
  children's food acceptance patterns, 303–322
  suckling and mother-infant bonding, 283–300
  taste and smell changes over life span, 341–361
  thirst and fluid balance in the elderly, 367–380
  hedonic aspects of, 6, 7, 69–167
  human salt appetite, 85–204
  hunger, hedonics, and satiation control, 127–145
  palatability in dietary obesity, 109–122
  the ponderostat, 149, 167
  sensory and hedonic aspects of sweet, high-fat foods, 69–81
  innate preference for protein, 270, 271
  mechanisms, not intake effects, 272, 273
  mediation by specific alterations in gustatory responsiveness, 447–461
    anatomy, 447, 448
    physiology, 448–461
  neural integration, 391–474
    CNS, taste, and the autonomic nervous system, 427–437

INDEX

metabolic influences on the gustatory system, 445–464
neuoranatomy and cephalic reflexes, 391–402
neuroendocrine activity during food intake, 405–422
protein-specific mechanisms, 270–273
   innate protein appetite, 270, 271
   learned protein appetite, 271, 272
regulatory aspects of, 1–65
   biological bases for sodium appetite, 3–15
   metabolic control of caloric intake, 19–36
   nutritional determinants of alcohol intake, 39–54
saliva secretion, 156–158
satieity, 269
   conditioned desatiation, 269
   conditioned satieties and noradrenergic feedings, 269, 270
social, sensory, and somatic factors in,
   multiple appetites and satieties, 264, 265
   social influences on eating, 264
sociology of, 261–264
   biological reductionism, 261, 262
   measurement of appetite mechanisms, 262–264
theory of, 265, 266
   development of, 266
   framework for appetite neuroscience, 265, 266
   regulation and function in, 266
Arteriosclerosis, primary pulmonary, depletion-specific salt appetite, 88
Artificial sweeteners, 158, 159
Autonomic nervous system, taste and the CNS, 427–437
   CNS mechanisms and pathways, 431–437
   gustatory, feeding, and general visceral interaction, 428–431
   insular cortex, 433, 434
AVP
   secretion during hypovolemia, 5
   thirst, 367

Barter's syndrome, depletion-specific salt appetite, 89
Bell's palsy, and taste, 345
β-casomorphine, role in infant quieting, 296, 388
$β_2$-adrenergic stimulation, 407, 408
Bilateral adrenalectomy, 3
Bilateral nephrectomy, 5
Bitterness, 130, 268, 451
   innate preferences, 267
   loss of in the elderly, 344
Blood pressure, depletion-specific salt appetite, 104
Body weight regulation, 405–422
   alliesthesia, 153–156
      obesity, 153
   artificial sweetners, 158, 159
   change in sensitivity to glucose, 417, 418
   eating disorders, 157, 158, 163
   food hoarding, 159, 165
   hypothalamic influences on, 152
   glucose, insulin, and body weight, 417–421
   hepatocytes and adipocytes of the islets of Langerhans, 410–412
   neuroendocrine activity during food intake, 405–422
   cephalic responses, 412–417
   connection between CNS and

Body weight regulation (*cont.*)
  islets of Langerhans, 406, 407
  regulation of pancreatic islet function, 407–410
  saliva evaluation, 156–158, 164
Brain stem, 455
  CNS, taste, and the autonomic nervous system, 431–434
  CNS pathways and mechanisms, 431–437
  convergence of gustatory and general visceral efferents, 436
Bulimia, 157, 158, 163, 164
  hedonic response profiles, 75, 76
  salivary response, 157
  smell, 361
  taste, 361

Calcitonin gene-related peptide, *see* CGRP
Calcium, 224, 230
Caloric intake
  deprivation-dependent effect on learning, 242
  dietary obesity, 112, 113
  fuel oxidation and control of calorie intake, 27–34
    change in response to homeostatic demands, 30, 31
    constancy of calorie intake and illusion of precision, 29, 30
    disturbances in calorie intake, 31–34
  metabolic stimulus and receptor, 23–27
  hepatic receptor, 25–27
  stimulus of free oxidation, 23, 24

postabsorptive control of, 20–23
  gastrointestinal signals, 20, 21
  metabolizable calories, 22, 23
  postabsorptive signals, 21, 22
regulatory aspects of appetite, 1–65
  metabolic control of caloric intake, 19–36
cAMP, mechism of olfactory losses, 349–352
Carbohydrate-specific cravings, *see also* Protein-specific cravings
Carbohydrates, 118, 119, 241
  cravings of, 77–79
  deficiency of, 77–79
  postingestional effects, 268
CCK, 408
  anorexia, 453
  insulin release, 406
  role in quieting infants, 295, 296
  satiation, postdecerebration, 453
Central oxytocinergic neurons, 9, 10
Cephalic phase insulin response, *see* CPIR
Cephalic phase reflexes and neuroanatomy, 391–402
  CPIR
    effects of preganglionic neurons, 395–397, 399–401
    mediation by the vagus, 391, 394, 401
    neural architecture, 400, 401
    neuron control of, 394, 395
    vagal efferents, 396–400
  during food intake, 412, 417
    preabsorption insulin release, 412–417
  gustatory, feeding, and general visceral interaction, 428–431
  saccharin, 473
CGRP

CNS, taste, and autonomic nervous system, 431–434
   insular cortex, 433
   ventral basal thalamus, 435, 436
Cholecystokinin, see CCK
Chorda tympani
  sodium deficiency, 221
  and taste, 345
CNS
  mechanisms and pathways, 431–437
   brain stem, 436, 437
   insular cortex, 431–434
   ventral basal thalamus, 434–436
  taste and the autonomic nervous system, 427–437
   gustatory, feeding, and general visceral interactions, 428–431
Congenital adrenal hyperplasia, and taste, 345
Corticosteroid insufficiency, depletion-specific salt appetite, 88
CPIR
  delineating neural mechanisms of, 391
  food intake preabsorption insulin release, 412–417
  mediation by the vagus, 391, 394, 395
  neural architecture of, 400, 401
  neuron control of, 394, 395
  postdecerebration, 453
  preganglionic neurons, 395–397, 399–401
  vagal efferents, 396–400
Cravings
  carbohydrate-specific, 261–273
  during the menstrual cycle, 327–335
  factors predisposing toward salt hunger, 211–231
  in pregnancy, 325–336, 386
  protein-specific, 261–273
Cretinism, and taste, 345
Crohn's disease, oral, and taste, 345
Cushing's syndrome
  smell, 343
  taste, 345
Cyclophosphamide, taste-directed changes in food selection physiological need, 446

Davis, Clara, 179, 180, 188
  children's eating behavior, 306, 307
Dehydration, in the elderly, 368, 376, 377
Deoxycorticosterone, taste-directed changes in food selection physiological need, 446
Deoxycorticosterone, see also DOC
Diabetes mellitus
  smell, 343
  taste, 345
Diet, selection and avoidance, 177–191
  ability to select a nutritionally adequate diet, 179–181
  cooking, 204
  food acquisition and processing, 202, 203
  in infants, 306–308, 310
  ingestion of toxins, 178
   Pavlovian conditioning process, 181
  measurement of appetite mechanisms, 262–264
  origins of food taboos, 205
  primate cuisine, 196–205
  social influence of adult animals on their young, 184–186

Diet (*cont.*)
    deposition of residual olfactory cues, 184, 185
    flavor cues in mother's milk, 185
    olfactory cues on breath of adult rats, 185, 186
    presence of adults at feeding site, 184
    social solutions to the problem of selecting a nutritionally adequate diet, 182–188
    taste
        directed changes in, 446
        palatability to a human of wild chimpanzee food, 197, 198
    total self-regulatory capacity, 179
    use of socially acquired information, 186–188
    identification and avoidance of toxins, 187, 188
    what to eat, 186, 187
    where to eat, 188
    vagotomy, 429
        gustatory, feeding, and general visceral interactions, 428–431

Dietary obesity, palatability in, 109–122, 161–163
    alliesthesia, 153–156
    caloric intake, 112, 113
    diet selection, 111, 112
    general considerations
        changes in energy expenditure, 110
        genetic obesity, 110
        nutritional manipulation, 110
    hyperphagia, 422
    overeating, 113, 114
    postingestional factors responsible for, 111–114, 118–121
    significance of sensory factors in foods, 114–118

DOC, 6, 7, 446
    eliciting sodium appetite, 3
    suppression of renin secretion, 7, 8
Dwarfism, 303
Dysautonomia, familial, and taste, 345

Elderly
    chemosensory loss in, 345–358
        age-related modification of smell, 343
        decreased sensitivity to acids, 348
        decreased sensitivity to sodium, 346, 347
        olfactory losses, 349–353
        reduced food recognition by odor cues, 349
    hypertension and taste perception, 387
    thirst and fluid balance in, 367–380
        dehydration, 376, 377
        mechanisms of thirst deficit, 373, 374
        neuroendocrine function, 375, 376
        renal function, 374, 375
Ethacrynic acid, induced sodium depletion, 93

Facial hypoplasia, and taste, 345
FAP, in response to taste stimuli, 135
Fat, 96, 97, 114, 115, 121, 298, 299
    attitudinal studies, 79, 80
    endogenous opioid peptide system, 79
    hedonic response profiles, eating-

INDEX

481

disordered groups, 75, 76
metabolic factors, 77–79
cravings, 77
sensory evaluation studies
oral sensations, 71, 72
relationship between perceived fat content and food acceptance, 71–75
FFA
connection between CNS and islets of Langerhans, 406, 407
regulation of pancreatic islet function, 409, 410
Fitness, diet, selection and avoidance, 177
5'-monophosphate, see cAMP
5'-triphosphate, see GTP
Fixed-action patterns, see FAP
Fluid balance, and thirst in the elderly, 367–380
dehydration, 376, 377
mechanisms of thirst deficit, 373, 374
neuroendocrine function, 375, 376
renal function, 374, 375
Food acceptance patterns, children, 303–322
acquisition of specific food likes and dislikes, 316–321
caretaker influence and physiological feedback, 316–321
conditioned aversions, 316
social learning effects on the formation of food preferences, 319
birth order, 385
control of food intake, 304–315
caloric density cues, 304, 305
depletion-based cues, 304, 305
interaction of caretaker influence and internal cues, 304–315

parental health misconceptions, 308, 309
Food intake
cardiovascular-associated changes, 430
control of, by liver, 429
CPIR, preabsorption insulin release, 412–417
destruction of SCN, 417
effect of vagus nerve, 429
vagotomy, 429
flavor-associated nutrient values, 242–256
hoarding, rats, 159, 165
hyperglycemia, 457, 464
hyperinsulinemia, 457, 464
learned preferences, 239–256
phenomenological studies, 240–242
site of unconditional stimulus transduction, 242–252
neuroendocrine activity during, 405–422
cephalic responses during, 412–417
connection between CNS and the islets of Langerhans, 406, 407
hypothalamic influences on hepatocytes and adipocytes of islets of Langerhans, 410–412
regulation of pancreatic islet function, 407–410
Formalin, taste-directed changes in food selection, physiological need, 446
Free fatty acids, see FFA
Fuel oxidation, 252–254, 278
Furosemide, induced sodium depletion, 93

Gastrin, insulin release, 406

Geniculate ganglion cells, sodium/
     lithium specificity, 103
Glossitis, and taste, 345
Glucagon
  promotion of gluconeogenesis,
       408
     stimulation of VMH, 411
  release
     modulation of D-cell function,
          406
     peptidergic mechanisms, 406,
          408
     stimulation of α_2-adrenoceptors,
          407
Gluconeogenesis, 408
Glucose, 473
  alteration of in central gustatory
       neuronal responsiveness,
       428
  hypothalamic influences on
     body weight, 417–421
     glucose, 417–421
     insulin, 417–421
     obesity, 417, 421
  regulation of pancreatic islet function, 408, 409
     FFA, 409, 410
Glycogenolysis, 408
  stimulation of VMH, 410
GTP, mechanism of olfactory
     losses, 349–352
Gustatory, feeding, and general visceral interaction, 428–431
  CPIR, 428–431
  effects of vagus nerve, 428–431
  influence of general visceral input, 428, 434, 435
     behavioral changes related to
          feeding, 428, 429
     blood glucose levels, 428
     extracellular sodium concentration, 428
     meal-related stimuli, 429

  metabolic influences on, 445–461
     anatomical considerations, 447,
          448
     establishment of preferences,
          445
     experience, 445
     physiological need and considerations, 446–461
     taste perceptions, 448, 449
  physiological welfare-based code,
       453, 454
  plasticity of, 453, 454

Halogeton, 182
hCG, 326
Hedonics, 95, 102, 111, 240
  alliesthesia and satiation, 137,
       138
  definition of, 127, 130
  diet-governed responses, 112
  eating disorders, 75, 76
  hunger and the caloric control of
       satiation, 140–142
  importance of hunger, 138–140
  location of "monitor" for, 453
  palatability, 130, 131
     relationship to hunger, 142–
          144
  satiety cascade, 131–133
     sensory-specific, 133–137
  sensory control of ingestion, 267
  site of unconditioned stimulus
       transduction, 243, 244
  social, sensory, and somatic factors in appetite, 264, 265
  synchronicity of physiology, behavior, and cognitions,
       144
  taste, 469, 470, 472, 473
     directed changes in food selection, 446
Histidine, feeding behavior, 446

INDEX

induced deficiencies of, 446
physiological need, 446
Human chorionic gonadotrophin, *see* hCG
Hunger, 242
  alliesthesia and satiation, 137, 138
  caloric control of satiation, 140–142
  definition of, 127–130
  the importance of, 138–140
  infant's cues for, 305
  palatability, 130, 131
  relationship between palatability and hunger, 142–144
    opioids, 142
  for salt, 211–231
    consequences, 222
    evolution of, 227–229
    innate reasons, 211–213
    latent learning studies, 213–220
    mineral deficiencies, 222–224
    salt lick, 224–227
    salty taste ingestion, 220–222
  satiety cascade, 131–133
  sensory-specific, 133–137
  synchronicity of physiology, behavior, and cognitions, 144
Hyperglycemia, 457, 464
Hyperinsulinemia, 457, 464
Hypernatremia and thirst, 373
Hyperphagia, 110, 114, 142
  insulin-induced, 453
    postdecerebration, 453
  obesity, 422
Hypertension, and taste, 345
Hypotension, 4
Hypothyroidism
  smell, 343
  taste, 345
Hypovolemia, 4

Insular cortex, 437
  CNS, taste, and the autonomic nervous system, 431–434
  CNA pathways and mechanisms, 431–437
  loss of taste discrimination, 432
  regulation of ingestion, 432
  CGRP, 432
Insulin, 412, 473
  hypothalamic influences of glucose, insulin, and body weight, 417–421
  CCK, 406
  somatostatin, 406
  obesity, 164
  regulation of pancreatic islets of Langerhans function, 407, 408
  release
    modulation of D-cell function, 406
    peptidergic mechanisms, 406, 408
    in response to meal-related stimuli, 430
    stimulation of 2- and 2-adrenoceptors, 407
Iron deficiency
  depletion-specific salt appetite, 90
Islets of Langerhans
  divergence of CPIR vagal efferents, 396
  insulin release, 405, 418
  hypothalamic influences on glucose, insulin, and body weight, 417–421
  hepatocytes and adipocytes, 410–412
  neuroendocrine activity during food intake, 406, 407
  regulation of pancreatic islet function, 407–410
    glucose, 408, 409

Islets of Langerhans (*cont.*)
role of CNS, 406, 407
norepinephrine and epinephrine, 406, 407
peptidergic mechanisms, 406, 408
PVN, 406, 407

Kallman's syndrome, and smell, 343
Korsakoff's syndrome, and smell, 343

Larkspur, 182
Laryngectomy
smell, 343
taste, 345
Learned food preferences, 239–256
phenomenological studies, 240–242
deprivation state, 241, 242
dissociation of conditioned and unconditioned stimuli, 240, 241
site of unconditional stimulus transduction, 242–252
cranial, 249, 250
gastric, 245, 246
hepatic, 247, 248
information transfer from transduction site to brain, 251, 252
intestinal, 246, 247
oral, 243–245
parenteral and systematic, 248, 249
what is transduced?, 250, 251
Legume anorexia, 303, 319
Leprosy
smell, 343
taste, 345
Lipolysis
glucagon release, 410

regulation of pancreatic islet function, 409
stimulation of VMH, 410, 411
Lithium, 229
activation of gustatory sodium transport system, 220, 221
Liver
alcohol, 50
control of food intake, 429
disease
smell, 343
taste, 345
fuel oxidation flavor-associated nutrient values, 252–256
regulation of pancreatic islet function, 408, 409
glycogen phosphorylase conversion, 408
Locoweed, 182
Low-pressure baroreceptors, thirst, 4, 5
Lupine, 182

Mineralocorticoids, 224
Mitochondrial myopathy, depletion-specific salt appetite, 89
Mother–infant bonding, suckling, 283–300
contact as a calming agent, 284–286, 288–297, 398
drug-versus contact-induced infant quieting, 288, 289
Multiple sclerosis
smell, 343
taste, 345

Nausea and vomiting during pregnancy, *see* NVP
Nausea, due to sodium depletion, 91
Neuoranatomy and cephalic phase reflexes, 391–402
CPIR

INDEX

effects of preganglionic neurons, 395–397, 399, 400
mediation by the vagus, 394
neural architecture, 400, 401
neuron control of, 394, 395
vagal efferents, 396–400
Neuroendocrine activity
during food intake, 405–422
cephalic responses, 412–417
connections between CNS and islets of Langerhans, 406, 407
hypothalamic influences on glucose, insulin, and body weight, 417–421
hypothalamic influences on hepatocytes and adipocytes of islets of Langerhans, 410–412
regulation of pancreatic islet function, 407–410
in the elderly, 375, 376
Normalkalemia, with periodic paralysis, depletion-specific salt appetite, 89
Nutrition
determinants of alcohol intake, 39–54
food acceptance patterns, children, 303–322
parenteral health misconceptions, 308, 309
NVP, 326, 332–335

Obesity, *see also* Dietary obesity
change in sensitivity to glucose, 417, 418
smell, 361
taste, 361
Olfaction
cAMP
mechanisms of olfactory losses, 349–352

diet, selection and avoidance, 184–186
obesity, 361
olfactory
cues, 184–186
sarcoidosis
smell, 343
Open-loop model, ponderostat, 151, 158, 163
Opioid and nonopioid processes, 283–300
mother–infant bonding, 284
OT
inhibition of sodium appetite, 9
relation between thirst, sodium, appetite, plasma, 7
Ozena, and smell, 343

Palatability, 109–122, 130–133, 264, 265
alteration of, 240
of primates to human food, 197
Pancreas, 395
neural innervation of, 401
regulation of islet function, 407, 410
Panhypopituitarism, and taste, 345
Paralysis, periodic with normalkalemia, depletion-specific salt appetite, 89
Paraventricular neurons, *see* PVN
Parkinson's disease, and smell, 343
PEG
sodium appetite, 4, 7
thirst stimulation, 8
Pica, 325
iron deficiency, 383
Pleoconial myopathy, depletion-specific salt appetite, 89
Polydipsia, 46
Polyethylene glycol, *see* PEG
Polysaccharides, 118, 298, 299

Ponderostat
  alliesthesia, 153–156
  artificial sweetners, 158, 159
  food hoarding, rats, 159
  open-loop model, 151
    actuation of negative feedback loop, 151
    manipulation of body weight, 152, 153
  positive feedforward, 159–163
    body weight manipulation, 159–163
  saliva secretion, 156–158
    as an index of appetite, 156, 157
  stability of body weight, 152, 153
  steady state versus regulated system, 149–151
    global perturbation, 150
Potassium, 221, 223, 224, 230
  activation of gustatory sodium transport system, 220
  chloride
    as a salty stimulus, 103
Predation risk, 266
Preganglionic neurons, 391, 392, 395–397, 399, 400
Pregnancy
  changes in appetitive variables during, 325–337
  color aversion, 383
  diet
    aversions of nutritional and sensory food groups, 331–333
    cravings of nutritional and sensory food groups, 328–330, 386
  functions of cravings and aversions, 325–327
  problems with past research, 327
  research needs, 335–337
    chemosensory changes and food intake, 336
  food intake and chemosensory changes as related to cravings, aversions, and hormonal status, 336
  hormone levels and chemosensory changes, 336
  maternal appetitive behavior and neonatal taste responsiveness or weight gain, 336
  symptoms of nausea and vomiting, 326, 332–335
  Yale Pregnancy study, 327–335
    aversions and cravings during the menstrual cycle, 327–335
Primary amenorrhea, and smell, 343
Primate gastronomy, 195–207
  gustatory sense, 195, 196
    similarities between humans and chimpanzees, 195, 196
  influence of satiety on taste-evoked activity, 457–461
  primate cuisine, 196–205
    cooking, 204
    dietary selection, 196–202
    food acquisition and processing, 202, 203
    food versus drugs, 203, 204
    origin of food taboos, 205
Protein- and carbohydrate-specific cravings, 261–273, 277–279
  immediate sensory control of ingestion, 266–270
    conditioned appetites and satieties, 269, 270
    innate preferences and aversion, 267
    learned preferences, 267, 268
    nutrient-specific sensory preferences, 270
  protein-specific appetite mecha-

nism, 270–273
  protein appetites, innate and learned 270–272
  social, sensory, and somatic factors in appetite, 264, 265
    multiple appetite and satieties, 264, 265
  research need, 272, 273
    appetite mechanisms, not intake effects, 272, 273
  sociology of appetite science, 261–264
    biological reductionism, 261, 262
    measurement of appetite mechanisms, 262–264
  theory of appetite
    framework for appetite neuroscience, 265, 266
      development of, 266
      regulation and function of, 266
Pseudohypoparathyroidism
  smell, 343
  taste, 345
PVN, 110, 132
  CNS control of islets of Langerhans function, 406, 407
  inhibition of sodium appetite, 9
  insulin release, 410

Quinine, 451, 455, 456
  caloric intake reduction, 113

Radiation therapy, and taste, 345
Raeder's paratrigeminal syndrome, and taste, 345
Regulated system, 149
  studying the ponderostat, 149, 150
Renal failure, chronic

  smell, 343
  taste, 345
Renal function, in the elderly, 374, 375
Renin–angiotensin system, 5
  blockade of, 7
    abolish salt consumption, 7
  induced sodium depletion, 95, 101
  thirst, 368
Richter, Curt, 179, 180, 211
  taste-directed changes in food selection, 446

Saccharin, 111, 117, 139, 158, 253, 268, 278, 454, 455, 471, 473
  CPIR, 473
  site of unconditioned stimulus transduction, 243
SAD, 70
Salivation, 164
  cephalic response, 430
  eating disorders, 157
  ponderostat, 156–158
Salt, *see* Sodium
Saltiness, 95–104, 268
  loss of in the elderly, 344
  pregnancy, 330
Satiation, 96, 456, 457
  alliesthesia, 137, 138
  caloric control of, and hunger, 140–142
  CCK-induced, 453
    postdecerebration, 453
  chemoreceptor-mediated, 429
  conditioned appetite, 269, 270
    in children, 310–315
    conditioned desatiation, 269
    noradrenergic feeding, 269, 270
  definition of, 127–130
  of hunger and thirst

Satiation (cont.)
  role of gastric mechanoreceptros, 432
  importance of hunger, 138–40
  infant's cues for, 305
  measurement of appetite mechanisms, 263
  multiple appetites, 264, 265
  palatability, 130, 131
  relationship between palatability and hunger, 142–144
  satiety cascade, 131–133
  sensory-specific satiety, 113, 133–137
  of sodium appetite, 428
    gustatory, feeding, and general visceral interaction, 428–431
    synchronicity of physiology, behavioral, and cognitions, 144
SCN, food intake, 417
Seasonal affective disorder, see SAD
Serotonin
  carbohydrate deficiency, 77–79
  learned protein preference, 271
Sickle cell hemoglobinopathies, depletion-specific salt appetite, 90
Sinusitis, and smell, 343
Sjögren's syndrome
  smell, 343
  taste, 345
Smell, 341–361
  age-related modification of signal transduction pathways, 343, 350–352
  anorexia, 361
  bulimia, 361
  chemosensory losses in the elderly, 345–358
    age of onset, 351, 353–355, 358
    age-related preference function for an odor, 295
    comparison of olfactory and trigeminal functioning in deficit smell detection, 352, 353
    mechanisms of olfactory losses, 349–352
    odorant identification, 353, 357
    reduced food recognition by odor cues, 349
  compounds affecting it, 342
  disorders affecting it, 343
  drugs affecting it, 344
  methods of enhancing chemosensory signals, 358–360
    to reverse the effects of malnutrition, 358
  olfactory cues and feeding, 184–186
  perceptual losses of it, 341, 342
SOA, 111, 471
Sodium appetite, 85–104, 471
  age of onset, 87–91, 103
  aldosterone, 3, 6, 228
  angiotensin, 6, 7
    regulation of extracellular fluid volume, 228
  biological bases for
    central oxytocinergic neurons, 9, 10
    excitatory stimuli, 6–8
    inhibitory stimuli, 8, 9
    research needs, 12–14
  blood pressure, 104
  clinical reports, 87–91
  cultural variations, 86, 87
  depletion-specific sodium appetite, 3, 85–87
    anorexia, 91
    nausea, 91

factors predisposing toward hunger for, 211–231
consequences of salt ingestion, 222
evolutionary factors, 227–229
innate, 211–213
latent learning studies, 213–220
mineral deficiencies, 222–224
salt lick, 224–227
salty taste ingestion, 220–222
human experimental approaches, 91, 92
inducing sodium depletion, 91
pituitary secretion of OT, 9, 10
physiological need, 456
in pregnant and lactating mammals, 227, 229
salt appetite or altered salt preference, 100–102
satiation of gustatory, feeding, and general visceral interaction, 428–431
taste
-directed changes in food selection, 446
and human sodium depletion, 92–100
Sodium bicarbonate, sodium hunger, 230
Sodium, excretion of, 368
Somatostatin, release, hypothalamus, influences on glucose, insulin, and body weight, 406, 408, 420
SON, inhibition of sodium appetite, 9
Sourness, 111, 112, 223
chimpanzees, 196
loss of in the elderly, 344
Steady-state system, 149
studying the ponderostat, 149
Steiniger, Fritz, 183, 184
Strangeness breeds contempt principle, human and chimpanzee aversion to the leopard, 205
Suckling, mother–infant bonding, 283–300
behavioral stress reduction, 285–299
analgesic action of tactile and milk component substances, 290–293
CCK, role of, 295, 196
contact as a calming agent, 284–286, 288–297, 398
definition, 285
exorphins, 296
infant stimulation, two classes, 296
interactions between opioid and nonopioid control system, 286, 288, 290, 291, 293, 294, 296, 297, 300
milk, 285, 290, 293–296, 298, 388
suckling and its consequences, 285, 290–292
suckling-related stimulation and reward, 284, 285, 293, 298
Sucrose octaacetate, *see* SOA
Sugar, 298, 299
attitudinal studies, 79, 80
endogenous opioid peptide system, 79
hedonic response profiles, 75–76
eating disordered groups, 75–76
metabolic factors, 77–79
cravings, 77
sensory evaluation studies
body weight, 71
population groups, 71
predictor for food preferences, 71

Suprachiasmatic nucleus, *see* SCN
Supraoptic nuclei, *see* SON
Sweetness, 96, 97, 115–117, 120,
    121, 130, 141, 142, 145,
    253, 268, 451, 456
  artificial, 158, 159
  children's eating pattern, 303
  chimpanzees, 196
  as an innate preference, 239, 267
  loss of in the elderly, 344
  satiety values, 263

Taste
  anatomy of hindbrain taste nuclei,
    447, 448
  anorexia, 361
  bulimia, 361
  CNS and the autonomic nervous
    system, 427–437
    CNS mechanisms and pathways, 431–437
    gustatory, feeding, and general visceral interaction, 428–431
  disorders affecting it, 345
  drugs affecting it, 346
  FAP, 135
  hypertension and the elderly, 387
  loss of insular cortex involvement in gustatory function, 432
  methods enhancing chemosensory signals, 358–360
    behavioral changes in eating, 360
    flavor amplification, 359, 360
    pharmacological modes of enhancement, 358, 359
    to reverse malnutrition, 358
  as a motivator and director of feeding, 445
  orofacial responses to stimuli, 453
  perceptual losses of, 342–345
    to acids, 348
    reduced food recognition by odor cues, 349
    to sodium salts, 346, 347
  primary function, 453
  relationship to metabolic state, 448–461
Thiamine, 224
  feeding behavior, 446
    induced deficiencies of, 446
    physiological need, 446
Thirst
  aldosterone, 7, 368, 375, 376
  fluid balance in the elderly, 367–380
  dehydration, 368, 376, 377
  mechanisms of thirst deficit, 373, 374
  neuroendocrine function, 375, 376
  renal function, 374, 375
  in response to heat stress and thermal dehydration, 372
  renin–angiotensin system, 368
  stimulation of
    low-pressure baroreceptors, 5
    PEG injection, 8
    pituitary secretion of AVP, 4, 5
Total self-regulatory capacity, 179
Toxicosis, 181, 182
Triglycerides, obesity, 164
Turner's syndrome
  and smell, 343
  and taste, 345

Urinary excretion values, 94

Vagus nerve, 396–400, 447
  as afferent projection for visceral information from gut and liver, 429

gustatory, feeding, and general visceral interaction, 428–431
mediation of CPIR, 391, 394, 395
motility of gastrointestinal tract and absorption processes, 417, 418
  glucagon release, 407
  insulin release, 407
  in response to stimulation of vagus nerve, 430
  separation from gustatory responsive neurons in brain stem, 436
SCN, 417
VMH, 410
Vasopressin, see AVP
Ventral basal thalamus, 432, 433

CNS, taste, and the autonomic nervous system, 431–434
  CGRP, 433
  CNS pathways and mechanisms, 431–437
  general visceral input, 434
Ventromedial hypothalamus, see VMH
Vitamin B_3 deficiency, and taste, 345
Vitamin B_{12} deficiency, and smell, 343
VMH, 110
  lipolysis, 410, 411
  pancreatic vagal nerves, 410

Zinc, and taste, 345